Uncertainty in Climate Change Research

Linda O. Mearns • Chris E. Forest • Hayley J. Fowler •
Robert Lempert • Robert L. Wilby
Editors

Uncertainty in Climate Change Research

An Integrated Approach

Editors
Linda O. Mearns *(deceased)*

Hayley J. Fowler
School of Engineering
Newcastle University
Newcastle upon Tyne, UK

Robert L. Wilby
Loughborough University
Loughborough, UK

Chris E. Forest
Department of Geosciences
Pennsylvania State University
University Park, PA, USA

Robert Lempert
RAND Corporation
Santa Monica, CA, USA

ISBN 978-3-031-85541-2   ISBN 978-3-031-85542-9   (eBook)
https://doi.org/10.1007/978-3-031-85542-9

This work was supported by UCAR University Corporation for Atmospheric Research.

© The Editor(s) (if applicable) and The Author(s) 2025. This book is an open access publication.

Open Access This book is licensed under the terms of the Creative Commons Attribution 4.0 International License (http://creativecommons.org/licenses/by/4.0/), which permits use, sharing, adaptation, distribution and reproduction in any medium or format, as long as you give appropriate credit to the original author(s) and the source, provide a link to the Creative Commons license and indicate if changes were made.

The images or other third party material in this book are included in the book's Creative Commons license, unless indicated otherwise in a credit line to the material. If material is not included in the book's Creative Commons license and your intended use is not permitted by statutory regulation or exceeds the permitted use, you will need to obtain permission directly from the copyright holder.

The use of general descriptive names, registered names, trademarks, service marks, etc. in this publication does not imply, even in the absence of a specific statement, that such names are exempt from the relevant protective laws and regulations and therefore free for general use.

The publisher, the authors and the editors are safe to assume that the advice and information in this book are believed to be true and accurate at the date of publication. Neither the publisher nor the authors or the editors give a warranty, expressed or implied, with respect to the material contained herein or for any errors or omissions that may have been made. The publisher remains neutral with regard to jurisdictional claims in published maps and institutional affiliations.

Cover illustration: Front cover art acknowledgment: Jeremy Cain.

This Springer imprint is published by the registered company Springer Nature Switzerland AG
The registered company address is: Gewerbestrasse 11, 6330 Cham, Switzerland

If disposing of this product, please recycle the paper.

This work is dedicated to the memories of

Linda O. Mearns, much missed friend, mentor, and scholar.
This book would not have happened without your enduring wit, vision, and cajoling.
Thank you.

and

Samuel Lepie Hallward, who at age 12 had to cope with the uncertainty of life and death much too soon.

Preface

Uncertainty is present in all phases of climate change research from the physical science (e.g., projections of future climate) to the impacts through to the effort to make decisions regarding mitigation and adaptation across different spatial scales. This theme will embrace all aspects of uncertainty in climate change research, providing a pedagogic whole for students, post-docs, and early-career scientists interested in any and all aspects of climate change. One central focus will be the need to understand the strands of uncertainty through-out the climate change problem in order to focus effectively in any one area. We aim to train the next generation of postgraduates in interdisciplinary thinking.

So said the background to the Advanced Study Program (ASP) Colloquium on *Uncertainty in Climate Change Research: An Integrated Approach*, 21 July–6 August 2014. The wider mission of the ASP at the US National Center for Atmospheric Research still includes stimulating the intellect of the research community; fostering the professional development of graduate students and postdoctoral fellows; promoting advanced scientific educational opportunities; focusing attention on emerging areas of science; and facilitating interactions across research communities. Simply put, this is all about developing people with the cutting-edge knowledge and technical skills needed to face an uncertain future. These aspirations were central to our colloquium and are just as valid today.

All five editors of this book have worked in climate change research and consulting for decades. We do not relish seeing the projections of our early models becoming lived realities in the shape of incessant weather disasters and climate impacts. Focusing on such problems rather than on their solutions is a route to despair. So, our greatest hope for the future is vested in next-generation policy-makers, researchers and practitioners. These colleagues face daunting challenges, but they are equipped with extraordinary tools like artificial intelligence and machine learning, big data and novel Earth observations. Moreover, we now recognize the importance of working *with* decision-makers to arrive at practicable solutions.

Many of the students who participated in the 2014 workshop are, by now, educators and mentors of early career scientists. And so our collective capacity builds from one generation to the next. We hope that the following chapters by colloquium speakers and experts will inspire others to embark on this pathway—whether as a physical or social scientist, someone who is interested in ethics or communication, or who is mathematically minded. We need more engineers and geographers working on climate change too!

We trust that within the *pedagogic whole* that follows there is something for everyone.

Boulder, CO, USA	Linda O. Mearns
University Park, PA, USA	Chris E. Forest
Newcastle upon Tyne, UK	Hayley J. Fowler
Santa Monica, CA, USA	Robert Lempert
Loughborough, UK	Robert L. Wilby

Acknowledgements

We thank the National Science Foundation and the National Center for Atmospheric Research for supporting the 2014 Advanced Study Program Colloquium on *Uncertainty in Climate Change Research: An Integrated Approach*. Without their sustained commitment to education, engagement and early career development, the event would not have happened.

We are also grateful to colleagues at Springer Nature for their continued patience and belief in this long publishing journey. To bring together such a wide arc of authors and topics in one place was always going to be challenging.

Finally, we acknowledge all the mentors, colleagues, family and friends that have encouraged us along the way.

Contents

1. **Integrated Uncertainty: An Introduction** 1
   Linda O. Mearns, Hayley J. Fowler, and Robert L. Wilby

2. **Laying the Policy Groundwork for Considering Integrated Uncertainty** 9
   Sarah Michaels

3. **Climate-Informed Decision Analysis Via Decision Scaling** 19
   Casey Brown

4. **Supporting Climate-Related Decisions Under Uncertainty** 31
   Robert Lempert

5. **The Policy Portfolio Problem** 47
   Rachel Warren

6. **Climate Change Adaptation in Practice: Navigating Uncertainty in the Real World** 61
   Linda A. Joyce, Laurna Kaatz, and Joel Smith

7. **Uncertainty of Climate Change Impacts on Crop Production** 71
   Daniel Wallach, Senthold Asseng, and Alex C. Ruane

8. **Uncertainty in Ecological Models** 81
   Dan L. Warren, Lukas Baumbach, Jamie M. Kass, and Alke Voskamp

9. **Uncertainty in Hydrologic and Water Resources Modelling** 93
   Robert L. Wilby and Geoff Darch

10. **Dimensions of Uncertainty in Mitigating Flooding** 105
    Sarah Michaels

11. **Uncertainty in Transportation Infrastructure** 115
    Jennifer M. Jacobs

12. **Science in Coastal Adaptation Decision-Making: Working Effectively with Persistent Uncertainties** 125
    Susanne C. Moser

13. **Uncertainty in Determining Impacts of Climate Change on Human Health** .. 137
    Kristie L. Ebi, Mary H. Hayden, Morgan E. Gorris, Christopher K. Uejio, and Jennifer Vanos

14. **Uncertainty and Socioeconomic Vulnerability to Climate Change** 145
    Kirstin Dow, Paty Romero-Lankao, and Olga Wilhelmi

15. **Climate/Earth System Projections and Their Uncertainties: An Overview** ... 155
    Chris E. Forest and William D. Collins

16. **Emissions and Concentration Scenarios** 163
    Jennifer Morris and John M. Reilly

17 **The Importance of Internal Variability for the Uncertainty in Climate Change Projections and Decision-Making** 177
Flavio Lehner and Clara Deser

18 **Downscaling Future Climate Projections: Compounding Uncertainty But Adding Value?** 185
Hayley J. Fowler, Linda O. Mearns, and Robert L. Wilby

19 **Characterizing the Uncertainty Surrounding Sea-Level Projections to Inform Decisions** 199
Tony E. Wong, Ryan L. Sriver, and Andra J. Garner

20 **Uncertainty Quantification: A Statistical Perspective** 207
Stephan R. Sain and William Kleiber

21 **Uncertainty and Extremes** 217
Mark D. Risser and Claudia Tebaldi

22 **How Uncertainty Interacts with Ethical Values in Climate Change Research** 229
Casey Helgeson, Wendy Parker, and Nancy Tuana

23 **Expert Judgment and Communication of Uncertainty** 237
Stephen B. Broomell, Emily Ho, Daniel M. Benjamin, and David V. Budescu

24 **Uncertainty in the Economic Appraisal of Adaptation** 247
Paul Watkiss

25 **Uncertainty Management in a Decision Context: What We Learned from the Decision Center for a Desert City** 255
Patricia Gober and Howard Wheater

26 **Acting with Uncertainty: Reflecting on a Decade of Rapid Progress in Climate Policy, Research and Practice** 263
Linda O. Mearns and Robert L. Wilby

Index 269

About the Editors

Linda O. Mearns (deceased) was a prominent climate scientist specializing in climate variability, climate change impacts, and the use of regional climate models. She held a Ph.D. in Geography and had decades of expertise in climate science and policy. She was a Senior Scientist at the National Center for Atmospheric Research (NCAR) and former Director of the Weather and Climate Impacts Assessment Science Program. Dr. Mearns contributed significantly to several Intergovernmental Panel on Climate Change (IPCC) reports and led multiple interdisciplinary research projects such as the North American Regional Climate Change Program (NARCCAP). Her work focused on improving the understanding of climate risks and informing adaptation strategies. In 2016, she received the American Association of Geographers (AAG) Distinguished Scholarship Award.

Chris E. Forest is Professor of Climate Dynamics in the Department of Meteorology and Atmospheric Science at The Pennsylvania State University, specializing in the evaluation of climate models to assess long-term climate projections, impacts, and uncertainties. He is the Director of the Penn State Center for Earth System Modeling, Analysis, and Data. His research focuses on basic understanding of climate dynamics, quantifying uncertainty in climate predictions, and understanding how to use climate information for assessing climate risks. He has contributed to multiple high-profile climate research initiatives and interdisciplinary projects aimed at enhancing climate resilience. He holds a Ph.D. in Meteorology from the Massachusetts Institute of Technology.

Hayley J. Fowler is Professor of Climate Change Impacts and directs the Centre of Climate and Environmental Resilience at Newcastle University. She advises United Kingdom Government on climate resilience, serving on the Department for Energy Security and Net Zero Science Expert Group. Her research interests lie in improving the physical understanding of changing weather extremes and delivering high-resolution projections for climate adaptation. She won the EGU's Sergey Soloviev Medal in 2024 and is an AGU Fellow. She is past-President of the British Hydrological Society. She is passionate about transdisciplinary working and regularly delivers lectures to engage the public on the climate crisis.

Robert Lempert is a Principal Researcher at the RAND Corporation and Director of the Frederick S. Pardee Center for Longer Range Global Policy and the Future Human Condition. His research focuses on climate risk management and decision-making under conditions of deep uncertainty. Dr. Lempert was a coordinating lead author for Working Group II of the United Nation's Intergovernmental Panel on Climate Change (IPCC) Sixth Assessment Report and the inaugural president of the Society for Decision Making Under Deep Uncertainty (http://www.deepuncertainty.org). A Professor of Policy Analysis in the Pardee RAND Graduate School, Dr. Lempert is an author of the book *Shaping the Next One Hundred Years: New Methods for Quantitative, Longer-Term Policy Analysis*.

Robert L. Wilby is a hydro-climatologist and Professor of Hydroclimatic Modelling in the Department of Geography and Environment at Loughborough University, UK. His research focuses on the forecasting and management of climate risks to freshwater systems and urban environments. He co-developed the Statistical DownScaling Model (SDSM)—a public domain climate scenario tool that has been used in numerous risk assessments, including for water resources, fluvial flooding and storm surge, air quality and urban heat stress. He was involved in the first four UK Climate Change Risk Assessments. He works extensively with Multilateral Development Banks and national agencies seeking to improve the resilience of built environments, water and energy infrastructure to climate change.

Integrated Uncertainty: An Introduction

Linda O. Mearns, Hayley J. Fowler, and Robert L. Wilby

*Le doute est un état mentale désagréable,
mais la certitude est ridicule.*
(Doubt is an uncomfortable condition, but
certainty is a ridiculous one).
Voltaire

1.1 What Is Uncertainty?

Uncertainty is a fact of life and not the sole "problem" of those researching or supporting action on climate change. There are many definitions of uncertainty that we could use to introduce the concept. However, for consistency, we apply the definition of uncertainty from Working Groups I, II, and III of the Intergovernmental Panel on Climate Change (IPCC) Sixth Assessment Report. The IPCC (2022) glossary says uncertainty is *"a state of incomplete knowledge that can result from a lack of information or from disagreement about what is known or even knowable."* We add that uncertainty applies to the past, present, and future. More specifically, uncertainty is a state of partial knowledge (such as not knowing the number of sides on some dice) combined with a random outcome (such as rolling this dice). Hence, total uncertainty is a blend of both this incomplete knowledge (epistemic) and the random (aleatory) behavior of a complex world.

There are many ways of representing and approaching uncertainty (Chaps. 14 and 23, this volume). Traditionally, probabilities are used as a measure of uncertainty, but this is only feasible when uncertainty can be clearly described and quantified. Although there are different types of uncertainty, the most important type in the context of integrated uncertainty in climate change is "deep uncertainty" (Marchau et al., 2019). This can arise in situations characterized by a lack of agreement among parties about aspects of a system that are crucial to the outcome of a decision. This might be the external context of the system, how the system works and its boundaries, and/or the system responses of interest, and/or their relative importance (weighting or values) (Lempert et al., 2004). For example, there is deep uncertainty around the thresholds and timing of climate tipping points for Antarctic ice sheet collapse and associated sea-level rise estimates (Chap. 19). Climate research and expert elicitation contribute to both reducing and managing deep uncertainty (Lempert et al., 2024).

Associated with the concept of deep uncertainty is that of "wicked problems" or "super wicked problems" (Lazarus, 2009). These are public policy problems that defy rational and optimal solutions. They are characterized by deep uncertainties, with many interdependencies and causes that interact, and where posited solutions can lead to new unintended consequences that are difficult to test. Climate change has been described as the quintessential super wicked problem. Essentially climate change becomes increasingly "super wicked" as time goes on as the need for greater integration across the different parts of the problem becomes more and more apparent. Of course, this further complicates the task of identifying, characterizing, and quantifying uncertainties in all parts of the climate change problem.

1.2 Our Approach to Uncertainty and Decision-Making

Our points of departure are around the approaches to decisions that must be followed to manage climate change

---

Linda O. Mearns has died before the publication of this book.

L. O. Mearns (deceased)

H. J. Fowler (✉)
School of Engineering, Newcastle University, Newcastle upon Tyne, UK
e-mail: hayley.fowler@newcastle.ac.uk

R. L. Wilby
Department of Geography and Environment, Loughborough University, Loughborough, UK

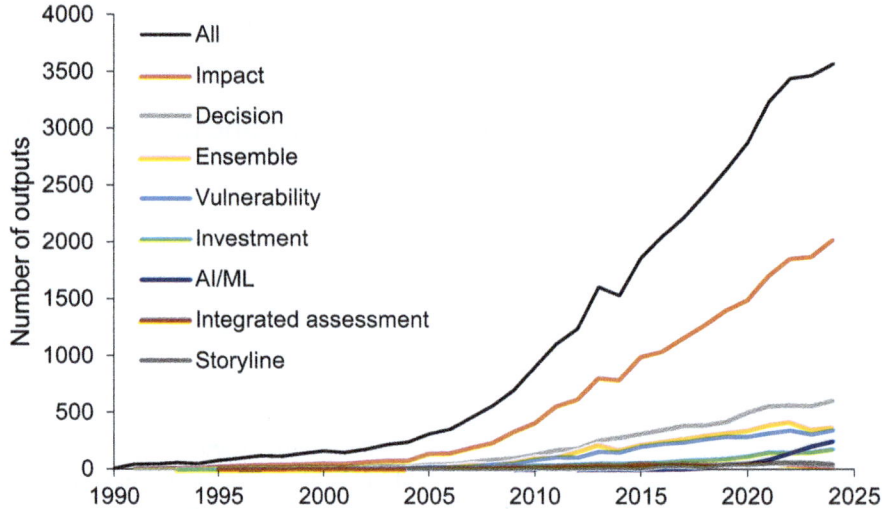

Fig. 1.1 Growth in the number of research outputs addressing various aspects of climate change uncertainty. The subtotals are not mutually exclusive as individual outputs may cover more than one theme. Counts for 2024 were estimated based on the number of papers accrued by mid-October. (Data source: Web of Science)

(via mitigation and/or adaptation) and their uncertainties. We describe this simply as "acting with uncertainty" (Lempert, *pers. comm.*). This contrasts with a more conventional three-step approach that (1) projects climate change—usually from climate models that tell us how the climate might change in the future; then (2) assesses what the impacts of climate change could be (across sectors); and finally (3) evaluates what actions can be taken to manage both negative and positive impacts—the so-called "uncertainty cascade" (Mitchell & Hulme, 1999), now characterized with varying degrees of sophistication (Smith et al., 2018). We assert that by starting with the decision context, one can more readily manage the many uncertainties that emerge in the various parts of the climate change problem, many of which cannot be reduced, or at least not within the decision timeframe.

It has long been recognized that uncertainty is an important consideration in risk and policy analysis (Morgan et al., 1990). We present several perspectives on decision-making under uncertainty (DMUU), including decision scaling (Chap. 3) and robust decision-making (RDM) approaches (Chap. 4). Both methods recognize that waiting to make decisions until critical uncertainties are reduced is a fool's errand. Indeed, some uncertainties related to climate change are irreducible—such as internal climate variability (Chap. 17). Given such realities, DMUU and RDM argue that one should now approach decision-making using methods that produce results that can manage the uncertainties about the future as they currently exist. Such methods affect how policies toward reducing climate change can be formulated (Chap. 2). In this way, the complicated task of reducing uncertainty before acting can be side-stepped. Otherwise, the uncertainty obstacle can be used by some as an excuse for not acting until the uncertainties are sufficiently "reduced" (Curry & Webster, 2011; Oreskes, 2015).

Concerns about uncertainties around future climate change have been with us for a long time. (Naturally, there are numerous uncertainties regarding most topics that concern the future.) Early studies tended to be about the climate changes themselves at various spatial and temporal scales. Initially, then, this concerned uncertainties regarding simulations by climate models, particularly of the future, but also around the data used to run and evaluate climate model experiments. Figure 1.1 shows how the volume of scientific outputs addressing climate uncertainty has grown over time. Annual output in 2024 on this topic approached 4000 papers, with over 37,000 in total since 1991. This represents about 8% of all climate change publications (listed in the Web of Science Core Collection to date).

Several overlapping uncertainty themes in climate change research have emerged since the 1990s (Fig. 1.1). These may be listed by output volume. First are *impact assessments*, some of which downscale climate model ensembles to drive multiple (crop, water, ecosystem, or human health) impact models, thereby generating envelopes of impact uncertainty (e.g., Wilby & Harris, 2006; Fowler et al., 2007). A second theme is around evaluating mitigation and adaptation options, within the *decision* space bounded by credible climate scenarios (e.g., Lempert et al., 2006; Brown et al., 2012). A third covers multiple GCMs, emissions scenarios, and internal climate variability, compiled in *ensembles* to show the relative contributions of these sources of climate uncertainty (e.g., Tebaldi & Knutti, 2007; Deser et al., 2012). Fourth, uncertainties have been framed around *vulnerability* and exposure in both top-down and bottom-up risk assessments (e.g.,

Conway et al., 2019; Sutton, 2019). Fifth are studies that use multiple-objective modeling to prioritize and sequence climate *investments* in the context of uncertain climate, land use, and technology scenarios (e.g. Müller & Robertson, 2013; Dunnett et al., 2018). A sixth and rapidly growing theme on *assistive technologies* (such as artificial intelligence and machine learning) covers the generation of super-ensembles of climate change at high resolution and hence better quantification of the likelihood of extreme weather events (e.g., Materia et al., 2024; Schneider et al., 2024). Seventh is research into *integrated assessment*, involving combinations of models for evaluating climate mitigation pathways in the context of other uncertain factors (e.g., van Vuuren et al., 2011; Dunford et al., 2015). Finally, the smallest theme—in terms of output volume—is about narrative and storyline approaches that deliberately avoid attaching probabilities to deeply uncertain futures but instead focus on the underlying drivers (e.g., Shepherd et al., 2018; Yates et al., 2015).

1.3 What Do we Really Know (i.e., What Is NOT Uncertain)?

One of the guiding principles when communicating climate change is that we should NOT lead with the uncertainties. This recognizes that an audience will likely anchor onto the uncertainties presented and assume that they are much greater than the certainties. But, of course, we already know a great deal about current and future climate change, and we are learning more every day (IPCC, 2021; USGCRP, 2023). For example, we know that the climate is significantly changed now and is continuing to change—this is not solely a future problem—and we know that the production of greenhouse gases by human activity is the cause. Detection studies have provided a considerable amount of evidence on changes to the climate. We know that temperatures are increasing throughout the world and that changes in precipitation regimes are occurring as a result. We know that oceans are warming and becoming more acidic and less oxygenated. We know that weather extremes important to human society and ecosystems, such as heatwaves and precipitation intensities, have increased and will likely continue to increase. We know that the number of intense tropical cyclones has increased over the past several decades and that these have been attended by increases in heavy precipitation (as witnessed during Hurricane Helene in 2024). We know that sea level is rising and will continue to rise for centuries to come. We know that glaciers and Arctic ice are melting and will continue to melt (IPCC, 2021).

We also continue to learn ever more details about what is driving these changes, and the extent to which human activities are responsible. Attribution science evaluates the relative contributions from multiple causal factors to a climate change or extreme weather event. Considerable research has been dedicated to attributing human causes to observed climate changes and their effects. We know, for example, that there is a human contribution to rising air temperature extremes (Hegerl et al., 2004), increased atmospheric moisture (Santer et al., 2007), Arctic moistening (Min et al., 2008), as well as mean (Zhang et al., 2007) and extreme (Min et al., 2011) precipitation changes. The IPCC Second Assessment Report (IPCC 1995) concluded that *"the balance of evidence suggests that there is a discernible human influence on global climate."* This statement was strengthened in 2001, 2007, and 2013. By the Sixth Assessment Report (IPCC, 2021), the assessment was that it is unequivocal that human activity is causing climate change. Attribution studies have advanced considerably over this period. Initially, attribution research focused on the climate changes themselves, such as *"most of the observed warming over the last 50 years is likely to have been due to the increase in greenhouse gas concentrations"* (SAR, 1995).

More recently, attention has shifted to attributing extreme events since these are strongly associated with damage to human societies and natural ecosystems (e.g., Stott et al., 2003; Coumou & Rahmstorf, 2012). This all began with a thought exercise after the 2003 floods in central England (Allen, 2003), but has since evolved into near-real-time attribution of extreme events by the World Weather Attribution (WWA) group.[1] Prominent examples of attribution studies linking increased severity and/or likelihood of extreme events to human greenhouse emissions include the 2003 European heatwave (Stott et al., 2003), 2010 Thailand floods and 2010–2011 Somalia drought (Otto et al., 2018), and 2017 Hurricane Harvey (Risser & Wehner, 2017). Other studies are advancing the attribution of the impacts of climate change, such as by using machine learning approaches (Callaghan et al., 2021). Impact examples include an analysis establishing that about 37% of heat-related deaths worldwide from 1991 to 2018 could be attributed to anthropogenic climate change (Vicedo-Cabrera et al., 2021). We know that the rise in the global concentration of CO_2 since the year 2000 is about 20 ppm per decade (and has increased by 25 ppm per decade since 2010). Another study found that with an average estimate of damages from Hurricane Harvey assessed at about US$90bn, the best estimate of damages attributable to human influence on the climate is US$67bn, with a likely lower bound of at least US$30bn; hence, one-third to three-quarters of the damages were due to human influence on global warming (Frame et al., 2020).

However, there remain numerous uncertainties regarding the details of climate change, especially at finer spatial and temporal scales. Global mean temperatures are rising, and this increase is expected to continue in the future, but the

---

[1] https://www.worldweatherattribution.org/analyses/

exact regional patterns and timings of change are uncertain. This is partially due to ambiguity about future emissions pathways, but also due to uncertainties in climate sensitivity and feedbacks (Chaps. 15 and 16). Uncertainty in changes to precipitation is even greater and may be compounded by the choice of downscaling method, so awareness of the value added by this extra work is important (Chap. 18). Nonetheless, it is clear that precipitation extremes at daily and shorter scales increase at the Clausius Clapeyron rate (6.5% per °C of warming) or greater (Fowler et al., 2021). We know that the Earth has warmed due to anthropogenic factors by around 1.3 °C compared to the pre-industrial period (during 2024, it was almost 1.5 °C warmer) and that it will continue to warm if emissions of greenhouse gases are not reduced. Initially, it was expected that global warming could reach 1.5 °C by the 2040s, but this was updated to the early 2030s (IPCC, 2021). However, as scientific understanding has advanced, it is now predicted that there is an 80% likelihood that at least 1 year will temporarily exceed 1.5 °C during 2024–2028 (WMO, 2024).

1.4 Calculation and Communication of Uncertainty

1.4.1 Quantifying Uncertainty

When a particular part of the climate change issue is clearly identified and has clear statistical properties, then the uncertainty may be described statistically. However, it should be noted that approaches to quantifying uncertainty vary between disciplines, largely due to differences in the perceived importance of various sources of uncertainty (Simmonds et al. 2022). Two contributions on the general statistics of uncertainty (Chap. 20) and the statistics of extremes (Chap. 21) provide the basics on how to quantify uncertainty under these conditions. However, it is not possible to provide simple statistical guidance on how to quantify deep uncertainty. By definition, deep uncertainty exists when experts and stakeholders cannot define probability distributions for important system variables and parameters (Lempert et al., 2024).

1.4.2 Communicating Uncertainty

Communication is a crucial aspect of uncertainty in climate change (Chap. 23). Communicating uncertainty appropriately for a particular audience—whether climate scientists, stakeholders, policymakers, or the general public—has received greater attention over the past decades as the urgency of action has intensified. Conveniently, the Intergovernmental Panel on Climate Change has acted as a bell weather for communicating uncertainty since its inception in 1990 as its three Working Groups cover most aspects of climate change (i.e., climate science, impacts and adaptation, and mitigation). Although the communication of uncertainty has evolved across the six IPCC reporting phases, the approaches have garnered criticism (e.g., Budescu et al., 2009, 2012, 2014). Many studies have been produced on the communication of uncertainty in the IPCC reports (e.g., Kandlikar et al., 2005; Curry, 2011; Budescu et al., 2014; Kause et al., 2022). For instance, some note that the verbal descriptors provided by the IPCC (such as "unlikely") may be misinterpreted by experts and nonexperts alike (Budescu et al., 2014; Kause et al., 2022). One solution is to supplement these verbal descriptors with quantitative ranges to reduce their misinterpretation. Other studies show that when scientists acknowledge uncertainty in their results there is increased trust and acceptance of their messages by the public (Howe et al., 2019). Improved dialogue is also needed between climate service providers and their clients (Wilby & Lu, 2022). This is best achieved through a collaborative and co-productive approach to knowledge creation (as in Chap. 6).

1.4.3 Integrating Uncertainty

One of the most common integrations across different types of uncertainty has been the blending of climate projection and impact model uncertainties. For example, an early agricultural study employed nine different economic models in combination with five crop models and two climate models (Nelson et al., 2013). Other forms of integration may involve connecting climate impacts and uncertainties that span different resource areas, such as at the water, food, and energy nexus (Yates et al., 2023). A consistent message is that climate and impact model uncertainty (due to parameters and structure) tends to dominate over emissions uncertainty to the 2050s (e.g., Wilby et al., 2009). However, other uncertainties around benchmarking and data quality for model evaluation apply too. There are also various sector-specific issues and approaches to uncertainty for agriculture (Chap. 7), ecosystems (Chap. 8), water resources (Chap. 9), flood risk management (Chaps. 10 and 12), transportation (Chap. 11), and human health (Chap. 13). For instance, water resource planners may use uncertainty in regional climate change scenarios to stress-test model representations of whole water supply systems (e.g., Fowler et al., 2022). This helps identify the most vulnerable elements and conditions under which the system could fail (with and without adaptation measures). In this case, the uncertainty is turned around to give deeper insight rather than to delay action. Political uncertainties can also be accommodated within this framework by working with stakeholders through long-term boundary organizations (Chap. 25).

1.4.4 Vulnerability Uncertainty

An important element that requires further development is the exploration of uncertainty in vulnerability (Chap. 14). Disentangling the various societal elements making up vulnerability is a difficult task, and research in this area has only come to the fore in recent years. But improved understanding of how vulnerability evolves is critical for creating climate-resilient societies. Integrated assessment modeling enables the testing of outcomes of various mitigation/adaptation options in "model worlds" to evaluate the effectiveness of interventions, as well as to expose any trade-offs and co-benefits. When seeking solutions to the climate change problem, one must consider both adaptation (Chap. 6) and mitigation (Chap. 16) together. Combining these solutions in a place-based way involves managing numerous uncertainties (Chap. 5). Developing aligned portfolios of adaptation and mitigation is a very wicked problem indeed, but making progress in this arena is critical for not just Net Zero planning but for increasing societal resilience to extreme weather events. Economic appraisal is a well-established technique for assessing the case for climate action and estimating the relative costs and benefits of different options. Economic risks due to climate uncertainty can be managed by prioritizing no- and low-regret options, as well as by devising multiple adaptation pathways that avoid lock-in and maximize future flexibility (Chap. 24), including resilient Net Zero Pathways.

1.5 Navigating Uncertainty for Climate Action

The overall goal of this book is to show how cutting-edge knowledge and technologies can help us navigate uncertainty and support climate action. This is an inherently multidisciplinary endeavor, but we are not saying that everyone should be an interdisciplinarian. Rather, we believe that individual uncertainty research activities should be informed by and benefit from awareness of the wider uncertainty context. This is because climate actions are seldom, if ever, one-dimensional. Chances are that even hardcore climate modelers are now talking to social scientists, policymakers, and engineers. We must also be mindful of the ethical dimensions of our research as there are different ways of treating uncertainty in our work (Chap. 22). This is because there are potential societal consequences of ignoring, misunderstanding, or miscommunicating uncertainty in climate change research to the public and policymakers.

The following 25 chapters reflect the expertise of 51 contributors. Author disciplines span agriculture, atmospheric science, business studies, climatology, earth and environmental science, ecology, economics, engineering, ethics, forestry, geography, information science, mathematics, meteorology, philosophy, politics, psychology, public health, sociology, and more. There has been considerable progress in climate policy, research, and practice in all these areas over the last decade (Chap. 26). We trust that our diverse perspectives will contribute to further innovation in climate research methods, interdisciplinary collaboration, and communication in order to better navigate the profound uncertainties that lie ahead.

References

Allen, M. (2003). Liability for climate change. *Nature, 421*, 891–892. https://doi.org/10.1038/421891a

Brown, C., Ghile, Y., Laverty, M., & Li, K. (2012). Decision scaling: Linking bottom-up vulnerability analysis with climate projections in the water sector. *Water Resources Research, 48*, W09537. https://doi.org/10.1029/2011WR011212

Budescu, D. V., Broomell, S., & Por, H. H. (2009). Improving communication of uncertainty in the reports of the Intergovernmental Panel on Climate Change. *Psychological Science, 20*, 299–308. https://doi.org/10.1111/j.1467-9280.2009.02284.x

Budescu, D. V., Por, H. H., & Broomell, S. B. (2012). Effective communication of uncertainty in the IPCC reports. *Climatic Change, 113*, 181–200. https://doi.org/10.1007/s10584-011-0330-3

Budescu, D., Por, H. H., Broomell, S., & Smithson, M. (2014). The interpretation of IPCC probabilistic statements around the world. *Nature Climate Change, 4*, 508–512. https://doi.org/10.1038/nclimate2194

Callaghan, M., et al. (2021). Machine-learning-based evidence and attribution mapping of 100,000 climate impact studies. *Nature Climate Change, 11*, 966–972. https://doi.org/10.1038/s41558-021-01168-6

Conway, D., et al. (2019). The need for bottom-up assessments of climate risks and adaptation in climate-sensitive regions. *Nature Climate Change, 9*, 503–511. https://doi.org/10.1038/s41558-019-0502-0

Coumou, D., & Rahmstorf, S. (2012). A decade of weather extremes. *Nature Climate Change, 2*, 491–496. https://doi.org/10.1038/nclimate1452

Curry, J. (2011). Reasoning about climate uncertainty. *Climatic Change, 108*, 723–732. https://doi.org/10.1007/s10584-011-0180-z

Curry, J. A., & Webster, P. J. (2011). Climate science and the uncertainty monster. *Bulletin of the American Meteorological Society, 92*, 1667–1682. https://doi.org/10.1175/2011BAMS3139.1

Deser, C., Phillips, A., Bourdette, V., & Teng, H. (2012). Uncertainty in climate change projections: The role of internal variability. *Climate Dynamics, 38*, 527–546. https://doi.org/10.1007/s00382-010-0977-x

Dunford, R., Harrison, P. A., & Rounsevell, M. D. A. (2015). Exploring scenario and model uncertainty in cross-sectoral integrated assessment approaches to climate change impacts. *Climatic Change, 132*, 417–432. https://doi.org/10.1007/s10584-014-1211-3

Dunnett, A., Shirsath, P. B., Aggarwal, P. K., Thornton, P., Joshi, P. K., Pal, B. D., Khatri-Chhetri, A., & Ghosh, J. (2018). Multi-objective land use allocation modelling for prioritizing climate-smart agricultural interventions. *Ecological Modelling, 381*, 23–35. https://doi.org/10.1016/j.ecolmodel.2018.04.008

Fowler, H. J., et al. (2021). Anthropogenic intensification of short-duration rainfall extremes. *Nature Reviews Earth and Environment, 2*, 107–122. https://doi.org/10.1038/s43017-020-00128-6

Fowler, H. J., Blenkinsop, S., & Tebaldi, C. (2007). Linking climate change modelling to impact studies: Recent advances in downscaling techniques for hydrological modelling. *International Journal of Climatology, 27*, 1547–1578. https://doi.org/10.1002/joc.1556

Fowler, K., Ballis, N., Horne, A., John, A., Nathan, R., & Peel, M. (2022). Integrated framework for rapid climate stress testing on a monthly timestep. *Environmental Modelling and Software, 150,* 105339. https://doi.org/10.1016/j.envsoft.2022.105339

Frame, D. J., Wehner, M. F., Noy, I., & Rosier, S. M. (2020). 2020: The economic costs of Hurricane Harvey attributable to climate change. *Climatic Change, 160,* 271–281. https://doi.org/10.1007/s10584-020-02692-8

Hegerl, G. C., Zwiers, F. W., Stott, P. A., & Kharin, V. V. (2004). Detectability of anthropogenic changes in annual temperature and precipitation extremes. *Journal of Climate, 17,* 3683–3700. https://doi.org/10.1175/1520-0442(2004)017<3683:DOACIA>2.0.CO;2

Howe, L. C., MacInnis, B., Krosnick, J. A., Markowitz, E. M., & Socolow, R. (2019). Acknowledging uncertainty impacts public acceptance of climate scientists' predictions. *Nature Climate Change, 9,* 863–867. https://doi.org/10.1038/s41558-019-0587-5

IPCC (1995), Climate Change 1995: A report of the Intergovernmental Panel on Climate Change, Second Assessment Report of the Intergovernmental Panel on Climate Change, https://www.ipcc.ch/assessment-report/ar2/

IPCC (2021). Climate change 2021: The physical science basis. Contribution of Working Group I to the Sixth Assessment Report of the Intergovernmental Panel on Climate Change [Masson-Delmotte, and Coauthors (eds.)]. Cambridge University Press. Cambridge University Press, Cambridge, UK and New York, NY, USA, 2391 pp., doi:https://doi.org/10.1017/9781009157896.

IPCC (2022) Annex II: Glossary: climate change 2022—Impacts, adaptation and vulnerability. Contribution of Working Group II to the Sixth Assessment Report of the Intergovernmental Panel on Climate Change [Pörtner, H.-O. and Coauthors (Eds.).] Cambridge University Press, Cambridge UK and New York NY, USA (2023), pp. 2897–2930, DOI: https://doi.org/10.1017/9781009325844.029.

Kandlikar, M., Risbey, J., & Dessai, S. (2005). Representing and communicating deep uncertainty in climate-change assessments. *Comptes Rendus Geoscience, 337,* 443–455. https://doi.org/10.1016/j.crte.2004.10.010

Kause, A., Bruine de Bruin, W., Persson, J., Thorén, H., Olsson, L., Wallin, A., Dessai, S., & Vareman, N. (2022). Confidence levels and likelihood terms in IPCC reports: A survey of experts from different scientific disciplines. *Climatic Change, 173,* 2. https://doi.org/10.1007/s10584-022-03382-3

Lazarus, R. J. (2009). Super wicked problems and climate change: Restraining the present to liberate the future. *Cornell Law Review, 94,* 5. Georgetown Public Law Research No. 1302623, https://ssrn.com/abstract=1302623

Lempert, R., Nakicenovic, N., Sarewitz, D., & Schlesinger, M. (2004). Characterizing climate-change uncertainties for decision-makers. *Climatic Change, 65,* 1–9. https://doi.org/10.1023/B:CLIM.0000037561.75281.b3

Lempert, R. J., Groves, D. G., Popper, S. W., & Bankes, S. C. (2006). A general, analytic method for generating robust strategies and narrative scenarios. *Management Science, 52,* 514–528. https://www.jstor.org/stable/20110530

Lempert, R. J., Lawrence, R., Kopp, R. E., Haasnoot, M., Reisinger, A., Grubb, M., & Pasqualino, R. (2024). The use of decision making under deep uncertainty in the IPCC. *Frontiers in Climate, 6,* 1380054. https://doi.org/10.3389/fclim.2024.1380054

Marchau, V. A., Walker, W. E., Bloemen, P. J., & Popper, S. W. (2019). *Decision making under deep uncertainty: From theory to practice* (p. 405). Springer Nature. https://doi.org/10.1007/978-3-030-05252-2

Materia, S., et al. (2024). Artificial intelligence for climate prediction of extremes: State of the art, challenges, and future perspectives. *Wiley Interdisciplinary Reviews: Climate Change,* e914. https://doi.org/10.1002/wcc.914

Min, S.-K., Zhang, X., & Zwiers, F. W. (2008). Human-induced Arctic moistening. *Science, 320,* 518–520. https://doi.org/10.1126/science.1153468

Min, S.-K., Zhang, X., Zwiers, F. W., & Hegerl, G. C. (2011). Human contribution to more-intense precipitation extremes. *Nature, 470,* 378–381. https://doi.org/10.1038/nature09763

Mitchell, T. D., & Hulme, M. (1999). Predicting regional climate change: Living with uncertainty. *Progress in Physical Geography, 23,* 57–78. https://doi.org/10.1177/030913339902300103

Morgan, M. G., Henrion, M., & Small, M. (1990). *Uncertainty: A guide to dealing with uncertainty in quantitative risk and policy analysis* (p. 332). Cambridge University Press.

Müller, C., & Robertson, R. D. (2013). Projecting future crop productivity for global economic modeling. *Agricultural Economics, 45*(1), 37–50. https://doi.org/10.1111/agec.12088

Nelson, G. C., et al. (2013). Climate change effects on agriculture: Economic responses to biophysical shocks. *Proceedings of the National Academy of Sciences, 111,* 3274–3279. https://doi.org/10.1073/pnas.1222465110

Oreskes, N. (2015). The fact of uncertainty, the uncertainty of facts and the cultural resonance of doubt. *Philosophical Transactions of the Royal Society A, 373,* 20140455. https://doi.org/10.1098/rsta.2014.0455

Otto, F. E., Philip, S., Kew, S., Li, S., King, A., & Cullen, H. (2018). Attributing high-impact extreme events across timescales—A case study of four different types of events. *Climatic Change, 149,* 399–412. https://doi.org/10.1007/s10584-018-2258-3

Risser, M. D., & Wehner, M. F. (2017). Attributable human-induced changes in the likelihood and magnitude of the observed extreme precipitation during hurricane Harvey. *Geophysical Research Letters, 44,* 12457–12464. https://doi.org/10.1002/2017GL075888

Santer, B. D., et al. (2007). Identification of human-induced changes in atmospheric moisture content. *Proceedings of the National Academy of Sciences, 104,* 15248–15253. https://doi.org/10.1073/pnas.0702872104

Schneider, T., Leung, L. R., & Wills, R. C. (2024). Opinion: Optimizing climate models with process knowledge, resolution, and artificial intelligence. *Atmospheric Chemistry and Physics, 24,* 7041–7062. https://doi.org/10.5194/acp-24-7041-2024

Shepherd, T. G., et al. (2018). Storylines: An alternative approach to representing uncertainty in physical aspects of climate change. *Climatic Change, 151,* 555–571. https://doi.org/10.1007/s10584-018-2317-9

Simmonds, E. G., et al. (2022). Insights into the quantification and reporting of model-related uncertainty across different disciplines. *iScience, 25,* 105512. https://doi.org/10.1016/j.isci.2022.105512

Smith, K. A., Wilby, R. L., Broderick, C., Prudhomme, C., Matthews, T., Harrigan, S., & Murphy, C. (2018). Navigating cascades of uncertainty—As easy as ABC? Not quite.... *Journal of Extreme Events, 1850007.* https://doi.org/10.1142/S2345737618500070

Stott, P., Stone, D., & Allen, M. (2003). Human contribution to the European heatwave of 2003. *Nature, 432,* 610–614. https://agupubs.onlinelibrary.wiley.com/doi/full/10.1029/2003GL017324

Sutton, R. T. (2019). Climate science needs to take risk assessment much more seriously. *Bulletin of the American Meteorological Society, 100,* 1637–1642. https://www.jstor.org/stable/27028499

Tebaldi, C., & Knutti, R. (2007). The use of the multi-model ensemble in probabilistic climate projections. *Philosophical Transactions of the Royal Society A, 365,* 2053–2075. https://doi.org/10.1098/rsta.2007.2076

USGCRP, 2023: *Fifth National Climate Assessment.* Crimmins, A.R., C.W. Avery, D.R. Easterling, K.E. Kunkel, B.C. Stewart, and T.K. Maycock, Eds. U.S. Global Change Research Program, Washington, DC, USA. https://doi.org/10.7930/NCA5.2023.

van Vuuren, D. P., et al. (2011). How well do integrated assessment models simulate climate change? *Climatic Change, 104,* 255–285. https://doi.org/10.1007/s10584-009-9764-2

Vicedo-Cabrera, A. M., et al. (2021). The burden of heat-related mortality attributable to recent human-induced climate change. *Nature Climate Change, 11*, 492–500. https://doi.org/10.1038/s41558-021-01058-x

Wilby, R. L., & Harris, I. (2006). A framework for assessing uncertainties in climate change impacts: Low-flow scenarios for the River Thames, UK. *Water resources Research, 42*, W02419. https://doi.org/10.1029/2005WR004065

Wilby, R., & Lu, X. (2022). Tailoring climate information and services for adaptation actors with diverse capabilities. *Climatic Change, 174*, 33. https://doi.org/10.1007/s10584-022-03452-6

Wilby, R. L., Troni, J., Biot, Y., Tedd, L., Hewitson, B. C., Smith, D. M., & Sutton, R. T. (2009). A review of climate risk information for adaptation and development planning. *International Journal of Climatology, 29*, 1193–1215. https://doi.org/10.1002/joc.1839

World Meteorological Organization (WMO) (2024). Global temperature is likely to exceed 1.5°C above pre-industrial level temporarily in next 5 years. Retrieved October 17, 2024, from https://wmo.int/news/media-centre/global-temperature-likely-exceed-15degc-above-pre-industrial-level-temporarily-next-5-years

Yates, D., Miller, K. A., Wilby, R. L., & Kaatz, L. (2015). Decision-centric adaptation appraisal for water management across Colorado's Continental Divide. *Climate Risk Management, 10*, 35–50. https://doi.org/10.1016/j.crm.2015.06.001

Yates, D., Szinai, J., & Jones, A. D. (2023). Modeling the water systems of the Western US to support climate-resilient electricity system planning. *Earth's Futures, 12*, e2022EF003220. https://doi.org/10.1029/2022EF003220

Zhang, X., Zwiers, F. W., Hegerl, G. C., Lambert, F. H., Gillett, N. P., Solomon, S., Stott, P., & Nozawa, T. (2007). Detection of human influence on 20th century precipitation trends. *Nature, 448*, 461–465. https://doi.org/10.1038/nature06025

Open Access This chapter is licensed under the terms of the Creative Commons Attribution 4.0 International License (http://creativecommons.org/licenses/by/4.0/), which permits use, sharing, adaptation, distribution and reproduction in any medium or format, as long as you give appropriate credit to the original author(s) and the source, provide a link to the Creative Commons license and indicate if changes were made.

The images or other third party material in this chapter are included in the chapter's Creative Commons license, unless indicated otherwise in a credit line to the material. If material is not included in the chapter's Creative Commons license and your intended use is not permitted by statutory regulation or exceeds the permitted use, you will need to obtain permission directly from the copyright holder.

Laying the Policy Groundwork for Considering Integrated Uncertainty

Sarah Michaels

2.1 Introduction

Policymaking under uncertainty has gained prominence with increasing concern about climate change (Jensen & Wu, 2016). Climate change is a long-term policy problem in which uncertainties arise from the long lags between policies and their consequences, our incomplete understanding of complex systems and climate change's interconnectedness with other issues (Dewulf & Biesbroek, 2018). Exacerbating the challenge is that for communities that do not perceive they have yet experienced climate change impacts firsthand; the threat appears remote temporally and spatially. Under these circumstances, it is particularly difficult to convey the urgency of acting (Moser, 2010).

While there are a range of reactions to uncertainty (Curry & Webster, 2011), one constructive way to think about uncertainty in the policy process is as "doubt that threatens to block action" (Schmitt & Klein, 1996 63). When doubt leads to a reluctance to act, the resulting hesitation can cause delay or the window of opportunity may close in which action is possible (Schmitt & Klein, 1996), or most effective as for climate change. This is a challenge for problem-solving, the core of making public policy (Birkland, 2016).

To lay the groundwork for considering uncertainty in the realm of climate change policy requires understanding the generic policy sphere in which decision-making impacting climate change research and societal response to climate change takes place. Consequently, in this chapter we explore the policy sphere using climate change policy examples.

S. Michaels (✉)
Department of Political Science and Nebraska Public Policy Center, University of Nebraska-Lincoln, Lincoln, NE, USA
e-mail: sarah.michaels@fulbrightmail.org

2.2 The Policy Sphere

It is novel and instructive to think about the policy sphere as consisting of three complementary segments (Fig. 2.1). The bottom segment consists of the conceptual foundations of the policy process. This is the realm in which policy scholars work to understand and explain how the policy process functions regardless of the specific policy topics they may use as testbeds for their ideas. It is where scholars work to advance the science of the policy process (Weible, 2017). The second segment is policy analysis, the systematic process of prescriptive activities used to aid decision-making (Patton et al., 2013). This is the domain of policy analysts who use analytical techniques applicable across specific policy topics to provide advice to decision-makers. The third segment is subject knowledge. This is the province of experts, often people who have acquired specialized training and work to advance knowledge in a specific area of knowledge, such as climate science. The boundaries between these segments are permeable, and working at the interface between segments is the source of new insights to those working in the adjoining segments.

2.2.1 Conceptual Foundations of the Policy Process

Accepting and working with uncertainty throughout the policy process is a necessity given uncertainty is the norm in public policy problems. The policy process is how politics, which Lasswell notes within his 1936 book title *Who Gets What*, is translated into policies, which Dye (1972 2) observes is "what government chooses to do or not to do." The conceptual foundations of the policy process consist of the efforts by scholars to explain how and why public policy originates, is shaped, and executed. This includes considering failure along with success and how that distinction transpires and is determined.

2.2.1.1 Stages of the Policy Process Heuristic

The stages of the policy process, also referred to as the textbook model, are widely used as a basis for analyzing and studying the policy process (Birkland, 2016; DeLeon, 1999) and as a foil for more sophisticated depictions of the policy process (Birkland, 2016). It breaks down the process into several stages, with different authors providing variations in demarcating the stages (Birkland, 2016; Bardach & Patashnik, 2020; Patton et al., 2013). Each stage in the decision-making process constitutes a potential point of entry for uncertainties (Sigel et al., 2010) and opportunities for policy analysis (Bardach & Patashnik, 2020; Patton et al., 2013). Below, the policy process is presented as eight stages (Fig. 2.2): (1) issue emergence (problem definition), (2) agenda setting, (3) generating alternatives, (4) selecting among alternatives, (5) enactment, (6) implementation, (7) monitoring, and (8) evaluation. Feedback in the form of monitoring and evaluation contributes to issue emergence and reiteration of the cycle. This eight-stage characterization is a variation of Patton et al.'s (2013) step-by-step depiction of policy analysis combined with Birkland's (2016) presentation of the stages model of the policy process.

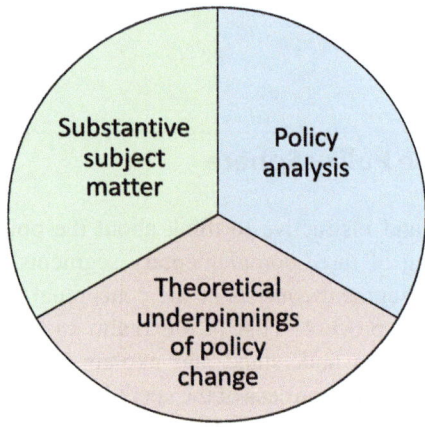

Fig. 2.1 The policy sphere consisting of three complementary segments

Issue Emergence or Problem Definition

In the policy domain, a distinction is made between conditions and problems. Conditions are understood as circumstances with which we can live, and problems as issues that must be addressed. Problems are defined in such a way they

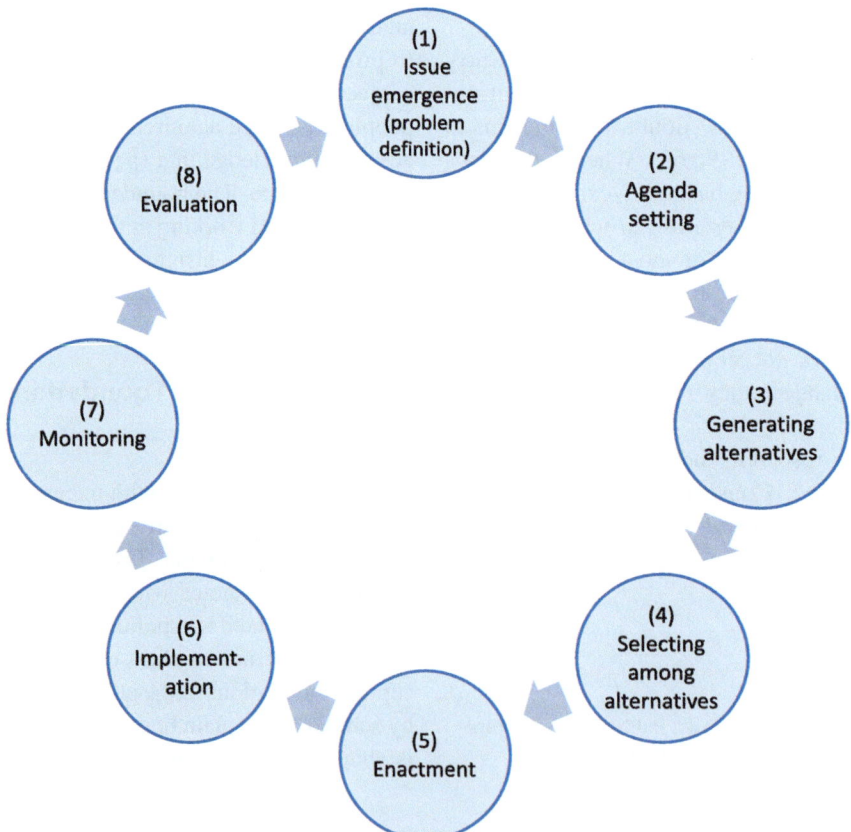

Fig. 2.2 Stages of the policy process

can be addressed by "solutions." Conditions become problems through social and economic changes and/or scientific and technological advances. For example, it became more plausible to think about effectively and economically reducing the carbon footprint with the development of commercial-scale renewable energy.

The policy process in the stages heuristic begins with defining what is the problem to be addressed. This is not straightforward as there may not be consensus about the nature and scope of what constitutes "the" problem, as is the situation with climate change. This may be a function of the range of social, political, and economic considerations that actors bring to the issue. Some definitions are more likely to be accepted and adopted than others. Advocates of a particular definition work hard to convince others of the merit of their chosen definition. The problem may well be defined in the context of the arena in which it is being addressed. For example, in climate change, a local government may define climate change in the context of what options it has to address a limited number of aspects of it. For example, municipalities can focus on decarbonizing local transport. This can include helping individuals make trips using more carbon-efficient means than personal vehicles and reducing how far and how many trips individuals make, as was done in Bath Riverside, England, as well as promoting zero tailpipe emission vehicles for trips requiring vehicles (Campbell et al., 2020). How a problem is defined often influences its treatment during the policy process (Schattschneider, 1960).

Problems generally become known in one of two ways (Kingdon, 2011). (1) There are changes in problem indicators (Birkland, 2016). For example, the IPCC shared the 2007 Nobel Peace Prize because of its meticulous approach to demonstrating the changes, notably human induced, in key indicators of climate change. At least in theory, considerable attention is paid to the issue if the data indicates a deteriorating situation (Kingdon, 2011), and these are interpreted and publicized by policy influencers (Birkland, 2016). Yet, those engaged in generating such evidence are often frustrated that making known monitoring demonstrating changes in indicators rarely leads directly and immediately to policy change (Michaels & Tyre, 2012). As climate change scientists have discovered, it does not even result in appropriate coverage in the politically influential US prestige press (Boykoff & Boykoff, 2004). What has been discernible is a correlation between increased concern about climate change and the release of new scientific reports, notably by the Nobel Prize-winning International Panel on Climate Change (IPCC) or routine political activity, such as US presidential campaigns (Hulme, 2008; Boykoff & Boykoff, 2004). (2) There is a focusing event, "a sudden, exceptional experience that, because of how it leads to harm or exposes the prospect for great devastation, is perceived as the impetus for policy change" (Michaels et al., 2006 983). Major weather events, such as Superstorm Sandy, have been presented as focusing events indicative of climate change (Feldpausch-Parker et al., 2018).

Agenda Setting

Agenda setting refers to how particular topics come to the attention of policymaking decision-makers (Cobb & Elder, 1983). This may occur through government officials sensing, often with the aid of opinion polls, changes in how a large number of people in a jurisdiction think about a topic. It may happen through the campaigning of interest groups and also through legislatures, notably when there are changes in their memberships (Herweg et al., 2017). Agenda setting has been one of the first and most heavily theorized stages of the policy process (Kingdon, 2011). An issue can reach the agenda in the same two ways, discussed above, of how problems become known through changes in indicators and through focusing events.

As agenda setting is an important expression of power, actors may work to keep an issue off the agenda while others work to get it on the agenda. Here is an example of the latter. On September 4, 2019, each of the then 10 qualifying Democratic presidential candidates answered questions individually about prospective climate change policies in 40-minute town hall segments broadcast on CNN. While the candidates may have had differing policy proposals, the overall intent of the seven-hour climate talkathon was to highlight the Democratic Party's agenda-setting commitment to climate change. Arguably, it was historic for two reasons. First, because the candidates made the cases as to why they should be the presidential nominee for their party based on their plans to address climate change. Second, the seven-hour presentation may well have been the longest stretch of programming up until then devoted to climate change by a US news network (Dickinson et al., 2019).

Agenda setting matters because there is limited agenda space for issues. For example, coverage of the Democrats' climate town hall was punctuated with updates on the progress of Hurricane Dorian. Arguably, in this case the main issue and the interrupting issue were complementary. Advances in extreme event attribution are leading to insights about the extent to which the intensity or probability of a particular form of weather event is affected by climate change in modelled worlds (National Academies of Sciences, Engineering, and Medicine., 2016).

In the Democratic Party town hall example, there is little variation in the construction of the climate change issue because of the desire to highlight the party's unanimity in its willingness to address climate change. Policy change advocates are aware that divisiveness among the positions of supporters and that more attention on an issue yields more negative attention and decreases the likelihood of making policy change on their chosen issue. "Policy change refers to a detectable deviation from the purposive course of action that is in place to address a perceived societal problem" (Michaels et al., 2006 983).

Generating Alternatives

Generating alternatives refers to coming up with different intervention schemes to ameliorate or solve a problem (Bardach & Patashnik, 2020). Thinking about how a given problem framing leads to an array of options may well reveal information-related uncertainties. Different problem framing, such as whether addressing climate change is a sufficiency or an efficiency problem, leads to considering different arrays of options. If climate change is framed as a sufficiency problem, the challenge is to pursue what is adequate to get the job done when more could be done. For example, it is using a bicycle rather than a car or hanging clothes out to dry on a clothing line rather than using a clothes dryer. If climate change is framed as an efficiency problem, the challenge is to pursue getting more out of resources expended. For example, it is increasing how far an automobile can travel on a tank of gas or reducing how much energy is necessary to power a clothes dryer. Consequently, when there are differences in problem framing there may not be agreement on what constitutes the appropriate array of policy options to consider. Some may think better options than those now proposed will be available in the near future because of anticipated new technologies (Hadorn et al., 2015). The range of policy alternatives generated can be wide or narrow, mutually compatible or redundant. For example, in addressing climate change, one option maybe to promote development of solar power, another to use carbon-based fuels more efficiently. More narrowly, a carbon tax maybe decided upon and so the alternatives relate to how much the tax should be, on whom, and on what activities it will be imposed. When alternatives are not mutually exclusive, the preferred option may well be a mix of alternatives.

Selecting Among Alternatives

It is rare one alternative is anticipated to produce better outcomes across all evaluation criteria than other alternatives. Usually, there are trade-offs between values expressed as projected outcomes produced by different alternatives. That these outcomes are projected means they are attached to a range of forecasting-associated uncertainties. Common trade-offs include between money and extent of desired output, such as risk reduction; and between costs carried privately, such as installing emissions abatement equipment, and social benefits, protecting forests (Bardach & Patashnik, 2020). A key criterion separating one alternative from another and action on the same alternative in two different time frames is political feasibility. For example, the United States signed the Paris Climate Agreement, committing signatory countries to reduce carbon dioxide emissions and similar emissions, during one presidential administration and then in another gave notice the United States would withdraw from it.

Choosing an option to implement may be seen as the concluding step in deciding how to cast the decision problem. Yet this may be premature if further learning is required because of outstanding, uncertain dimensions of the decision problem. "Learning may include identifying or recategorizing uncertainties of information on the basis of scientific analysis and new information, revising information because of changing circumstances, reconsidering ethical positions and reassessing the feasibility or efficiency of policies" (Hadorn et al., 2015 115). In the case of climate change policy, not only are there outstanding, uncertain dimensions of the decision problem, there is also recognition that future policies will need to be bolder, more malleable, and expansively scoped than what is in place (National Research Council, 2010).

Enactment

Enactment refers to when a governing authority sets out an action or a rule of conduct. The term is often associated with legislative action. For example, in general, federal statute laws in the United States are enacted when the president signs legislation presented by Congress or when Congress overrides a presidential veto (Birkland, 2016). The president can act through other means. For example, on November 1, 2013, President Obama issued an Executive Order—Preparing the United States for the Impacts of Climate Change. Enactments can be and are reversed, as occurred on March 28, 2017, when President Trump signed an Executive Order revoking the above President Obama's Order. Enactment can also occur at a subnational scale. For example, the State of California adopted the Low Carbon Fuels Standard (LCFS) as a regulatory effort to reduce the carbon content of transportation fuel used in the State.

Implementation

Enactment of laws is not simply translated into actions by agencies or complied with by citizens. Government agencies have to ascertain what legislation requires or enables them to do. For example, while implementing the iterative EU Water Framework Directive, responsible entities sought opportunities to embed adaptation measures within it (Wilby et al., 2006). Translating legislation into rules and regulations can be difficult and contentious. Also, it is often during implementation attention to detail flags because people think the problem has been addressed by the passage of legislation (Birkland, 2016).

Some policies are implemented successfully while others are not. Most fall short of what their most enthusiastic proponents claim they will achieve (Birkland, 2016). Failure may be attributed to poor policy design or theory or to disagreements between federal, state, local, and street-level implementers or to insufficient resources (National Research Council, 2010).

Since the Paris Agreement, the role of climate science is more about aiding in implementing and monitoring policy actions than previously, when arguably its key role had been to provide scientific evidence global warming was taking place (Beck & Mahony, 2018).

Monitoring

Monitoring involves routinely collecting data on policy performance indicators (Schoenefeld et al., 2018). It is integral to the bottom-up system of international climate governance that began to be emphasized in the 2009 Copenhagen Summit, officially known as the United Nations Climate Change Conference (COP15), because such a system is predicated on sound monitoring for assessing both past performance and estimating future contributions. The 2015 Paris Agreement committed signatory countries to report regularly on their emission reduction pledges known as Nationally Determined Contributions (NDCs) pledges to tackle climate change. Public policies in many countries provide the means for converting pledges into actions. Salient insights of how this might work come from the European Union's experience, as a self-declared climate leader, of monitoring national climate policies. The EU's decision to employ a predominantly technical interpretation of four international reporting quality criteria—completeness, consistency, comparability, and transparency—constrained knowledge production and debate on how individual climate policies performed. Impediments to more in-depth reporting include political concerns about the costs and burdens of reporting, who determines what should be monitored, how useful the reporting information is perceived to be, and the political control emanating from policy knowledge (Schoenefeld et al., 2018).

Evaluation

Evaluation involves producing information about the worth of policy outcomes, a function of the extent to which they contribute to goals and objectives. It does so through assessing policy performance and clarifying and critiquing values underpinning the choice of goals and objectives. Commonly used evaluation criteria include effectiveness, efficiency, adequacy, equity, responsiveness, and appropriateness (Dunn, 2012). A change in outcome is not necessarily attributable to a particular policy. For example, various policy measures may produce improvements in energy efficiency and macroeconomic shifts may influence energy prices, incentivizing or disincentivizing energy efficiency (European Environmental Agency, 2016).

In more than one variant of the idealized stages of the policy process, feedback based on monitoring and evaluation leads to refining any of the stages and may suggest the emergence of new issues (Birkland, 2016). For example, documenting poor policy performance may lead to reconfiguring the policy problem or suggesting the preferred alternative be dropped (Dunn, 2012).

2.2.1.2 Problems with Stages Heuristic

The stages heuristic is appealing for its simplicity. It intuitively makes sense, conforming with stages familiar to anyone exposed to linear project planning. It is a heuristic and as such not every step always happens, the sequence of steps does not always happen in the same order, the process is not always completed (Birkland, 2016). Even more fundamentally, Campbell et al. (2019) argue climate change may be unframeable because of three of its ontological dimensions: unboundedness, incalculability, and unthinkability.

More sophisticated theories provide a deeper understanding of the policy process. While beyond the scope of this chapter to explore, it is important to acknowledge the theories and frameworks of the policy process literature that have eclipsed, at least in policy scholarship, the stages heuristic. They share a recognition of the importance of individual actors, such as policy entrepreneurs, groups, including coalitions of groups, and the role of ideas and reject the stages heuristic based on its lack of defensible conceptual foundations (Weible & Sabatier, 2017). Yet the stages heuristic continues to be employed in practice. It fits with the idealized step-by-step planning process, has an easy-to-grasp logic, and perceived widespread applicability.

2.2.2 Policy Analysis

The second segment of the policy sphere is policy analysis. Intended to improve policymaking (Dunn, 2012 53), policy analysis involves providing informed rationales for making decisions to solve collective, real-world problems (Dunn, 2012; Patton et al., 2013). The purpose of policy analysis is to help policymakers choose among complex options under uncertain conditions. Policy analysis is not an open-ended, unconstrained exercise in exploring the dimensions of a pressing issue. Rather it is characterized by a search conducted in a limited time frame, directed at a particular issue, constrained set of alternatives, preparation of document leading to action, client alignment, problem orientation, time horizon influenced by the terms of elected office and uncertainty, and politically sensitive strategy (Patton et al., 2013).

Policy analysts use multiple methods of inquiry often drawn from applied social sciences. Typically they do so to address five types of questions: (1) What is the problem we are trying to solve? (defining the problem), (2) On what basis should we choose which policies to execute? (establishing evaluation criteria), (3) What are the different bundles of anticipated results of our policy decisions? (designing policy alternatives), (4) What are the on-the-ground consequences of the policy implemented? (monitoring), and (5) To what extent do these consequences fulfill the intent of the policy and/or mitigate the current form of the problem? (evaluation) (Dunn, 2012). These questions inform and highlight individual stages of the stages heuristic outlined above.

Policy analytic methods are intended to structure problems, forecast expected policy outcomes, construct alternatives, prescribe preferred policies, and monitor and evaluate observed policy outcomes (Dunn, 2012; Bardach & Patashnik, 2020). At the agenda-setting stage of policymaking, problem structuring may reveal hidden assumptions underpinning problem definition, identify problem origins, identify possible objectives, reconcile conflicting perspectives, and generate new policy choices. Anticipated consequences of adopting preferred policies may be revealed through forecasting. Forecasting contributes to examining possible future scenarios, estimating the consequences of future policies, and specifying likely constraints on fulfilling intended objectives (Dunn, 2012). For example, the United Kingdom Committee on Climate Change (2015) based its recommendation for the United Kingdom's fifth carbon budget for 2028–2032 on a number of scenarios it built for reducing emissions leading up to 2050.

Ambiguously, an alternative may refer to an intervention involving foregoing an option or it may refer to an option that can be undertaken in combination with one or more other options (Bardach & Patashnik, 2020). In developing an initial list of prospective options, policy analysts may begin by noting those being considered by critical political actors and then trying to design alternatives to the ones already being floated (Bardach & Patashnik, 2020). In suggesting preferred policies, analysts may specify criteria for policy selection, indicate which administrative unit should have the authority and responsibility to implement a policy, detect externalities, and approximate risk and uncertainty levels. Monitoring indicators of policy outcomes and impacts assist in policy implementation by measuring the consequences of adopted policies. Extent of compliance, unintended consequences, obstacles and constraints to implementation, and encouraging administrative accountability may be a product of monitoring. Discrepancies between anticipated and actual policy performance is how evaluation aids policy assessment and adaptation. Doing so may lay the groundwork for restructuring problems, modifying or reconstituting policies, clarifying and critiquing values, underpinning policies, and establishing the extent to which the original problem has been alleviated (Dunn, 2012).

While some, if not all, of the above methods may be familiar to climate scientists from their own work, policy analysis and research are not always interchangeable. Compared to research, policy analysis is conducted under more severe time constraints, can be less concerned with generating new data, explicitly considers alternatives, including the impact of predicted outcomes on various interests, may indicate a favored option, is not intended to discuss disciplinary contributions (Lindquist, 1988), or give primacy to informing theory. These differences may well lead to discrepancies in what uncertainties are considered and how they might be treated.

Since the presence of uncertainty is the norm in public policy problems, accepting and working with uncertainty is a necessity in policy analysis. While sensitivity testing and decision-analytic methods can be helpful when values for certain variables cannot be obtained (Patton et al., 2013), uncertainty continues to be a challenging dimension for climate change policy analysis as it does for the valuation of adaptation outcomes. For example, differences in cost–benefit valuations result from how to value uncertain outcomes, notably high-consequence events for which probabilities are thought to be low at low levels of warming and become more pronounced with greater climate forcing increase. Plausibly such outcomes could lead to significant irreversibilities in climate-impacted systems and in the climate itself. An example of such irreversibilities is the release of large amounts of greenhouse gases from warming permafrost (National Research Council, 2010).

2.2.3 Subject Knowledge

Subject matter expertise, the third segment of the policy sphere, is highly valued in rational-based approaches to policymaking. The insights of experts may spur policy change, reinforce existing choices, and be used to validate what decision-makers want to do, notwithstanding competing evidence. Those educated in climate sciences are likely to find themselves engaged in policy through their substantive, subject matter knowledge. This contrasts with professionals working out of the other two policy sphere segments, who may have developed less subject matter expertise and more capability, relevant across subjects, related to politics and policy analysis.

While subject matter expertise is needed, some experts do not know how the policy process works, worry their work may be "misused," and/or fear participating in the policy process will not advance their careers. The latter may become less of a concern with increasing interdisciplinarity, heightened understanding of the needs of information users (Dow & Carbone, 2007), and mounting pressure for policy-relevant research. For example, this includes the United States' National Science Foundation considering broader impacts in evaluating funding proposals and the United Kingdom's Research Excellence Framework incentivizing "impactful" research.

Employing expertise increases in importance as understanding the connections between science and policy becomes more refined (McNie, 2007; Michaels, 2009). Recognizing societal reliance on nonstationary natural systems makes forecasting integral to decision-making about the future (Michaels & Tyre, 2012). Projecting climate change and impacts, based on scenarios and storylines, links climate change science and policy (Feichter & Gramelsberger, 2011). The IPCC has gone to great lengths to establish protocols of how to inform policymaking based on rigorous, peer-reviewed scientific findings. Integral to the IPCC Assessment Reports' extended review process is identifying and assessing climate change projections uncertainty to support policymaking under uncertainty (Feichter & Gramelsberger, 2011).

Natural scientists confront uncertainty as a research process attribute and thus are familiar with the shortcomings of the models and tools employed in representing the external world. Gathering more data or refining models are widespread means of addressing epistemic uncertainty (Kundzewicz et al., 2018). Natural scientists also recognize aleatory uncertainty, a function of a phenomenon's intrinsic randomness (Kundzewicz et al., 2018), or ontological uncertainty, a function of innately complex structure performance (Dewulf & Biesbroek, 2018), which cannot be reduced (Kundzewicz et al., 2018).

Climate change scientists engaging in the policy process need to recognize human activity involves at least three attributes leading to social indeterminism: random process, myopicness, and intentionality (Pielke Jr., 2007). The first is random process. While people may be clear on what they wish to achieve, they may employ a haphazard series of actions in their efforts to obtain a result. The second is short-sightedness; people consider only the short term or weigh heavily the immediate over the long term. The third is intentionality, usually a function of behavior shaped by contingencies, in which people learn directly through their own experiences of the consequences of their actions.

Detecting the presence of social indeterminism, such as when there is value-based disagreement about whether to reduce greenhouse gas emissions, may help scientists distinguish when their findings are unlikely to influence decision-making (Michaels & Tyre, 2012). Bringing climate change science to bear in policy does not reduce social indeterminism. Categorizations of uncertainty suggest when science may or may not influence decision-making (Michaels & Tyre, 2012) and that different strategies are required to address different sources and forms of uncertainty (Jensen & Wu, 2016).

2.3 Conclusion

This chapter has laid out the policy groundwork for considering integrated uncertainty in the context of climate change research. To do so, it has employed a depiction of three segments constituting the policy sphere: (1) conceptual foundations of the policy process, (2) policy analysis, and (3) subject knowledge (Fig. 2.1). While this depiction is applicable to disparate policy topics, in this chapter it has been used to help structure thinking about climate change policy, including how uncertainty factors into its policymaking. In the first segment, conceptual foundations of the policy process, the stages of the policy process heuristic were presented. While there are theoretical and pragmatic limitations to this heuristic, by breaking down the policy process into different steps, it becomes apparent how uncertainty is a consideration throughout the process. The discussion of the second segment, policy analysis, brings to the fore how generic tools, such as cost–benefit analysis, used across policy domains are applicable and can be tailored specifically to climate change policy. It is worth noting how policy analysis, with its emphasis on providing timely input and its lack of emphasis on originality, constitutes a different form of advice than traditional, scientific research, which prizes meticulous investigative process, thorough documentation, and originality. The third segment, subject knowledge, highlights what is distinct about the contribution of those with substantive expertise, such as in investigating climate change. It is important to recognize when scientific expertise is more and less applicable to advancing climate change policy. This may well be a function of what uncertainty is present at a particular point in the process of climate change policymaking. The challenge for incorporating uncertainty into climate change policymaking is to do so in such a way that acknowledges the origins, actuality, and implications of uncertainty without paralyzing decision-making.

Acknowledgments The author thanks Robert Wilby and Linda Mearns for their valuable remarks on an earlier draft of this chapter and Seth McGinnis for help in modifying the presentation of the figures.

References

Bardach, E., & Patashnik, E. M. (2020). *A practical guide for policy analysis* (6th ed., p. 190). Sage CQ Press.

Beck, S., & Mahony, M. (2018). The IPCC and the new map of science and politics. *WIREs Climate Change, 9*, e547. https://doi.org/10.1002/wcc.547. 16 pp.

Birkland, T. A. (2016). *An introduction to the policy process: Theories, concepts, and models of public policy making* (4th ed., p. 399). Routledge.

Boykoff, M. T., & Boykoff, J. M. (2004). Balance as bias: Global warming and the US prestige press. *Global Environmental Change, 14*, 125–136. https://doi.org/10.1016/j.gloenvcha.2003.10.001

Campbell, N., McHugh, G., & Ennis, P. J. (2019). Climate change is not a problem: Speculative realism at the end of organization. *Organization Studies, 40*(5), 725–744. https://doi.org/10.1177/0170840618765553

Campbell, M., Walker, R., Marsden, G., McCulloch, S., Jenkinson, K., & Anable, J. (2020). *Decarbonising transport: The role of land use, localisation and accessibility* (p. 14). Local Government Association.

Cobb, R. W., & Elder, C. D. (1983). *Participation in American politics: The dynamics of agenda-building* (2nd ed., p. 196). Johns Hopkins University Press.

Committee on Climate Change (2015). *The fifth carbon budget*: The next step towards a low-carbon economy. Presented to the Secretary of State pursuant to section 34 of the Climate Change Act 2008 [United Kingdom], 130 pp.

Curry, J. A., & Webster, P. J. (2011). Climate science and the uncertainty monster. *Bulletin of American Meteorological Society, 92*, 1667–1682. https://doi.org/10.1175/2011BAMS3139.1

DeLeon, P. (1999). The stages approach to the policy process: What has it done? Where is it going? In P. A. Sabatier (Ed.), *Theories of the policy process* (1st ed., pp. 19–34). Westview Press.

Dewulf, A., & Biesbroek, R. (2018). Nine lives of uncertainty in decision-making: Strategies for dealing with uncertainty in environmental governance. *Policy and Society, 37*(4), 441–458. https://doi.org/10.1080/14494035.2018.1504484

Dickinson, T., Bort, R., & Reis, P. (2019). 10 takeaways from the Democrats' historic climate town hall. *Rolling Stone, 5*, 2019. https://www.rollingstone.com/politics/politics-news/cnn-climate-crisis-town-hall-democrats-sanders-warren-harris-biden-880176/

Dow, K., & Carbone, G. (2007). Climate science and decision making. *Geography Compass, 1*(3), 302–324. https://doi.org/10.1111/j.1749-8198.2007.00036.x

Dunn, W. N. (2012). *Public policy analysis: An introduction* (5th ed., p. 460). Routledge.

Dye, T. R. (1972). *Understanding public policy* (p. 305). Prentice-Hall.

European Environmental Agency. (2016). *Environment and climate policy evaluation*. EEA Report No 18/2016. European Environmental Agency, pp. 24 https://doi.org/10.2800/68508.

Feichter, J., & Gramelsberger, G. (2011). Chapter 1 introduction to the volume. In G. Gramelsberger & J. Feichter (Eds.), *Climate change and policy* (8). Springer-Verlag. https://doi.org/10.1007/978-3-642-17700-2_1

Feldpausch-Parker, A. M., Peterson, R., Rai, T., Stephens, J. C., & Wilson, E. J. (2018). Smart grid electricity system planning and climate disruptions: A review of climate and energy discourse post-superstorm Sandy. *Renewable and Sustainable Energy Reviews, 82*, 1961–1968. https://doi.org/10.1016/j.rser.2017.06.015

Hadorn, G. H., Brun, G., Saliva, C. R., Stenke, A., & Thomas, P. (2015). Decision strategies for policy decisions under uncertainties: The case of mitigation measures addressing methane emissions from ruminants. *Environmental Science & Policy, 52*, 110–119. https://doi.org/10.1016/j.envsci.2015.05.011

Herweg, N., Zahariadis, N., & Zohlnhofer, R. (2017). The multiple streams framework: Foundations, refinements, and empirical applications. In C. Weible & P. Sabatier (Eds.), *Theories of the policy process* (4th ed., pp. 17–53). Westview Press.

Hulme, M. (2008). The conquering of climate: Discourses of fear and their dissolution. *The Geographical Journal, 174*(1), 5–16. https://doi.org/10.1111/j.1475-4959.2008.00266.x

Jensen, O., & Wu, X. (2016). Embracing uncertainty in policy-making: The case of the water sector. *Policy and Society, 35*(2), 115–123. https://doi.org/10.1016/j.polsoc.2016.07.002

Kingdon, J. W. (2011). *Agendas, alternatives, and public policies* (2nd ed., p. 273). Longman.

Kundzewicz, Z. W., Krysanova, V., Benestad, R. E., Hov, Ø., Piniewski, M., & Otto, I. M. (2018). Uncertainty in climate change impacts on water resources. *Environmental Science & Policy, 79*, 1–8. https://doi.org/10.1016/j.envsci.2017.10.008

Lasswell, H. D. (1936). *Politics: Who gets what, when, how*. McGraw-Hill.

Lindquist, E. A. (1988). What do decision models tell us about information use? *Knowledge in Society, 1*(2), 86–111. https://doi.org/10.1007/BF02687215

McNie, E. C. (2007). Reconciling the supply of scientific information with user demands: An analysis of the problem and review of the literature. *Environmental Science & Policy, 10*, 17–38. https://doi.org/10.1016/j.envsci.2006.10.004

Michaels, S. (2009). Matching knowledge brokering strategies to environmental policy problems and settings. *Environmental Science & Policy, 12*, 994–1011. https://doi.org/10.1016/j.envsci.2009.05.002

Michaels, S., & Tyre, A. J. (2012). How indeterminism shapes ecologists' contributions to managing socio-ecological systems. *Conservation Letters, 5*, 289–295. https://doi.org/10.1111/j.1755-263X.2012.00241.x

Michaels, S., Goucher, N., & McCarthy, D. (2006). Policy windows, policy change and organizational learning: Watersheds in the evolution of watershed management. *Environmental Management, 38*, 983–992. https://doi.org/10.1007/s00267-005-0269-0

Moser, S. C. (2010). Communicating climate change: History, challenges, process and future directions. *WIREs Climate Change, 1*, 31–53. https://doi.org/10.1002/wcc.11

National Academies of Sciences, Engineering, and Medicine. (2016). *Attribution of extreme weather events in the context of climate change* (p. 186). The National Academies Press. https://doi.org/10.17226/21852

National Research Council. (2010). *Advancing the science of climate change* (p. 526). The National Academies Press. https://doi.org/10.17226/12782

Patton, C. V., Sawicki, D. S., & Clark, J. (2013). *Basic methods of policy analysis and planning* (3rd ed., p. 464). Pearson.

Pielke, R., Jr. (2007). *The honest broker* (p. 186). Cambridge University Press.

Schattschneider, E. E. (1960). *The semi-sovereign people: A realist's view of democracy in America* (p. 147). Holt, Rinehart and Winston.

Schmitt, J. F., & Klein, G. (1996). Fighting in the fog: Dealing with battlefield uncertainty. *Marine Corps Gazette, 80*, 62–69.

Schoenefeld, J. J., Hildén, M., & Jordan, A. J. (2018). The challenges of monitoring national climate policy: Learning lessons from the EU. *Climate Policy, 18*, 118–128. https://doi.org/10.1080/14693062.2016.1248887

Sigel, K., Klauer, B., & Pahl-Wostl, C. (2010). Conceptualising uncertainty in environmental decision-making: The example of the EU water framework directive. *Ecological Economics, 69*, 502–510. https://doi.org/10.1016/j.ecolecon.2009.11.012

Weible, C. M. (2017). Introduction: The scope and focus of policy process research and theory. In C. Weible & P. Sabatier (Eds.), *Theories of the policy process* (4th ed., pp. 1–13). Westview Press.

Weible, C. M., & Sabatier, P. A. (Eds.). (2017). *Theories of the policy process* (4th ed., p. 402). Westview Press.

Wilby, R. L., Orr, H. G., Hedger, M., Forrow, D., & Blackmore, M. (2006). Risks posed by climate change to delivery of water framework directive objectives. *Environment International, 32*, 1043–1055. https://doi.org/10.1016/j.envint.2006.06.017

Open Access This chapter is licensed under the terms of the Creative Commons Attribution 4.0 International License (http://creativecommons.org/licenses/by/4.0/), which permits use, sharing, adaptation, distribution and reproduction in any medium or format, as long as you give appropriate credit to the original author(s) and the source, provide a link to the Creative Commons license and indicate if changes were made.

The images or other third party material in this chapter are included in the chapter's Creative Commons license, unless indicated otherwise in a credit line to the material. If material is not included in the chapter's Creative Commons license and your intended use is not permitted by statutory regulation or exceeds the permitted use, you will need to obtain permission directly from the copyright holder.

Climate-Informed Decision Analysis Via Decision Scaling

Casey Brown

3.1 The Need for Climate-Informed Decision Analysis

A climate-informed decision analysis is a tailoring of the methods of decision analysis to the opportunities and constraints inherent to the state of the science in climate prediction and projection. The goal of a climate-informed decision analysis is to identify the optimal choice when climate uncertainty is a dominant factor affecting the outcomes of the decision and where climate information, such as derived from instrumental records, palaeoclimatological reconstructions, or climate simulations, is used within the analysis. Optimality is defined by the decision-maker and implemented via a corresponding decision rule. The premise is that the use of climate information may provide value by improving the outcomes of decisions when climate information is included in the decision analysis. However, there are two problems that often arise in the use of climate information, and climate projections in particular, within a decision analytic framework. The first is the mismatch between climate projections and the attributes of the "states of the world" or scenarios that are used in decision analysis, which must be mutually exclusive and collectively exhaustive. We call this challenge scenario ambiguity. The proposed solution is the introduction of the climate stress test, which is a statistical climate scenario generator specifically designed to create climate scenarios that are congruent with the adaptation decision analysis. The second challenge is the incorporation of useful climate information from Global Climate Model (GCM) projections if it is not used as the "states of the world" or scenarios in the decision analytic framework. The proposed solution follows the concepts of pre-posterior analysis, robustness, and ex post scenario analysis (Bryant and Lempert, 2010) and combines them with the concept of categorical climate forecasts to provide a method for including climate information as a weighting factor for assessing vulnerability or selecting a robust decision option.

The result of these two proposed solutions comprises the design of a climate-informed decision analysis previously described as "decision scaling." The name derives from the idea that the climate information is scaled in the sense of the spatial, temporal, and categorical scales to meet the needs of the scenarios that emerge from the decision analysis. This methodology was originally developed for applications of seasonal climate forecasting (e.g., Brown et al., 2006). The need of operational agencies, specifically, those that manage or invest in water infrastructure, for practical guidance on vulnerability analysis and adaptation planning revealed the applicability of decision scaling to these issues and they soon became a focal point.

There are many studies that focus on the impacts of climate change on a human interest. These studies are distinct from climate studies that focus solely on the climate variables themselves, uncovering trends and change. These impact analyses typically include the modeling of an entity's response to climate change, and the entity may include ecosystem response, public health responses, or, very commonly, infrastructure system response. The prevalence of climate-related impacts on water systems contributes to the prominence of this focus, and it serves as the departure point for this chapter. Initially, there was general interest in how these systems responded to the possible impacts of climate changes as depicted by a sampling of projections of future climate, typically derived from GCMs. Increasingly, however, those tasked with long-term decisions potentially affected significantly by possible climate changes sought more understanding than could be provided by a sampling of possible impacts. They sought information and guidance that they needed in order to act and make decisions. For example,

C. Brown (✉)
Hydrosystems Research Group, Civil and Environmental Engineering, University of Massachusetts, Amherst, MA, USA
e-mail: casey@engin.umass.edu

the World Bank Group, which includes the World Bank and International Finance Corporation, was an early advocate for "actionable" climate information. A study by their Independent Evaluation Group (World Bank Independent Evaluation Group, 2012:69) summarized the challenge: "*In retrospect, the Bank Group has pioneered—often in innovative ways—the use of climate models but has discovered that they often have relatively low value-added for many of the applications*" for which climate information is needed. This report recommended a redirection of effort toward decision-making processes that performed well under conditions of irreducible uncertainty.

Our response was to depart from the focus on improving the credibility of climate simulations or attempting to increase the resolution of raw simulations and instead to focus on the decision process itself. Decision analysis presents an appealing quantitative framework, but application is complicated by the mismatch between the typical assumptions of the framework and the nature of GCM-derived climate simulations. In particular, GCM projections do not satisfy the assumptions of "states of the future" in the typical statistical decision analytic framework that they be collectively exhaustive and mutually exclusive (Stainforth et al., 2007). In addition, the climate projections from GCMs are not presented as predictions yet contain some indication of what might happen in the future for plausible scenarios of greenhouse gas emissions and presumably what might not happen. Consequently, an ensemble of GCM projections may not be well-suited for use in expected utility analysis nor for use in nonprobabilistic decision rules, such as Maxi-Min (maximize the minimum utility across collectively exhaustive states of nature).

The goal here is to design a decision framework specifically to make the best use of the available climate projections. The key innovation is to decouple the scenarios used in the decision framework from GCM projections, instead using a statistical climate scenario generator for this purpose. The GCM projections are then used as a sensitivity factor when weighing alternative decisions. Note that here we use the term scenarios as the future states of the world used in decision analysis, whereas projections refer to simulations of future climate from GCMs. There are now many examples of decision scaling (DS) applications that illustrate its various aspects, and particular methods for individual steps continue to evolve (e.g., Whateley et al., 2014; Turner et al., 2014; Steinschneider et al., 2015a, b; Poff et al., 2016; Taner et al., 2019; Ray et al., 2019). However, there has not been the opportunity to provide an articulation of why it was developed and the motivation for the advances that it represented. This chapter attempts to provide background motivation and brief review of the basic concepts, introducing the problems that climate projections posed for decision analysis and explaining the DS solutions to these problems.

The chapter concludes with an explication of the science and the methodological challenges that need further study.

3.2 The Analysis of Decisions

Formal presentation of mathematical approaches to decision analysis was developed in the twentieth century as a systematic, repeatable approach to making decisions under uncertainty. The standard problem formulation for a decision analysis framework (c.f., Schlaifer & Raiffa, 1961) consists of the formal structuring of decisions in terms of choices, unknown future states of the world, "rewards" or the benefit (or disbenefit) the decision-maker receives as a result of selecting a particular choice, and the decision rule, which is the algorithm for determining the optimal decision. The decision rule defines optimality. There has been a rich history of the application of decision analysis to the use of weather forecasts and seasonal to interannual (often called climate forecasts) forecasts of precipitation and temperature, especially in agriculture and water management. Seasonal to interannual climate forecasts became a focus as a result of the recognition of deterministic aspects of the climate system at these timescales. In particular, the emergence of skillful prediction of the El Niño/Southern Oscillation resulted in a great deal of attention of the use of these forecasts for societal benefit. Applications at these forecast time scales benefit from the ability to assess the predictive error of the forecast, that is, given a forecast of X, did X happen or not. This can be usefully summarized as a conditional probability: given a forecast of X, the probability of getting X is p. The conditional probability can be used with a decision analysis framework that maximizes expected utility. Assuming an accepted utility function for measuring the benefit of outcomes of different decision alternatives, the expected utility maximizing decision alternative can be identified.

An early application of decision analysis to climate change adaptation on the Great Lakes was described in Hobbs et al. (1997). Their application presents the maximizing expected utility decision framework but avoids the issue of estimating probabilities for the future states of the world. The authors present a decision analysis framework but do not specify the probability of future lake levels. Unlike in the case of weather or seasonal climate forecasts, the needed conditional probability of success is not available for climate change projections. Without probabilities, the maximization of expected utility framework was no longer practicable.

However, alternative decision rules exist within decision analysis, and there are many that do not depend on probability. For example, to maximize the minimum reward decision rule, select the alternative without regard to the probabilities of future conditions. It simply selects the alternative that avoids the worst possible outcome (note that defining the worst outcome requires a judgment as to the existence of a scenario under which that outcome occurs, and thus assigning it a nonzero probability of existence—more below). Further, it is possible that the value of decision analysis for applicants might not be the ability to identify an optimal decision alternative. Instead, there was value in the organization of the information that is useful for evaluating alternatives, that is, the articulation of the future states of the world or scenarios and the investigation of the performance of different alternatives in those scenarios. Such analyses offer an example of decision analysis used for structured exploration and comparison of the consequences of alternative decisions rather than decision analysis used to rank alternative decisions.

3.2.1 Decision Analysis: An Illustrative Climate Adaptation Example

In a typical statistical decision analysis application, a decision-maker seeks the best choice among several possibilities, where "best" is determined by the performance of the selected option in the future. There are various ways to define "best," including the use of utility functions but with no loss of generality. Let us assume that best is defined in terms of the option with the highest net present value. If there are four possible choices, say options A–D and the key uncertain variable that determines their future performance can be categorized as four mutually exclusive and collectively exhaustive futures, then the problem can be summarized by Table 3.1. The table is completed by calculating the net present value of each of the options in each of the possible futures. The resulting decision matrix is used in combination with a decision rule to select the best option.

Now, consider the case of a water supply utility that needs to expand their water supply to meet the needs of a growing population. Table 3.2 summarizes this case, with the four possible water supply expansion options, and the net present value of each of those options in five mutually exclusive and collectively exhaustive futures. Here, the future variable of most importance to the performance of each option is the future precipitation. Other uncertain variables, such as the demand for water, population growth, or cost overruns, may also be influential, but for the purposes of this example only precipitation uncertainty is considered. Specifically, the future scenarios are defined by the change in 30-year mean annual precipitation from current conditions. For this strictly illustrative example, the net benefit for each option in each future is indicated in terms of $M. Notice that to this point we have not assigned probabilities to the futures.

Determination of the best option depends on the decision rule that is used. We begin with two decision rules that do not require probabilities given that we have none. The first is the maximize the maximum (MaxiMax) reward rule. This rule states that we simply choose the option with the highest net benefit in any scenario. In this case, the desalination plant is the optimum choice, with a reward of $500 M if the future is greater than 20% dryer than current conditions. An

Table 3.1 Decision matrix showing the options (row headings) from which the "best" will be selected based on their net present value in each future state of the world (column headings)

Options	Future 1	Future 2	Future 3	Future 4
Option A	Net Present Value (A|Future 1)	Net Present Value (A|Future 2)	Net Present Value (A|Future 3)	Net Present Value (A|Future 4)
Option B	Net Present Value (B|Future 1)	Net Present Value (B|Future 2)	Net Present Value (B|Future 3)	Net Present Value (B|Future 4)
Option C	Net Present Value (C|Future 1)	Net Present Value (C|Future 2)	Net Present Value (C|Future 3)	Net Present Value (C|Future 4)
Option D	Net Present Value (D|Future 1)	Net Present Value (D|Future 2)	Net Present Value (D|Future 3)	Net Present Value (D|Future 4)

Table 3.2 Example decision matrix as described in the text. The matrix entries are the net present value of each option (row headings) in each future state of the world (column headings). The blue cell indicates the optimal choice using the "MaxiMax" decision rule, and the green shading indicates the optimal choice using the "MaxiMin" decision rule

Options	>20% dryer	5 to 20% dryer	Little change	5 to 20% wetter	>20% wetter
Surface reservoir	−100	100	200	300	200
Groundwater	250	200	150	150	−150
Desalination plant	500	250	100	−200	−600
Demand reduction	60	60	60	60	60

Table 3.3 As in Table 3.2 but the red-shaded cell indicates the optimal option for the maximum likelihood decision rule and green shading indicates the optimal choice for the maximize expected utility approach

Options	>20% dryer	5 to 20% dryer	Little change	5 to 20% wetter	>20% wetter
Surface reservoir	−100	100	200	300	200
Groundwater	250	200	150	150	−150
Desalination plant	500	250	100	−200	−600
Demand reduction	60	60	60	60	60
Probability	0.2	0.25	0.35	0.1	0.1

Table 3.4 As in Table 3.2 but green shading indicates for each option the climate scenarios for which it satisfies the threshold on "acceptable performance" as defined in the text. This is a definition of robustness for each option

Options	>20% dryer	5 to 20% dryer	Little change	5 to 20% wetter	>20% wetter
Surface reservoir	−100	100	200	300	200
Groundwater	250	200	150	150	−150
Desalination plant	500	250	100	−200	−600
Demand reduction	60	60	60	60	60

alternative and more cautionary approach is to maximize the minimum reward ("MaxiMin"). In this case, the decision-maker seeks to avoid the worst outcomes by selecting the option that does best in a worst case. With this decision rule, the demand reduction option is optimum as its worst-case reward ($60 M) is higher than every other option's worst-case reward.

Now consider the case where a bold analyst has assigned probabilities to the future scenarios, as indicated in Table 3.3. This allows us to introduce commonly used decision rules that depend on probabilities. The first is called the maximum likelihood decision rule. This decision rule selects the option that has the best reward in the most likely future. While this may seem like a risky decision approach, the commonly heard request for decision-makers to analysis of "Tell me what will happen, and I'll make the best decision for that case" actually follows this decision rule. In this case, the surface reservoir has the highest net present value in the most likely future, which is the "little change" future with a probability of 0.35. Finally, the maximize expected value (or utility) decision rule is perhaps the most commonly applied. In this case, the expected value of the net benefit of each option is calculated over all the futures and the option with the highest expected reward is selected. In our example, the groundwater option is the optimal choice.

Another approach to selecting the preferred decision uses the concept of "satisficing" coined by Herbert Simon by combining satisfy and suffice. This means selecting a solution that provides an acceptable level of performance rather than seeking the ideal solution. This can be extended to the concept of robustness. Robustness can be defined in many ways but in this analysis the term is defined as providing satisfactory performance over a wide range of possible futures. The analyst may satisfy by selecting the option that provides acceptable performance over the widest range of possible futures and thus avoid being too cautious (maximin) or too dependent on predicted probabilities (max expected value). For example, Table 3.4 now shows the robustness range of each of the water supply options assuming a threshold of acceptability of $150 M.

Robustness provides a balance between caution due to overwhelming uncertainty and dependence on predicted probabilities that lack credibility. This is a promising decision analysis framing to address climate change. However, it is not yet clear how to use climate information with this framework. The challenge begins with the required scenarios.

3.3 Incongruence of Climate Projections and Decision Analysis

In the classic statistical decision analytic framework, the states of the world are defined as mutually exclusive and collectively exhaustive, and these conditions are required to guarantee a clear indication of the optimal choice(s) (Edwards et al., 2007). However, climate projections are not designed with this framework in mind and unsurprisingly are difficult to incorporate with it. Due to the increasing role of internal variability at the scale of adaptation decisions (Whateley & Brown, 2016), a typical ensemble of projections is particularly lacking in this regard when analyzed at the scale of adaptation decisions (Brown & Wilby, 2012). Collectively exhaustive climate scenarios would span the range of plausible climate changes. That is, there would be a scenario that represents the full range of plausible climate change. In practice, it is difficult to specify and simulate all such scenarios and so the changes analyzed could be restricted to those that occur in variables that are most influential in terms of decision outcomes. However, it is an open question as to whether an ensemble of climate projections could represent the full range of climate uncertainty. Stainforth et al. (2007) described the range presented by an ensemble

Table 3.5 As in Table 3.2 described in the text. Future states of the world as represented by GCM simulations

Options	GCM#1	GCM#2	GCM#3	GCM#4	GCM#5
Surface reservoir	100	−100	150	100	200
Groundwater	150	200	250	−150	150
Desalination plant	−200	150	150	−250	300
Demand reduction	90	60	80	60	50

of climate projections as the "minimum range of maximum uncertainty." For example, a recent convening of scientists specified a range of plausible climate change that exceeded the range of a multi-GCM, multirun ensemble at an expert elicitation workshop for a California water utility (Lempert et al., 2023). Furthermore, typical risk analysis for water resources systems involves precise simulation of stochastic hydrologic variability to sample the range of possible realizations, something that is not a focus of a typical GCM ensemble.

Table 3.5 illustrates the problem by updating the decision matrix with GCM runs as states of the world. The analyst is confronted with selecting the optimal strategy based on the sampling of results that the GCM projections provide. However, the resulting optimal decision (as defined by the decision rule) is likely to be highly dependent on the selection of climate projections used. A different selection of GCMs or a different method for processing the raw GCM outputs (i.e., downscaling and bias correction) would potentially produce different results for each option, which undermines repeatability of the analysis and perhaps the confidence of the analyst in ranking performance.

The challenge then for the decision analyst is matching the needs of the decision framework with the useful information that could be gleaned from an ensemble of GCMs. Perhaps the generation of climate scenarios for adaptation decisions may be improved by not directly using the GCM scenarios. In particular, stochastic weather models can be modified for use in climate change studies to provide the mutually exclusive and collectively exhaustive climate scenarios for applying the classic decision analysis framework to adaptation decisions. The result is a "climate stress test," further explained below.

3.4 A Solution to Scenario Ambiguity

We define decision analysis-ready climate scenarios as scenarios that meet the collectively exhaustive and mutually exclusive criteria for the traditional decision analysis framework. Such scenarios can be generated by designing the scenarios to sample the desired range of change of the selected relevant climate variables. The result is a climate stress test (Brown & Wilby, 2012). There are likely a large number of methods possible for creating climate stress test scenarios; the first is described in Steinschneider and Brown (2013). Some aspects of designing and creating a climate stress test bear special consideration: physical consistency, range of sampling, variability, and skill of prediction. A common question is how to define the range of possible climate changes to use in the climate stress test. Because the generation of additional scenarios requires insignificant additional effort, there are a few limitations to the range that can be tested. However, often the computational requirements of subsequent modeling steps do require a constraint on the number of scenarios tested. Methods to adaptively select the next climate scenario based on the simulated impact of the previous scenario could reduce the computational burden. Otherwise, strategic selection of scenarios to effectively sample the desired range can be used (e.g., Whateley et al., 2016).

Defining the plausible range of climate change ultimately is a subjective choice informed by historical trends and GCM projections. The analyst can be reassured that the goal is to characterize the response of the system to changes and not to predict what is likely to happen. There should be no hard choices here. If there is any doubt whether a given endpoint is plausible, expand the range an increment further. The goal here is to leave no stone unturned, not to guess what is more or less likely. In making these determinations, three sources of climate information should be informed. The first is any trends in the historical weather time series for the location of interest. Scenarios should include the range implied if historical trends were to continue or accelerate. Second, climate projections from GCMs should be consulted to include the range of changes that the models indicate. Finally, theoretical implications of increasing greenhouse gas concentrations in the atmosphere provide a basis for the direction of change in some variables. For example, in addition to expectations of increased temperatures, there is also an accompanying theoretical case for expectations of increased precipitation intensity due to the enhanced moisture capacity of the warmer atmosphere and rising sea levels, even if observations or climate simulations do not reflect these changes in all cases.

The next consideration is the representation of the internal variability of the climate system and the implications for decision analysis. For many systems, the variability of climate may be more influential in terms of outcomes than mean changes. Thus, sampling variability in the scenarios is important for them to be collectively exhaustive. Fortunately,

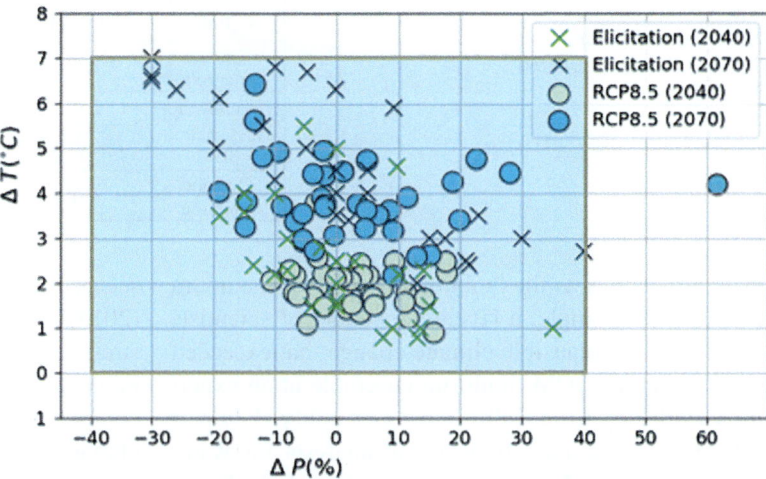

Type of uncertainty	Sampling range	Sample size
Natural climate variability	Stochastic realizations	10 realizations
Changes in mean annual precipitation (%)	−40 % to 40 % with 5% increments	17 change factors
Changes in mean annual temperature (°C)	0 to 7 °C with 1 °C increments	8 change factors
TOTAL		**1360 climate scenarios**

Fig. 3.1 Climate stress test design for a water resources planning application. Blue box encompasses the climate changes included in the analysis, which includes the majority of the GCM-based climate change projections and expert elicitations of expected climate change range

simulation of weather and hydrologic climate variability has long been a topic of study in water resources and agricultural modeling, and a wide range of statistical tools exist to generate scenarios of variability that are consistent with the statistics of the historical record. In the typical design of a stress test, a small number of realizations of variability (10) generated, for example, via Monte Carlo sampling of the weather generator are selected to be representative of the possible variability that will be experienced in the future. Alternatively, initial condition ensembles could be used (Piani et al., 2005). Rank statistics on appropriate measures of the variability in a trace can be used to ensure a wide range of variability is in fact used (Whateley et al., 2016). The climate changes are then applied to these realizations. This creates the ensemble of weather time series that together create a climate stress test designed to systematically sample the system of interest and generate a comprehensive understanding of its response to climate change.

Figure 3.1 is a depiction of a climate stress test sampling strategy that was developed for analysis of the San Francisco water supply system. It includes the climate simulations that were considered in designing the range of climate change explored in the analysis. Interestingly, the decision for the sampling range of climate change also included input gathered through an expert elicitation session with climate scientists who were familiar with California climate. Their estimates are also included in the figure.

3.5 Ex Post Scenario Analysis and Use

The methodology of DS can be described in three broad steps: (1) decision framing; (2) climate stress test; and (3) estimating climate-informed risks. Decision framing consists of the organization of information and materials to begin the analysis and is consistent with the various methods of stakeholder engagement and planning that are available (c.f., Winston & Goldberg, 2004). The key is that mission objectives and metrics to measure performance relative to those objectives are agreed. In addition, this step entails enumerating the uncertain factors that affect the decision, including future climate, and the computational models that will be used for the system of interest in the analysis. Finally, the options that are being considered if the process includes an options analysis should be identified to ensure the model has the capability to include them.

The second step is the use of the climate stress test to provide the mutually exclusive and collectively exhaustive scenarios needed for the traditional decision analysis framework. The climate stress test is described above. The stress test approach is designed to reveal insights regarding the system without concern regarding the probability of any of the particular scenarios to be realized. It is inspired by the analysis technique known as *pre-posterior analysis* (Schlaifer & Raiffa, 1961). Pre-posterior analysis involves identifying optimal decisions that you would make given

Table 3.6 As in Table 3.4, with the robustness of each option indicated as a robustness index (the demand reduction option is not robust to any of the possible futures)

Options	>20% dryer	5 to 20% dryer	Little change	5 to 20% wetter	>20% wetter
Surface reservoir	−100	100	200	300	200
Groundwater	250	200	150	150	−150
Desalination plant	500	250	100	−200	−600
Demand reduction	60	60	60	60	60
Robustness index	\multicolumn{5}{l|}{Desal Plant = 0.40 — Groundwater = 0.8 — Surface Reservoir = 0.60}				

a forecast (posterior to receiving the forecast) before you actually know what information the forecast will provide (thus, "pre"-posterior). To assess the value of the forecast, you have to assess the value of every optimal decision that would be made with the forecast relative to the optimal decision without the forecast, for every possible forecast (Brown et al., 2020). The climate stress test enables a similar process to determine the optimal adaptation choice for every possible future climate before we receive the information as to which of those climates is more likely.

The result of the climate stress test is a multidimensional dataset of climate conditions and corresponding values of the performance metrics of interest. This is used in the final step of the methodology. Here, ex post scenario analysis (Bryant and Lempert, 2010) is used to define scenarios in terms of the values of the performance metrics. A cluster analysis tool like the Patient Rule Induction Method (PRIM) is used to select values of the climate variables that define scenarios of acceptable performance and unacceptable performance. Because the cluster analysis is performed on the results of the climate stress test, the resulting ex post scenarios can be considered collectively exhaustive (to the degree possible as argued above) and mutually exclusive (via an appropriate cluster analysis tool with non-fuzzy set membership). By defining the ex post scenarios based on a performance threshold of acceptability, the scenarios leverage the concepts of satisficing and robustness. Table 3.6 illustrates the identification of ex post scenarios for our simple water supply example with the same acceptability threshold of $150 M. The ex post scenarios that define robust performance are indicated with green shading. For example, the surface reservoir option is robust for climate futures of "little change" through precipitation increases of greater than 20%. It is, however, vulnerable for climate changes of 5% or greater reductions in precipitation. This information can be further summarized by calculating the fraction of future conditions for which a particular option provides acceptable performance, or as we have defined it, is robust. Table 3.6 shows that the groundwater option is the most robust as it provides acceptable performance for 80% of the futures considered. The fraction can be summarized for each of the options as a simple robustness index.

Up to this point in the methodology, climate projections have been used only to inform the design of the climate stress test. The ex post scenarios present another important entry point. Presumably, the projections might be able to provide additional information to judge the options. For example, for options with similar robustness scores, it would be interesting to consider whether the projections provided information regarding the relative likelihood of the futures that are problematic. For example, if deciding between the desalination plant and the surface reservoir, it would be interesting to investigate whether the climate simulations could provide an indication of which of the two ex post climate futures that are problematic for each option is more likely. That is, the surface reservoir fails the robustness criteria if precipitation decreases by greater than 5%, while the desalination plant fails if precipitation does not decrease by at least 5%. These ex post scenarios provide the entry point to the final methodological activity using the best available climate information to assess the robustness and/or risk of each option. Fortunately, the ex post scenarios change the challenge from a point estimate of probability to something akin to a categorical forecast, which may improve the credibility of the estimates.

3.6 Probably or Not: The Use of Probabilities

A question that has long interested adaptation specialists is whether probabilities of climate change are needed for decision-making (e.g., Katzav et al., 2021). Decision rules used in decision analysis require probabilities in some cases but not in others. The decision rules that do not require probabilities explicitly and provide clear optimal selections such as minimizing the maximum regret criterion, making them attractive alternatives that seemingly sidestep the question of whether climate probabilities should be used for adaptation decision-making. However, the selection of the scenarios themselves can be highly influential on the indicated optimal decision, and this selection is inevitably based on judgments of probability. If, as has been argued, climate projections

Table 3.7 As in Table 3.6, with climate-informed robustness indices as defined in the text

Options	>20% dryer	5 to 20% dryer	Little change	5 to 20% wetter	>20% wetter
Surface reservoir	−100	100	200	300	200
Groundwater	250	200	150	150	−150
Desalination plant	500	250	100	−200	−600
Demand reduction	60	60	60	60	60
Probability	0.2	0.25	0.35	0.1	0.1
Robustness index	\multicolumn{5}{c}{Surface Reservoir = 0.55; Groundwater = 0.9; Desal Plant = 0.45}				

delineate the "minimum range of maximum uncertainty," the use of a climate simulation as the worst-case scenario leaves open the possibility that the realized climate outcomes could be worse. Selecting the range of plausible climate changes requires judgments of which futures are plausible or not in order to not bias the analysis of the options.

Probabilistic forecasts are the norm for seasonal to interannual prediction. However, simulations of climate change are forced with scenarios of greenhouse gas emissions that do not have assigned probabilities. Nonetheless, the use of a probabilistic framework has advantages for using climate simulations to inform adaptation decisions. For example, it provides a repeatable analytical means to summarize the information in a multimodel ensemble of GCM simulations. A probabilistic framework may also provide a more representative summary of the information within an ensemble by accounting for model relatedness and GCM sample size (Steinschneider et al., 2015a, b). The question is whether the information within an ensemble of climate projections can be incorporated into the decision analysis without introducing bias.

There are a number of methods that have been developed for estimating pdfs from observations and GCM simulations. Details of these approaches address critical issues including accounting for the potential sample bias due to different ensemble sizes per GCM, accounting for the closely related GCMs, accounting for differential skill if desired, and optimal merging of trends and projections. Ideally, the predictive error associated with the resulting pdfs could be estimated. Alas, because the observed future climate is not available for comparison, that error cannot be calculated. Thus, while the probabilistic framework is helpful for summarizing the information content in an ensemble of GCM simulations, it does imply that the resulting distributions are ready for use in a probability-based decision framework. Unlike seasonal climate forecast forecasts or weather forecasts, we cannot easily evaluate the predictive skill of a climate projection. For weather and seasonal climate forecasts, the prediction can be compared with the observed outcome as it occurs over the next days or months. We have a long wait until we know the observed outcome of future climate change. Therefore, it is difficult to quantify the reliability of predicted probabilities, and this is typically a barrier for trusting them for use in a decision framework where the optimal choice is a product of those probabilities. And yet, it would seem that the simulations provide information that could be useful for informing decisions.

The final step in DS is to investigate the best available climate projections for possible inclusion in the decision process. Table 3.7 illustrates one possible use. Here, the ex post scenarios of robustness now have assigned probabilities. The probabilities are derived in a responsible way from expert judgment, climate projections, or a combination of each. The framework is independent of the particular method used to derive the probabilities and indeed can accommodate multiple methods (and the concept of "belief dominance" (Baker et al., 2016) can be used to select optimal decisions). The climate change probabilities can be thought of as a "level of concern" regarding vulnerabilities or as an input to the robustness on the other hand. Table 3.7 shows the climate-informed robustness index that has now weighted the robustness index by the probability of the scenarios over which each option is robust. Here, the climate projection information has entered the analysis at the final stage and with abundant flexibility to recalculate as new information becomes available (e.g., IPCC WGI Report (2021) CMIP6), without need to repeat the climate stress test.

3.7 Process-Based Decision Scaling

The original design of decision scaling entailed another objective not yet discussed. The approach was designed to maximize the skill of the signal from GCM projections. At the time, it was clear that projections of the mean of a variable over larger spatial areas (multiple grid cells) and longer time periods (e.g., annual) were less biased than variables from single cells and shorter temporal periods (e.g., daily) for variables such as precipitation and temperature. Thus, by designing scenarios based on the mean of a variable over large areas (e.g., river basins) and longer time periods (annual mean precipitation) the GCM information used to inform the decision would be less biased than if other approaches were used (such as downscaling methods that disaggregate single grid cells to higher resolution grids they overlay, without consideration of skill). The climate-weather generator approach (e.g., Steinschneider & Brown, 2013) is designed to achieve exactly that by creating realistic high-resolution data that is consistent with a variable mean over a large spatial area and averaging time period.

The basic idea of creating scenarios that fit a decision framework while also maximizing the potential skill of climate simulations can be extended to focus on specific physical processes. Recent research has advanced the concept of "process-based" decision scaling. The premise underpinning these methods is that the skillful signal from climate simulations might be improved by focusing on well-simulated processes as the predictand. Then this predictand could serve as the basis for scenarios, assuming there is a strong influence of the predictand on the system of interest. Steinschneider et al. (2019) provide an example. A study of California water resources under climate change revealed a vulnerability to mean precipitation reductions. The mean annual precipitation total in California is the outcome of multiple precipitation processes, and the major portion of variability arises from the frequency of atmospheric river (AR) events (Dettinger, 2011). Investigation of changes in the frequency of ARs is potentially more tractable than mean annual precipitation by studying this physical initiation and development of the events and especially if the driving factors, including sea surface temperatures, might be more credibly simulated than precipitation itself.

Using AR-based scenarios requires the generation of mutually exclusive AR scenarios. This can be done by designing the climate stress test to be a function of the AR frequency and modifying that frequency as desired. Schlef et al. (2018) applied a similar approach to flood frequency analysis. In this case, the variables driving the stress test were soil moisture and the location of the Pacific North American atmospheric index. Since direct projection of extreme precipitation that causes flooding is difficult, the study focused on variables that caused floods and conditions conducive to extreme precipitation. GCM projections of the variables were used to inform rather than to project the change in flood risk. These approaches are developing and provide a promising pathway to both reduce bias in projections of decision-relevant variables, while also producing needed insights for decisions when incorporated with the decision scaling framework.

3.8 Conclusion

Decision scaling was created as an attempt to bring available climate change information to a decision analytic framework. For that purpose, the concept of the climate stress test was created by developing a climate scenario generator. In addition, information from GCM projections was used not for scenarios but rather to assess the relative likelihood of the ex post scenarios identified in the analysis. Doing so required the pre-posterior decision analysis framing, and this required creation of the climate stress test. The climate stress test proved useful for methods like robust decision-making, which at the time used GCM simulations as scenarios but were neither mutually exclusive nor collectively exhaustive. In turn, DS adopted scenario discovery or ex post scenario analysis for linking to climate information at the end of the analysis.

The climate stress test approach also overcame a major challenge that practitioners faced—the choices involved with selection of GCM projections and downscaling method that preceded every impact study at the time. They were in an impossible position, making choices that were potentially highly influential in terms of the study outcome but without any basis or understanding of the consequences for making the choice. Because DS begins with a climate stress test, the choice of projections is both less consequential and initially unnecessary. As a result, a number of operational agencies have developed guidelines for their climate analysis based on DS, including the World Bank (Ray & Brown, 2015), the International Hydropower Association (IHA, 2019), the U.S. Army Corps of Engineers (Mendoza et al., 2018), and the Millennium Challenge Corporation (Brown et al., 2024) or otherwise applied the methodology (Ray et al., 2020; François et al., 2024).

Ultimately, DS was designed as a general framework to improve adaptation decision-making through novel incorporation of climate information. As such, there are continuous opportunities to improve the application of DS and enhance adaptation decision-making. The first is continued research into the generation of probabilities of key climate variables that are influential for decisions, especially at time scales relevant for planners, which is often a decade or two ahead. Such advances will increase the confidence in acting for adaptation and help overcome the barrier that climate uncertainty so often causes for decision-makers, such as in cases where the decision to take action is highly dependent on the assumptions of future climate conditions. Second, analytical tools that reduce the cost and complexity of applying DS are needed to enable scaling and aiding adaptation by a wider range of practitioners who cannot afford the financial and labor cost of a typical study. Finally, further research into the sequencing of investments and the effects of current choices in constraining future opportunities or risks is needed. An integration of climate science with traditional sequential decision analysis has potential.

References

Baker, E., Valentina, B., & Ahti, S. (2016). Finding common ground when experts disagree: Belief dominance over portfolios of alternatives. *FEEM Working Paper No. 46.2016*, https://doi.org/10.2139/ssrn.2815365.

Brown, C., & Wilby, R. L. (2012). An alternate approach to assessing climate risks. *Eos, Transactions American Geophysical Union, 93*, 401–402. https://doi.org/10.1029/2012EO410001

Brown, C., Rogers, P., & Lall, U. (2006). Demand management of groundwater with monsoon forecasting. *Agricultural Systems, 90*, 293–311. https://doi.org/10.1016/j.agsy.2006.01.003

Brown, C., Boltz, F., Freeman, S., Tront, J., & Rodriguez, D. (2020). Resilience by design: A deep uncertainty approach for water systems in a changing world. *Water Security, 9*, 100051.

Brown, C., et al. (2024). Achieving robust project design. Report prepared for The Millennium Challenge Corporation, 59 pp, https://assets.mcc.gov/content/uploads/guidance-achieving-robust-project-design.pdf

Bryant, B. P., & Lempert, R. J. (2010). Thinking inside the box: A participatory, computer-assisted approach to scenario discovery. *Technological Forecasting and Social Change, 77*(1), 34–49.

Dettinger, M. (2011). Climate change, atmospheric rivers, and floods in California–a multimodel analysis of storm frequency and magnitude changes. *Journal of the American Water Resources Association, 47*, 514–523. https://doi.org/10.1111/j.1752-1688.2011.00546.x

Edwards, W., Miles, R. F., & Von Winterfeldt, D. (2007). *Advances in decision analysis*. Cambridge.

François, B., Dufour, A., Nguyen, T. N. K., Bruce, A., Park, D. K., & Brown, C. (2024). From many futures to one: Climate-informed planning scenario analysis for resource-efficient deep climate uncertainty analysis. *Climatic Change, 177*(7), 111. https://doi.org/10.1007/s10584-024-03772-9

Hobbs, B. F., Chao, P. T., & Venkatesh, B. N. (1997). Using decision analysis to include climate change in water resources decision making. *Climatic Change, 37*(1), 177–202.

International Hydropower Association. (2019). *Hydropower sector climate resilience guide*. London, UK, 63 pp, https://www.hydropower.org/publications/hydropower-sector-climate-resilience-guide

Katzav, J., Thompson, E. L., Risbey, J., Stainforth, D. A., Bradley, S., & Frisch, M. (2021). On the appropriate and inappropriate uses of probability distributions in climate projections and some alternatives. *Climatic Change, 169*, 15. https://doi.org/10.1007/s10584-021-03267-x

Lempert, R. J., Berry, S. H., & Tanverakul, S. A. (2023). *Elicitation of climate information to support water agency planning: Report of a workshop with the San Francisco Public Utilities Commission*. RAND, 63 pp, https://www.rand.org/pubs/conf_proceedings/CFA2138-1.html.

Mendoza, G., Jeuken, A., Matthews, J. H., Stakhiv, E., Kucharski, J., & Gilroy, K. (2018). *Climate Risk Informed Decision Analysis (CRIDA): Collaborative water resources planning for an uncertain future*. UNESCO Publishing.

Piani, C., Frame, D. J., Stainforth, D. A., & Allen, M. R. (2005). Constraints on climate change from a multi-thousand member ensemble of simulations. *Geophysical Research Letters, 32*, L23825. https://doi.org/10.1029/2005GL024452

Poff, L., et al. (2016). Sustainable water management under future uncertainty with eco-engineering decision scaling. *Nature Climate Change, 6*, 25–34. https://doi.org/10.1038/nclimate2765

Ray, P., & Brown, C. (2015). *Confronting climate uncertainty in water resources planning and project design—the decision tree framework* (p. 128). World Bank Group Press. http://documents.worldbank.org/curated/en/516801467986326382/Confronting-climate-uncertainty-in-water-resources-planning-and-project-design-the-decision-tree-framework

Ray, P. A., Taner, M. Ü., Schlef, K. E., Wi, S., Khan, H. F., Freeman, S. S. G., & Brown, C. M. (2019). Growth of the decision tree: Advances in bottom-up climate change risk management. *Journal of the American Water Resources Association, 55*, 920–937. https://doi.org/10.1111/1752-1688.12701

Ray, P., Wi, S., Schwarz, A., Correa, M., He, M., & Brown, C. (2020). Vulnerability and risk: Climate change and water supply from California's Central Valley water system. *Climatic Change, 161*, 177–199. https://doi.org/10.1007/s10584-020-02655-z

Schlaifer, R., & Raiffa, H. (1961). *Applied statistical decision theory* (p. 356). Clinton Press.

Schlef, K. E., François, B., Robertson, A. W., & Brown, C. (2018). A general methodology for climate-informed approaches to long-term flood projection—Illustrated with the Ohio river basin. *Water Resources Research, 54*, 9321–9341. https://doi.org/10.1029/2018WR023209

Stainforth, D. A., Downing, T. E., Washington, R., Lopez, A., & New, M. (2007). Issues in the interpretation of climate model ensembles to inform decisions. *Philosophical Transactions of the Royal Society A, 365*, 2163–2177. https://doi.org/10.1098/rsta.2007.2073

Steinschneider, S., & Brown, C. (2013). A semiparametric multivariate, multisite weather generator with low-frequency variability for use in climate risk assessments. *Water Resources Research, 49*, 7205–7220. https://doi.org/10.1002/wrcr.20528

Steinschneider, S., McCrary, R., Mearns, L. O., & Brown, C. (2015a). The effects of climate model similarity on probabilistic climate projections and the implications for local, risk-based adaptation planning. *Geophysical Research Letters, 42*, 5014–5022. https://doi.org/10.1002/2015GL064529

Steinschneider, S., McCrary, R., Wi, S., Mulligan, K., Mearns, L. O., & Brown, C. (2015b). Expanded decision-scaling framework to select robust long-term water-system plans under hydroclimatic uncertainties. *Journal of Water Resources Planning and Management, 141*, 04015023. https://doi.org/10.1061/(ASCE)WR.1943-5452.0000536

Steinschneider, S., Ray, P., Rahat, S. H., & Kucharski, J. (2019). A weather-regime-based stochastic weather generator for climate vulnerability assessments of water systems in the western United States. *Water Resources Research, 55*, 6923–6945. https://doi.org/10.1029/2018WR024446

Taner, M. Ü., Ray, P., & Brown, C. (2019). Incorporating multidimensional probabilistic information into robustness-based water systems planning. *Water Resources Research, 55*, 3659–3679. https://doi.org/10.1029/2018WR022909

Turner, S. W., Marlow, D., Ekström, M., Rhodes, B. G., Kularathna, U., & Jeffrey, P. J. (2014). Linking climate projections to performance: A yield-based decision scaling assessment of a large urban water resources system. *Water Resources Research, 50*, 3553–3567. https://doi.org/10.1002/2013WR015156

Whateley, S., & Brown, C. (2016). Assessing the relative effects of emissions, climate means, and variability on large water supply systems. *Geophysical Research Letters, 43*, 11329–11338. https://doi.org/10.1002/2016GL070241

Whateley, S., Steinschneider, S., & Brown, C. (2014). A climate change range-based method for estimating robustness for water resources supply. *Water Resources Research, 50*, 8944–8961. https://doi.org/10.1002/2014WR015956

Whateley, S., Steinschneider, S., & Brown, C. (2016). Selecting stochastic climate realizations to efficiently explore a wide range of climate risk to water resource systems. *Journal of Water Resources Planning and Management, 142*, 06016002. https://doi.org/10.1061/(ASCE)WR.1943-5452.00006

Winston, W. L., & Goldberg, J. B. (2004). *Operations research: Applications and algorithms* (p. 1418). Thomson Brooks/Cole.

World Bank Independent Evaluation Group. (2012). *Adapting to climate change: Assessing World Bank Group experience* (p. 193). World Bank Group.

Open Access This chapter is licensed under the terms of the Creative Commons Attribution 4.0 International License (http://creativecommons.org/licenses/by/4.0/), which permits use, sharing, adaptation, distribution and reproduction in any medium or format, as long as you give appropriate credit to the original author(s) and the source, provide a link to the Creative Commons license and indicate if changes were made.

The images or other third party material in this chapter are included in the chapter's Creative Commons license, unless indicated otherwise in a credit line to the material. If material is not included in the chapter's Creative Commons license and your intended use is not permitted by statutory regulation or exceeds the permitted use, you will need to obtain permission directly from the copyright holder.

Supporting Climate-Related Decisions Under Uncertainty

Robert Lempert

4.1 Introduction

Quantitative, evidence-based analysis often proves indispensable for making good policy choices. To inform such choices, decision-makers often seek predictions about the future. This seems natural because prediction is the bedrock of science enabling researchers to test their hypotheses and demonstrate understanding of complicated systems. Decision-makers find predictions attractive because when good ones are available they unquestionably provide valuable input toward better choices. Society's vast enterprise in the physical, biological, and social sciences continually improves its predictive capabilities in order to test and refine scientific hypotheses. Recent analytic innovations, such as big data and machine learning, enhance decision-relevant predictive capabilities. New processes, such as crowd-sourcing, prediction markets, and super-forecasting (Tetlock & Gardner, 2015), provide new and more reliable means to aggregate human judgments into probabilistic forecasts.

But the quest for predictions—and a reliance upon decision and risk analysis methods that require them—can prove counterproductive and sometimes dangerous in a fast-changing, complex world. Prediction-focused analysis risks overconfidence in organizations' decision-making and in their internal and external communications (Sarewitz & Pielke, 2000). Prediction-focused policy debates can also fall victim to the strategic uses of uncertainty. Opponents may attack a proposed policy by casting doubt on the predictions used to justify it, rather than engage with the merits of a policy itself, knowing that the policy may be sounder than the predictions (Herrick & Sarewitz, 2000; Lempert & Popper, 2005; Rayner, 2000; Weaver et al., 2013).

A reliance on prediction can also skew the framing of a decision challenge. President Eisenhower (reportedly) advised, "If a problem cannot be solved, enlarge it." But science often reduces uncertainty by narrowing its focus, prioritizing questions that can be resolved by prediction, not necessarily on the most decision-relevant inquiries. So-called "wicked problems" (Rittel & Webber, 1973) present this contrast most starkly. In addition to their irreducible uncertainty and nonlinear dynamics, wicked problems are not well-bounded, are framed differently by various stakeholders, and are not well-understood until after formulation of a solution. Climate change and how to deal with it is quintessentially a wicked problem. Using predictions to adjudicate such problems skews attention toward the proverbial lamp post, not the true location of the keys to a policy solution.

This chapter describes robust decision making (RDM), a set of concepts, processes, and enabling tools designed to reimagine the role of quantitative models and data in informing decisions when prediction is perilous (Lempert et al., 2003, 2006).[1] As discussed in more detail below, RDM is one of a class of methods called decision making under deep uncertainty (DMDU) (Marchau et al., 2019).[2] Rather than regard models as tools for prediction and the subsequent prescriptive ranking of decision options, models and data become vehicles for systematically exploring the consequences of assumptions; expanding the range of futures considered; crafting promising new responses to dangers and opportunities; and sifting through a multiplicity of scenarios, options, objectives, and problem framings to identify the most important tradeoffs confronting decision-makers. That

---

R. Lempert (✉)
Pardee Center for Longer Range Global Policy, RAND, Santa Monica, CA, USA
e-mail: lempert@rand.org

[1] This chapter is a reprint, with some adapting and updating, of Lempert, R., 2019: Robust Decision Making (RDM). In: V. A. W. J. Marchau, W. E. Walker, P. J. T. M. Bloemen and S. W. E. Popper (eds.), *Decision Making under Deep Uncertainty: From Theory to Practice*. Springer: 329. Cham, Switzerland.

[2] Also see Chap. 3 in this volume by Casey Brown.

is, rather than making better predictions, quantitative models and data can inform better decisions (Popper et al., 2005).

RDM rests on a simple concept (Lempert et al., 2013c). Rather than using computer models and data as predictive tools, the approach runs models myriad times to stress test proposed decisions against a wide range of plausible futures. Analysts then use visualization and statistical analysis of the resulting large database of model runs to help decision-makers identify the key features that distinguish those futures in which their plans meet and miss their goals. This information helps decision-makers identify, frame, evaluate, modify, and choose robust strategies—ones that meet multiple objectives over many scenarios.

RDM provides decision support under conditions of deep uncertainty, defined as the condition in which parties to a decision do not know or agree on the model(s) that relate their actions to consequences, the prior probability distributions for key parameters to those models, and the importance of various objectives their actions seek to achieve (Lempert et al., 2003; Walker et al., 2013). As described Sect. 4.2, RDM builds on strong foundations of relevant theory and practice, providing an operational and newly capable synthesis through the use of today's burgeoning information technology. As one motivation, RDM notes that the consideration that the most commonly used analytic methods for predictive decision and risk analysis have their roots in the 1950s and 1960s when relative computational poverty made a virtue of analytics recommending a single best answer based on a single best estimate prediction. Today's ubiquitous and inexpensive computation enables analytics better suited to more complex problems, many of them "wicked" and thus poorly served by the approximation that there exists such an ideal solution.

4.2 RDM Foundations

How can quantitative, evidence-based analysis best inform our choices in today's fast-paced and turbulent times, particularly concerning climate change. RDM in general aims to answer this question. In particular, RDM does so by providing a new synthesis of four key concepts: decision analysis, assumption-based planning, scenarios, and exploratory modeling.

4.2.1 Decision Analysis

A large body of empirical research makes clear that people, acting as individuals or in groups, often make better decisions when using well-structured decision aids. The discipline of decision analysis (DA) comprises the theory, methodology, and practice that inform the design and use of such aids. RDM represents one type of quantitative DA method, drawing, for instance, on the field's decision structuring frameworks, a consequentialist orientation in which alternative actions are evaluated in each of several alternative future states of the world, a focus on identifying tradeoffs among alternative decision options, and tools for comparing decision outcomes addressing multiple objectives.

As one key contribution, DA and related fields help answer the crucial question: What constitutes a good decision? No universal criterion exists. Seemingly reasonable decisions can turn out badly, but seemingly unreasonable decisions can turn out well. Good decisions tend to emerge from a process in which people are explicit about their goals, use the best available evidence to understand the potential consequences of their actions, carefully consider the tradeoffs, contemplate the decision from a wide range of views and vantages, and follow agreed-upon rules and norms that enhance the legitimacy of the process for all those concerned (Jones et al., 2014).[3]

While broad in principle, in practice the DA community often seeks to inform good decisions using an expected utility framework for characterizing uncertainty and comparing decision options (Morgan & Henrion, 1990). This expected utility framework characterizes uncertainty with a single joint probability distribution over future states of the world. Such distributions often reflect Bayesian (i.e., subjective) rather than frequentist probability judgments. The framework then uses optimality criteria to rank alternative options. RDM, in contrast, regards uncertainties as deep and thus either eschews probabilities or uses sets of alternative distributions drawing on the concepts of imprecise probabilities (Walley, 1991). RDM uses decision criteria based on robustness rather than optimality.

DA based on expected utility can be usefully termed "agree-on-assumptions" (Kalra et al., 2014) or "predict-then-act" (Lempert et al., 2004) approaches because they begin by seeking agreement regarding the likelihood of future states of the world and then use this agreement to provide a prescriptive ranking of policy alternatives. In contrast, RDM and other DMDU methods follow an "agree-on-decisions" approach, which inverts these steps.[4] They begin with one

---

[3] These attributes follow from a broadly consequentialist, as opposed to rule-based (deontological) view of decision-making. March, J. G., 1994: *A Primer on Decision Making: How Decisions Happen*. The Free Press.

[4] The DMDU literature often uses different names to describe this inverted analytic process, including "backwards analysis." Lempert, R. J., and Coauthors, 2013c: Making Good Decisions Without Predictions: Robust Decision Making for Planning Under Deep Uncertainty. "Bottom up." Ghile, Y. B., M. Ü. Taner, C. Brown, J. G. Grijsen, and A. Talbi, 2014: Bottom-up climate risk assessment of infrastructure investment in the Niger River Basin. *Climatic Change*, **122**, 97–110. "Context first." Ranger, N., A. Millner, S. Dietz, S. Fankhauser, A. Lopez, and G. Ruta, 2010: Adaptation in the UK: a decision making process. "Assess risk of policy." Carter, T. R., and Coauthors, 2007: New Assessment Methods and the Characterisation of Future Conditions. *Climate Change 2007: Impacts, Adaptation and Vulnerability. Contribution of Working Group II to the Fourth Assessment Report of the Intergovernmental Panel on*

or more strategies under consideration, use models and data to stress test the strategies over a wide range of plausible paths into the future, and then use the information in the resulting database of runs to characterize vulnerabilities of the proposed strategies and identify and evaluate potential response to those vulnerabilities. Such approaches seek to expand the range of futures and alternatives considered and, rather than provide a prescriptive ranking of options, often seek to illuminate tradeoffs among not-unreasonable choices. As summarized by Helgeson (2018), agree-on-assumptions approaches generally focus on identifying a normative best choice among a fixed menu of decision alternatives, while agree-on-decision approaches focus on supporting the search for an appropriate framing of complex decisions.

4.2.2 Assumption-Based Planning

As part of an "agree-on-decisions" approach, RDM draws on the related concepts of stress testing and red teaming. The former, which derives from engineering and finance (Borio et al., 2014), subjects a system to deliberately intense testing to determine its breaking points. The latter, often associated with best practice in US and other militaries' planning (Zenko, 2015), involves forming an independent group to identify means to defeat an organization's plans. Both stress testing and red teaming aim to reduce the deleterious effects of overconfidence in existing systems and plans by improving understanding of how and why they may fail (Lempert, 2007).

In particular, RDM draws on a specific form of this concept, a methodology called assumption-based planning (ABP) (Dewar et al., 1993). Originally developed to help the US Army adjust its plans in the aftermath of the Cold War, ABP begins with a written version of an organization's plans and then identifies load-bearing assumptions—that is, the explicit and implicit assumptions made while developing that plan that, if wrong, would cause the plan to fail. Planners can then judge which of these load-bearing assumptions are also vulnerable—that is, could potentially fail during the lifetime of the plan.

ABP links the identification of vulnerable, load-bearing assumptions to a simple framework for adaptive planning that is often used in RDM analyses. Essential components of an adaptive strategy include a planned sequence of actions, the potential to gain new information that might signal a need to change this planned sequence, and actions to be taken in response to this new information, that is, contingent actions (Walker et al., 2001). After identifying the vulnerable, load-bearing assumptions, ABP considers *shaping actions* (those designed to make the assumptions less likely to fail), *hedging actions* (those that can be taken if assumptions begin to fail), and *signposts* (trends and events to monitor in order to detect whether any assumptions are failing).

4.2.3 Scenarios

RDM draws from scenario analysis the concept of a multiplicity of plausible futures as a means to characterize and communicate deep uncertainty (Lempert et al., 2003). Scenarios represent internally consistent descriptions of future events that often come in sets of two or more. Most simply, scenarios are projected futures that claim less confidence than probabilistic forecasts. More generally, a set of scenarios often seeks to represent different ways of looking at the world without an explicit ranking of relative likelihood (Wack, 1985).

Scenarios are often developed and used in deliberative processes with stakeholders. Deemphasizing probabilistic ranking—focusing on a sense of possibility, rather than probability—helps stakeholders expand the range of futures they consider, allowing them to contemplate their choices from a wider range of views and vantages, thus helping participants consider uncomfortable or unexpected futures (Gong et al., 2017; Schoemaker, 1993). The sense of possibility rather than probability can also help scenarios to communicate a wide range of futures to audiences not necessarily eager to have their vantage expanded. By representing different visions of the future without privileging among them, scenarios can offer a comfortable entry into an analysis. Each person can find an initially resonant scenario before contemplating ones that they find more dissonant.

RDM draws from scenario analysis the concept of organizing information about the future into a small number of distinct cases that help people engage with, explore, and communicate deep uncertainty. In particular, the Intuitive Logics school of scenario analysis (Schwartz, 1996) uses qualitative methods to craft a small number of scenarios, distinguished by a small number of key uncertain determinants that differentiate alternative decision-relevant paths into the future. As described below, RDM uses quantitative "Scenario Discovery" algorithms to pursue the same ends. The resulting scenarios summarize the results of the ABP-style stress tests and can link to the development of adaptive strategies (Groves & Lempert, 2007; Groves et al., 2014). Note that while the scenario literature traditionally distinguishes between proba-

---

Climate Change, M. L. Parry, O. F. Canziani, J. P. Palutikof, P. J. v. d. Linden, and C. E. Hanson, Eds., Cambridge University Press, 1, 33–171, Dessai, S., and M. Hulme, 2007: Assessing the Robustness of Adaptation Decisions to Climate Change Uncertainties: A Case Study on Water Resources Management in the East of England. *Global Environmental Change*, **17**, 59–72, Lempert, R., N. Nakicenovic, D. Sarewitz, and M. Schlesinger, 2004: Characterizing climate-change uncertainties for decision-makers - An editorial essay. *Climatic Change*, **65**, 1–9.

bilistic and nonprobabilistic treatments, RDM often employs a third alternative—entertaining multiple views about the likelihood of the scenarios.

The scenario literature also describes a process for seeking robust strategies that includes choosing a set of scenarios that include the most important uncertainties facing the users and then identifying strategies that perform well across all of them (van der Heijden, 1996). This process provides an animating idea for RDM.

4.2.4 Exploratory Modeling

RDM integrates these concepts—DA, ABP, and scenarios—through exploratory modeling (EM). Bankes (1993) encapsulated the 1980s RAND debates on useful and predictive models by dividing computer simulations into two types: (1) consolidative models, which gather all known facts together into a single package that, once validated, can serve as a surrogate for the real world, and (2) exploratory models, which map a wide range of assumptions onto their consequences without privileging one set of assumptions over another. Exploratory models are useful when no single model can be validated because of missing data, inadequate or competing theories, or an irreducibly uncertain future.

Running a model many times is not profound. But as perhaps its key insight, EM notes that when used with an appropriate experimental design—that is, appropriate questions and a well-chosen set of cases designed to address those questions—the large database of results generated from nonpredictive, exploratory models can prove surprisingly useful toward informing policy choices. Bankes (1993) describes several types of questions one may address with EM, including hypothesis generation, reasoning from special cases, and assessing properties of the entire ensemble (Weaver et al., 2013, Box 4.2). RDM uses them all but focuses in particular on robustness as a property of the entire ensemble. That is, identifying and evaluating robust strategies become key capabilities one can elicit with EM.

EM provides RDM with a quantitative framework for stress testing and scenario analysis. While consolidative models most usefully support deductive reasoning, exploratory models serve best to support inductive reasoning—an iterative cycle of question and response. As described in Sect. 4.3, RDM also aims to support a decision-analytic, human/machine collaboration that draws upon what each partner does best.

RDM also exploits another EM advantage: the focus on the simple computational task of running models numerous times in the forward direction. This facilitates exploration of futures and strategies by reducing the requirements for analytic tractability on the models used in the analysis, relative to approaches that rely on optimization or dynamic programming. In addition, EM enables truly global sensitivity explorations since it privileges no base case or single future as an anchor point.

> **Box 4.1 Key Elements of RDM**
>
> RDM meets its goals by proceeding in multiple iterations as humans and computers alternatively test each other's conclusions about futures and strategies. Four key elements govern these interactions (Lempert et al., 2003):
>
> - Consider a multiplicity of plausible futures. The ensemble of futures should be as diverse as possible to adequately stress test proposed policies. The ensemble can also facilitate group processes by including futures that correspond to different groups' worldviews.
> - Seek robust, rather than optimal strategies. Robust strategies perform well compared to the alternatives, over a wide range of plausible futures.
> - Employ adaptive strategies to achieve robustness. Adaptive strategies are designed to evolve over time in response to new information. Generally, such strategies reflect decision-making rules and in practice are often organized around near-term actions, signposts to monitor, and contingency actions to take in response to those signposts.
> - Use the computer to facilitate human deliberation over explorations, options, and tradeoffs, not as device for recommending a particular ordering of strategies.

4.3 RDM Process

RDM explicitly follows a learning process called "deliberation with analysis" in which parties to a decision deliberate on their objectives and options; analysts generate decision-relevant information using system models; and the parties to the decision revisit their objectives, options, and problem framing influenced by this quantitative information (NRC, 2009). Among learning processes, deliberation with analysis proves most appropriate for situations with diverse decision-makers who face a changing decision environment and whose goals can evolve as they collaborate with others. Deliberation with analysis also supports continuous learning based on indicators and monitoring (NRC, 2009, p. 74), a process important to the literature on adaptive policymaking (Swanson & Bhadwal, 2009; Walker et al., 2010). The RDM process is composed of five steps (Fig. 4.1).

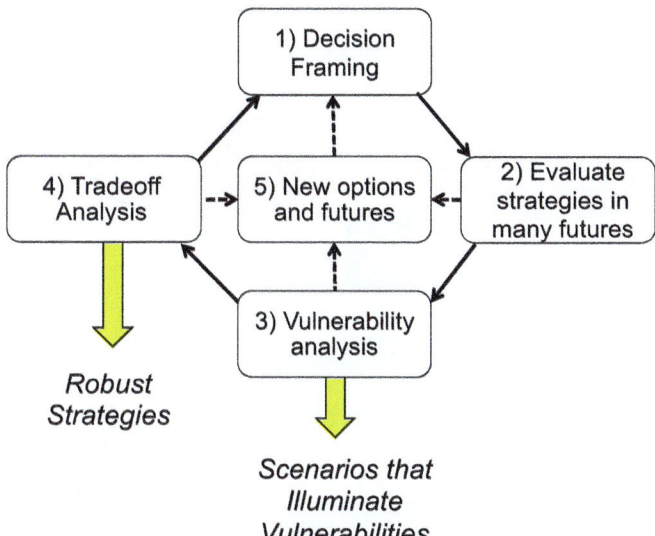

Fig. 4.1 Steps in an RDM analysis. (Source: Lempert et al. (2013c), with the earlier versions in Lempert and Groves (2010) and Lempert and Kalra (2011)

Step 1: The RDM process starts with a decision-framing exercise in which stakeholders define the key factors in the analysis: the decision-makers' objectives and criteria; the alternative actions they can take to pursue those objectives; the uncertainties that may affect the connection between actions and consequences; and the relationships, often instantiated in computer simulation models, between actions, uncertainties, and objectives. This information is often organized in a 2x2 matrix called "XLRM" (Lempert et al., 2003), for uncertainties ("X" factors), policy levers (L), relationships (R), and measures of performance (M).

Step 2: As an "agree-on-decision" approach, RDM next uses simulation models to evaluate proposed strategies in each of many plausible paths into the future, which generates a large database of simulation model results. The proposed strategies can derive from a variety of sources. In some cases, an RDM analysis might start with one or more specific strategies drawn from the relevant public debate. For instance, an RDM analysis for a water agency might begin with that agency's proposed plan for meeting its supply requirements (Groves et al., 2014) or water quality requirements (Fischbach et al., 2015). In other cases, optimization routines for one or more expected futures or decision criteria might yield the initial proposed strategies (Hall et al., 2012). Additionally, an analysis might begin with a wide span of simple strategies covering the logical spectrum and then refine, select, and modify to yield a small group of more sophisticated alternatives (Popper et al., 2009). Often, such as in the example below, an application uses a combination of these approaches.

Step 3: Analysts and decision-makers next use visualization and data analytics on these databases to explore for and characterize vulnerabilities. Commonly, RDM analyses use statistical "Scenario Discovery" algorithms (see below) to identify and display for users the key factors that best distinguish futures in which proposed strategies meet or miss their goals. These clusters of futures are usefully considered policy-relevant scenarios that illuminate the vulnerabilities of the proposed policies. Because these scenarios are clearly, reproducibly, and unambiguously linked to a policy stress test, they can avoid the problems of bias and arbitrariness that sometimes afflict more qualitative scenario exercises (Lempert, 2013; Parker et al., 2015).

Step 4: Analysts and decision-makers may use these scenarios to display and evaluate the tradeoffs among strategies. For instance, one can plot the performance of one or more strategies as a function of the likelihood of the policy-relevant scenarios (e.g., see Fig. 4.2) to suggest the judgments about the future implied by choosing one strategy over another. Other analyses plot multiobjective tradeoff curves—for instance, comparing reliability and cost (Groves et al., 2013a)—for each of the policy-relevant scenarios to help decision-makers decide how to best balance among their competing objectives.

Step 5: Analysts and decision-makers could then use the scenarios and tradeoff analyses to identify and evaluate potentially more robust strategies—ones that provide better tradeoffs than the existing alternatives. These new alternatives generally incorporate additional policy levers, often the components of adaptive decision strategies: short-term actions, signposts, and contingent actions to be taken if the predesignated signpost signals are observed. In some analyses, such adaptive strategies are crafted using expert judgment (e.g., see Lempert & Groves, 2010; Lempert et al., 1996, 2000, 2003; Popper et al., 2009). In other analyses, optimization algorithms may help suggest the best combination of near-term actions, contingent actions, and signposts for the new adaptive strategies (Herman et al., 2014; Kasprzyk et al., 2013; Lempert & Collins, 2007; Lempert et al., 2006).

RDM uses both absolute and relative performance measures to compare strategies in the vulnerability and tradeoff analyses. Absolute performance measures are useful when decision-makers are focused on one or more outcomes, such as profit, energy produced, or lives saved. Absolute performance measures are also useful when decision-makers are focused on some invariant standard—for instance, a regulatory requirement on reliability or environmental quality, a required threshold for an economic rate of return, or a requirement that benefits exceed costs. Relative performance measures are often useful when uncertainties create a wide

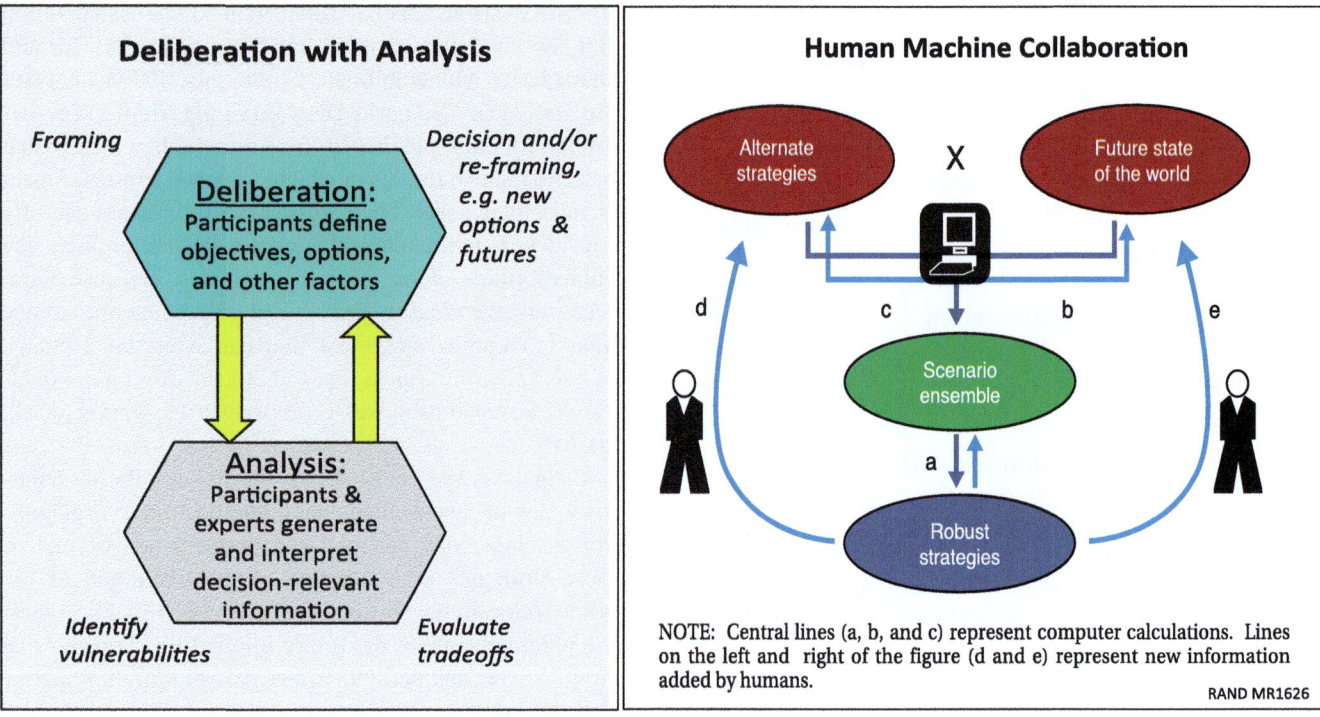

Fig. 4.2 Two views of RDM as a deliberative process. (Source: Lempert et al. 2003)

range of outcomes, so decision-makers seek strategies that perform well compared to alternatives over a wide range of futures. RDM often uses regret to represent relative performance.

At each of the RDM steps, information produced may suggest a reframing of the decision challenge. The process produces key deliverables, including (1) the scenarios that illuminate the vulnerabilities of the strategies and (2) potential robust strategies and the tradeoffs among them.

The left panel of Fig. 4.2 shows these RDM steps in support of a process of deliberation with analysis. Stakeholders begin by deliberating over the initial decision framing. In the vulnerability and tradeoff analysis steps, stakeholders and analysts produce decision-relevant information products. Using these products, stakeholders deliberate over the choice of a robust strategy or return to problem framing, for instance, seeking new alternatives or stress testing a proposed strategy over a wider range of futures. In practice, the process often moves back and forth between problem framing, generating scenarios that illuminate vulnerabilities, identifying new alternatives based on those scenarios, and conducting a tradeoff analysis among the alternatives.

People teaming with computers—each doing what they do best—are more capable than computers or people alone.[5] RDM uses EM to support deliberation with analysis in a process of human/machine collaboration (Lempert et al., 2003). As shown in the right panel of Fig. 4.2, people use their creativity and understanding to pose questions or suggest solutions—for instance, candidate robust strategies. Computers consider numerous combinations of strategies and futures to help users address their questions, search for initially unwelcome counterexamples to proposed solutions, and help people find new candidate robust strategies to propose (Lempert & Popper, 2005; Lempert et al., 2002).

The RDM steps and deliberative processes are consistent with others in the DMDU literature. For instance, multiobjective RDM (MORDM) offers a similar iterative process but with the major advance of more articulation of the step of generating alternative strategies (Kasprzyk et al., 2013). Among related literature, Many-Objective Visual Analytics uses interactive visualizations to support problem framing and reframing (Kollat & Reed, 2007; Woodruff & Reed, 2013), often with a posteriori elicitation of preferences (Cohon & Marks, 1975; Maass et al., 1962).

Overall, the RDM process aims to provide quantitative decision support that helps meet the criteria for good decisions even in the presence of deep uncertainty and the other attributes of wicked problems. The process encourages participants to be explicit about their goals and consider the most important tradeoffs. The process uses scenario concepts

---

[5] For example, Thompson, C., 2013: *Smarter Than You Think: How Technology Is Changing Our Minds for the Better.* Penguin notes that competent chess players teamed with computers can defeat both grandmasters without computers and computers without human assistants.

linked to the idea of policy stress tests, along with computer-assisted exploration, to encourage and facilitate consideration of the decision from a wide range of views. It helps recognize the legitimacy of different interests, values, and expectations about the future by using models as exploratory, rather than predictive, tools within an evidence-based, multi-scenario, multiobjective decision-making process.

> **Box 4.2 How Is a Robust Strategy Operationalized Within the Context of RDM**
>
> A robust strategy is one that performs well, compared to the alternatives, over a wide range of plausible futures (Lempert et al., 2003; Rosenhead et al., 1972). Other definitions exist, including trading some optimal performance for less sensitivity to assumptions (Lempert & Collins, 2007) and keeping options open (Rosenhead, 1990). These definitions can be implemented via a variety of decision analytic criteria (Lempert, 2019). Choosing a specific quantitative criterion to judge robustness can, however, prove complicated because many robustness criteria exist, and in some cases they can yield a different ordering of strategies (Giuliani & Castelletti, 2016). No robustness criterion is best in all circumstances and, as befits a decision support methodology designed to facilitate problem framing, RDM often includes the choice of a robustness criterion as part of its problem-framing step.

4.4 Tools

The concepts underlying RDM—scenario thinking, robustness decision criteria, stress testing proposed plans, and the use of exploratory models—have long pedigrees. But over the last decade new computer and analytic capabilities have made it possible to combine them in practical decision analyses. In particular, RDM often relies on Scenario Discovery and visualization, robust multiobjective optimization, integrated packages for EM, and high-performance computing.

Scenario Discovery algorithms often implement the RDM vulnerability analysis step (step 3) in Fig. 4.1. Scenario Discovery begins with a large database of model runs in which each model run represents the performance of a strategy in one future. The Scenario Discovery cluster-finding algorithms then offer concise descriptions of those combinations of future conditions that best distinguish the cases in which the implementation plan does or does not meet its goals. The requisite classification algorithms—often Patient Rule Induction Method (PRIM) (Friedman & Fisher, 1999) or Classification and Regression Tree (CART) (Breiman et al., 1984), combined with a principal component analysis (Dalal et al., 2013)—seek to balance between the competing goals of simplicity and accuracy in order to describe sets of strategy-stressing futures as concise, understandable, and decision-relevant scenarios (Bryant & Lempert, 2010; Groves & Lempert, 2007). Software to implement the PRIM algorithm is available, both in standalone routines and embedded in the EM software packages described below. Overall, Scenario Discovery replicates analytically the ideas of qualitative Intuitive Logics scenario analysis and provides information products that can prove compelling in stakeholder deliberations (Lempert, 2013). RDM analyses also use computer visualization of the database of runs to support the vulnerability and tradeoff analyses. Tableau, a commercially available platform, has proven particularly useful in much RDM work (Cervigni et al., 2015; Groves et al., 2013a).

The "new alternatives" step (step 5) in Fig. 4.1 can employ a variety of methods. While some RDM analyses use only stakeholder and/or expert judgment to craft responses to potential vulnerabilities (e.g., Groves et al., 2020; Lempert & Groves, 2010), many applications use some multiobjective robust optimization tool. Some such applications have used constrained optimization to trace out a range of potentially robust solutions, both in single-objective (Lempert & Collins, 2007) and multiobjective (Groves et al., 2013b) cases. The latter instance involved a planning tool that allowed analysts to trace the Pareto tradeoff curves in each scenario for any two objectives using constrained optimizations over the other objectives (Groves et al., 2012) and has been widely used. Other applications, such as the Colorado Basin Supply and Demand Study (Groves et al., 2013a), run large portfolio optimizations for many futures; note the individual actions that occur in the optimal set for most, some, and few of the futures; and use this information to craft adaptive strategies that begin with the actions that occur in most of the futures' optimal sets, and implement the others depending on which future comes to pass (Bloom, 2015; Groves et al., 2013a). Multiobjective RDM (MORDM) tools provide a more general solution, using evolutionary algorithms to identify the regions of a Pareto surface over many objectives that are most robust to uncertainty (Kasprzyk et al., 2013). MORDM has been used to identify, through a process called direct policy search, adaptive strategies modeled as controllers in the control theory sense (Quinn et al., 2017).

While many RDM analyses use stand-alone software to generate and analyze many model runs, several integrated EM packages exist that can greatly facilitate such analysis, such as the exploratory modeling and analysis workbench,[6] open MORDM (Hadka et al., 2015), and rhodium.[7]

While many RDM analyses may be conveniently run on a laptop or desktop computer (e.g., Lempert et al., 2013b), re-

---

[6] http://simulation.tbm.tudelft.nl/ema-workbench/contents.html
[7] https://github.com/Project-Platypus/Rhodium/wiki/Philosophy

cent studies have also used high-performance computation—either large-scale cluster computing (Groves et al., 2016; Zeff et al., 2014) or cloud-based computer services (Cervigni et al., 2015; Isley, 2014)—to quickly and inexpensively conduct a very large number of runs.

4.5 Example: Carrots and Sticks for New Technology

An early RDM application provides an ideal example of the approach, the types of tools used, and how EM can draw together concepts from decision analysis, scenarios, and ABP. The example, called "carrots and sticks for new technology," focused on determining the most robust combination of two policy instruments—carbon prices and technology subsidies—to reduce climate-altering greenhouse gas emissions (Lempert, 2002; Robalino & Lempert, 2000). We describe this study using the framework offered in this book's opening chapter (the major subheadings below), while also relating the framework to the RDM steps shown in Fig. 4.1.

4.5.1 Frame the Analysis

4.5.1.1 Formulate Question (RDM Step 1)

It is well-understood that an economically ideal greenhouse gas emission reduction policy should include an economy-wide carbon price implemented through mechanisms such as a carbon tax or a cap-and-trade system. This early RDM study addressed the question of whether and under what conditions technology incentives, such as tax credits or subsidies for clean energy technologies, also prove necessary and important as part of a greenhouse gas reduction strategy. Many national and regional jurisdictions worldwide employ such incentives in their climate policies because they prove politically popular and have a compelling logic, if for no other reason than significant technology innovation will prove crucial to limiting climate change. Recent climate legislation in the United States, such as the Inflation Reduction Act, follows this approach. But technology incentives have a mixed record of success (Cohen & Noll, 2002), and sometimes link to larger debates about the appropriate role of government (Wolf, 1993). Standard economic analysis proves a poor platform to adjudicate such questions because the extent to which technology incentives prove economically important may depend on coordination failures that occur in the presence of increasing returns to scale, imperfect information, and heterogeneous preferences—deeply uncertain factors not well-represented in standard economic models. Addressing the technology-incentive question with RDM thus proved useful due to this deep uncertainty and because people's views on technology incentives can be strongly affected by their worldviews.

4.5.1.2 Identify Alternatives (RDM Step 1)

The study was organized as a head-to-head comparison between strategies incorporating two types of policy instruments: an economy-wide carbon tax and a technology price subsidy for low-carbon-emitting technologies. We considered four combinations: (1) no carbon reduction policy, (2) tax only, (3) subsidies only, and (4) a combination of both taxes and subsidies. As described below, both the tax and subsidy were configured as adaptive strategies designed to evolve over time in response to new information.

4.5.1.3 Specify Objectives (RDM Step 1)

This study compared the strategies with two output measures: the present value of global economic output (pvGDP) and the mid-twenty-first-century level of greenhouse gas emissions. We focused on pvGDP to facilitate comparison of this work with other analyses in the climate change policy literature using more standard economic formulations. As described below, we calculated the regret for each strategy in each future and used a domain criterion for robustness (see Box 4.2), thus looking for strategies with low regret over a wide range of plausible futures.

4.5.1.4 Specify System Structure (RDM Step 1)

To compare these adaptive strategies, we employed an agent-based model of technology diffusion, linked to a simple macro model of economic growth that focused on the social and economic factors that influence how economic actors choose to adopt, or not to adopt, new emissions-reducing technologies. The agent-based representation proved useful because it conveniently represents key factors potentially important to technology diffusion, such as the heterogeneity of technology preferences among economic actors and the flows of imperfect information that influence their decisions. Considering this tool as an exploratory, rather than predictive, model proved useful because it allowed the study to make concrete and specific comparisons of price- and subsidy-based strategies even though, as described below, available theory and data allowed the model's key outputs to vary by over an order of magnitude.

The model was rooted in the microeconomic understanding of the process of technology diffusion (Davies, 1979). Each agent in our model represents a producer of a composite good, aggregated as total GDP, using energy as one key input. Each period the agents may switch their choice of energy-generation technology, choosing among high-, medium-, or low-emitting options. Agents choose technology to maximize their economic utility. The agents estimate utility based on their expectations regarding each technology's cost and performance. Costs may or may not decline significantly due to increasing returns to scale as more agents choose to adopt. The agents have imperfect information and current and future technology cost and performance but can gain information based on their own experience, if any, and by querying other agents who have used the technology. Thus, the model's technology diffusion rates depend reflexively on themselves since by adopting a technology each agent generates new information that may influence the adoption decisions of other potential users. The model also used simple, standard, but deeply uncertain relationships from the literature on the connections between greenhouse gas emissions and the economic impacts due to climate change (Nordhaus, 1994).

4.5.2 Perform Exploratory Uncertainty Analysis

4.5.2.1 Specify Uncertainties or Disagreements (RDM Step 2)

The agent-based model had 30 input parameters representing the deeply uncertain factors, including the macroeconomic effects of potentially distortionary taxes and subsidies on economic growth, the microeconomic preferences that agents use to make technology adoption decisions, the future cost and performance of high- (e.g., coal), low- (e.g., gas), and nonemitting (e.g. solar) technologies, the way information about new technologies flows through agent networks, and the impacts of climate change.

We employed three sources of information to constrain our EM. First, the agent-based model embodied the theoretical economic understanding of technology diffusion. Second, we drew plausible ranges for each individual parameter using estimates from the microeconomics literature. Third, we required the model to reproduce macroeconomic data regarding the last 50 years of economic growth and market shares for different types of energy technology. We also constrained future technology diffusion rates in the model to be no faster than the fastest such rates observed in the past.

Consistent with these constraints, the model nonetheless was able to generate a vast range of plausible futures. To choose a representative sample of futures from this vast set, the study launched a genetic algorithm over the model inputs, searching for the most diverse set of model inputs consistent with the theoretical, macroeconomic, and microeconomic constraints (Miller, 1998). This process yielded an ensemble of 1611 plausible futures, with each future characterized by a specific set of values for each of the 30 uncertain model input parameters. Each member of the ensemble reproduced the observed history from 1950 to 2000 but differed by up to an order of magnitude in projected mid-twenty-first-century emissions.

4.5.2.2 Stress Test Strategies Against Futures (RDM Steps 3–5)

The study represented its carbon tax and technology subsidies as adaptive strategies using a single set of parameters for each to describe their initial conditions, and how they would evolve over time in response to new information.

As with many carbon price proposals, the study's carbon tax would start with an initial price per ton of CO_2 at time zero and rise at a fixed annual rate, subject to two conditions meant to reflect political constraints. First, we assumed that the government would not let the tax rise faster than the increase in the observed social cost of carbon. Second, we assumed that if the economy dropped into recession (defined as a global growth rate below some threshold), the government would drop the carbon tax back to its original level. Three parameters defined the tax strategy: the initial tax rate, the annual rate of increase, and the minimum threshold for economic growth required not to repeal the tax.

The study's technology subsidy reduces the cost to users of a technology to a fixed percentage of its unsubsidized cost. The subsidy stays in effect until the policy either succeeds in launching the technology (defined as its market share rising above some threshold value) or the policy fails (defined as the market share failing to reach a minimum threshold after a certain number of years). Meeting either of these conditions permanently terminates the subsidy. Four parameters defined the subsidy strategy: the subsidy level, the market share defining success, the market share defining failure, and the number of years before the subsidy can be judged a failure.

The study chose a single set of parameter values to define the tax and subsidy policies, each set chosen to optimize the pvGDP for the future represented by the average value for each of the 30 model parameters. The results of the study were relatively insensitive to this simplification.

We used the agent-based model to calculate the pvGDP and mid-century greenhouse gas emissions for each strategy in each of the 1611 plausible futures. This ensemble of runs made it immediately clear that the *Taxes Only* and *Combined* strategies consistently perform better than the *No Policy* and *Subsidies Only* strategies. The remainder of the analysis thus focused on the first two.

4.5.3 Choose Short-Term Actions and Long-Term Contingencies

4.5.3.1 Illustrate Tradeoffs (RDM Step 4)

Lacking (not-yet-developed) Scenario Discovery algorithms and faced with too many dimensions of uncertainty for an exhaustive search, the study used importance sampling to find the five uncertain input parameters most strongly correlated with mid-century GHG emissions. Four of these key uncertainties related to the potential for coordination failures to slow technology diffusion—the rate of cost reductions for nonemitting technologies caused by learning by doing, the rate at which agents learn from one another about the performance of new technologies, the agents' risk aversion, and the agents' price-performance preferences for new technologies—and one related to the damages from climate change.

We then examined the regret in pvGDP across the cases for the *Taxes Only* and *Combined* strategies as a function of all ten two-dimensional combinations of the five key uncertainties.[8] Each visualization told a similar story—that the *Combined* strategy cases had lower mean and variability in pvGDP regret than those for *Taxes Only* except in the corner of the uncertainty space with low potential for coordination failures and/or low impacts from climate change.

This analysis provides the study's basic comparison among the strategies: *Taxes Only* performs best when the potential for coordination failures and the impacts from climate change are small, and *Combined* performs best otherwise. Both strategies perform better than *No Policy* or *Subsidies Only*.

4.5.3.2 Select and Plan for Adaptation (RDM Step 4)

To help decision-makers understand the conditions under which the *Taxes Only* and *Combined* strategies would each be favored over the other, we first collapsed the 10 visualizations into a single two-dimensional graph by combining all four key uncertain parameters relating to potential coordination failures into a single variable. Figure 4.3 shows the sets of expectations in which pvGDP for the *Taxes Only* strategy exceeds that of *Combined*, and vice versa, as a function of the probability assigned to high rather than low values of the four uncertain parameters related to potential coordination failures, which we labeled "probability of a nonclassical world" and "probability of high damages" due to climate change. The region dominated by *Combined* is larger than that of *Taxes Only* for two reasons. First, *Combined* dominates *Taxes Only* over larger regions of the state space. Second, in the regions of the uncertainty space where *Taxes Only* is the better strategy, *Combined*'s regret is relatively small, while in the regions where *Combined* is better, *Taxes Only*'s regret is relatively large. Existing scientific understanding proves insufficient to define with certainty where the future lies in Fig. 4.3. Different parties to the decision may have different views. But the boundary between the two regions is consistent with increasing returns to scale much smaller than those observed for some energy technologies—such as natural gas turbines, wind, and solar—and the middle of the figure is consistent with relatively small levels of risk aversion,

---

[8]Ten combinations because $\binom{5}{2} = 10$.

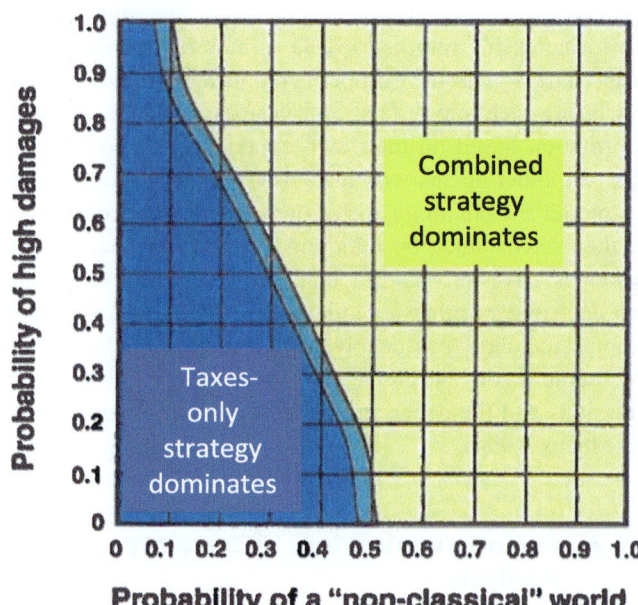

Fig. 4.3 Scenario map comparing "Taxes Only" and "Combined" strategies. Note: Taxes-only strategy uses carbon pricing mechanisms. Combined strategy uses carbon pricing and technology incentives. High damages refers to significant impacts from climate change. A nonclassical world is one in which the diffusion rate of new technologies is dominated by factors such as learning by doing and network learning effects not well-captured in many economic models

learning rates, and heterogeneity of preferences compared to those seen in various literatures. These results suggest that a combination of price instruments and technology subsidies may prove the most robust strategy over a wide range of plausible futures.

4.5.3.3 Implementation, Monitoring, and Communication

This study addressed a high-level question of policy architecture—the best mix of policy instruments for decarbonization. While the study did not provide detailed implementation plans, it does suggest how a national or state/provincial government might pursue the study's recommendations. The study envisions policymakers choosing a strategy that includes the rules by which the initial actions will be adapted over time (Swanson et al., 2007). The carbon price, presumably set by the legislature, would follow the social cost of carbon as periodically updated by executive agencies (National Academies of Sciences, 2016) whenever the economy was not in recession. The legislature would also set the technology subsidy and terminate it when the subsidized technologies either succeeded or failed based on market share data gathered by executive agencies. The study did not examine pre-commitment issues—that is, how the current legislature could ensure that future legislatures would in fact follow the adaptive strategy. These interesting issues of political economy have, however, been recently explored using RDM methods (Isley et al., 2015).

4.5.4 Iterate and re-Examine (RDM Steps 2 and 3)

The study's results are based on an examination of only 6 of the 30 dimensions of uncertainty in the model, representing a small subset of the full range of plausible futures. As a key final step, we tested the policy recommendations by launching a genetic search algorithm across the previously unexamined dimensions looking for additional futures that would provide counterexamples to our conclusions. This process represents the computer feedback loop in the right panel of Fig. 4.2. The genetic algorithm ran for most of the time the authors spent writing their manuscript and found no plausible counterexamples.

Overall, this study suggests that if decision-makers hold even modest expectations that market imperfections are likely to inhibit the diffusion of new, emissions-reducing technologies or that the impacts of climate change will turn out to be serious, then strategies combining both carbon taxes and technology incentives may be a promising component of a robust strategy for reducing greenhouse gas emissions.

4.6 Recent Advances and Future Challenges

The "Carrots and Sticks for New Technology" example includes all the steps of an RDM analysis shown in Fig. 4.1. It used optimization algorithms to define alternative, adaptive strategies, and generated its futures using genetic algorithms to perform what has more recently been called scenario diversity analysis (Carlsen et al. 2016a, b). The study employed the process of human/machine collaboration shown in Fig. 4.2, in particular, in the computer search for counterexamples to the human-derived patterns that constitute its policy conclusions.

Since this early example, the methods and tools for multiobjective RDM analyses have approached maturity, now reaching the point at which one can describe with some specificity how to conduct multiscenario, multiobjective RDM for many wicked problems. For instance, a recent study used RDM on a topic similar to the "Carrots and Sticks" example—examining how international finance institutions such as the Green Climate Fund (GCF) can best craft long-term investment strategies to speed decarbonization in the face of deep technological and climate uncertainty (Molina Perez, 2016). This more recent study was made possible by powerful new Scenario Discovery algorithms and visualization tools.

Recent work for four North Carolina cities illustrates the power of MORDM, a combination of RDM with new evolutionary algorithms for multiobjective robust optimization (Herman et al., 2014, 2016; Zeff et al. 2014, 2016). The study helped the neighboring cities of Raleigh, Durham, Chapel Hill, and Cary link their short-term operational and long-term investment water plans by shifting the former from rule-based procedures to new dynamic risk-of-failure triggers, and the latter from static to adaptive policy pathways. The study also helped the four independent cities coordinate their plans in the presence of different objectives and deep uncertainty.

Work for the US Bureau of Reclamation and the parties to the Colorado Compact showcases RDM's ability to facilitate deliberation with analysis, helping contesting parties to agree on the vulnerabilities they face and adaptive strategies for addressing them (Bloom, 2015; Groves et al. 2013a, 2019).

Some important technical hurdles still remain before these capabilities fully mature. First, an approach is needed that provides full Pareto satisficing surfaces. Current MORDM analyses identify Pareto *optimal* surfaces for best-estimate cases and measure the robustness of alternative strategies, represented by different regions on the Pareto surface, to the deep uncertainties (Kasprzyk et al., 2013). In the future, MORDM could produce sets of strategies chosen specifically because their performance across multiple objectives was largely insensitive to the deep uncertainties. Recent work has taken steps toward providing such Pareto satisficing surfaces (Watson & Kasprzyk, 2017), but more needs to be done. Furthermore, despite the availability of ubiquitous computation on the cloud, and through high-performance computation facilities, it still remains difficult in many cases to conduct a full MORDM analysis using realistic system models, which would require running many thousands of cases to perform the multiobjective robust optimization over each of many thousands of scenarios. Research is needed on what we might call adaptive sampling approaches to help navigate more efficiently through the set of needed runs. In addition, research could usefully provide guidance on when to use alternative robustness criteria, as well as the conditions under which RDM's iterative analytic process is guaranteed to converge independent of the initial problem framing or when path dependence may lead analyses to different answers (Kwakkel et al., 2016).

Finally, the cost of developing the needed system models often puts RDM analyses out of reach for many decision-makers. Research on "RDM-lite"—means to quickly develop such models through approaches such as expert elicitation and participatory modeling—could greatly increase the use of these methods (e.g., see O'Mahony et al., 2018; Popper, 2019).

Evaluation plays a crucial role in the design and use of any successful decision support system (NRC, 2009; Pidgeon & Fischhoff, 2011; Wong-Parodi et al. 2016, 2020). Some evaluations of RDM tools, visualizations, and processes exist, both in the laboratory (Budescu et al., 2013; Gong et al., 2017; Parker et al., 2015) and through field experiments

(Groves et al., 2008). Recent work has proposed frameworks for evaluating the impacts of RDM-based decision support in urban environments (Knopman & Lempert, 2016). But much more such evaluation work is required to improve the practical application of RDM decision support (Bartels et al., 2018).

More broadly, as DMDU methods reach technical maturity, they offer the opportunity to reshape the relationship between quantitative decision analytics and the way in which organizations use this information with their internal and external audiences and processes. The potential for such reshaping presents a rich menu of research needs to understand the organizational, anthropological, political, and ethical implications.

As one example, the concept of *risk governance* embeds risk management, which often has a narrow, more technocratic perspective, in a broader context that considers institutions, rules conventions, processes, and mechanisms through which humans acting as individuals and groups make choices affecting risk (Renn, 2008). Recent work has explored how to embed RDM methods and tools in a risk governance framework (Knopman & Lempert, 2016). For instance, RDM can help decision-makers working within a multiagent and multijurisdictional system organize their strategies into "tiers of transformation," which derives from the ideas of triple-loop learning. Lower tiers represent actions the decision-makers can address on their own, while the outer tiers represent large-scale, transformative system changes that only the decision-makers can help catalyze. Any understanding of how to implement and use such capabilities, and the extent to which they would prove useful, remains nascent.

Future work can also usefully situate the types of moral reasoning and social choice embodied in alternative approaches to decision support. In his treatise, the *Idea of Justice*, Amartya Sen (2009) describes two classes of moral reasoning—the transcendental and the relational. The former, represented by Sen's teacher John Rawls (1971), seeks to inform ethical societal choices by first envisioning a common vision of a perfectly just world. People can then use that vision to inform their near-term choices. The latter, Sen's preferred alternative, rests on the assumption that irreducible uncertainty about the consequences of our actions, and a diversity of priorities, goals, and values, are fundamental attributes of our world. Thus, no such transcendental vision of the type envisioned by Rawls is possible because the level of agreement and commonality of values it presupposes does not, and should not, exist in a diverse society in which people are free to pursue their lives according to their own, often very different, visions of what is good. In addition, Sen argues, even if a common transcendental vision were possible, it would prove insufficient to inform near-term choices, because human knowledge is too fallible and the uncertainties too deep to chart an unambiguous path to the ideal. But humans can obtain sufficient knowledge to craft near-term options and differentiate the better from the worse.

Relational reasoning thus involves an iterative process of debating, choosing, learning, and revisiting choices, always trying to move in the direction of more justice in the face of imperfect knowledge and conflicting goals.

"Agree-on-assumptions" approaches to decision support reflect transcendental reasoning, while "agree-on-decisions" approaches reflect relational reasoning (Lempert et al., 2013a). Sen emphasizes the importance of deliberation in a relational process of social choice. His framework provides attributes for judging what constitutes an ethical process of deliberation with analysis. In particular, such deliberations work best when they recognize the inescapable plurality of competing views; facilitate re-examination and iterative assessments; demand clear explication of reasoning and logic; and recognize an "open impartiality" that accepts the legitimacy and importance of the views of others, both inside and outside the community of interest to the immediate policy discussion.

Recent work has pioneered methods for conducting ethical–epistemological analysis on the extent to which decision support products, methods, and systems meet such ethical criteria (Bessette et al., 2017; Jafino et al., 2021; Lempert & Turner 2021, Mayer et al. 2017; Tuana, 2013), but much more remains to be done. Today's world presents numerous, complex decision challenges—from climate change and sustainability to national security—that require quantitative decision support to successfully address. But "agree-on-assumptions" methods often lure decision-makers toward overconfidence and can make it difficult to engage and promote consensus among participants with diverse expectations and interests. Such methods—built on the assumption that the decision analytics aim to provide a normative ranking of decision options—have their foundations in a time of computational poverty and rest on a narrow understanding of how quantitative information can best inform decisions. Recent years have seen an explosion of computational capabilities and a much richer understanding of effective decision support products and processes. RDM—a multiobjective, multiscenario "agree-on-decision" approach—exploits these new capabilities and understanding to facilitate deliberative processes in which decision-makers explore, frame, and reach consensus on the "wicked" problems that today's decision-makers increasingly face.

References

Bankes, S. C. (1993). Exploratory modeling for policy analysis. *Operations Research, 41*, 435–449. https://doi.org/10.1287/opre.41.3.435

Bartels, E., Mikolic-Torreira, I, Popper, S. W., & Predd, J. (2018). *What is the value proposition of analysis for decision making?* Santa Monica: RAND Corporation, PR-3485-RC.

Bessette, D. L., Mayer, L. A., Cwik, B., Vezer, M., Keller, K., Lempert, R., & Tuana, N. (2017). Building a values-informed mental model for New Orleans climate risk management. *Risk Analysis, 37*, 1993–2004. https://doi.org/10.1111/risa.12743

Bloom, E., 2015: Changing midstream: Providing decision support for adaptive strategies using robust decision making, 271 pp. https://www.rand.org/content/dam/rand/pubs/rgs_dissertations/RGSD300/RGSD348/RAND_RGSD348.pdf.

Borio, C., Drehmann, M., & Tsatsaronis, K. (2014). Stress-testing macro stress testing: Does it live up to expectations? *Journal of Financial Stability, 12,* 3–15. https://doi.org/10.1016/j.jfs.2013.06.001

Breiman, L., Friedman, J. H., Olshen, R. A., & Stone, C. J. (1984). *Classification and regression trees. Wadsworth Statistics/ Probability Series.* Monterey: Wadsworth. https://doi.org/10.1201/9781315139470

Bryant, B. P., & Lempert, R. J. (2010). Thinking inside the box: A participatory, computer-assisted approach to scenario discovery. *Technological Forecasting and Social Change, 77,* 34–49. https://doi.org/10.1016/j.techfore.2009.08.002

Budescu, D. V., Lempert, R., Broomell, S., & Keller, K. (2013). Aided and unaided decisions with imprecise probabilities. *European Journal of Operational Research, 2,* 31–62. https://doi.org/10.1007/s40070-013-0023-4

Carlsen, H., Lempert, R., Wikman-Svahn, P., & Schweizer, V. (2016a). Choosing small sets of policy-relevant scenarios by combining vulnerability and diversity approaches. *Environmental Modelling & Software, 84,* 155–164. https://doi.org/10.1016/j.envsoft.2016.06.011

Carlsen, H., Eriksson, E. A., Dreborg, K. H., Johansson, B., & Bodin, Ö. (2016b). Systematic exploration of scenario spaces. *Foresight, 18,* 59–75. https://doi.org/10.1108/FS-02-2015-0011

Carter, T. R., et al. (2007). New assessment methods and the characterisation of future conditions. *Climate change 2007: Impacts, adaptation and vulnerability.* In M. L. Parry, O. F. Canziani, J. P. Palutikof, P. J. v. D. Linden, & C. E. Hanson (Eds.), *Contribution of Working Group II to the Fourth Assessment Report of the Intergovernmental Panel on Climate Change* (Vol. 1, pp. 33–171). Cambridge University Press. https://www.ipcc.ch/site/assets/uploads/2018/03/ar4_wg2_full_report.pdf

Cervigni, R., Liden, R., Neumann, J. E., & Strzepek, K. M. (Eds.). (2015). *Enhancing the climate resilience of Africa's infrastructure: The water and power sectors.* World Bank. https://doi.org/10.1596/978-1-4648-0466-3

Cohen, L. R., & Noll, R. G. (2002). Technology pork barrel. https://doi.org/10.2307/20045032

Cohon, J., & Marks, D. (1975). A review and evaluation of multiobjective programing techniques. *Water Resources Research, 11,* 208. https://doi.org/10.1029/wr011i002p00208

Dalal, S., Han, B., Lempert, R., Jaycocks, A., & Hackbarth, A. (2013). Improving scenario discovery using Orthogonol rotations. *Environmental Modeling and Software, 48,* 1–16. https://doi.org/10.1016/j.envsoft.2013.05.013

Davies, S. (1979). *The diffusion of process innovations.* Cambridge, MA: Cambridge University Press.

Dessai, S., & Hulme, M. (2007). Assessing the robustness of adaptation decisions to climate change uncertainties: A case study on water resources management in the East of England. *Global Environmental Change, 17,* 59–72. https://doi.org/10.1016/j.gloenvcha.2006.11.005

Dewar, J. A., C. H. Builder, W. M. Hix, and M. H. Levin, 1993: Assumption-based planning—A planning tool for very uncertain times, 88 pp. .

Fischbach, J. R., Lempert, R. J., Molina-Perez, E., Tariq, A., Finucane, M. L., & Hoss, F. (2015). *Managing water quality in the face of uncertainty: A robust decision making demonstration for EPA's National Water Program.* https://doi.org/10.7249/rr720

Friedman, J. H., & Fisher, N. I. (1999). Bump hunting in high-dimensional data. *Statistics and Computing, 9,* 123–143. https://doi.org/10.1023/a:1008894516817

Ghile, Y. B., Taner, M. Ü., Brown, C., Grijsen, J. G., & Talbi, A. (2014). Bottom-up climate risk assessment of infrastructure investment in The Niger River Basin. *Climatic Change, 122,* 97–110. https://doi.org/10.1007/s10584-013-1008-9

Giuliani, M., & Castelletti, A. (2016). Is robustness really robust? How different definitions of robustness impact decision-making under climate change. *Climatic Change, 135,* 409–424. https://doi.org/10.1007/s10584-015-1586-9

Gong, M., et al. (2017). Testing the scenario hypothesis: An experimental comparison of scenarios and forecasts for decision support in a complex decision environment. *Environmental Modeling and Software, 91,* 135–155. https://doi.org/10.1016/j.envsoft.2017.02.002

Groves, D. G., & Lempert, R. J. (2007). A new analytic method for finding policy-relevant scenarios. *Global Environmental Change, 17,* 73–85. https://doi.org/10.1016/j.gloenvcha.2006.11.006

Groves, D. G., D. Knopman, R. Lempert, S. Berry, and L. Wainfan, 2008: Presenting uncertainty about climate change to water resource managers—Summary of workshops with the inland empire utilities agency. .

Groves, D. G., C. Sharon, and D. Knopman, 2012: Planning tool to support Louisiana's decision making on coastal protection and restoration. https://www.rand.org/pubs/technical_reports/TR1266.html.

Groves, D. G., Fischbach, J. R., Bloom, E., Knopman, D., & Keefe, R. (2013a). Adapting to a changing Colorado River: Making future water deliveries more reliable through robust management strategies. *RAND Corporation.* https://doi.org/10.7249/rr242

Groves, D. G., et al. (2013b). *Addressing coastal vulnerabilities through comprehensive planning.* RAND Corporation. https://www.rand.org/pubs/research_briefs/RB9696-1.html

Groves, D. G., Bloom, E. W., Lempert, R. J., Fischbach, J. R., Nevills, J., & Goshi, B. (2014). Developing key indicators for adaptive water planning. *Journal of Water Resources Planning and Management, 141*(7). https://doi.org/10.1061/(asce)wr.1943-5452.0000471

Groves, D. G., Lempert, R. J., May, D. W., Leek, J. R., & Syme, J. (2016). Using high-performance computing to support water resource planning. A workshop demonstration of real-time analytic facilitation for the Colorado River Basin, 12 pp. https://doi.org/10.7249/cf339

Groves, D. G., Molina Perez, E., Bloom, E., & Fischbach, J. R. (2019). In V. A. W. J. Marchau, W. E. Walker, P. J. T. M. Bloemen, & S. W. E. Popper (Eds.), *Robust decision making (RDM): Application to water planning and climate policy. Decision making under deep uncertainty: From theory to practice* (p. 329). Springer. https://doi.org/10.1007/978-3-030-05252-2_7

Groves, D. G., et al. (2020). *The benefits and costs of decarbonizing Costa Rica's economy: Informing the implementation of Costa Rica's national decarbonization plan under uncertainty.* https://doi.org/10.7249/rra633-1

Hadka, D., Herman, J., Reed, P., & Keller, K. (2015). An open source framework for many-objective robust decision making. *Environmental Modelling & Software, 74,* 114–129. https://doi.org/10.1016/j.envsoft.2015.07.014

Hall, J. M., Lempert, R., Keller, K., Hackbarth, A., Mijere, C., & McInerney, D. (2012). Robust climate policies under uncertainty: A comparison of info-gap and RDM methods. *Risk Analysis, 32,* 1657–1672. https://doi.org/10.1111/j.1539-6924.2012.01802.x

Helgeson, C. (2018). Structuring decisions under deep uncertainty. *Topoi, 1-13,* 257. https://doi.org/10.1007/s11245-018-9584-y

Herman, J., Zeff, H., Reed, P., & Characklis, G. (2014). Beyond optimality: Multistakeholder robustness tradeoffs for regional water portfolio planning under deep uncertainty. *Water Resources Research, 50,* 7692–7713. https://doi.org/10.1002/2014wr015338

Herman, J., Zeff, H., Lamontagne, J., Reed, P., & Characklis, G. (2016). Synthetic drought scenario generation to support bottom-up water supply vulnerability assessments. *Journal of Water Resources Planning and Management, 142.* https://doi.org/10.1061/(ASCE)WR.1943-5452.0000701

Herrick, C., & Sarewitz, D. (2000). Ex post evaluation: A more effective role for scientific assessments in Environmental policy. *Science, Technology, and Human Values, 25*, 309–331. https://doi.org/10.1177/016224390002500303

Isley, S. (2014). Evaluating the political sustainability of emission control policies in an evolutionary economics setting. *Pardee Rand Graduate School*. https://doi.org/10.7249/tr1308

Isley, S. C., Lempert, R. J., Popper, S. W., & Vardavas, R. (2015). The effect of near-term policy choices on long-term greenhouse gas transformation pathways. *Global Environmental Change, 34*, 147–158. https://doi.org/10.1016/j.gloenvcha.2015.06.008

Jafino, B. A., Kwakkel, J. H., & Taebi, B. (2021). Enabling assessment of distributive justice through models for climate change planning: A review of recent advances and a research agenda. *Wiley Interdisciplinary Reviews: Climate Change., 12*. https://doi.org/10.1002/wcc.721

Jones, R. N., et al. (2014). *Foundations for decision making. Climate Change 2014: Impacts, Adaptation, and Vulnerability*. Intergovernmental Panel on Climate Change (IPCC), Ed. https://www.ipcc.ch/site/assets/uploads/2018/02/WGIIAR5-Chap2_FINAL.pdf.

Kalra, N., Hallegatte, S., Lempert, R., Brown, C., Fozzard, A., Gill, S., & Shah, A. (2014). *Agreeing on robust decisions: A new process of decision making under deep uncertainty*. https://doi.org/10.1596/1813-9450-6906

Kasprzyk, J. R., Nataraj, S., Reed, P. M., & Lempert, R. J. (2013). Many-objective robust decision making for complex environmental systems undergoing change. *Environmental Modeling and Software, 42*, 55–71. https://doi.org/10.1016/j.envsoft.2012.12.007

Knopman, D., & Lempert, R. (2016). *Urban responses to climate change: Framework for decisionmaking and supporting indicators* (156 pp). https://doi.org/10.7249/rr1144.

Kollat, J., & Reed, P. (2007). A framework for Visually Interactive Decision-making and Design using Evolutionary Multi-objective Optimization (VIDEO). *Environmental Modeling and Software, 22*, 1691–1704. https://doi.org/10.1016/j.envsoft.2007.02.001

Kwakkel, J. H., Haasnoot, M., & Walker, W. E. (2016). Comparing robust decision-making and dynamic adaptive policy pathways for model-based decision support under deep uncertainty. *Environmental Modelling & Software, 86*, 168–183. https://doi.org/10.1016/j.envsoft.2016.09.017

Lempert, R. J. (2002). A new decision sciences for complex systems. *Proceedings of the National Academy of Sciences, 99*, 7309–7313. https://doi.org/10.1073/pnas.082081699

Lempert, R. J. (2007) Can scenarios help policymakers be both bold and careful? Blindside: How to anticipate forcing events and wild cards in global politics. https://www.jstor.org/stable/10.7864/j.ctt6wpff7.12.

Lempert, R. J. (2013). Scenarios that illuminate vulnerabilities and robust responses. *Climatic Change, 117*, 627–646. https://doi.org/10.1007/s10584-012-0574-6

Lempert, R. (2019). In V. A. W. J. Marchau, W. E. Walker, P. J. T. M. Bloemen, & S. W. E. Popper (Eds.), *Robust decision making (RDM). Decision making under deep uncertainty: From theory to practice* (p. 329). Springer. https://doi.org/10.1007/978-3-030-05252-2_2

Lempert, R. J., & Collins, M. (2007). Managing the risk of uncertain threshold responses: Comparison of robust, optimum, and precautionary approaches. *Risk Analysis, 27*, 1009–1026. https://doi.org/10.1111/j.1539-6924.2007.00940.x

Lempert, R., & Groves, D. G. (2010). Identifying and evaluating robust adaptive policy responses to climate change for water management agencies in the American West. *Technological Forecasting and Social Change, 77*, 960–974. https://doi.org/10.1016/j.techfore.2010.04.007

Lempert, R., and N. Kalra, 2011: Managing climate risks in developing countries with robust decision making. https://www.rand.org/pubs/external_publications/EP201100254.html.

Lempert, R. J., & Popper, S. W. (2005). In R. Klitgaard & P. Light (Eds.), *High-performance government in an uncertain world. High performance government: Structure, leadership, and Incentives*. RAND. https://www.rand.org/pubs/monographs/MG256.html

Lempert, R. J., & Turner, S. (2021). Engaging multiple worldviews with quantitative decision support: A robust decision-making demonstration using the Lake model. *Risk Analysis, 41*, 845–865. https://doi.org/10.1111/risa.13579

Lempert, R. J., Schlesinger, M. E., & Bankes, S. C. (1996). When we don't know the costs or the benefits: Adaptive strategies for abating climate change. *Climatic Change, 33*, 235–274. https://doi.org/10.1007/bf00140248

Lempert, R. J., Schlesinger, M. E., Bankes, S. C., & Andronova, N. G. (2000). The impact of variability on near-term climate-change policy choices. *Climatic Change, 45*, 129–161. https://doi.org/10.1007/978-94-017-3010-5_8

Lempert, R. J., Popper, S. W., & Bankes, S. C. (2002). Confronting surprise. *Social Science Computer Review, 20*, 420–440. https://doi.org/10.1177/089443902237320

Lempert, R. J., Popper, S. W., & Bankes, S. C. (2003). Shaping the next one hundred years: New methods for quantitative, long-term policy analysis. *RAND Corporation, xxi*, 187. https://doi.org/10.7249/mr1626

Lempert, R., Nakicenovic, N., Sarewitz, D., & Schlesinger, M. (2004). Characterizing climate-change uncertainties for decision-makers—An editorial essay. *Climatic Change, 65*, 1–9. https://doi.org/10.1023/b:clim.0000037561.75281.b3

Lempert, R. J., Groves, D. G., Popper, S. W., & Bankes, S. C. (2006). A general, analytic method for generating robust strategies and narrative scenarios. *Management Science, 52*, 514–528. https://doi.org/10.1287/mnsc.1050.0472

Lempert, R., D. G. Groves, and J. Fischbach, 2013a: Is it ethical to use a single probability density function?. https://www.rand.org/content/dam/rand/pubs/working_papers/WR900/WR992/RAND_WR992.pdf.

Lempert, R. J., Kalra, N., Peyraud, S., Mao, Z., Tan, S. B., Cira, D., & Lotsch, A. (2013b). *Ensuring robust flood risk management in Ho Chi Minh City: A robust decision making demonstration*. https://doi.org/10.1596/1813-9450-6465

Lempert, R. J., et al. (2013c). *Making good decisions without predictions: Robust decision making for planning under deep uncertainty*. https://doi.org/10.7249/rb9701

Maass, A., et al. (1962). *Design of Water Resources Systems; new techniques for relating economic objectives, engineering analysis, and governmental planning*. Harvard University Press.

March, J. G. (1994). *A primer on decision making: How decisions happen*. The Free Press.

Marchau, V. A. W. J., Walker, W. E., Bloemen, P. J. T. M., & Popper, S. W. E.. (2019). *Decision making under deep uncertainty: From theory to practice* (329 pp). Springer. doi:https://doi.org/10.1007/978-3-030-05252-2.

Mayer, L. A., et al. (2017). Understanding scientists' computational modeling decisions about climate risk management strategies using values-informed mental models. *Global Environmental Change, 42*, 107–116. https://doi.org/10.1016/j.gloenvcha.2016.12.007

Miller, J. H. (1998). Active Nonlinear Tests (ANTs) of complex simulations models. *Management Science, 44*(6), 820–830. https://doi.org/10.1287/mnsc.44.6.820

Molina Perez, E. (2016). Directed international technological change and climate policy: New methods for identifying robust policies under conditions of deep uncertainty. *Pardee RAND Graduate School*.https://doi.org/10.7249/rgsd369

Morgan, M. G., & Henrion, M. (1990). Uncertainty: A guide to dealing with uncertainty in quantitative risk and policy analysis. *Cambridge University Press*.https://doi.org/10.1017/cbo9780511840609

National Academies of Sciences, E., and Medicine. 2016 (NAS). (2016). *Assessment of approaches to updating the social cost of carbon: Phase 1 report on a near-term update*. https://nap.nationalacademies.org/catalog/21898/assessment-of-approaches-to-updating-the-social-cost-of-carbon.

National Research Council. (2009). Informing decisions in a changing climate. https://doi.org/10.17226/12626

Nordhaus, W. D. (1994). Managing the global commons: The economics of climate change. *MIT Press., 1*, 381. https://doi.org/10.1017/s1355770x00000735

O'Mahony, A., et al. (2018). *Assessing, monitoring and evaluating army security cooperation: A framework for implementation*. https://doi.org/10.7249/rr2165

Parker, A. M., Srinivasan, S., Lempert, R. J., & Berry, S. (2015). Evaluating simulation-derived scenarios for effective decision support. *Technological Forecasting and Social Change, 91*, 64–77. https://doi.org/10.1016/j.techfore.2014.01.010

Pidgeon, N., & Fischhoff, B. (2011). The role of social and decision sciences in communicating uncertain climate risks. *Nature Climate Change, 1*, 35–41. https://doi.org/10.1038/nclimate1080

Popper, S. W. (2019). Robust decision making and scenario discovery in the absence of formal models. *Futures & Foresight Science, 1*, e22. https://doi.org/10.1002/ffo2.22

Popper, S. W., Lempert, R. J., & Bankes, S. C. (2005). Shaping the future. *Scientific American, 292*, 66–71. https://doi.org/10.1038/scientificamerican0405-66

Popper, S. W., Berrebi, C., Griffin, J., Light, T., Min, E. Y., & Crane, K. (2009). Natural gas and Israel's energy future: Near-term decisions from a strategic perspective. https://doi.org/10.7249/mg927

Quinn, J. D., Reed, P. M., & Keller, K. (2017). Direct policy search for robust multi-objective management of deeply uncertain socio-ecological tipping points. *Environmental Modelling & Software, 92*, 125–141. https://doi.org/10.1016/j.envsoft.2017.02.017

Ranger, N., A. Millner, S. Dietz, S. Fankhauser, A. Lopez, and G. Ruta, 2010: Adaptation in the UK: A decision making process. https://www.lse.ac.uk/granthaminstitute/wp-content/uploads/2014/03/PB-Ranger-adaptation-UK.pdf.

Rawls, J. (1971). A theory of justice. *Harward University Press*. https://doi.org/10.4159/9780674042605

Rayner, S. (2000). Prediction and other approaches to climate change policy. In D. Sarewitz (Ed.), *Prediction: Science, decision making, and the future of nature* (pp. 269–296). Island Press.

Renn, O. (2008). Risk governance: Coping with uncertainty in a complex world. *Earth*. https://doi.org/10.4324/9781849772440

Rittel, H., & Webber, M. (1973). Dilemmas in a general theory of planning. *Policy Sciences, 4*, 155–169. https://doi.org/10.1007/bf01405730

Robalino, D. A., & Lempert, R. J. (2000). Carrots and sticks for new technology: Abating greenhouse gas emissions in a heterogeneous and uncertain world. *Integrated Assessment, 1*, 1–19. https://doi.org/10.1023/a:1019159210781

Rosenhead, J. (1990). In J. Rosenhead & J. Mingers (Eds.), *Rational analysis: Keeping your options open. Rational analysis for a problematic world: Problem structuring methods for complexity, uncertainty and conflict*. Wiley. https://doi.org/10.1002/sres.491

Rosenhead, M. J., Elton, M., & Gupta, S. K. (1972). Robustness and optimality as criteria for strategic decisions. *Operational Research Quarterly, 23*, 413–430. https://doi.org/10.2307/3007957

Sarewitz, D., & Pielke, R. A. (2000). *Prediction: Science, decisionmaking, and the future of nature*. Island Press.

Schoemaker, P. J. H. (1993). Multiple scenario development: Its conceptual and behavioral foundation. *Strategic Management Journal, 14*, 193–213. https://doi.org/10.1002/smj.4250140304

Schwartz, P. (1996). *The art of the long view—Planning for the future in an uncertain world*. 1996 edition ed. Currency-Doubleday.

Sen, A. (2009). The idea of justice. *Belknap Press*. https://doi.org/10.4159/9780674054578

Swanson, D., & Bhadwal, S. (2009). Creating adaptive policies: A guide for policy-making in an uncertain world. *Sage Publications*. https://doi.org/10.4135/9788132108245

Swanson, D., H. Venema, S. Barg, S. Tyler, J. Drexage, P. Bhandari, and U. Kelkar, 2007: Initial conceptual framework and literature review for understanding adaptive policies.

Tetlock, P. E., & Gardner, D. (2015). *Superforecasting: The art and science of prediction*. Broadway Books.

Thompson, C. (2013). *Smarter than you think: How technology is changing our minds for the better*. Penguin.

Tuana, N. (2013). Embedding philosophers in the practices of science: Bringing humanities to the sciences. *Synthese, 190*, 1955–1973. https://doi.org/10.1007/s11229-012-0171-2

van der Heijden, K., 1996: Scenarios: The art of strategic conversation.

Wack, P. (1985). The gentle art of Reperceiving - scenarios: Uncharted waters ahead (part 1 of a two-part article). *Harvard Business Review*, 73–89. https://hbr.org/1985/09/scenarios-uncharted-waters-ahead

Walker, W. E., Rahman, S. A., & Cave, J. (2001). Adaptive policies, policy analysis, and policy-making. *European Journal of Operational Research, 128*, 282–289. https://doi.org/10.1016/s0377-2217(00)00071-0

Walker, W., Marchau, V., & Swanson, D. (2010). Addressing deep uncertainty using adaptive policies. *Technology Forecasting and Social Change, 77*, 917–923. https://doi.org/10.1016/j.techfore.2010.04.004

Walker, W. E., Robert J. Lempert, and J. H. Kwakkel, 2013: Deep uncertainty. *Encyclopedia of Operations Research and Management Science*, Springer US., 395-402. doi:https://doi.org/10.1007/978-1-4419-1153-7_1140.

Walley, P. (1991). Statistical reasoning with imprecise probabilities. *Chapman and Hall*.https://doi.org/10.1007/978-1-4899-3472-7

Watson, A. A., & Kasprzyk, J. R. (2017). Incorporating deeply uncertain factors into the many objective search process. *Environmental Modeling and Software, 89*, 159–171. https://doi.org/10.1016/j.envsoft.2016.12.001

Weaver, C. P., Lempert, R. J., Brown, C., Hall, J. A., Revell, D., & Sarewitz, D. (2013). Improving the contribution of climate model information to decision making: The value and demands of robust decision frameworks. *WIREs Climate Change, 4*, 39–60. https://doi.org/10.1002/wcc.202

Wolf, C., 1993: Markets or governments: Choosing between imperfect alternatives.https://www.rand.org/pubs/notes/N2505.html.

Wong-Parodi, G., Krishnamurti, T., Davis, A., Schwartz, D., & Fischhoff, B. (2016). A decision science approach for integrating social science in climate and energy solutions. *Nature Climate Change, 6*, 563–569. https://doi.org/10.1038/nclimate2917

Wong-Parodi, G., Mach, K. J., Jagannathan, K., & Sjostrom, K. D. (2020). Insights for developing effective decision support tools for environmental sustainability. *Current Opinion in Environmental Sustainability, 42*, 52–59. https://doi.org/10.1016/j.cosust.2020.01.005

Woodruff, M., & Reed, P. (2013). Many objective visual analytics: Rethinking the design of complex engineered systems. *Structural and Multidiciplinary Optimization, 48*, 201–219. https://doi.org/10.1007/s00158-013-0891-z

Zeff, H. B., Kasprzyk, J. R., Herman, J. D., Reed, P. M., & Characklis, G. W. (2014). Navigating financial and supply reliability tradeoffs in regional drought management portfolios. *Water Resources Research, 50*, 4906–4923. https://doi.org/10.1002/2013wr015126

Zeff, H., Herman, J., Reed, P., & Characklis, G. (2016). Cooperative drought adaptation: Integrating infrastructure development, conservation, and water transfers into adaptive policy pathways. *Water Resources Research, 52*, 7327–7346. https://doi.org/10.1002/2016wr018771

Zenko, M. (2015). *Red team: How to succeed by thinking like the enemy*. Basic Books.

Open Access This chapter is licensed under the terms of the Creative Commons Attribution 4.0 International License (http://creativecommons.org/licenses/by/4.0/), which permits use, sharing, adaptation, distribution and reproduction in any medium or format, as long as you give appropriate credit to the original author(s) and the source, provide a link to the Creative Commons license and indicate if changes were made.

The images or other third party material in this chapter are included in the chapter's Creative Commons license, unless indicated otherwise in a credit line to the material. If material is not included in the chapter's Creative Commons license and your intended use is not permitted by statutory regulation or exceeds the permitted use, you will need to obtain permission directly from the copyright holder.

5. The Policy Portfolio Problem

Rachel Warren

5.1 Introduction

Given the urgent need to prevent further changes in global climate, and also to adapt to the climate change that has already occurred and the further change that may occur in the future, decision-makers are faced with a "Policy Portfolio Problem": that is, the following question: "With what combinations of adaptation and mitigation policies should we address climate change?" This question needs to be addressed in the context of climate-resilient, sustainable development pathways, and in combination with the equally urgent need to address other global issues such as land degradation, biodiversity loss, and pollution. These problems often have common drivers and their effects also interact strongly.

Decision-makers are diverse: some will be considering global, regional, national, or local climate change action across various sectors of the economy, while others might be focused on individual sectors, nature conservation, or small villages. Decision-makers often assume that there is a balance or trade-off between mitigation and adaptation, and often fail to consider the linkages between them. It is now widely accepted, however, that both adaptation and mitigation are essential. Given this, decision-makers are still confronted with a variety of mitigation and adaptation options to consider, as well as solar radiation management. However, as the need to address climate change becomes more and more urgent, the range of available portfolios of action to mitigate and adapt to climate change that is consistent with the UN Paris Agreement shrinks and the decision space becomes more and more constrained. For example, mitigation portfolios that limit globally averaged warming to 1.5 °C above pre-industrial levels with "limited" overshoot almost all include carbon dioxide removal (CDR), variously deploying Bioenergy with Carbon Dioxide Capture and Storage (BECCS) and/or Direct Air Carbon Dioxide Capture and Storage (DACCS). That is, because of the delay in deploying global climate change mitigation, it is now extremely difficult to identify Paris-compliant mitigation pathways that are not dependent on CDR. Excluding CDR now requires a low global population, the majority of whom adopt sustainable lifestyles, and optimistic assumptions about agricultural intensification and dietary change, alongside rapid renewable electrification and high technological efficiency (Van Vuuren et al., 2018) to achieve the Paris Agreement goals without these methods. Early action might have avoided the need for carbon dioxide removal at all. The longer action is delayed, the more carbon dioxide removal is required to avoid an overshoot of the global temperature targets. An overshoot, even if temporary, would significantly increase climate change risks (Schleussner et al., 2024) and any attempt to reverse such an overshoot would require still larger amounts of CDR, the deployment of which is of uncertain efficacy and feasibility and would undermine sustainability.

This chapter explores the issues that need to be considered when making such decisions. *Most* mitigation and adaptation actions in a given sector affect climate change-related risk in other sectors or interact with mitigation and adaptation actions in other sectors (Warren, 2010). It can be difficult to quantify the magnitude of such interactions, owing to both uncertainties about the effects of mitigation or adaptation within a single sector considered independently, and also further uncertainties about the degree to which they will interact. Despite this, it is vital that these interactions be used as a basis for designing portfolios of action across sectors, considering both adaptation and mitigation together. If actions are designed within silos, and efforts are afterward made to reduce unintended negative effects in other sectors, this is much less likely to be effective. The chapter also highlights the promising role of nature-based solutions (NBS),

R. Warren (✉)
Tyndall Centre for Climate Change Research, University of East Anglia, Norwich, UK
e-mail: r.warren@uea.ac.uk

which can contribute to both mitigation and adaptation, in these portfolios.

"Mitigation" is defined as "a human intervention to reduce the sources or enhance the sinks of greenhouse gases." This includes the avoidance of greenhouse gas emissions, either by reducing fossil fuel burning or by preventing deforestation, or alternatively by enhancing carbon sinks, for example, through restoration of natural systems. Use of biomass to produce bioenergy and then capture and store the CO_2 released ("BECCS"), and direct air capture and storage (DACCS) are also included here. "Adaptation" is defined as "the process of adjustment to actual or expected climate and its effects" (IPCC WGII AR5 Glossary). However, some actions may have both a strong mitigation effect and a strong adaptation effect (such as NBS).

"Geoengineering" techniques are only briefly covered here since they generally have adverse side effects (McCormack et al., 2016) including on agriculture and ecosystems, potentially compromising Article 2 of the UN Convention on Climate Change, which sets out to limit global warming so as to allow "ecosystems to adapt naturally" and ensure that food production is "not threatened."

Mitigation and adaptation action can focus on addressing the supply side (i.e., the way a commodity such as energy, food, or water is produced) or on addressing the demand side (i.e., reducing the demand for these commodities). Actions can include considering types of new technology to deploy in mitigation and adaptation, for example, to change how the supply (of energy, water, food) is produced, or to do so more efficiently; and also how changes in human behavior might reduce the demand. Since adaptation and mitigation actions can interact, when selecting a portfolio of these actions, it is vital to consider these linkages, including the uncertainties therein, in order to avoid examples of unintended adverse outcomes.

In order to actually deliver these changes, policies are required to alter the "status quo." Not only is there a decision to make about the combination of these supply or demand-side actions to take, but also a decision to make about the most effective way to implement them. A portfolio of policies will be required to deliver the portfolio of technical and behavioral change. There is not a one-to-one relationship, that is, a single policy made by a single actor often is not sufficient to effect a change in technology or behavior. Each one of these new technologies or behavioral changes is likely to require a portfolio of policies to be implemented by different actors, typically ensuring that government, business, finance, and public support are aligned. Not only national governments, but also local governments, businesses, investors, financial organizations, and regulators may all need to become involved. Their ultimate need is not only for advice on what portfolio of policies each decision-maker should adopt, but also there is a need to consider how to integrate policies across different actors, who are making decisions and setting policies for different sectors of society (and may be operating on different spatial and temporal scales). This advice should be based on both a qualitative and quantitative understanding of policy implications, including an understanding of the uncertainty budget in terms of the quantification of the potential outcomes of different policy combinations.

5.2 Adaptation and/or Mitigation: The Big Picture

A question that is often asked is whether there are trade-offs between adaptation and mitigation, or whether they are complementary. For a number of reasons, they are generally largely complementary. Firstly, a global warming of 1.1 °C has already occurred (IPCC, 2021). Limiting global warming to below 1.5 °C is now very challenging yet still feasible (IPCC, 2018; IPCC 2022 (WGIII report)), which means that adaptation to this level of 1.5 °C warming will almost certainly now be necessary in order to reduce the loss and damage that this level of warming is projected to entail. At the other extreme, if no further efforts are made to mitigate climate change beyond policies in place today, a global warming of up to 3.6 °C might emerge by 2100 (https://climateactiontracker.org/global/cat-thermometer/), leading to high risks to ecosystems and their services, human systems (including agriculture and water security), human lives, and livelihoods (Oppenheimer et al., 2014). Under such a situation, limits to adaptation would be exceeded in many systems (IPCC AR6 WGII SPM, 2022), and large-scale loss and damage would be inevitable.

Indeed, it is argued that there is a "human imperative" to limiting warming to 1.5 °C owing to the much greater risks at 2 °C, to both human and natural systems, including the Arctic (due to the loss of Arctic sea ice and permafrost), and in tropical regions (due to high mortality of coral reefs, persistent heat stress in livestock, and forest dieback) (Hoegh-Guldberg et al., 2018, 2019). The work of the Intergovernmental Panel on Climate Change (IPCC) Working Group II is to assess and review literature on the subject of climate change risks. As a part of this assessment, the IPCC considers five categories of risk called "Reasons for Concern (RFC)" and assesses how the level of risk is projected to change as the globe warms. Levels of risk due to extreme weather events (RFC2) become very high already with 2 °C of warming (IPCC, 2022a) while levels of risk in unique and threatened systems (RFC1) (including biodiversity hotspots) become very high already at 1.5 °C (IPCC, 2022a). Between 2 and 4.5 °C warming, levels of risk become very high for all other "reasons for concern," including global aggregate economic impacts and large-scale singular events (examples of these include sustained melting of the Antarctic and Greenland ice sheets, which are already losing mass presently). Estimates of aggregate economic impacts of global warming vary widely but reach hundreds of trillions of dollars (Warren et al., 2021). IPCC (2022b) finds that the benefits of climate change mitigation to limit warming to 2 °C exceed the costs of climate change mitigation. This statement is given medium confidence as it is qualified by the following exceptions, that is, that future damages are not discounted at high rates (which would have devalued the rights of future generations to a healthy planet), and also the unlikely possibility that climate change damages might be very low.

In general, the greater the investment in global climate change mitigation, the lower the level of global warming, and the lower the residual climate risks and the less of an adaptation challenge remains. Had there been action on climate change decades earlier, a solely mitigation-based response would have been feasible (except in the case of risks associated with sea-level rise, which reduces relatively slowly in response to mitigation action compared to other risks). But the world is now in a situation where a very great deal of both adaptation and mitigation is necessary to avoid the worst impacts of climate change, and, hence, they are now complementary, especially as synergies between the two types of policy exist and some actions can contribute to both processes.

Owing to the urgency of the need for climate change action, there is presently increased interest in the potential role of solar radiation management (SRM) in the portfolio of human responses to climate change. SRM relies on the continuous injection of sulfate aerosols into the stratosphere to reflect incoming solar radiation and thus reduce the greenhouse effect. However, this (a) only reduces global warming and fails to address increasing carbon dioxide concentrations and hence ocean acidification, (b) can reduce precipitation (Tilmes et al., 2013), and (c) if terminated, for financial or political reasons (including "black swan" events such as global pandemics or wars), a sudden increase in global warming would take place at a much faster rate than that presently occurring, which would present even greater challenges to adaptation than the present rate (Jones et al., 2013; McCormack et al., 2016). Ongoing research is exploring how spatial design of SRM injection might improve the outcome for precipitation (Wells et al., 2024). However, owing to very large uncertainties in the potential effect of SRM on regional precipitation patterns and other potential problems (National Academies of Sciences, Engineering and Medicine, 2021), SRM is not discussed further in this chapter.

Having established that both adaptation and mitigation are now required, there remains the question of whether there is a trade-off in how much of each is implemented at the global scale. Firstly, since there are limits to adaptation, in some cases at levels of warming as low as 1.5 or 2 °C, risks that are not avoided by mitigation cannot simply be avoided by adaptation instead. For example, many coral reef ecosystems have been assessed to be unable to persist above 1.5 °C warming, while above 2 °C warming almost all would be unable to persist, owing to the projected associated increase in sea surface temperatures and ocean heatwaves, which will increase the frequency of coral bleaching episodes so that the interval between episodes is too short for reef recovery (O'Neill et al., 2022), leading to the loss of ecosystem services since many communities are dependent on reefs for their livelihoods and reefs provide a breeding ground for marine animals that roam the world's oceans. Similarly, many biodiversity hotspots around the world are at risk of irreversible loss due to their inability to adapt to a global warming of 1.5–2 °C (O'Neill et al., 2022). Around 10% of species globally are projected to become endangered at 1.6 °C compared with >20% at 2.1 °C (median). Extinction is irreversible and forms a hard limit to adaptation. Species loss undermines the provisioning and regulating ecosystem services upon which humans depend. In human systems, beyond 2 °C, cultivar changes are projected to be unable to offset global production losses in agriculture, while beyond 3 °C, water shortages in Western and Central Europe are projected to be impossible to be offset by geophysical and technological adaptation methods (O'Neill et al., 2022), putting European agricultural production at risk. At 4.5 °C global warming, regional maximum temperatures are expected to exceed the human survivability threshold across most of South Asia.

Considerations such as this informed the UNFCCC Paris Agreement to "pursue efforts to limit warming to well below 2 °C" and "pursue efforts" to limit it to 1.5 °C. Integrated assessment models that represent the global economic system and incorporate our scientific understanding of the earth's climate system can provide examples of global mitigation portfolios that have at least a 50% chance of limiting warming to 1.5 or 2 °C. However, owing to uncertainties in climate sensitivity (Chap. 15, this volume), there is still a chance that global warming could be greater than 1.5 or 2 °C even if one of these portfolios is implemented fully, and hence planning adaptation for higher levels of warming is still prudent. Further, such a global mitigation portfolio requires global cooperation, and should some nations fail to deliver their contribution to the portfolio, again it would result in higher levels of warming. Finally, since adaptation always has a local context, the concept of adapting to a particular level of global warming needs to consider the pattern of regional climate change associated with different levels of global warming. The regional warming on land almost always exceeds the global average warming (owing to the slower warming response of the oceans) and can be very much greater in the center of large landmasses and also varies considerably from year to year (Seneviratne et al.,

2018). Furthermore, the precise regional pattern projected is uncertain and varies considerably across the ensemble of CMIP6 models (IPCC AR6 WGI Atlas). In some regions, it is even unclear whether precipitation will increase or decrease with warming, and hence adaptation actions robust to either outcome need to be prepared. However, in general, adapting to higher potential levels of global warming provides a buffer against the uncertainty in regional climate projection. These arguments led to an oft-quoted statement "mitigate to 2 °C and adapt to 4 °C" (R. Watson, pers. comm.). However, note that this statement is not realistic: adaptation to a level of 4 °C warming is not in fact possible, owing to the aforementioned extensive breaching of limits to adaptation.

The concept of a global trade-off between investments into adaptation and mitigation assumes a single decision-maker with a single budget, rather than a set of actors working at different scales in different systems. It also ignores the potential for synergies between adaptation and mitigation actions, and for actions that contribute to both. Single decision-makers in individual organizations might have limited budgets and may want to consider whether it is more appropriate or effective for them to meet their organization's goals by investing in mitigation, or adaptation, or both. However, at larger spatial scales, such a trade-off is not appropriate: mitigation has a global effect over a long time scale of decades, while adaptation has only a local effect. In practice, however, investments in mitigation and adaptation are often made separately by individual nations or organizations, potentially leading to considerable challenges as far as global coordination of both mitigation and adaptation is concerned.

The UNFCCC acts as an umbrella organization and, through its processes, attempts to bring nations' climate-related policies together so that the global community collaborates to reach the Paris Agreement goals and adapt to the residual impacts. At present, global action on both mitigation and adaptation has been assessed as insufficient by both the United Nations and the IPCC.

The adaptation gap report (United Nations Environment Programme, 2022) finds that, although adaptation efforts are increasing and 84% of countries have established adaptation plans, strategies, laws, or policies, many of these have not been implemented. In developing countries, this is largely due to the lack of finance, in particular given an expectation of international financial flows to support these actions. Yet, there are still large adaptation gaps in developed countries such as the United States and the United Kingdom, where there are particularly large gaps in introducing adaptation policies into building regulations, healthcare, and the management of the natural environment (CCRA3; HM Government, 2022). If the situation is not addressed, the adaptation gap will widen as the climate changes, affecting agriculture, water resources, and the natural environment. The size of the gap is also related to the amount of mitigation effort: the more effort is made to reduce greenhouse gas emissions, the smaller the adaptation gap will be. Further, monitoring the size of the adaptation gap is difficult because of the lack of inventories of existing adaptation action and its efficacy.

Presently, there is insufficient action on climate change mitigation to reach the goals of the Paris Agreement, with the current policies and action at the time of writing sufficient to limit warming to approximately 2.7 °C (range 2.2–3.4 °C, Meinshausen et al., 2022) with CO_2 emissions around 6–13% above 2010 levels in 2030 compared with the IPCC assessed 45% reduction below 2010 levels to constrain warming to 1.5 °C (see also https://climateactiontracker.org/global/cat-thermometer/). Insufficient action on the part of many nations leaves an implementation gap of 23–27 $GtCO_2e$. Nations have submitted nationally determined contributions for emission reductions in 2030: if these were to be achieved, global temperature rise could be limited to 2.4 °C (range 1.9–2.9 °C), while if in addition there was the full implementation of all announced targets, including net-zero targets, it would be limited to 1.8 °C (range 1.5–2.3 °C), corresponding to a greenhouse gas emission reduction of 32–34% by 2050 relative to 2010 levels, suggesting that the portfolio of net-zero targets would be sufficient to limit warming to 2 °C (but not 1.5 °C, which requires around 45% reduction of CO_2 emissions by 2030 relative to 2010). Importantly, the issue is that many nations are not putting in place the policies and measures necessary to achieve their announced long-term targets, and many of the 2030 targets are insufficient and not in line with these long-term targets (Meinshausen et al., 2022). Some countries have not put forward net-zero targets, while others would need to bring forward in time their net-zero targets, or even put forward net negative emission targets. Owing to this situation, emissions are still increasing (Friedlingstein et al., 2022) and yet this trend needs to be reversed within this decade to keep the Paris Agreement target within reach.

The modest levels of the NDCs put forward so far mean that mitigation costs in the short term remain relatively small, but this implies accelerated action after 2030 to reach the PA target, and this overall would be a more costly mitigation pathway in the long term, than pathways in which there is more stringent action by 2030 than in the present NDCs: in these pathways, the higher upfront costs are compensated for by economic recovery after net zero is obtained (Riahi et al., 2021, 2022).

Trading off risks and costs on different times cales using a cost–benefit approach requires placing a value on risks in the future compared with the present, a practice referred to by economists as the "discount rate," the value of which is a matter of great controversy. The common business practice of using discount rates to value investments over short periods is arguably not appropriate for use in problems such as climate change with century-long time scales since use of

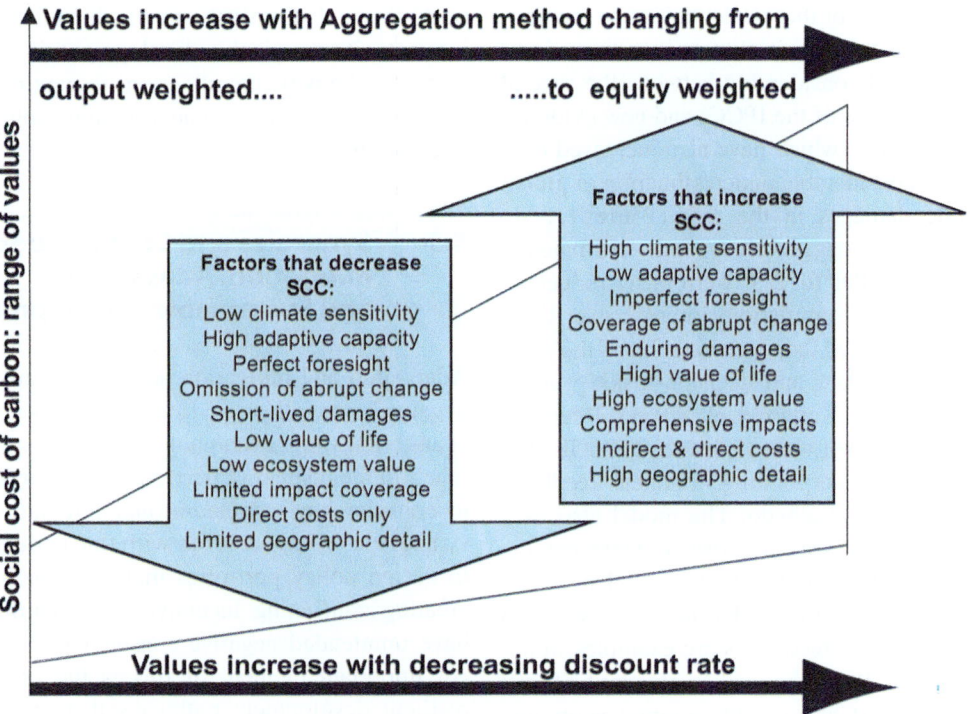

Fig. 5.1 The influence of integrated assessment model design, and of the values of uncertain and/or subjective parameters, upon the calculated value of the social cost of carbon. (Reproduced with permission from Fisher et al., 2007)

a high discount rate inevitably condemns future generations to large and unacceptable risks. This is a particularly important source of uncertainty in calculations of climate change damages and strongly affects the outcome of cost–benefit analyses that attempt to derive an "optimal" level of global warming at which costs and benefits of climate change mitigation equate. Overall, the values placed on future climate change damages can take a wide range of values depending on subjective issues such as the placing of monetary values on lives and ecosystems. Many studies do not include indirect costs or co-benefits, and often use very simple equations to estimate damages at a global scale, tending to focus on market risks, with little regional detail. Lastly, the approach assumes that the decision-maker is completely rational and has perfect foresight. The marginal damage done by the climate change caused by the release of one additional ton of carbon into the atmosphere is commonly referred to as the "social cost of carbon" (SCC). Figure 5.1 illustrates the uncertainties in estimating it. SCC values have often been used to justify larger or smaller investments in climate change mitigation.

However, IPCC (2022b) finds that over the twenty-first century, integrated assessment models currently show that the cost of limiting warming to 2 °C is "lower than the global economic benefits of reducing warming" unless low estimates of climate damages are used (which are unlikely to be valid, see below); or a high discount rate is used in which future damages are discounted at high rates.

Because of the large uncertainties in discount rates, values of (particularly nonmarket) assets, as well as in climate science itself, Stern (2007) used the PAGE2002 integrated assessment model in its cost–benefit analysis work because this model is based on a probabilistic representation of these uncertain parameters. This was one of the first attempts to include an economic analysis of the potential for large-scale discontinuities in the earth system, that is, the effects of feedback processes not well captured in global climate models, such as the potential for accelerated release of methane from permafrost as the global warms. It estimated potential damages of global warming over the next two centuries to be at least 5% and potentially 14% GDP, which at the time was considered an outlier in the literature but is now looking more consistent with more recent work in which values of 7–8% GDP have appeared (see below) while acknowledging that not all risks have yet been captured.

In fact, the UNFCC Paris Agreement was largely informed by scientific evidence about the risks at different levels of global warming. However, some countries, such as the United States, are legally required to compare the costs and benefits of their (climate) policies, and hence such analysis is still important in some nations and also can help support and incentivize climate change action. Over time, estimates of the aggregate global economic damages (often expressed as %GDP loss) have been steadily increasing and now also provide an economic argument for limiting global warming to 1.5 or 2 °C. There are several reasons for this.

Inevitably, it takes time for these calculations to incorporate new evidence about the earth's climate sensitivity (estimates of which increased considerably between the second and third assessment reports of the IPCC) and new evidence about climate change risks (which have also increased over time). This has led to considerable underestimation of global aggregate economic damages in the past (Warren et al., 2010). Global aggregate damages were typically estimated to fall in the range of 1–3% GDP for 2 °C warming (Arent et al., 2015; Tol 2018). Inevitably simulations improve over time, and recent work has improved the representation of the earth system, climate change risks, and decision-making under uncertainty (Warren, 2014; Keppo et al., 2021). Recently, the PAGE09 model has been refined to create PAGE-ICE (Yumashev et al., 2019) incorporating nonlinear feedback in the Arctic permafrost and albedo. The model also uses a fat-tailed distribution for sea-level rise to represent the effects of potential melting of the Greenland Ice Sheet, while its climate sensitivity values and a carbon cycle match IPCC AR5. These nonlinear feedbacks are examples of so-called "discontinuities." When applied to standard Shared Socioeconomic Pathways (Chap. 16, this volume), economic damages of 6% GDP are estimated for 2.7C warming above pre-industrial as opposed to 1–2% GDP in the absence of these improvements (Chen et al., 2020; Warren et al., 2021). Similarly, the integrated model DICE2016R2 now includes a blanket 25% uplift to damages to account for discontinuities in general (Nordhaus & Sztorc, 2013) and produces the year 2100 damage estimates of 2.0% of income at 3 °C and 7.9% of global income at a global temperature rise of 6 °C (Nordhaus, 2018). Further work on this model (Hänsel et al., 2020) to update the carbon cycle, making it consistent with the IPCC Special Report on 1.5 °C warming (Rogelj et al., 2018) and other improvements, has led to a revised estimate of damages of 6.69% of global GDP for 3 °C, leading to an optimal limit to global warming of 1.77 °C in 2100, consistent with the Paris Agreement.

Despite these improvements, there is empirical evidence for even larger damages (Burke et al., 2015), while a large number of potentially high-risk climate change outcomes are still not incorporated or only poorly represented. These include the potential conversion of rainforests to savannas, escalation of fire risks in forest, loss of coral reefs, ocean acidification, risk cascades due to the loss of ecosystem services, and the potential for climate change-induced migration and conflict.

The evidence that significant investment in *both* mitigation and adaptation portfolios is needed is clear. How do we decide which actions to implement? How should uncertainties in effectiveness, costs, benefits, risks, or unintended consequences of actions be handled and what are the most important ones (i.e., the "uncertainty budget")? The next sections will address this. A critical element in designing the portfolios is that mitigation and adaptation interact, and also interact with sustainable development: it is therefore critical to design the portfolio of actions to maximize synergies and minimize trade-offs. The next section explores some of these interactions.

5.3 Synergies and Conflicts Between Mitigation, Adaptation, and Sustainable Development

Adaptation and mitigation actions can impact positively or negatively on other environmental risks or other climate change risks, on sustainable development, or also on each other. In the context of portfolios, it is particularly important to consider how adaptation and mitigation actions can be synergistic or may conflict with one another. It is important to design policy portfolios that are synergistic, rather than creating conflicting incentives, or creating incentives that have unintended negative consequences for other systems. An important element of this is the creation of climate-resilient development pathways that encompass clean development while increasing adaptative capacity to climate change. Conversely, unsustainable development tends to increase vulnerability to the climate change risks. This is not a minor issue: *most* adaptation and mitigation actions in a given sector have an effect on another sector (Berry et al., 2015).

Examples of synergies and conflicts between adaptation and mitigation actions can be classified into various types, and some examples are considered here: the portfolio problem requires a careful consideration of these various issues.

1. Adaptation to climate change in one sector can influence climate change-related risks in other sectors or systems. For example, protecting coastlines by building sea defenses to protect cities can increase erosion in adjacent natural habitats.
2. Adaptation can contribute to, or detract from, greenhouse gas emissions. For example, the installation of air conditioning as an adaptation to heat stress increases energy demand and hence greenhouse gas emissions unless the energy is produced from renewable sources. However, even in that case, increasing energy demand increases the challenge of producing a larger fraction of energy from renewable sources. Conversely, adapting to a drier climate by using no-till agriculture to reduce crop water use also prevents soil erosion and helps retain carbon in the soil. Adaptation in the water and urban sectors [sustainable urban drainage (SUDS)] also contributes to mitigation through carbon storage and has positive effects on biodiversity conservation as well as reducing the urban heat island (Gill et al., 2007; Chance, 2009). Bosello et al. (2013) review a number of synergies.

3. Mitigation can contribute to, or detract from, adaptation or adaptive capacity. For example, large hydroelectric power schemes can result in the inundation of large swathes of land, displacing human settlements and drowning ecosystems. They are also vulnerable to increasing climate variability, in particular, drought.
4. Climate change mitigation and adaptation involving changes in land use can be either synergistic or in conflict. A clear example of potential conflict related to the use of bioenergy with carbon capture and storage (BECCS) is used to achieve CDR. Some studies estimate that 1.1–1.5 Gha of land would be necessary to use BECCS to reduce global warming from 2.5 °C to 1.7 °C (Boysen et al., 2017). Thus, if deployed on large scales, BECCS would have "far reaching consequences for land and water availability" (IPCC, 2018). This would inevitably create competition for land and water use with both food production and biodiversity conservation, conflicting with the goals of Article 2 of the UNFCCC, which states that the goal of the Convention is to limit global warming in order to protect food production and natural ecosystems. However, in some regions, more limited use of BECCS might be carefully governed to minimize these conflicts (IPCC, 2018).
5. On the other hand, a clear example of synergy is using ecosystem restoration (and prevention of deforestation) to sequester (or retain) stores of carbon on land or in the coastal zone. This also increases climate resilience since these ecosystems provide services such as air and water purification, pollination, and protection against extreme weather events. On land, ecosystem restoration and forest protection not only stores carbon but prevents soil erosion and land degradation, reducing the risks of downstream flooding and maintaining soils for agricultural use. In coastal wetlands, protection and restoration of saltmarshes, seagrass meadows, and coral reefs (commonly referred to as "blue" carbon) protect the coast from storm surges and preserve the breeding grounds of marine fish and mammals. These are examples of "nature-based solutions" (NBS). United Nations Environment Programme (2021) states that 6.6 (range 2–11) $GtCO_2e$ per year could be removed from the atmosphere between 2020 and 2050 by halting land degradation and land use change (including deforestation); while a further 18.6 (range 1.8–35.5) $GtCO_2e$ per year could be removed by ecosystem restoration on land over the same period.

There is also enormous potential for "blue" carbon sequestration. Conservation of mangroves, saltmarshes, and seagrass meadows can avoid emissions of 304 (141–466) Tg/yr. CO_2e, while potential restoration of 0.2–3.2 million ha tidal marshes, 8.3–25.4 million ha seagrasses, and 9–13 million ha mangroves could additionally sequester 841 (621–1064) Tg CO_2e per year by 2030, which is equivalent to 3% of global emissions (Macreadie et al., 2021). Mangrove protection and/or restoration could provide the greatest carbon-related benefits. Overall, NBS illustrate strong synergies between climate change adaptation, mitigation, and the SDGs, with multiple benefits for both human and natural systems.

IPCC (2018) explores synergies and trade-offs between climate change mitigation and the sustainable development goals more generally, finding a strong majority of synergies. IPCC (2022b) now concludes that "accelerated and equitable climate action in mitigating, and adapting to, climate change impacts is critical to sustainable development" and also finds that "policies that shift development pathways towards sustainability can broaden the portfolio of available mitigation responses and enable the pursuit of synergies with development objectives."

Figure 5.2 summarizes the linkages between a range of climate change mitigation actions and the SDGs.

5.4 Portfolios of Mitigation Policy

Global mitigation portfolios that satisfy particular climate change constraints can be simulated by integrated assessment models (IAMs) that combine at minimum a representation of the economy, combined with a simplified representation of the global climate system. Some contain a detailed bottom-up representation of the energy system and a range of technical solutions for reducing emissions. *They all show that a major transition in the energy system is required to achieve the Paris Agreement goals.* This includes rapid reductions in fossil fuel use and rapid increases in low-carbon energy sources. The range of technical solutions covers (a) decarbonization of energy supply side, typically including renewable energy such as solar, wind, geothermal, nuclear, and bioenergy; (b) demand-side measures such as energy efficiency measures; and (c) a range of land-based measures to remove carbon dioxide from the atmosphere (CDR). Direct air capture with storage (DACCS) is starting to become included in some IAMs. Whilst DACCS has the advantage of a low land use footprint, there are presently significant obstacles preventing its being operated at scale (Realmonte et al., 2019) and hence it is considered that it should be considered only as a useful addition to mitigation portfolios, rather than as a replacement for other techniques. A key element of all portfolios is the avoidance of continuing to build fossil fuel-based infrastructure as this would lock society into their continued use. Another key element is that a net-zero target is only reached if any remaining fossil fuel use is offset by CDR. However, large scale deployment of CDR may be ineffective in reversing global warming as it may be undermined by earth system feedbacks (IPCC, 2021), and its feasibility is also

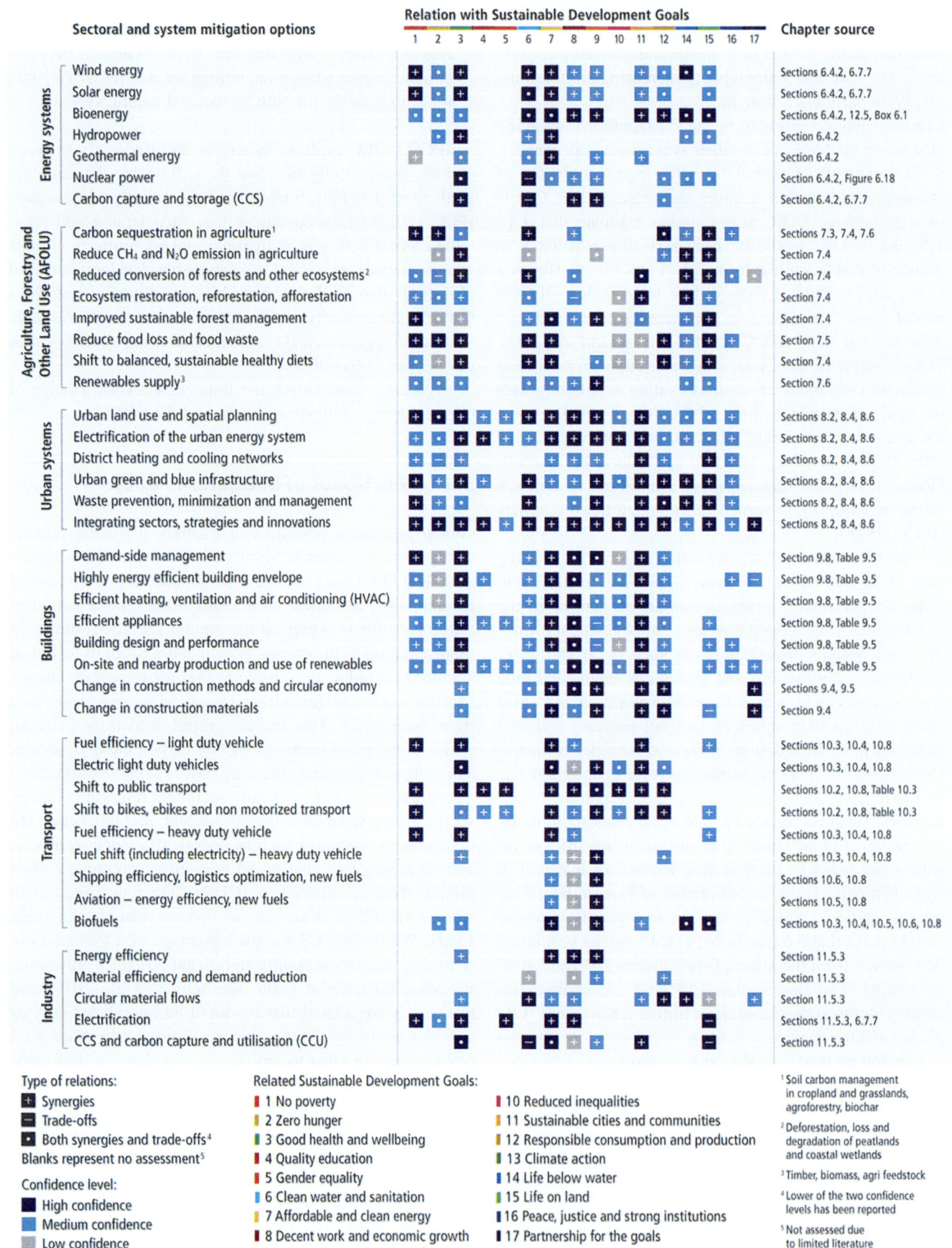

Fig. 5.2 Synergies and trade-offs between sectoral and system mitigation options and the SDGs. (Reproduced with permission from IPCC, 2022b)

technically and economically unproven (Schleussner et al., 2024). Thus CDR is not an effective method of reaching the goals of the Paris Agreement, which can only be achieved with rapid near-term greenhouse gas emission reductions (Schleussner et al., 2024). This can be more easily achieved under a scenario in which there is a low future global population, the majority of whom adopt sustainable lifestyles, and optimistic assumptions about agricultural intensification and dietary change, involving rapid renewable electrification and high technological efficiency (Van Vuuren et al., 2018).

There are trade-offs between these technical solutions, for example, failure to reduce greenhouse gas emissions deeply by the 2030s leads to a greater need for land-based measures to remove carbon dioxide from the atmosphere [referred to as carbon dioxide removal (CDR)] in order to avoid "overshooting" the Paris Agreement goals (IPCC, 2022b). These emission reductions come from a combination of supply- and demand-side measures: greater efforts to reduce energy demand also reduce the need for CDR and also the need to use nuclear energy, which has high initial costs (IPCC, 2022b). IPCC (2022b) assesses modeled pathways to global net-zero emissions and finds that 74% (54–90%) of emission reductions come from supply and demand measures, 13% (4 = 20%) by CO_2 mitigation options in the AFOLU sector, and 13% (10–18%) through the reduction of non-CO_2 emissions.

There is also a trade-off in terms of the trajectories of declining fossil fuel use over time, with some portfolios reducing coal, oil, and gas use at different rates. IPCC (2022b) assesses modeled pathways that limit warming to 1.5 °C with no or limited overshoot. Across these simulations, IPCC (2022b) finds that the use of these three fossil fuels variously declines by 95% (interquartile range 80–100)%, 60% (40–75%), and 45% (20–60%) by 2050. However, their continued use at these levels is only consistent with the Paris 1.5 °C target with carbon capture and storage technologies: without this, their use declines by 100% (95–100%), 60% (45–75%), and 70% (60–80%) by 2050 relative to 2019. In all of these pathways, electricity production is almost completely decarbonized. A wide range of potential portfolios for so doing exists, including various combinations of demand-side management, smart grids, electrolytic hydrogen, sustainable biofuels, and many others.

However, it is not only the global energy system that needs to be decarbonized: it is also the global food production system, which is responsible for 21–37% of current greenhouse gas emissions. Thus, IAMs such as IMAGE (van Vuuren et al., 2017) are also used to explore the allocation of land to growing food, biofuels, or biodiversity conservation, and consideration of the demand for food in relation to human population growth and dietary preferences. There is potential to avoid 0.7–8 GtCO_2e of greenhouse gas emissions by 2050 (2–20% of current emissions) so that a greater proportion of protein is derived from plants (United Nations Environment Programme, 2021).

IPCC, 2018 and, more recently, IPCC, 2022b explored various pathways that limit warming to 1.5 °C with no or limited overshoot, drawing on a large database of IAM outputs. All of these pathways use CDR, but the amount varies and is much lower in scenarios where energy demand is greatly reduced. CDR methods include bioenergy with carbon capture and storage (BECCS), and actions in the Agriculture, Forestry and Other Land Use (AFOLU) sector such as "Afforestation." Some pathways avoid BECCS deployment entirely through a combination of demand-side measures and use of AFOLU-related CDR measures. In modeled scenarios that include CDR, and limit warming to 1.5C with no or limited overshoot, the potential reported BECCS over 2020–2100 ranges from 30 to 780 GtCO_2, while the potential DACCS is 0–310 GtCO_2. As previously mentioned, large-scale deployment of BECCS has a high land and water footprint and creates a conflict with several SDGs including biodiversity conservation of life on land and eliminating hunger.

Most of the IAMs do not distinguish, though, between the types of "Afforestation." Some types of afforestation, such as the planting of monocultures of conifers, especially in areas where the natural habitat would have been savannah, grassland, or deciduous woodland, can, however, be detrimental to biodiversity conservation and the provision of some ecosystem services. In practice, within mitigation portfolios, "Afforestation" should be generalized to "Ecosystem restoration" since large amounts of carbon are also stored in other types of ecosystems, including natural grasslands and coastal wetlands. This creates a synergy with both adaptation and the sustainable development goals relating to life on land and below water.

The precise choice of mitigation portfolios in a given location is obviously influenced by the national, regional, and local context. In particular, the relative costs of alternatives will be a key factor, with the costs of wind, solar, and storage having fallen greatly in recent years. More generally, IPCC, 2022b finds that a portfolio of mitigation options "costing USD100 tCO_2-eq-1 or less could reduce global GHG emissions by at least half the 2019 level by 2030." An important issue is the relative emphasis on reducing emissions of non-CO_2 greenhouse gases, in particular methane. IPCC (2022b) finds that deep reduction of methane emissions during the period when CO_2 emissions are being reduced to net zero is an effective method to reduce peak global warming. The Glasgow Climate Pact, in which 104 countries pledged to reduce methane emissions by 30% by 2030, is a step in the right direction.

The question of in which sectors mitigation efforts are concentrated also arises: IPCC (2022b) explores how, if the goals are to be kept in reach, reduced effort in one sector requires compensation by increased efforts in another. IPCC, 2022b finds that in assessed net-zero scenarios, models indicate that net zero is reached earlier in the AFOLU and energy supply sectors than it is for the buildings, industry, and transport sectors. Decarbonization of the transport sector is facilitated by urban planning measures to reduce demand (e.g., active transport), electric vehicles, and efficiency measures.

Another important issue to consider when designing portfolios is the sustainability of materials required, for example, mining of materials required for batteries, where efforts to make mining more sustainable (e.g., avoid damaging high carbon/high biodiversity ecosystems) and to recycle minerals, can reduce undesirable side effects (IPCC, 2022b).

Finally, an emerging issue is the effects of climate change on portfolios, for example, changing climate will affect the present-day regional energy generation potential of renewable energy technologies in positive or negative ways. Examples include projected reduction in solar power potential in Europe and in hydropower potential in Latin America (Hou et al., 2021; Wasti et al., 2022).

5.5 Portfolios of Adaptation Policy

The United Kingdom has been a lead player in developing a prioritization system designed to inform its adaptation planning. It was one of the first countries to organize a national Climate Change Risk Assessment (CCRA) and its Climate Change Act 2008 established statutory requirements to assess risks to inform the National Adaptation Plan. The Netherlands and the United States rapidly followed suit with their own national assessments. Here, however, the focus is on the UK experience as its successive CCRA approaches provide interesting lessons in adaptation prioritization methods.

Typically, as in the first CCRA, Climate Change Risk Assessment is conducted using a set of climate change scenarios to drive impact models in a harmonized fashion, to determine the magnitude of risks to human and natural systems in the country. However, in the United Kingdom, it was found that this approach did not deliver what was needed to inform the adaptation plan due to the need to make decisions about which adaptation to invest in as a priority. It was recognized that factors other than the magnitude of future risks (in which there are large uncertainties) were important, for example, the vulnerability of the affected systems, the implications of damage to those systems, and the adaptive capacity.

A lack of incorporation of existing adaptation and inconsistent treatment of socioeconomic change was also problematic. It was also noted that decision-makers would prefer a "receptor-based" assessment, that is, one more focused on sectors at risk rather than academic classifications of risk types. Reviewers recommended an "adaptation-first" approach be taken in CCRA2.

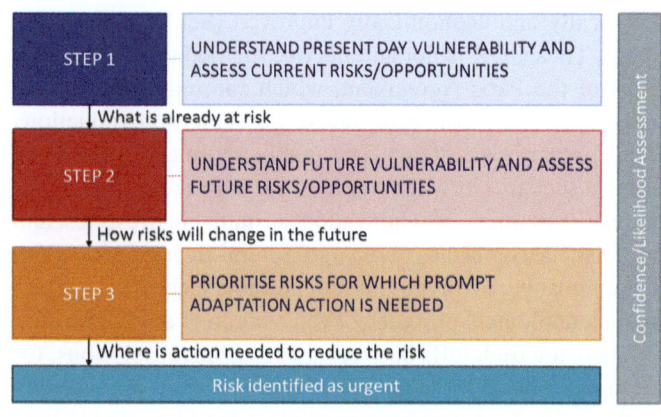

Fig. 5.3 Steps in the CCRA2 prioritization process. (Source: Warren et al., 2016)

In CCRA2, instead of performing a harmonized, top-down risk assessment to inform adaptation priorities, a new three-step, bottom-up process was developed (Warren et al., 2018, Fig. 5.3): (1) assess climate vulnerability in the present day, (2) assess future vulnerability due to both climate and non-climate drivers, and (3) prioritize adaptation options based on effectiveness, feasibility, and other criteria. In step (3), an "urgency scoring" framework was used as a basis for prioritization, in which the following attributes would increase the urgency for implementation: (a) rapid speed of onset of an impact/consequence; (b) a large spatial scale of the consequences, for example, national scale; (c) a large magnitude of the social, environmental, and/or economic impact; (d) a time-limited opportunity—or window of intervention—for early adaptation; (e) a benefit in the next 5 years if action taken; (f) long lead times for adaptation (i.e., a process that takes a long time to complete); (g) socioeconomic trends that will drive high consequences; and (h) a potential for learning, to inform adaptation decisions in the near future. By including all these factors, the urgency scoring process was able to tease out priorities more clearly despite uncertainties inherent in precisely quantifying the size of future risks. This process resulted in the classification of potential adaptation actions into categories as follows: (1) more urgent issues where adaptation action is needed; (2) more urgent issues for which further investigation is needed; and (3) less urgent issues where current action is to be sustained or a watching brief is to be kept.

CCRA3 further develops the CCRA2 method by elaborating in step 3 analysis of the potential for lock-in of inappropriate or maladaptive responses, and the potential to exceed critical thresholds that impact on the effectiveness of adaptation, as well as looking at interactions between individual risks (Betts & Brown, 2021). The rest of the methodology remains similar but was further improved using a quantitative magnitude-scoring process and a formal quality of evidence process.

5.6 A Note on Uncertainties

Scientific evidence can greatly assist decision-makers facing the portfolio problem, and co-production of research on climate change risks and adaptation benefits can be important. Since decision-makers may need advice on how to handle uncertainty, where possible scientists can provide information in a way that minimizes uncertainty. For example, quantifying precisely risks in the 2080s is much more difficult than estimating the percentage of risks avoided by limiting warming to 1.5 °C rather than 3 °C, or by implementing a particular adaptation policy. That is more tractable and is a mitigation-related question but still with large uncertainty. However, it is not always possible to reformulate questions so as to minimize uncertainty either because decision-makers need answers to questions that have more uncertain answers, or because uncertainties might be inherently large. This is discussed more fully in Chaps. 3 and 4 (this volume).

5.7 Conclusion

United Nations Environment Programme (2021) emphasizes how important it is to solve the environmental problems of climate change, biodiversity loss, land degradation, and pollution together. This means that the "portfolio problem" is not only a problem of climate change policy: it is one of creating climate-resilient development pathways that address all four issues, while contributing to the sustainable development goals. It is also one of maximizing synergies, and avoiding conflicts, between policies put in place in different sectors, and managing the myriad uncertainties present in the portfolio problem. Many synergies involve biodiversity or water, and hence, there is a high potential for designing integrated policy portfolios in these areas, including via nature-based solutions. Pathways will involve a multitude of actors in governments, businesses, financial organizations, NGOs, academia, as well as individuals. Interdisciplinary knowledge and communication between organizations responsible for financial flows and incentives in different systems will be critical in the effort to increase synergies and minimize trade-offs between goals.

References

Arent, D. J., Tol, R. S., Faust, E., Hella, J. P., Kumar, S., Strzepek, K. M., Tóth, F. L., & Yan, D. (2015). Key economic sectors and services. In C. B. Field et al. (Eds.), *Climate change 2014: Impacts, adaptation, and vulnerability. Part a: Global and sectoral aspects. Contribution of working group II to the fifth assessment report of the intergovernmental panel on climate change* (pp. 659–708). Cambridge University Press.

Berry, P. M., Brown, S., Chen, M., Kontogianni, A., Rowlands, O., Simpson, G., & Skourtos, M. (2015). Cross-sectoral interactions of adaptation and mitigation measures. *Climatic Change, 128*, 381–393. https://doi.org/10.1007/s10584-014-1214-0

Betts, R. A., & Brown, K. (2021). Introduction. In R. A. Betts, A. B. Haward, & K. Pearson (Eds.), *The third UK climate change risk assessment technical report* (pp. 1–30). Prepared for the Climate Change Committee.

Bosello, F., Carraro, C., & De Cian, E. (2013). Adaptation can help mitigation: An integrated approach to post-2012 climate policy. *Environment and Development Economics, 18*, 270–290. https://doi.org/10.1017/S1355770X13000132

Boysen, L. R., Lucht, W., Gerten, D., Heck, V., Lenton, T. M., & Schellnhuber, H. J. (2017). The limits to global-warming mitigation by terrestrial carbon removal. *Earth's Future, 5*, 463–474. https://doi.org/10.1002/2016EF000469

Burke, M., Hsiang, S. M., & Miguel, E. (2015). Global non-linear effect of temperature on economic production. *Nature, 527*, 235–239. https://doi.org/10.1038/nature15725

Chance, T. (2009). Towards sustainable residential communities; the Beddington zero energy development (BedZED) and beyond. *Environment and Urbanization, 21*, 527–544. https://doi.org/10.1177/0956247809339007

Chen, Y., Liu, A., & Cheng, X. (2020). Quantifying economic impacts of climate change under nine future emission scenarios within CMIP6. *Science of the Total Environment, 703*, 134950. https://doi.org/10.1016/j.scitotenv.2019.134950

Fisher, B., et al. (2007). Issues related to mitigation in the long-term context. In B. Metz et al. (Eds.), *Climate change 2007: Mitigation of climate change. Contribution of working group III to the fourth assessment report of the intergovernmental panel on climate change (IPCC)* (pp. 169–250). Cambridge University Press.

Friedlingstein, P., et al. (2022). Global carbon budget 2022. *Earth System Science Data, 14*, 4811–4900. https://doi.org/10.5194/essd-14-4811-2022

Gill, S. E., Handley, J. F., Ennos, A. R., & Pauleit, S. (2007). Adapting cities for climate change: The role of the green infrastructure. *Built Environment, 33*, 115–133. https://doi.org/10.2148/benv.33.1.115

Hänsel, M. C., Drupp, M. A., Johansson, D. J., Nesje, F., Azar, C., Freeman, M. C., Groom, B., & Sterner, T. (2020). Climate economics support for the UN climate targets. *Nature Climate Change, 10*, 781–789. https://doi.org/10.1038/s41558-020-0833-x

HM Government (2022). *UK climate change risk assessment 2022*. 49 pp.

Hoegh-Guldberg, O., et al. (2018). Impacts of 1.5°C global warming on natural and human systems. Global warming of 1.5°C. An IPCC special report on the impacts of global warming of 1.5°C above pre-industrial levels and related global greenhouse gas emission pathways, in the context of strengthening the global response to the threat of climate change, sustainable development, and efforts to eradicate poverty, Masson-Delmotte, V. et al., Eds., Cambridge University Press, 175–312.

Hoegh-Guldberg, O., et al. (2019). The human imperative of stabilizing global climate change at 1.5 C. *Science, 365*, eaaw6974. https://doi.org/10.1126/science.aaw6974

Hou, X., Wild, M., Folini, D., Kazadzis, S., & Wohland, J. (2021). Climate change impacts on solar power generation and its spatial variability in Europe based on CMIP6. *Earth System Dynamics, 12*, 1099–1113. https://doi.org/10.5194/esd-12-1099-2021

IPCC (2018). Summary for policymakers Global warming of 1.5°C. An IPCC special report on the impacts of global warming of 1.5°C above pre-industrial levels and related global greenhouse gas emission pathways, in the context of strengthening the global response to the threat of climate change, sustainable development, and efforts to eradicate poverty, V. Masson-Delmotte et al., Eds., Cambridge University Press, 3–24

IPCC (2021). Summary for policymakers. In V. Masson-Delmotte et al. (Eds.), *Climate change 2021: The physical science basis. Contribution of working group I to the sixth assessment report of the intergovernmental panel on climate change* (pp. 3–32). Cambridge University Press.

IPCC (2022a). Summary for policymakers. In H.-O. Pörtner et al. (Eds.), *Climate change 2022: Impacts, adaptation and vulnerability. Contribution of working group II to the sixth assessment report of the intergovernmental panel on climate change* (pp. 3–33). Cambridge University Press.

IPCC (2022b). Summary for policymakers. In P. R. Shukla et al. (Eds.), *Climate change 2022: Mitigation of climate change. Contribution of working group III to the sixth assessment report of the intergovernmental panel on climate change.* Cambridge University Press.

Jones, A., et al. (2013). The impact of abrupt suspension of solar radiation management (termination effect) in experiment G2 of the geoengineering model intercomparison project (GeoMIP). *Journal of Geophysical Research, [Atmospheres], 118*, 9743–9752. https://doi.org/10.1002/jgrd.50762

Keppo, I., et al. (2021). Exploring the possibility space: Taking stock of the diverse capabilities and gaps in integrated assessment models. *Environmental Research Letters, 16*, 053006. https://doi.org/10.1088/1748-9326/abe5d8

Macreadie, P. I., et al. (2021). Blue carbon as a natural climate solution. *Nature Reviews Earth and Environment, 2*, 826–839. https://doi.org/10.1038/s43017-021-00224-1

McCormack, C. G., et al. (2016). Key impacts of climate engineering on biodiversity and ecosystems, with priorities for future research. *Journal of Integrative Environmental Sciences*, 1–26.

Meinshausen, M., Lewis, J., McGlade, C., Gütschow, J., Nicholls, Z., Burdon, R., Cozzi, L., & Hackmann, B. (2022). Realization of Paris agreement pledges may limit warming just below 2° C. *Nature, 604*, 304–309. https://doi.org/10.1038/s41586-022-04553-z

National Academies of Sciences, Engineering, and Medicine. (2021). *Reflecting sunlight: Recommendations for solar geoengineering research and research governance.* The National Academies Press. https://doi.org/10.17226/25762

Nordhaus, W. (2018). Projections and uncertainties about climate change in an era of minimal climate policies. *American Economic Journal: Economic Policy, 10*, 333–360.

Nordhaus, W., & Sztorc, P. (2013). DICE 2013R: Introduction and user's manual. Retrieved October 9, 2019, from http://www.econ.yale.edu/~nordhaus/homepage/homepage/documents/DICE_Manual_100413r1.pdf .

O'Neill, B., et al. (2022). Key risks across sectors and regions. In H.-O. Pörtner et al. (Eds.), *Climate change 2022: Impacts, adaptation and vulnerability. Contribution of working group II to the sixth assessment report of the intergovernmental panel on climate change* (pp. 2411–2538). Cambridge University Press.

Oppenheimer, M., Campos, M., Warren, R., Birkmann, J., Luber, G., O'Neill, B. C., & Takahashi, K. (2014). Emergent risks and key vulnerabilities. In C. B. Field et al. (Eds.), *Climate change 2014: Impacts, adaptation, and vulnerability. Part A: Global and sectoral aspects. Contribution of working group II to the fifth assessment report of the intergovernmental panel of climate change* (pp. 1039–1099). Cambridge University Press.

Realmonte, G., Drouet, L., Gambhir, A., Glynn, J., Hawkes, A., Koberle, A. C., & Tavoni, M. (2019). An inter-model assessment of the role of direct air capture in deep mitigation pathways. *Nature Communications, 10*, 3277.

Riahi, K., et al. (2021). Cost and attainability of meeting stringent climate targets without overshoot. *Nature Climate Change, 11*, 1063–1069. https://doi.org/10.1038/s41558-021-01215-2

Riahi, K., et al. (2022). Mitigation pathways compatible with long-term goals. In P. R. Shukla et al. (Eds.), *Climate change 2022: Mitigation of climate change. Contribution of working group III to the sixth assessment report of the intergovernmental panel on climate change.* Cambridge University Press.

Rogelj, J., et al. (2018). Mitigation pathways compatible with 1.5°C in the context of sustainable development. Global warming of 1.5°C. An IPCC special report on the impacts of global warming of 1.5°C above pre-industrial levels and related global greenhouse gas emission pathways, in the context of strengthening the global response to the threat of climate change, sustainable development, and efforts to eradicate poverty, V. Masson-Delmotte et al., Eds., Cambridge University Press, 93–174.

Seneviratne, S. I., et al. (2018). The many possible climates from the Paris Agreement's aim of 1.5 C warming. *Nature, 558*, 41–49. https://doi.org/10.1038/s41586-018-0181-4

Schleussner, C., et al. (2024). Overconfidence in overshoot. *Nature 634*, 366–373.

Stern, N. (2007). *The economics of climate change: The Stern review* (p. 692). Cambridge University Press.

Tilmes, S., et al. (2013). The hydrological impact of geoengineering in the geoengineering model Intercomparison project (GeoMIP). *Journal of Geophysical Research, [Atmospheres], 118*, 11–036. https://doi.org/10.1002/jgrd.50868

Tol, R. S. (2018). The economic impacts of climate change. *Review of Environmental Economics and Policy, 12*(1), 4–25. Oxford University Press. https://doi.org/10.1093/reep/rex027

United Nations Environment Programme (2021). *Making peace with nature: A scientific blueprint to tackle the climate, biodiversity and pollution emergencies.* https://www.unep.org/resources/making-peace-nature.

United Nations Environment Programme (2022). *Adaptation gap report 2022: Too little, too slow—Climate adaptation failure puts world at risk.* https://www.unep.org/adaptation-gap-report-2022.

van Vuuren, D. P., et al. (2017). Energy, land-use and greenhouse gas emissions trajectories under a green growth paradigm. *Global Environmental Change, 42*, 237–250.

Van Vuuren, D. et al. (2018). Alternative pathways to the 1.5C target reduce the need for negative emission technologies, *Nature Climate Change, 8*, 391–397. https://doi.org/10.1038/s41558-018-0119-8

Warren, R. (2010). The role of interactions in a world implementing adaptation and mitigation solutions to climate change. *Philosophical Transactions of the Royal Society A, 369*, 217–241. https://doi.org/10.1098/rsta.2010.0271

Warren, R. (2014). Optimal carbon tax doubled. *Nature Climate Change, 4*, 534–535. https://doi.org/10.1038/nclimate2288

Warren, R., Mastrandrea, M. D., Hope, C., & Hof, A. F. (2010). Variation in the climatic response to SRES emissions scenarios in integrated assessment models. *Climatic Change, 102*, 671–685. https://doi.org/10.1007/s10584-009-9769-x

Warren, R., Watkiss, P., Wilby, R. L., Humphrey, K., Ranger, N., Betts, R., Lowe, J., & Watts, G. (2016). *UK climate change risk assessment evidence report*: Chapter 2, Approach and context. Report prepared for the Adaptation Sub-Committee of the Committee on Climate Change, London.

Warren, R. F., Wilby, R. L., Brown, K., Watkiss, P., Betts, R. A., Murphy, J. M., & Lowe, J. A. (2018). Advancing national climate change risk assessment to deliver national adaptation plans. *Philosophical Transactions of the Royal Society of London, Series A, 376*, 20170295. https://doi.org/10.1098/rsta.2017.0295

Warren, R., Hope, C., Gernaat, D. E. H. J., Van Vuuren, D. P., & Jenkins, K. (2021). Global and regional aggregate damages associated with global warming of 1.5 to 4 °C above pre-industrial levels. *Climatic Change, 168*, 24. https://doi.org/10.1007/s10584-021-03198-7

Wasti, A., Ray, P., Wi, S., Folch, C., Ubierna, M., & Karki, P. (2022). Climate change and the hydropower sector: A global review. *Wiley Interdisciplinary Reviews: Climate Change, 13*, e757. https://doi.org/10.1002/wcc.757

Wells, A. F., Henry, M., Bednarz, E. M., MacMartin, D. G., Jones, A., Dalvi, M., & Haywood, J. M. (2024). Identifying climate impacts from different stratospheric aerosol injection strategies in UKESM1. *Earth's Futures, 12*, e2023EF004358. https://doi.org/10.1029/2023EF004358

Yumashev, D., et al. (2019). Climate policy implications of nonlinear decline of Arctic land permafrost and other cryosphere elements. *Nature Communications, 10*, 1–11. https://doi.org/10.1038/s41467-019-09863-x

Open Access This chapter is licensed under the terms of the Creative Commons Attribution 4.0 International License (http://creativecommons.org/licenses/by/4.0/), which permits use, sharing, adaptation, distribution and reproduction in any medium or format, as long as you give appropriate credit to the original author(s) and the source, provide a link to the Creative Commons license and indicate if changes were made.

The images or other third party material in this chapter are included in the chapter's Creative Commons license, unless indicated otherwise in a credit line to the material. If material is not included in the chapter's Creative Commons license and your intended use is not permitted by statutory regulation or exceeds the permitted use, you will need to obtain permission directly from the copyright holder.

Climate Change Adaptation in Practice: Navigating Uncertainty in the Real World

Linda A. Joyce, Laurna Kaatz, and Joel Smith

6.1 Introduction

Climate change, though not a new challenge (Houghton et al., 1990), is still considered an emerging and significant issue for practitioners and resource managers. As many chapters in this book discuss, analytic approaches have been developed to assist in making sound decisions in the face of uncertainty about climate change (see Chaps. 3, 4, and 9 in this volume). These contributions either address how to approach uncertainty or offer tools to aid decision-making. How stakeholders are navigating climate change considerations in natural resource-related planning and decision-making efforts is the focus of this chapter.

The challenge of anticipating climate change is that we know the climate is changing and will continue to change in the coming decades, but we cannot forecast precisely when and how this change will come to fruition on the ground. In the past, stakeholders have made decisions considering climate and weather with imbedded assumptions that the future would not be substantially different from past variability in weather and climate. With respect to future climate, we know temperatures and sea levels will continue to rise, but we do not know by how much, by when, and what path—sudden or smooth—the changes in climate will take.

Additionally, we are more certain about the change in direction in some climatological variables than in others. Figure 6.1 displays the relative certainty about the direction of change in key variables. We have the most confidence that temperature and sea levels will rise in the future. Temperature increases will lead to earlier peak snowmelt and more intense precipitation. In contrast, there is less certainty about how precipitation patterns and climate variability will change, although knowledge is improving (Doblas-Reyes et al., 2021; Hayhoe et al., 2018; Seneviratne et al., 2021). This differentiated knowledge does give some direction for adaptation. We know we need to prepare for higher temperatures, sea levels, etc., but the magnitude and manifestation of these changes, especially at localized scales, is less clear (Fig. 6.1).

The uncertainty associated with future climate ripples through the physical, economic, social, and political networks in which stakeholders make decisions (Dewulf & Biesbroek, 2018; Stults & Larsen, 2020). A survey of local adaptation plans across the United States found that most plans acknowledged future climate uncertainty, whereas *few efforts incorporated uncertainty* about local coping capacity, effectiveness of future strategies, or uncertainty related to actions of other entities or governmental policies. Our experience with climate change adaptation also sees the planning approach spectrum described by Stults and Larsen (2020) as ranging from deterministic *Predict and Plan* to more flexible *Adapt and Monitor* approaches.

Based on our experience and that of others working with stakeholders, we offer these observations on uncertainty and the decision-making process through our case studies drawn from the United States:

1. Uncertainty has always been and will continue to be a critical part of the decision-making process. Stakeholders and decision-makers seek to make the best decisions possible in light of uncertainty. A key part of this process is the recognition that actual changes in climate may differ from projections. Decisions should, to the extent practicable, incorporate flexibility and adaptive capacity as information and understanding increases. *Adaptive Capacity* is a "built-in" ability to rapidly adjust, take advantage of new

L. A. Joyce
USDA Forest Service, Rocky Mountain Research Station, Fort Collins, CO, USA
e-mail: linda.joyce@usda.gov

L. Kaatz (✉)
Formerly at Denver Water, Denver, CO, USA

J. Smith
Independent Researcher, Boulder, CO, USA

© The Author(s) 2025
L. O. Mearns et al. (eds.), *Uncertainty in Climate Change Research*,
https://doi.org/10.1007/978-3-031-85542-9_6

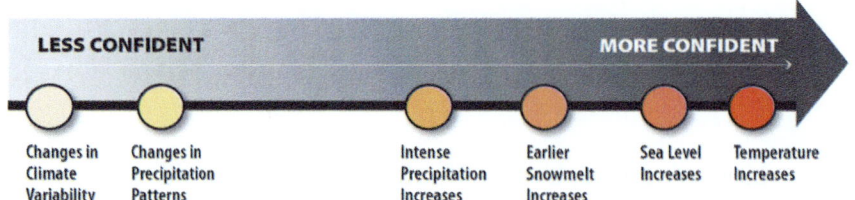

Fig. 6.1 Continuum of certainty on climate change. (Source: U.S. Climate Resilience Toolkit, 2023)

opportunities, or cope with change (California Adaptation Forum, 2016).

2. The desire for certainty and predictive skill when approaching a decision is understandable. User needs for quantitative tools encompass current tools that are predictive, as well as unmet needs for which predictive tools are on the horizon and still yet another area where the needs may not yet be met by existing or in-development tools. Predictive skill comes in the form of accuracy and precision. Precision of climate change projections (i.e., model resolution) does not necessarily result in improvement of accuracy (predictive skill). While estimates of climate sensitivities have narrowed and overall understanding of important dynamics and process-interactions in climate have progressed, uncertainties about future human-generated greenhouse emissions create a fundamental uncertainty (Hausfather et al., 2020; see also Chap. 16 in this volume). Projections should be selected with relevance to the natural resource or environmental question (Snover et al., 2013), treated as plausible futures (Langner et al., 2020; Woodhouse et al., 2021), and not forecasts or predictions.

3. The type and magnitude of response will also depend on the coping capacity of the natural resource being managed. For example, a forest fire in a highly managed, densely forested watershed that would have historically burned regularly will likely be far more disastrous, that is, have a lower coping capacity, than a forest allowed to burn regularly or one subjected to controlled fires. Increasing the coping capacity of the managed natural resource will further help address an inability to decrease the uncertainty in how the future will come to fruition. Decision-makers apply techniques to enhance coping capacity, and increase adaptive capacity while minimizing the need for certainty and prediction.

4. Stakeholder involvement is critical. Stakeholders need to be closely involved in decision-making and buy into outcomes. Decisions on managing climate-sensitive resources have to be made by those who will be affected by and manage such resources. Scientists are needed to advise stakeholders on the state of science and its proper applications to decision-making, but stakeholders need to have confidence in the decisions. Though involving multiple parties is time-consuming, the process of arriving at a shared understanding of goals, constraints, and outcomes results in unpaired success (Yates et al., 2015). Such cooperation between stakeholders and scientists is often referred to as "coproduction" of knowledge (Lemos et al., 2019).

5. Sound decisions that embrace uncertainty about climate change can and are being made (Bierbaum et al., 2013; Halofsky et al., 2015). Examples of climate adaptation in practice are illustrated in the next section.

For each case study, we draw out the particular management or decision focus, the nature of stakeholder engagement, and whether and how uncertainty was addressed. The following examples span private and federal land ownership, as well as a variety of natural resources.

6.2 Collaborative Efforts to Address Grazing Management Decisions under Drought in the Great Plains

Climate change brings the possibility of future sustained droughts in the Great Plains (Frankson et al., 2022). On the short-grass prairie, environmental and socioeconomic systems are tightly connected; drought affects ecological resilience of the shortgrass ecosystem, as well as economic resilience of the private ranch (Joyce & Marshall, 2017). In response to a severe drought, a collaborative effort of federal and private land managers was initiated to explore management decisions on federal land that private ranchers seasonally grazed their livestock. Domestic livestock grazing on Pawnee National Grassland typically starts in May and ends in October. However, very dry weather occurred over a series of years preceding 2009, and these conditions forced seasonal grazing activity to be vacated earlier than planned. Consequently, unplanned decisions about alternate forage sources or the need to move or sell cattle were made quickly. After this experience, federal managers and private rangeland managers began to explore decision-making processes that would incorporate risks associated with weather (Maczko et al., 2019). Federal managers, in this case the United States Department of Agriculture Forest Service, must manage the impact of weather on the availability and use of ecosystem services from federal land, such as Pawnee National Grassland. Private land ranchers must manage the economic risk arising from environmental conditions that affect forage quantity and quality on both federal and private land. The collaborative effort shared available historical and current information on climate and resource conditions. This effort identified the need for an early season look at current conditions, with the potential need mid-season to address declining conditions. This collaborative decision-making process, providing two known points for decision-making by federal managers and ranchers, continues to the present.

In another example of the coproduction of knowledge, the Collaborative Adaptive Rangeland Management project, on the nearby Central Plains Experiment Range, brought together a diverse set of scientists (ecologists, biologists, climate scientists, sociologists, economists) and an equally diverse stakeholder group (ranchers, conservationists, state and federal agencies) to manage 10 herds grazing 3200 acres (1295 hectares) over the last 10 years (Wilmer et al., 2022). Working together and learning to listen to each other, scientists and stakeholders continually evaluated multiple objectives, including grazing management, wildlife habitat provisions, and ranch profitability and economic sustainability (Ganguli & O'Rourke, 2022).

6.3 Coproduction of Climate Change Information to Assess Future Risk to Roads in a National Forest

In this example, several entities organically came together to facilitate coproduction of data to assess future risk to roads in an area where snow/rain patterns were likely to change in the future. In 2013, the United States Department of Agriculture Forest Service Mt. Baker-Snoqualmie National Forests (MBS NF) initiated a travel analysis as required by the 2005 Travel Management Rule using a science-based process called the Sustainable Roads Strategy. A public engagement process asked stakeholders which roads mattered the most (Cerveny et al., 2022). At the same time, the MBS NF was a collaborator in the North Cascadia Adaptation Partnership (NCAP) that brought together resource management agencies from the United States Department of Agriculture and the United States Department of the Interior as well as federal and university research scientists. This partnership, along with numerous stakeholders, explored the current science related to climate change and the effects on natural resources within 6,177,635 acres (2.5 million hectares) of north-central Washington (Raymond et al., 2014).

Using climate change information developed through NCAP and the travel analysis, a University of Washington PhD student explored the impacts to watersheds across MBS NF (Strauch et al., 2015). Staff at Conservation Northwest, building on Strauch et al. (2015), assigned each road segment a composite climate hazard score determined from the combined potential peak 2080 flood-level increase and potential 2080 winter soil moisture increase (Fig. 6.2) (Wooten, 2016). The analysis also considered potential shifts in the timing of shoulder seasons, allowing earlier or later user access and road maintenance (Fig. 6.2).

The final travel analysis product, including the work of Strauch et al. (2015) and Wooten (2016), provided information that the MBS NF can use to minimize future risk of climate hazards in the maintenance of road systems and presents a variety of opportunities for making changes to current road management practices (USDA, 2015). This information will serve as a basis for long-term maintenance cost reductions, prioritizing scarce resources to maintain the desired forest transportation system that meets the access needs for public or administrative purposes. The MBS has completed two access and travel management documents with a climate change roads analysis. This coproduction example shows how bringing scientists and managers together can assist in adding climate change information to a standard planning process on a National Forest.

6.4 Denver Water Embraces Uncertainty through Scenario Planning

Denver Water, the largest drinking water utility in Colorado, has a long legacy of planning for the future. The utility began Integrated Resource Planning (IRP) in the 1990s, and each iteration has built upon the last, factoring in new challenges and taking advantage of new techniques and resources.

In 2002, the utility experienced simultaneous natural disasters. Colorado experienced the single worst drought year in the state's recorded history. Then in June of this same year, the fourth largest wildfire the state has experienced burned across Denver's largest and hardest working watershed. Weeks after the fire was contained, a summer rainstorm brought sediment and debris into streams and reservoirs, significantly impacting water quality and reservoir capacity. Nearly 20 years later, the watershed is still noticeably scarred.

This experience presented Denver Water with an opportunity to reflect on its IRP planning assumptions and approach. Foundational questions such as "Does planning with the observed hydrologic record sufficiently prepare Denver Water for the variability the region has experienced prior to the observed record and may experience in the future?" and "How might local hydroclimatic conditions change over time as anthropogenic warming accelerates?" emerged as critical new inquiries.

Denver Water reconstructed past hydrology using paleohydrology, a technique that uses tree rings to recreate past hydrology, to address the first inquiry. The utility now plans

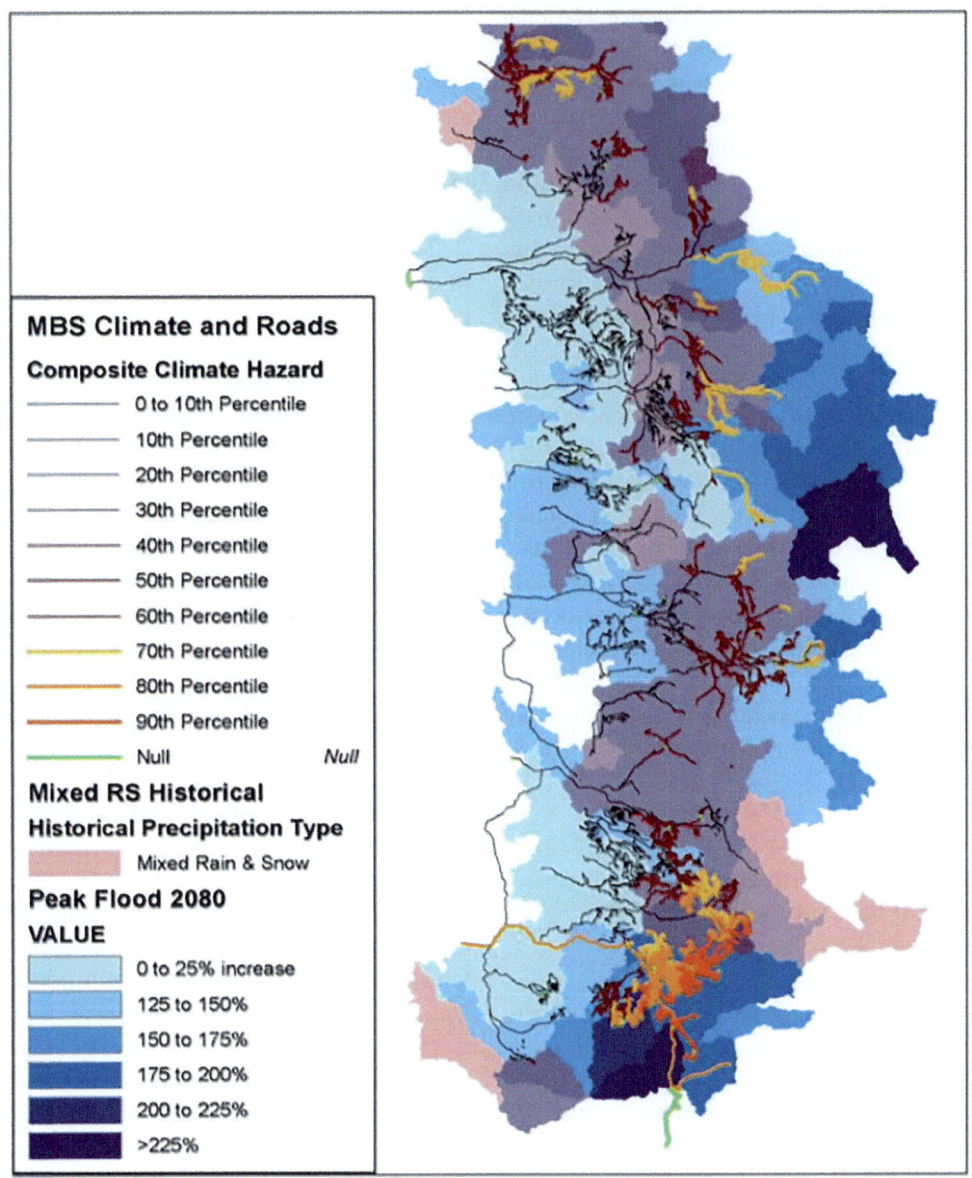

Fig. 6.2 Composite road hazard scores overlaid on the potential 2080 peak flood level percent increase for each sub-watershed, with mixed rain-and-snow sub-watersheds shaded darker. (Source: Wooten, 2016)

with data going back to 1634 and found droughts in the 1600s and 1800s that are much more challenging than seen in the observed record. Next, Denver Water turned to downscaled General Circulation Models coupled with hydrology models to understand potential future hydroclimatic conditions. The utility learned that modeling precipitation is incredibly complicated, its region may not see model agreement for precipitation, it should not expect skillful precipitation projections, and there is significantly more skill and confidence in temperature projections. As a result of these findings, Denver Water shifted its climate analysis focus to higher temperatures and, through simple sensitivity assessments, found its supply and demand is significantly vulnerable to warming. Basic assessments are informative and have proved to be an incredibly valuable mechanism in helping the utility understand and evaluate its vulnerability to climate change. The most important outcomes of these inquiries were the understanding and acceptance that history cannot fully portray the range of possible future conditions and that a *Predict then Plan* approach will not adequately prepare Denver Water for the future.

Following these lessons and seeking to avoid its experience in 2002, Denver Water embraced uncertainty as a critical aspect of planning and adopted Scenario Planning in 2008. Scenarios incorporate combinations of assumptions around external drivers of change (parameters outside of the utility's control including the regional economy, global greenhouse gas emissions, and community values) to evaluate impacts to the water system and prepare for a range of future challenges. A key element of the scenario planning process is to challenge embedded assumptions that staff have about how the water system can operate and what the future will look like, and by envisioning different futures, encourage a broad range of thinking and discussion within the organization. In its current IRP, Denver Water is planning for both a warmer and a hotter future. The utility recognizes climate change is here and now, and that the region will continue to warm for the foreseeable future, though exactly how much it warms and how the warming manifest on the ground is uncertain.

6.5 Adaptation in Practice: Denver Water's "From Forests to Faucets" Program

In this example, a partnership across federal, state, and local agencies depicts the effectiveness of coproduction and value of adaptation investments. The *From Forests to Faucets* program is a multimillion-dollar partnership between Denver Water, the U.S. Forest Service, the Colorado State Forest Service, the Natural Resources Conservation Service, the Colorado Forest Restoration Institute at the Colorado State University, and other agencies. It is creating resilient watersheds by mitigating the impacts of fire by treating overly dense forestland and continuing reforestation efforts in areas ravaged by wildfires.

Between 2010 and 2028, the program will have invested more than $96 million and hundreds of labor hours to conduct fuel reduction and forest management actions. Through the beginning of 2023, over 150,000 acres (60,700 hectares) within Denver Water's priority watersheds have been treated. Program treatments include thinning and hazardous fuels removal, prescribed fire, clear cuts, noxious weed treatments, and tree planting. Treatment plans are individually designed and implemented by program partners based on the characteristics and conditions of each watershed and are intended to decrease the risk of catastrophic wildfires, not mitigate wildfires completely. Wildfires are important occurrences in healthy watersheds and will happen in the future, likely heightened by climate change (Abatzoglou & Williams, 2016; Higuera et al., 2021; Rodrigues et al., 2023) (Fig. 6.3).

Program success, which became evident during the Buffalo Mountain fire above Dillon Reservoir in 2018 (Fig. 6.3), illustrates the effectiveness of coproduced treatment investments. Specific fuel reduction treatments to the 900-acre (364 hectare) Buffalo Mountain watershed were developed and implemented by the partnership and resulted in saved lives, property, ecosystems, and minimized impacts to reservoirs and water quality. Estimates indicate the treatment approach saved nearly $1 billion in damages. As fires increase in frequency, duration, and intensity, and occur outside of the traditional season and across higher elevation bands, coproduced investments in watershed health will save money and decrease impacts on increasingly vulnerable communities and natural resources.

Fig. 6.3 Photograph of burn line near residential homes immediately following the Buffalo Mountain fire in 2018. (Source: Denver Water)

6.6 Incorporating Climate in Culvert Size Selection

Culverts, which are designed to channel streamflow, often below roads, come in standard sizes ranging from 3-inch (0.080 m) diameter increments for small culverts to 1-foot (0.305 m) for large culverts. Each size culvert (Fig. 6.4) (U.S. Department of Transportation, 1972) can handle a range of high stream flows. Therefore, in selecting a culvert size, it is not necessary to know the exact amount of peak flow, but it is important to have an idea about which range of flow is most likely to adequately account for potential increase in future high flows. In the case of culverts, the marginal cost, and downsides of selecting a larger size is relatively small (Fig. 6.4).

Given this situation, it may be possible to readily anticipate future increases in peak flow by installing incrementally larger culverts as they are put in or reach the end of their design life. For example, using Fig. 6.4, assume that a 42-inch (1.067 m) culvert was sufficiently sized for observed peak flows. Perhaps peak flow was 130 cfs (cubic feet per second; 3.7 m^3/s). As a result of climate change, the peak flows could increase. The exact amount by which it would

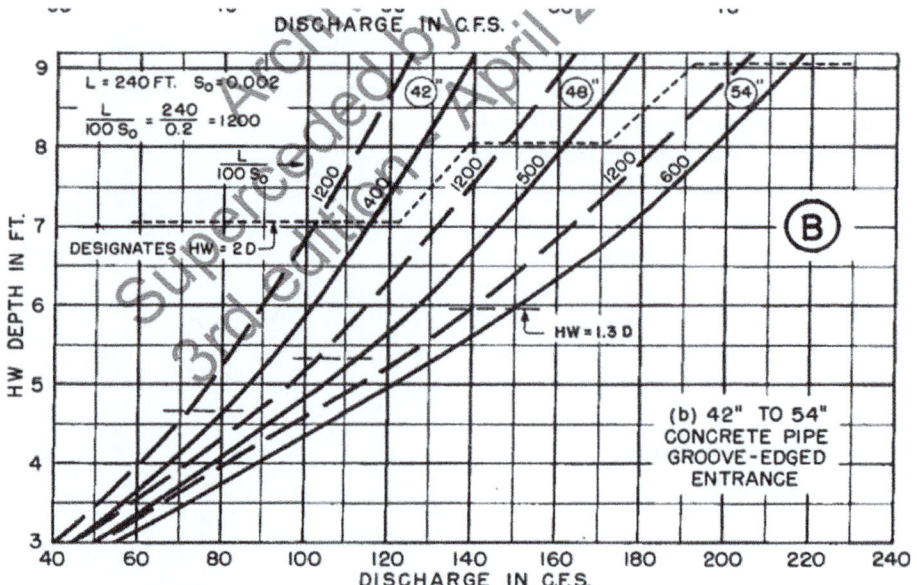

Fig. 6.4 Culvert sizes and capacities. Typical culvert capacity chart. (Source: Fig. 1 in "Capacity Charts for Hydraulic Design of Highway Culverts." Hydraulic Engineer Circular No. 10. 1972. U.S. Department of Transportation. https://www.fhwa.dot.gov/engineering/hydraulics/pubs/hec/hec10.pdf)

increase is uncertain but increases of 10–20% or more over coming decades seem plausible (see Martel et al., 2021). Under this assumption, peak flows of approximately 145–160 cfs (4.1–4.5 m^3/s) could be anticipated. In such a case, the next higher size of culvert, 48-inch (1.2192 m), would be needed. What this suggests is that if a new culvert is being installed or an older one being replaced, where observed peak flows might have supported installation of a 42 inches (1.067 m) culvert, anticipation of climate change could justify installation of an incrementally larger culvert, in this case 48-inch (1.219 m). This simple calculation may only be cost-effective if the marginal cost differences between installing a 42-inch (1.067 m) and a 48-inch (1.219 m) culvert are small and there are no other complicating factors, such as a larger culvert necessitating changing design of the road above.

The sizing of culverts in anticipation of climate change was analyzed by McCurdy and Travis (2017). They analyzed culverts across Colorado examining costs of installing culverts and benefits such as avoiding flooding and traffic diversions. The article found that increasing the "resilience factor," that is, the capacity of a culvert to withstand a flow above design levels, is advantageous. This appears to be consistent with the idea of installing somewhat larger culverts that can withstand larger flows. However, McCurdy and Travis (2017) found that replacing existing culverts with larger ones before their design lifetime was reached would not be cost-effective.[1] So going to a larger-size culvert may only be justified when new ones are being installed for reasons other than anticipation of climate change.

6.7 Boston Waterfront Planning

Spurred on by increasing climate disasters since the 1990s and a near miss in the Boston area from Superstorm Sandy in 2012, the City of Boston undertook a comprehensive planning effort to prepare for increases in sea levels, temperatures, precipitation, and storm intensity.

Boston used a single scenario of sea level rise, assuming 9 inches (0.229 m) of sea level rise by 2030, 21 inches (0.533 m) by 2050, and 36 inches (0.914 m) by 2070. A uniform scenario was needed to align all the roadways, walkways, and landscapes. The scenario, which is referred to as the "40 inches scenario," includes the 1:100 storm event on top of the sea level rise projections.[2] The plan also allows for the possibility that sea level rise will exceed 40 inches (1.067 m) this century (Bosma et al., 2015).

A key part of the planning included extensive local stakeholder consultations. This was important to develop and sustain community support for the long-term implementation of the plan and enable stakeholders to identify local concerns and develop creative solutions (Walsh, 2016).

The planning focused on eight areas across the city. Here, we describe one area: the "Coastal Resilience Solutions for Downtown Boston and the North End." This is a $200–300-million 50-year project to protect Boston waterfront including Downtown, the North End, and the eastern edge

---

[1] Design lives of culverts can be as long as 50–100 years (CCPPA, 2023).

[2] The 40-inch scenario was based on projections of sea level rise by the US Government in 2012 and analysis of change in storm surge undertaken for the State of Massachusetts (Bosma et al., 2015).

of the city's West End from the 40-inch scenario by late this century (City of Boston, 2020). The integrated plan, which was developed with extensive stakeholder input, relies on a combination of natural (green infrastructure) defenses, breakwaters, sea walls, harbor walks, and raised land to protect the waterfront and inland areas from increases in coastal flooding and sea level rise.

The Downtown Boston and North End plan was developed with extensive input from local stakeholders, including property owners, neighborhood groups, state and federal agencies, and others. Two open houses were held for the public to give input into the plan. The stakeholders strongly preferred a combination of effectiveness, adaptability, feasibility, and consideration of environmental impacts when prioritizing options for protecting the waterfront.

Boston's resilience plan is aimed at reducing flood risks in an area going from Rowes Wharf in Downtown, the Wharf District, North End waterfront, and the eastern part of the West End not including the New Charles River Dam.

There are three timeframes for implementing specific components of the plan:

- Near-term out to 2030. This includes immediate measures to address current flood risks.
- Mid-term actions out to about the 2040s.
- Long-term measures from about 2050–2070 and beyond.

The various time horizons allowed areas with different elevations to take sufficient measures to protect against the uniform scenario. Areas with lower elevation will build defensive measures in the shorter timeframes while areas with higher elevations can put off implementing measures until later time periods. New construction will be built for the 2070 scenario (Paul Kirshen, University of Massachusetts Boston, *pers. comm.*). The resilience plan proposes an integrated coastal defense system, including specific options that provide protection from flooding caused by sea level rise and more intense coastal storms. The defense system includes combinations of

- Green infrastructure (living shorelines).
- Offshore breakwater structures.
- Raising land such as Puopolo Park in the North End.
- Building and raising hard structures such as elevated waterfronts, harbor walks, sea walls, and bulkheads.
- Where necessary to provide additional defense against elevated flood risks, raised roadways, intersections, and bike paths.

The plan incorporates parks and bikeways and maximizes public access to the city's waterfront. Property owners will need to take additional actions to protect their assets. Specific plans are included to defend against the "40-inch scenario," but provisions are made to put in more protections should sea level rise exceed that scenario.

Capital costs are estimated to be between $189 and $315 million, with estimated annual maintenance costs of $3–five million. The plan reports that net benefits over its lifetime (applying discount rates of 3 and 7%) will be $2–6 billion.

6.8 Conclusions

This chapter demonstrates—through selected examples of natural resource management—that added uncertainty about climate change need not paralyze decision-makers or the decision-making process. What the examples show is that sound climate change-informed decisions *can be made* to manage natural resources despite uncertainty about climate change.

- Federal and private land managers collaborate to anticipate and manage grassland production in the Pawnee National Grassland.
- Federal forest managers and stakeholders worked together to anticipate potential impacts of climate change on roads in the Mt. Baker-Snoqualmie National Forests.
- Denver Water uses scenario planning, including scenarios not just of change in climate but also in socioeconomic conditions such as population, economy, and consumer preferences, to improve its decision-making about long-term water management.
- Denver Water works collaboratively with state and federal forest managers to reduce fire risks in watersheds in ways that reduce risks of fire and harm to water supplies.
- Culverts incorporate a safety margin for a range of high flows. Using larger size culverts is a way of incorporating potential increases in high flow in a manner that can be effective over a range of possible future conditions.
- The City of Boston worked with scientists and local stakeholders to plan for sea level by selecting an individual scenario that serves as a minimum for its waterfront but also allows for raising the waterfront above that scenario should sea level rise be higher than currently planned for.

These decisions recognize that while we know (and have known for some decades) that the climate is changing, and for many key variables we know the direction of change (e.g., higher temperatures and sea levels), skillful projections of exactly how climate change will manifest locally are not available now and are not expected to be available in the near future. Despite this, sound decisions that reduce risks and improve resilience to future climate change are being made.

These examples also show that the process of -making is as, if not more, important than the decisions themselves. Stakeholder involvement is critical to arriving at management actions or decisions under climate change. Collaborative and coproduction efforts build trust among the stakeholders, scientists, and decision-makers and facilitate shared understanding of scenarios, effects, and possible tradeoffs. A diversity of perspectives is needed to challenge embedded assumptions about the future. Risk depends on the perspective/interest. Multiple objectives, such as ecological resilience and economic resilience, are needed. These partnerships also provide opportunities to integrate scientific knowledge and experiential knowledge.

What has long been desired to help those making decisions on long-term investments is to provide forecasts or guidance on what exact changes in climate and extreme events can be expected to happen in coming decades. In other words, the perception has been that science was not providing sufficient information, so a major focus has been on improving science. This perception is sometimes expressed as a desire for "actionable science," that is, provide projections with high enough spatial resolution and narrow enough of a range in magnitude to be useful for decision-making.

While improvement in the science of projecting future changes in climate would be helpful, it is not reasonable to expect such a change in the science of projecting climate change will emerge in the near future (Barsugli et al., 2009; Pielke Sr & Wilby, 2012). Yet, given the way we make decisions about long-term climate-sensitive investments, there are ways to productively incorporate what is known about climate change into such decisions. We need to use science in more creative ways than we have in the past to enable us to be better prepared for climate change impacts than the old assumptions of climate stationarity would have yielded.

References

Abatzoglou, J. T., & Williams, A. P. (2016). Impact of anthropogenic climate change on wildfire across western US forests. *Proceedings of the National Academy of Sciences, 113*(42), 11770–11775. https://doi.org/10.1073/pnas.1607171113

Barsugli, J., Anderson, C., Smith, J., & Vogel, J. M. (2009). *Options for improving climate modeling to assist water utility planning for climate change* (p. 146). Water Utility Climate Alliance. Retrieved May 22, 2023, from https://www.researchgate.net/profile/Jason-Vogel-5/publication/252503020_Options_for_Improving_Climate_Modeling_to_Assist_Water_Utility_Planning_for_Climate_Change/links/0deec52f13326797100000000/Options-for-Improving-Climate-Modeling-to-Assist-Water-Utility-Planning-for-Climate-Change.pdf

Bierbaum, R., et al. (2013). A comprehensive review of climate adaptation in the United States: More than before, but less than needed. *Mitigation and Adaptation Strategies Global Change, 18*, 361–406. https://doi.org/10.1007/s11027-012-9423-1

Bosma, K., Douglas, E., Kirshen, P., McArthur, K., Miller, S., & Watson, C. (2015). *MassDOT-FHWA pilot project report: Climate change and extreme weather vulnerability assessments and adaptation options for the central artery*. University of Massachusetts Boston. Retrieved June 7, 2023, from http://eea-nescaum-dataservices-assets-prd.s3.us-east-1.amazonaws.com/resources/production/Pilot_Project_Report_MassDOT_FHWA.pdf

California Adaptation Forum. (2016). *Pre-forum drought exercise: Situation manual* (p. 19) Retrieved May 22, 2023, from http://cdn2.hubspot.net/hubfs/472557/Situation_Manual_CAF_Drought_Exercise_0902_Cadmus.pdf?__hssc=121325015.1.1592530318019&__hstc=121325015.88669e8617d1e4e944ffa7fe08790257.1591140862832.1591922957115.1592530318019.3&__hsfp=204825244&hsCtaTracking=a4ffba1d-d34e-446c-8b79-4ef3daea37c1%7C85783f6c-35d3-4941-a295-e7daaa33d659

CCPPA (Canadian Concrete Pipe and Precast Association) (2023). Estimated material service life of drainage pipes. Retrieved May 22, 2023, from https://ccppa.ca/estimated-material-service-life-of-drainage-pipes/

Cerveny, L. K., McLain, R. J., Banis, D., & Todd, A. (2022). The use of socio-spatial data for sustainable roads planning: A national forest case study. *Journal of Environmental Planning and Management, 8*, 157. https://doi.org/10.3390/f8050157

City of Boston. (2020). *Coastal resilience solutions for downtown Boston and North End* (p. 117) Retrieved May 22, 2023 from https://www.boston.gov/sites/default/files/file/2020/10/Climate%20Ready%20North%20End%20Downtown%20Final_EMBARGO%20102820.pdf

Dewulf, A., & Biesbroek, R. (2018). Nine lives of uncertainty in decision-making: Strategies for dealing with uncertainty in environmental governance. *Policy and Society, 37*(4), 441–458. https://doi.org/10.1080/14494035.2018.1504484

Doblas-Reyes, F. J., et al. (2021). Linking global to regional climate change. In V. Masson-Delmotte et al. (Eds.), *Climate change 2021: The physical science basis* (pp. 1363–1512). Cambridge University Press. https://doi.org/10.1017/9781009157896.012

Frankson, R., Kunkel, K. E., Stevens, L. E., Easterling, D. R., Umphlett, N. A., Stiles, C. J., Schumacher, R., & Goble, P. E. (2022). Colorado state climate summary 2022. NOAA Technical Report NESDIS 150-CO. NOAA/NESDIS, 5 pp. https://statesummaries.ncics.org/chapter/co/

Ganguli, A. C., & O'Rourke, M. E. (2022). How vulnerable are rangelands to grazing? *Science, 378*(6622), 834. https://doi.org/10.1126/science.add4278

Halofsky, J. E., Peterson, D. L., & Marcinkowski, K. W. (2015). Climate change adaptation in United States Federal Natural Resource Science and Management Agencies: A Synthesis. Adaptation Science Interagency Working Group, Interagency Land Management Adaptation Group, U.S. Department of Agriculture Forest Service. 89 pp.

Hausfather, Z., Drake, H. F., Abbott, T., & Schmidt, G. A. (2020). Evaluating the performance of past climate model projections. *Geophysical Research Letters, 47*, e2019GL085378. https://doi.org/10.1029/2019GL085378

Hayhoe, K., et al. (2018). Our changing climate. In D. R. Reidmiller et al. (Eds.), *Impacts, risks, and adaptation in the United States: Fourth National Climate Assessment, volume II* (pp. 72–144). U.S. Global Change Research Program. https://doi.org/10.7930/NCA4.2018.CH2

Higuera, P. E., Shuman, B. N., & Wolf, K. D. (2021). Rocky Mountain subalpine forests now burning more than any time in recent millennia. *Proceedings of the National Academy of Sciences, 118*(25), e2103135118. https://doi.org/10.1073/pnas.2103135118

Houghton, J. T., Jenkins, G. J., & Ephraums, J. J. (Eds.). (1990). *Climate change. The intergovernmental panel on climate change scientific assessment*. Cambridge University Press.

Joyce, L. A., & Marshall, N. (2017). Managing climate change risks in rangeland systems. In D. Briske (Ed.), *Rangeland Systems* (pp. 491–526). Springer Series on Environmental Management. Springer. https://doi.org/10.1007/978-3-319-46709-2_15.491-526

Langner, L. L., Joyce, L. A., Wear, D. N., Prestemon, J. P., Coulson, D., & O'Dea, C. B. (2020). Future scenarios: A technical document supporting the USDA Forest Service 2020 RPA Assessment. Gen. Tech. Rep. RMRS-GTR-412. U.S. Department of Agriculture, Forest

Service, Rocky Mountain Research Station. 34 p., https://doi.org/10.2737/RMRS-GTR-412.

Lemos, M. C., Wolske, K. S., Rasmussen, L. V., Arnott, J. C., Kalcic, M., & Kirchhoff, C. J. (2019). The closer, the better? Untangling scientist–practitioner engagement, interaction, and knowledge use. *Weather, Climate, and Society, 11*(3), 535–548. https://doi.org/10.1175/WCAS-D-18-0075.1

Maczko, K., Hidinger, L., Tanaka, J. A., Morgan, J. A., Mitchell, J. E., Fox, W. E., Joyce, L., & Duke, C. S. (2019). *Climate change on the range: Monitoring and adaptation for sustainability. Sustainable rangelands roundtable publication no. 6. MP-139* (46). University of Wyoming. https://www.wyoextension.org/agpubs/pubs/MP-139.pdf

Martel, J.-L., Brissette, F. P., Lucas-Picher, P., Troin, M., & Arsenault, R. (2021). Climate change and rainfall intensity-duration-frequency curves: Overview of science and guidelines for adaptation. *Journal of Hydrologic Engineering, 26*, 10. https://doi.org/10.1061/(ASCE)HE.1943-5584.0002122

McCurdy, A. D., & Travis, W. R. (2017). Simulated climate adaptation in stormwater systems: Evaluating the efficiency of adaptation strategies. *Environment Systems and Decisions, 37*, 214–229. https://doi.org/10.1007/s10669-017-9631-z

Pielke, R. A., Sr., & Wilby, R. L. (2012). Regional climate downscaling: What's the point? *Eos, Transactions American Geophysical Union, 93*(5), 52–53. https://doi.org/10.1029/2012EO050008

Raymond, C. L., Peterson, D. L., & Rochefort, R. M., eds., (2014). Climate change vulnerability and adaptation in the North Cascades region, Washington. Gen. Tech. Rep. PNW-GTR-892. Portland, OR: U.S. Department of Agriculture, Forest Service, Pacific Northwest Research Station. 279 p.

Rodrigues, M., Camprubí, A. C., Balaguer-Romano, R., Coco Megía, C. J., Castañares, F., Ruffault, F., Fernandes, P. M., & Resco de Dios, V. (2023). Drivers and implications of the extreme 2022 wildfire season in Southwest Europe. *Science of the Total Environment, 859*, 160320. https://doi.org/10.1016/j.scitotenv.2022.160320

Seneviratne, S. I., et al. (2021). Weather and climate extreme events in a changing climate. In V. Masson-Delmotte et al. (Eds.), *Climate change 2021: The physical science basis* (pp. 1513–1766). Cambridge University Press. https://doi.org/10.1017/9781009157896.013

Snover, A. K., Mantua, N. J., Littell, J. S., Alexander, M. A., McClure, M. M., & Nye, J. (2013). Choosing and using climate change scenarios for ecological-impact assessments and conservation decisions. *Conservation Biology, 27*, 1147–1157. https://doi.org/10.1111/cobi.12163

Strauch, R., Raymond, C. L., Rochefort, R. M., Hamlet, A. F., & Lauver, C. (2015). Adapting transportation to climate change on federal lands in Washington State, U.S.A. *Climatic Change, 130*, 185–199. https://doi.org/10.1007/s10584-015-1357-7

Stults, M., & Larsen, L. (2020). Tackling uncertainty in US local climate adaptation planning. *Journal of Planning, 40*(4), 416–431. https://doi.org/10.1177/0739456X18769134

U.S. Climate Resilience Toolkit (2023). Climate Science for water professionals, Online Training for Water Utilities/WUCA, https://toolkit.climate.gov/course-lessons/climate-science-water-professionals.

USDA Forest Service. (2015). Mt. Baker-Snoqualmie National Forest: Forest-wide Sustainable Roads Report (Accessed 4/9/2025). 42p. https://www.fs.usda.gov/Internet/FSE_DOCUMENTS/fseprd486757.pdf

U.S. Department of Transportation. (1972). Capacity charts for the hydraulic design of highway culverts. Federal Highway Administration. Hydraulic Engineering Circular No. 10. https://www.fhwa.dot.gov/engineering/hydraulics/pubs/hec/hec.10.pdf.

Walsh, M. (2016). Climate ready Boston: Executive summary. https://www.boston.gov/sites/default/files/file/2023/03/2016_climate_ready_boston_executive_summary_1.pdf.

Wilmer, H., et al. (2022). Social learning lessons from collaborative adaptive rangeland management. *Rangelands, 44*(5), 316–326. https://doi.org/10.1016/j.rala.2021.02.002

Woodhouse, C. A., Smith, R. M., McAfee, S. A., Pederson, G. T., McCabe, G. J., Miller, W. P., & Csank, A. (2021). Upper Colorado River Basin 20th century droughts under 21st century warming: Plausible scenarios for the future. *Climate Services, 21*, 100206. https://doi.org/10.1016/j.cliser.2020.100206

Wooten, G. (2016). Analysis of climate change impacts to roads on the mount-baker Snoqualmie National Forest. *Conservation Northwest*. http://cascadiapartnerforum.org/wp-content/uploads/2016/04/Climate-Risks-on-MBSNF-Roads_2016_GWooten.pdf

Yates, D. N., Miller, K. A., Wilby, R. L., & Kaatz, L. (2015). Decision-centric adaptation appraisal for water management across Colorado's continental divide. *Climate Risk Management, 10*, 35–50. https://doi.org/10.1016/j.crm.2015.06.001

Open Access This chapter is licensed under the terms of the Creative Commons Attribution 4.0 International License (http://creativecommons.org/licenses/by/4.0/), which permits use, sharing, adaptation, distribution and reproduction in any medium or format, as long as you give appropriate credit to the original author(s) and the source, provide a link to the Creative Commons license and indicate if changes were made.

The images or other third party material in this chapter are included in the chapter's Creative Commons license, unless indicated otherwise in a credit line to the material. If material is not included in the chapter's Creative Commons license and your intended use is not permitted by statutory regulation or exceeds the permitted use, you will need to obtain permission directly from the copyright holder.

7. Uncertainty of Climate Change Impacts on Crop Production

Daniel Wallach, Senthold Asseng, and Alex C. Ruane

7.1 Effect of Climate Change on Agriculture

Climate changes could affect all aspects of the food system, beginning with impacts on average crop production and the stability of yields and additionally shaped by socioeconomic and geopolitical forces (Porter et al., 2014). Society demands that future agricultural systems provide sufficient nutritious food for the consumption patterns of a growing and developing population, adapt to climate changes, play a role in the mitigation of greenhouse gas emissions, and provide an economic incentive that keeps farms in business (Ruane et al., 2017). The agricultural sector needs to advance by considering crop quantity, quality, efficiency of resource use, and environmental and sociopolitical footprint. This would be a substantial challenge if faced in isolation, but cropping systems are only a component of larger societal systems and thus compete under additional constraints for resources such as public and private investment, labor, land, water, and energy (Rosenzweig et al., 2020). Here, we focus on uncertainties in projections of climate impacts on field-level agricultural production (the fundamental engine of food systems through pastoral systems is also of fundamental importance and could be severely impacted) (Soussana et al., 2010).

D. Wallach (✉)
Institute of Crop Science and Resource Conservation, University of Bonn, Bonn, Germany

S. Asseng
Technical University of Munich, Digital Agriculture, HEF World Agricultural Systems Center, Freising, Germany
e-mail: senthold.asseng@tum.de

A. C. Ruane
Climate Impacts Group, NASA Goddard Institute for Space Studies, New York, NY, USA
e-mail: alexander.c.ruane@nasa.gov

A number of climate factors affect agricultural systems, including long-term trends and episodic extreme events (IPCC, 2014). The most widespread climate impacts stem from increases in mean temperature, particularly during portions of the growing season that correspond to key crop development stages. The increase in atmospheric CO_2 concentrations from continuing greenhouse gas emissions can have a beneficial effect on crop growth and water use efficiency. A third major effect is the expected change in rainfall patterns, which could impact agricultural production either directly for rain-fed agriculture, or indirectly through the availability of water for irrigation. Changes in extreme conditions, most notably heat waves and droughts given their ability to cause severe regional yield losses, are also important. Warmer conditions may also lead to a reduction in cold season hazards while those areas with increased rainfall could be forced to manage wet hazards such as flooding, water logging, and soil runoff degradation. Additional agroclimatic hazards are less well studied, including severe storms (e.g., tropical cyclones and hail), changes in ozone concentrations, and climate-driven alterations to the likelihood and extent of outbreaks of pests and diseases. In recent years, there has also been enhanced attention on approaches that capture the exacerbating role of connected extremes for agricultural impacts, which may be compound (e.g., heatwave and drought at the same time and place), concurrent (e.g., multiple breadbaskets suffering bad harvests in the same year), or sequential (e.g., an early season drought followed by a later drought).

Contextualized information about future climate impacts on regional and crop system-specific agricultural production can play an important role in supporting food systems decision-making. Rather than assuming static farming systems, it is important to recognize the agency of farmers and other food system actors who may take actions in the face of changing climate and socioeconomic conditions (Valdivia et al., 2015). Climate impact information can inform reactive interventions that respond to agroclimatic risks that have already occurred or are ongoing (e.g., supplemental irrigation, pesticide application, farm bailouts) or proactive interventions that reduce or eliminate the likelihood of agroclimatic risks so that disasters are more manageable (e.g., shifting growing seasons, selecting cultivars suited to changing climate conditions, building infrastructure for alternative field amendments, utilizing new agronomic technologies, establishing crop insurance programs, moving farmland away from hazard-prone areas, providing improved forecasts and early-warning systems). Foresight and advanced planning are needed to ensure that these preventive options are in place to guard against future climate trends and extremes, and information that corresponds to the scale of regional systems is most useful to stakeholder decision-making.

7.2 Methods for Understanding and Quantifying the Effect of Climate Change

7.2.1 Forecasting by Analogy

Anticipated future climate in some specific location may be similar to current climate in a different location. Future crop production under climate change can then be inferred by analogy. However, use of this approach is limited by two major difficulties, namely that the climate analogy is in general only approximate, and there may be important differences between the two locations other than climate (Bos et al., 2015).

7.2.2 Observation and Experimentation

Observations of field results, assuming they include both environmental and crop growth data, give information about crop response to the environment. Specially designed field and controlled environment experiments provide more targeted information about how crops respond to weather and can thus be used as the basis for simulating the effects of climate change. Of particular usefulness are experiments that are specifically designed to explore conditions that mimic in some way expected future weather. This is more easily done in controlled environment experiments, but there is always uncertainty because of the question of how far, and in what ways, controlled environment chambers differ from normal growing conditions. This has led to the creation of free-air carbon dioxide enrichment (FACE) experiments (Fig. 7.1), which can be combined with supplemental heating or increased ozone (Ewert & Porter, 2000; Kimball et al., 1995). These experimental results are also subject to uncertainties, related to both the imposed treatments (e.g., how homogeneous is the enriched CO_2 concentration) and

Fig. 7.1 A free air CO_2 enrichment (FACE) experiment with wheat at Braunschweig, Germany. (Courtesy of Remy Manderscheid copyright)

to the heterogeneous nature of a crop. While observation and experimentation are the foundation for predicting the effect of climate change on crop production, they are insufficient. It is only possible to observe or test a few environments from a much larger number of environments of interest. Tools are then needed to generalize or extrapolate to other situations from those observed. Mathematical dynamic crop simulation models provide those tools.

7.2.3 Statistical/Empirical Models

These models are based on historic weather and yield data that are analyzed to evaluate the separate and interacting effects of temperature and water (growing season mean and select extreme event thresholds), sometimes incorporating additional factors to represent increasing CO_2 concentration (Urban et al., 2015). These models have the advantage of being directly based on observations, but are limited to conditions within the range that have been observed (Lobell & Asseng, 2017). Most statistical models do not include responsive management or genetics (Chenu et al., 2017).

7.2.4 Process-Based Crop Models

Process-based crop models are based on our understanding of the biological and physical processes that determine crop production (Chenu et al., 2017). They can be used to simulate how changes in temperature and rainfall regimes, as well as CO_2 concentration or other forcing variables, might affect crop production and yield. These models are based on biophysical equations and a set of parameters used to configure processes not directly resolved. Their accuracy is heavily dependent on proper configuration to local conditions, including soils, management, and cultivars (Asseng et al., 2013). Experimentation as described above is important for developing and testing such models (Asseng et al., 2015; Maiorano et al., 2017). Being based on biological and physical processes, it is expected that crop models should be able to extrapolate to conditions outside those observed (Iizumi et al., 2017; Ruane et al., 2017). Process-based models are the major method for evaluating the impact of climate change, so the focus here is on such models.

7.3 Model Uncertainty

Different models, faced with the same prediction problem, in general obtain different results. The variability between models is a measure of model uncertainty.

7.3.1 Model Structure Uncertainty

Soil–crop–atmosphere systems are very complex, and process-based mathematical models can only be, and are only meant to be, simplifications of the true systems. It is not surprising then that there are multiple crop model structures (different sets of equations) for the same crop species, which have been developed to focus on various aspects of the system or ranges of environments. For example, Liu et al. (2019) compared more than 30 models for wheat, and Kumudini et al. (2014) compared eight different thermal functions that have been used to model maize phenology.

The usual way of estimating model structure uncertainty is from ensemble studies, where multiple models are all provided with the same input variables and the same data for calibration. The variability among simulated outputs is then an indication of the uncertainty that results from uncertainty in model structure. It is only fairly recently that crop modelers have begun to do such studies. A major impetus was the launching of the Agricultural Model Intercomparison and Improvement Project (AgMIP, Rosenzweig et al., 2013) that led to numerous multimodel studies of the effect of climate change, including the effect of increased temperature or increased temperature with CO_2 concentration on yield or grain protein content of wheat (Asseng et al., 2015, 2019), yield of corn (Basso et al., 2018), yield of rice (Hasegawa et al., 2017), and potatoes (Fleisher et al., 2017). Other multimodel studies have targeted the effect of extreme weather events (high temperatures over a limited time in particular) (Liu et al., 2016b; Troy et al., 2015), the interaction of high temperatures with increased CO_2 (Thomey et al., 2019), and the effect of ozone, which is the most damaging air pollutant for crops (Choquette et al., 2019; Guarin et al., 2019). All these studies show that the variability between models is large. For example, Asseng et al. (2015) showed that simulated yields of wheat can vary by over two tons per hectare between models.

The implicit assumption when estimating uncertainty from multimodel studies is that the models that participate are a random sample from some theoretical distribution of "plausible" models. But this is a simplification. In most cases, the collections are ensembles of opportunity. They are not specifically designed to cover the range of plausible models, but rather simply represent the modeling groups that volunteered to participate in the study. This may lead to an exaggerated estimate of variability if the ensemble includes models that are not adapted to the study in question and so give unreasonable results. On the other hand, several related models may participate in the study, leading to an underestimation of variability. In consequence, the estimates of model structure uncertainty based on multimodel ensembles are themselves somewhat uncertain due to the design of the intercomparison experiment.

We have simplified the treatment of uncertainty by regarding model structure uncertainty as a single source of uncertainty, but one could go to a finer level of granularity and examine uncertainty related to specific processes in crop models. For example, Alderman and Stanfill (2017) focus just on the uncertainty due to the fact that several different functions are possible for modeling temperature effect on wheat phenology.

There are two main pathways toward reducing uncertainty due to model structure. The first is to use as the simulated value not the value simulated by an individual model, but rather an ensemble value, often the mean or median of the values simulated by the models in an ensemble. It is well known in statistics that the variance of the mean of a sample is the variance of the population, divided by the sample size. In exactly the same way, the variance of the mean of multiple simulated values is the variance in the population divided by the size of the ensemble, assuming that the models in the ensemble are a random sample of plausible models (Wallach et al., 2018; Martre et al., 2015). The larger the ensemble size then, the smaller the model structure uncertainty when predicting with the mean of simulated values. The second path to reducing model uncertainty is through shared model improvement. For example, Maiorano et al. (2017) found that improving the temperature response functions decreased the 10th–90th percentile range of simulated grain yields on average by 26% in an independent evaluation dataset (Fig. 7.2).

7.3.2 Model Parameter Uncertainty

Crop models usually contain a large number of parameters. In general, most of these parameters are considered fixed, and only a relatively small number need to be estimated by calibration, that is, by fitting the model to a set of field data. There are uncertainties in both categories of parameters, but attention is usually focused on the uncertainty in calibrated parameters.

Estimating uncertainty of estimated parameters is a major topic in statistics, and standard statistical approaches have been applied to crop models. Often the amount of data is insufficient to unequivocally determine the parameter values, leading to equifinality, meaning that various combinations of parameter values give equally good fits to the data (Bevin & Freer, 2008). However, even in the absence of exact equifinality, calibrated parameter values are uncertain because only a finite amount of data are ever available for calibration. It is generally found that parameter uncertainty is substantially smaller than structure uncertainty (e.g., Tao et al., 2018). In addition to parameter uncertainty due to limited data, simulated results can also depend on the calibration method. Parameter uncertainty is the result of uncertainties in multiple parameters. It can also be of interest to look specifically at uncertainty due to certain parameters or groups of parameters.

Parameter uncertainty can be reduced by increasing the amount and/or improving the quality of the calibration data. It can also be reduced by shared improvements in the calibration approach. The way forward is better data availability, in particular through more effective data sharing, and through guidelines for calibration specifically adapted to mechanistic models like crop models.

7.3.3 Model Input Uncertainty

7.3.3.1 Weather and Climate

Models that project climate impacts on crop production are driven by a set of weather or climate information that captures the changing field environment under specified scenarios. Uncertainties related to the observation and projection of regional climate changes are well covered by other chapters in this volume, but here we concentrate on the elements of climate uncertainty represented within the selected weather or climate-forcing dataset. The gold standard for climate datasets used in crop modeling remains comprehensive, precise, and long-term meteorological observations from the field of interest. As high-quality observations are not available in the vast majority of agricultural impact assessments and future conditions cannot be observed, uncertainties are introduced when climate data are drawn from observational products and future climate model simulations. These uncertainties may result in biases critical for agricultural impacts assessment, including those related to the mean climate, the distribution of extremes (e.g., intensity, duration, frequency, and timing), and relational aspects (e.g., associations between weather variables).

Historical climate data can be obtained from nearby stations, remote sensing products, retrospective analyses, or hybrid products that combine these into climatic forcing datasets suited for agricultural model application. Ruane et al., 2021 analyzed the implications of historical climate data uncertainty for an ensemble of global gridded crop models driven with 11 climatic forcing datasets, finding large uncertainties related to growing season temperature and (especially) precipitation biases in regions with sparse meteorological station networks (particularly in tropical countries) and complex topography (where agricultural lands are likely not well represented by coarse datasets). Biases were also larger for distributional aspects such as the number of heavy precipitation days and extreme heat days.

Future climate forcing for crop models needs to be identified according to a specific emission scenario, climate model, downscaling technique (or lack thereof), and bias adjustment (or lack thereof); each of which introduces uncertainties as discussed elsewhere in this volume and summarized here.

7 Uncertainty of Climate Change Impacts on Crop Production

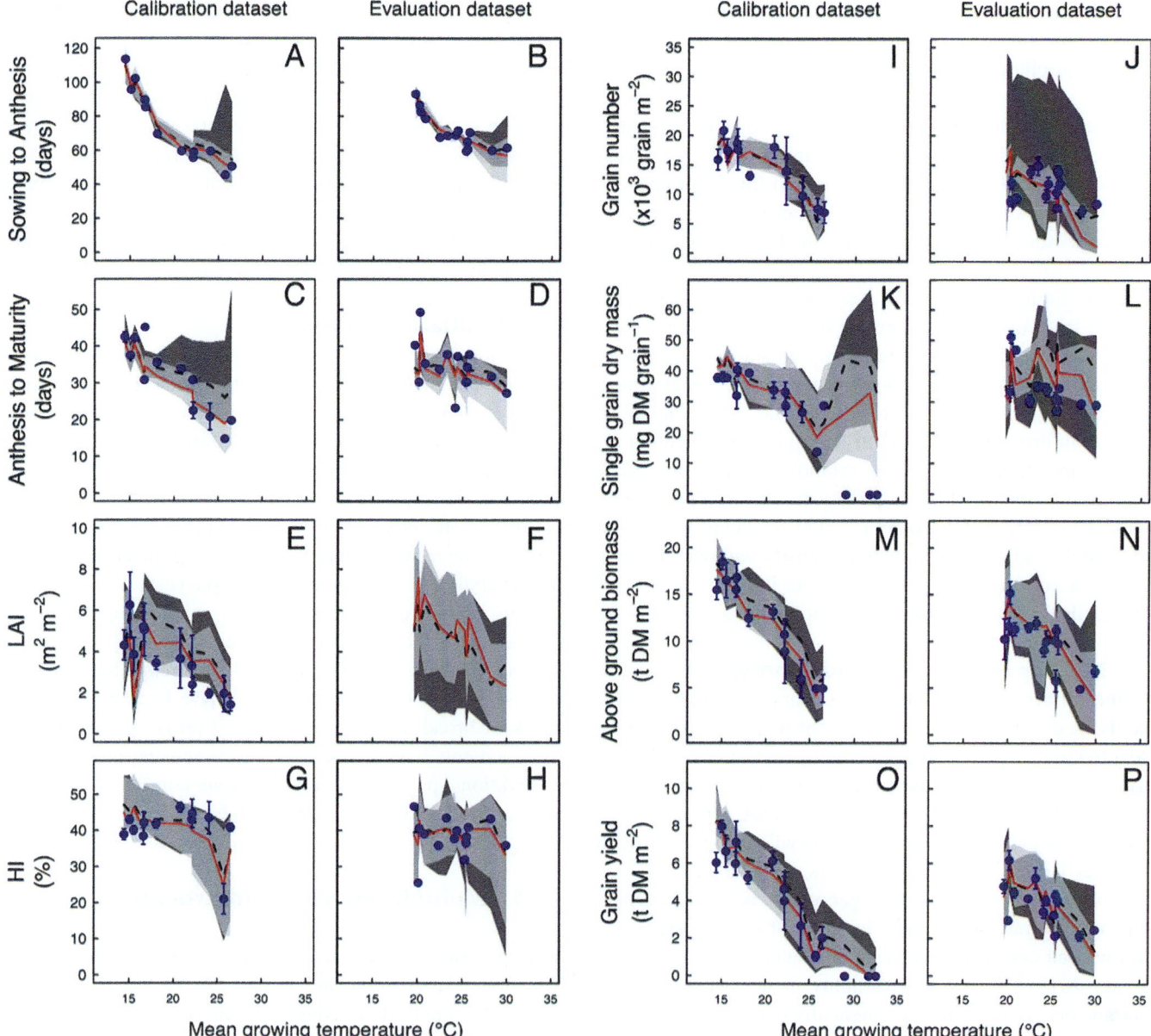

Fig. 7.2 Simulated and measured days from sowing to anthesis (**a, b**), days from anthesis to maturity (**c, d**), leaf area index (LAI) (**e, f**), harvest index (HI) (**g, h**), grain number (**i, j**), single grain dry mass (**k, l**), final total aboveground biomass (M and N), final grain yield (**o, p**), versus mean growing season temperature for the calibration (**a, c, e, g, i, k, m, o**) and evaluation (**b, d, f, h, j, l, n, p**) datasets. Black dotted lines and dark grey areas are e-median (ensemble median) and the 10th–90th percentile range of the 15 original (unimproved) models, respectively. Solid red lines and light gray areas are e-median and the 10th–90th percentile range of the 15 improved models, respectively. Symbols are measured mean ± 1 s.d. for $n = 3$ independent replicates. (Source: Maiorano et al., 2017)

First, uncertainty in future scenarios reflects differences in radiative forcing associated with aerosol and greenhouse gas emissions represented by nonprobabilistic shared socioeconomic pathways and representative concentration pathways (SSP-RCPs). These may take the form of an SSP-RCP combination and time slice, such as the SSP370 mid-century (2040–2069) period, or a specific global warming level determined by global mean surface air temperature compared to pre-industrial conditions (e.g., the +2.0 °C world). Second, for a given scenario there are multiple general circulation models (GCMs), which represent different possible predictions of future climate for a given pathway (Eyring et al., 2016). Agricultural applications may pay particular attention to uncertainties introduced by the range of models' overall equilibrium climate sensitivities and their regional pattern of climatic changes. Additionally, the sequence of extreme events could be highly dependent on internal climate variability that may be further understood using large ensembles of a single climate model. Third, the resolution of GCMs is typically hundreds of kilometers, requiring downscaling to capture climatic phenomena and local geographies important to agricultural applications. Finally, discrepancies introduced by these climate simulations may still be further reduced through bias adjustment to more decision-relevant scales (Galmarini et al., 2019), although this can introduce methodological uncertainties and further uncertainties from the observational datasets that form the target of bias adjustment.

Both future and historic agroclimatic impact analyses benefit from the use of ensemble approaches that are clear about sources of climatic uncertainty. In some cases, use of multiple models or methods may not reduce uncertainties as methods used to simulate, downscale, and bias-adjust climate information each has structural components that affect the mean, distribution, and relational aspects of climate forcing information that can propagate into crop production impact uncertainty. For example, a statistical bias-adjustment method that adjusts precipitation intensity without adjusting precipitation frequency can miss shifts in the frequency of dry spells that are important to irrigation management.

7.3.3.2 Crop Management and Genotype

In the future, cultivars adapted to changed conditions will likely be developed, and management will also probably be adapted to those cultivars and future conditions. Simulations for future climate often use current management, which may lead to unrealistically low simulated yields. However, predicting how cultivars and management will change is difficult and has large uncertainty (Rosenzweig et al., 2013). One approach to predicting future genotypes is via ideotyping where crop traits are defined in terms of model parameters to find trait combinations adapted to future climate (Asseng et al., 2019). However, uncertainty arises because model parameters are not directly genetically determined, so modifying parameters does not accurately represent the range of possible genotypes (Chenu et al., 2017). Furthermore, even accepting that parameter combinations represent possible genotypes, the possible ranges and correlations between parameters are uncertain.

Adaptation via altered crop management is another option. However, in practice, crop management depends in complex ways on farmer situation and the socioeconomic environment influencing decision-making on a farm. One approach is to define scenarios for future socioeconomic conditions relevant to farming, called representative agricultural pathways (RAPs) (Antle et al., 2017a). As with RCPs, these are possible scenarios rather than probabilistic forecasts. One approach is to search for crop management that maximizes yield under future conditions, but this is of course subject to uncertainties. There could also be disruptive changes, such as vertical farming (Asseng et al., 2020), robotic management (Asseng & Asche, 2019), or massive genetic engineering affecting future agriculture. One approach to better understand uncertainty due to crop management is to explore this with the current range of crop management practices, but this has not been done yet. More information on the effect of crop management on uncertainty is needed. Crop modelers also need cultivar parameters that represent the range of current and possible future genotypes grown in the field. Better understanding of genetics and new approaches in gene-based crop models could be ways forward (Chenu et al., 2017).

7.3.3.3 Initial Conditions

Simulations of cropping systems depend on initial soil water, N, and C pools. While "typical" initial values are often used and reset each year, a carryover effect, which is more realistic, from year to year can result in very different simulated climate change impacts on cropping system outcomes (Basso et al., 2018). Hence, there is uncertainty due to initial conditions and due to the practice of resetting initial conditions each year. To overcome the initial value setting uncertainty, crop models could be executed across multiple years without reinitializing (Basso et al., 2018). However, this might require simulation of other crops in a crop rotation or a fallow period, each adding new uncertainties to the exercise. Simulations of these initial effects at large scale have been hindered by a lack of broad measurements of soil properties. However, improvements in soil moisture remote sensing provide new observational perspectives on field water conditions throughout the season (Mladenova et al., 2020).

7.3.4 Multiple Sources of Uncertainty

It is important to study uncertainties using multiple model structures, parameters, and inputs (Wallach & Thorburn, 2017). This can be done using super ensembles. For example, simulations could be performed using multiple crop models, with weather data generated by multiple GCMs (Liu et al., 2019), or using multiple crop models, each with multiple parameter values generated from the distribution of parameter values for that model (Alderman & Stanfill, 2017), or using multiple crop models with multiple GCMs and multiple parameter values (Tao et al., 2018). The variability in simulated values would represent uncertainty due to structure and weather uncertainty, or structure and parameter uncertainty, or structure, weather, and parameter uncertainty, respectively. Wallach et al. (2016a, b) showed how to decompose the overall variance from such ensemble studies into separate contributions from each source of uncertainty. They showed that the combined uncertainty from structure and inputs is not simply a sum of uncertainties due to each of those sources but rather the interaction between them. There have to date been relatively few studies of multiple sources of uncertainty, so it is difficult to generalize about the results, but there is some indication that structure uncertainty is larger than parameter uncertainty (Zhang et al., 2017). Note that multiple uncertainty studies are also of interest in other sectors such as water (Clark et al., 2016).

7.3.5 Upscaling Uncertainty

Crop models simulate a single homogeneous field, but often the interest is in regional or global production (Ruane et al., 2017). Crop model application at a regional or global scale requires extrapolation methods, including aggregation and upscaling, each introducing additional uncertainty (Ewert et al., 2015). One common way of obtaining regional yields is to divide the region into uniform grid cells, to simulate for a single point in each cell, and then to aggregate the results. The choice of resolution is a problem. Higher resolution (smaller grid cells) reduces the variability within the cell, and therefore the uncertainty due to choosing a single point, but requires input data from more locations, which may be difficult to obtain. An alternative is to simulate for a limited number of representative points in the region, chosen because the necessary input data are available for those points. The problem with this upscaling approach is in how to weigh the individual points to get the regional value. It has been found that aggregation of grid cells and upscaling representative points give similar results with a similar range of uncertainties (Liu et al., 2016b; Zhao et al., 2017, Fig. 7.3). In addition, there is also uncertainty in future land use and future irrigation for individual crops (Yu et al., 2019) and how, for instance, bioenergy and future land conservation might affect cropland.

Improvement in both aggregation and upscaling would result from having more data. For aggregation, one requires more accurate model inputs, including soil characteristics, initial soil conditions, and crop management, for each grid cell (Ray et al., 2015). For upscaling, one requires detailed data for more representative points.

7.4 Prediction Uncertainty

Given a simulated value, true crop response is uncertain because models are not perfect predictors. Prediction uncertainty is evaluated by comparing simulated with observed values. A common summary value for prediction uncertainty is mean squared error of prediction (MSEP).

Prediction uncertainty is clearly different from model uncertainty. For example, given simulations with multiple model structures, one can have close agreement between models (small model structure uncertainty), but all the models could be wrong (large prediction uncertainty). On the other hand, the two types of uncertainty are not totally disconnected. If there are large differences between models, then at least some of the models must have large prediction errors. Thus, large model uncertainty is necessarily associated with large MSEP, when averaged across models (Wallach et al., 2016b).

Multimodel studies have indeed found that errors vary between models and can be quite large (Martre et al., 2015; Wallach et al., 2018). The variability between models might suggest that a way to limit prediction error is to choose the best predicting model. However, studies have shown that the ranking of models with respect to model error is not stable, but depends on the specific output of interest and on the environment simulated (Martre et al., 2015; Wallach et al., 2018).

It has been observed empirically that the mean or median of simulated values from an ensemble of models tend to have smaller errors than many or even all of the individual models in the ensemble (Asseng et al., 2013). The reason is that at least some of the model errors cancel out when averaging over models. In fact, it can be proven that the MSEP of an ensemble mean must be as small or smaller than the average MSEP of the individual models (Wallach et al., 2018). That is, prediction uncertainty is reduced by averaging over an ensemble of crop models compared to the average prediction uncertainty of the models in the ensemble. This has led, in practice, to the use of the ensemble median (often found to be slightly better than the mean) to draw conclusions about the effect of climate change on crop production (Liu et al., 2019). The path to further reduction in prediction error is through model improvement, that is, a more accurate description of the way weather, soil, and management impact crop growth and development.

7.5 The Road Ahead to Reduce and Better Characterize Uncertainty

A key requirement for reducing uncertainty in climate change impact assessment is to better characterize climate change at the local level, which is in the domain of climate models. Another major uncertainty concerns the adaptation of crop cultivars and cropping systems to climate change. Cultivar characteristics and crop management are drivers of crop models, but crop models are also a major tool for projecting their evolution. Further studies on the use of crop models here are necessary. Crop models are generally designed for projections of a single homogeneous field, while impact studies are more concerned with regional and global scales. Upscaling crop models is a major source of uncertainty (Ewert et al., 2015), which could be reduced by improving global databases for the inputs required by crop models.

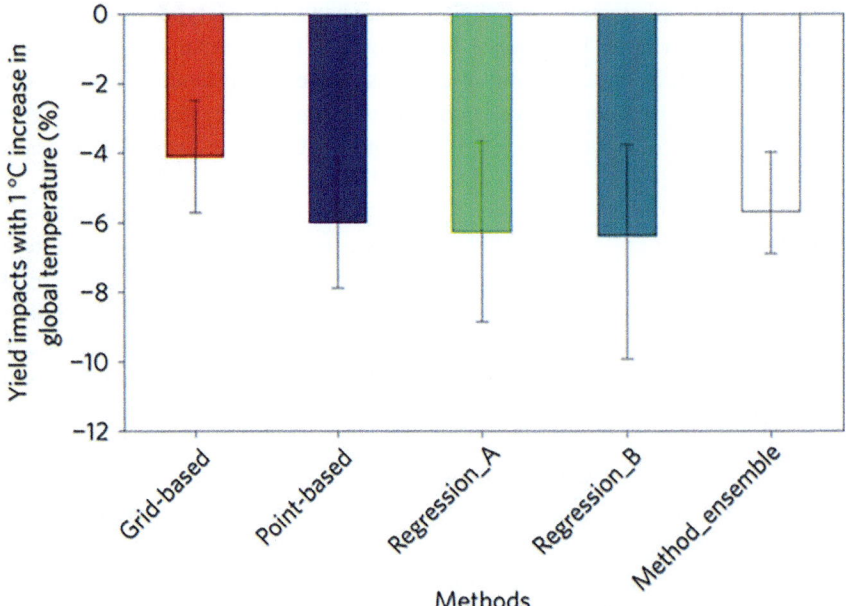

Fig. 7.3 Impacts of 1 °C global temperature increase on global wheat yield estimated by different assessment methods. The grid-based (0.5° × 0.5° grid cells) method is an ensemble median from seven global gridded crop models, averaged over 30 years and aggregated over all simulated grid cells. The point-based method is an ensemble median from 30 models, averaged over 30 years and aggregated over 30 global locations. Regression_A is based on a country-level statistical regression. Regression_B is based on a global-level statistical regression. The error bars for the four different methods indicate the 95% confidence intervals based on multimodel ensembles in the simulations and bootstrap resampling in the statistical regressions. **The mean of the method_ensemble is shown with error bars indicating the 95% confidence intervals based on medians of individual methods.** (**Source:**Liu et al.,2016a, 2016b)

Increasing the data for crop calibration and testing would be a major way to better characterize and then reduce uncertainty. To some extent, this is a problem of making existing data more readily available to all modeling groups, which would require a new paradigm for data collection and sharing (Antle et al., 2017b). Additional measurements of initial soil conditions and in-season crop and soil dynamics are necessary to understand crop systems dynamics, better characterize the uncertainty in crop model simulations, and reduce this uncertainty through model improvement. Of particular interest is experimentation that mimics climate change, including increased CO_2 with respect to ambient, increased temperature with periods of extreme temperatures, and interactions of CO_2 and temperature with crop cultivar, crop nutrition, water availability, pests and diseases, and other factors.

Predictors based on multimodel ensembles can reduce structure and prediction uncertainty, and shared improvement can further reduce prediction uncertainty. Collaboration between modeling groups is therefore essential. Improved modeling techniques, including better use of data and interoperable model components, in next-generation crop models (Antle et al., 2017b) will help make collaboration more efficient. Cross-fertilization between disciplines is also important, for instance, crop modeling has much to learn from work on multimodel ensembles in climatology (Wallach et al., 2016a).

Uncertainty of climate change impacts on crop production is a vast field, and our vision of uncertainty is based on numerous individual studies, which are necessarily limited by the range of situations studied. To facilitate the generalization of this research, it is important to develop standardized procedures for evaluating and reporting uncertainty. Uncertainty in uncertainty estimates, though a second-order effect, also requires more attention. While it will not be possible to eliminate uncertainty, reducing uncertainty and providing improved estimates of uncertainty are possible and will enable more informed responses to the challenge of climate change.

References

Alderman, P. D., & Stanfill, B. (2017). Quantifying model-structure- and parameter-driven uncertainties in spring wheat phenology prediction with Bayesian analysis. *European Journal of Agronomy, 88,* 1–9. https://doi.org/10.1016/j.eja.2016.09.016

Antle, J. M., et al. (2017a). Design and use of representative agricultural pathways for integrated assessment of climate change in U.S. Pacific Northwest Cereal-Based Systems. *Frontiers in Ecology and Evolution, 5,* 99. https://doi.org/10.3389/fevo.2017.00099

Antle, J. M., Jones, J. W., & Rosenzweig, C. E. (2017b). Next generation agricultural system data, models and knowledge products: Introduction. *Agricultural Systems, 155,* 186–190. https://doi.org/10.1016/j.agsy.2016.09.003

Asseng, S., & Asche, F. (2019). Future farms without farmers. *Science robotics, 4*, eaaw1875. https://doi.org/10.1126/scirobotics.aaw1875

Asseng, S., et al. (2013). Uncertainty in simulating wheat yields under climate change. *Nature Climate Change, 3*, 827–832. https://doi.org/10.1038/nclimate1916

Asseng, S., et al. (2015). Rising temperatures reduce global wheat production. *Nature Climate Change, 5*, 143–147. https://doi.org/10.1038/nclimate2470

Asseng, S., et al. (2019). Climate change impact and adaptation for wheat protein. *Global Change Biology, 25*, 155–173. https://doi.org/10.1111/gcb.14481

Asseng, S., Guarin, J. R., Raman, M., Monje, O., Kiss, G., Despommier, D. D., Meggers, F. M., & Gauthier, P. P. G. (2020). Wheat yield potential in controlled-environment vertical farms. *Proceedings of the National Academy of Sciences, 117*, 19131–19135. https://doi.org/10.1073/PNAS.2002655117

Basso, B., et al. (2018). Soil organic carbon and nitrogen feedbacks on crop yields under climate change. *Agricultural & Environmental Letters, 3*, 180026. https://doi.org/10.2134/ael2018.05.0026

Bevin, K., & Freer, J. (2008). Equifinality, data assimilation, and uncertainty estimation in mechanistic modelling of complex environmental systems using the GLUE methodology. *Journal of Hydrology, 249*, 11–29.

Bos, S. P. M., Pagella, T., Kindt, R., Russell, A. J. M., & Luedeling, E. (2015). Climate analogs for agricultural impact projection and adaptation—A reliability test. *Frontiers in Environmental Science, 3*, 65. https://doi.org/10.3389/fenvs.2015.00065

Chenu, K., Porter, J. R., Martre, P., Basso, B., Chapman, S. C., Ewert, F., Bindi, M., & Asseng, S. (2017). Contribution of crop models to adaptation in wheat. *Trends in Plant Science, 22*, 472–490. https://doi.org/10.1016/j.tplants.2017.02.003

Choquette, N. E., et al. (2019). Uncovering hidden genetic variation in photosynthesis of field-grown maize under ozone pollution. *Global Change Biology, gcb.14794*. https://doi.org/10.1111/gcb.14794

Clark, M. P., et al. (2016). Characterizing uncertainty of the hydrologic impacts of climate change. *Current Climate Change Reports, 2*, 55–64. https://doi.org/10.1007/s40641-016-0034-x

Ewert, F., & Porter, J. R. (2000). Ozone effects on wheat in relation to CO_2: modelling short-term and long-term responses of leaf photosynthesis and leaf duration. *Global Change Biology, 6*, 735–750. https://doi.org/10.1046/j.1365-2486.2000.00351.x

Ewert, F., et al. (2015). Uncertainties in scaling-up crop models for large-area climate change impact assessments. In *Handbook of climate change and agroecosystems* (pp. 261–277). https://doi.org/10.1142/9781783265640_0010

Eyring, V., Bony, S., Meehl, G. A., Senior, C. A., Stevens, B., Stouffer, R. J., & Taylor, K. E. (2016). Overview of the coupled model intercomparison project phase 6 (CMIP6) experimental design and organization. *Geoscientific Model Development, 9*, 1937–1958. https://doi.org/10.5194/gmd-9-1937-2016

Fleisher, D. H., et al. (2017). A potato model intercomparison across varying climates and productivity levels. *Global Change Biology, 23*, 1258–1281. https://doi.org/10.1111/gcb.13411

Galmarini, S., et al. (2019). Adjusting climate model bias for agricultural impact assessment: How to cut the mustard. *Climate Services, 13*. https://doi.org/10.1016/J.CLISER.2019.01.004

Guarin, J. R., Emberson, L., Simpson, D., Hernandez-Ochoa, I. M., Rowland, D., & Asseng, S. (2019). Impacts of tropospheric ozone and climate change on Mexico wheat production. *Climatic Change, 155*, 157–174. https://doi.org/10.1007/s10584-019-02451-4

Hasegawa, T., et al. (2017). Causes of variation among rice models in yield response to CO_2 examined with free-air CO_2 enrichment and growth chamber experiments. *Scientific Reports, 7*. https://doi.org/10.1038/s41598-017-13582-y

Iizumi, T., Furuya, J., Shen, Z., Kim, W., Okada, M., Fujimori, S., Hasegawa, T., & Nishimori, M. (2017). Responses of crop yield growth to global temperature and socioeconomic changes. *Scientific Reports, 7*, 7800. https://doi.org/10.1038/s41598-017-08214-4

IPCC. (2014). Climate change 2014: Impacts, adaptation, and vulnerability. Part A: Global and sectoral aspects. In C. B. Field, V. R. Barros, D. J. Dokken, & K.J (Eds.), *Contribution of working group II to the fifth assessment report of the intergovernmental panel on climate change* (p. 1132). Cambridge University Press.

Kimball, B., et al. (1995). Productivity and water use of wheat under free-air CO2 enrichment. *Global Change Biology, 1*, 429–442. https://doi.org/10.1111/j.1365-2486.1995.tb00041.x

Kumudini, S., et al. (2014). Predicting maize phenology: Intercomparison of functions for developmental response to temperature. *Agronomy Journal, 106*, 2087–2097. https://doi.org/10.2134/agronj14.0200

Liu, B., et al. (2016a). Similar estimates of temperature impacts on global wheat yield by three independent methods. *Nature Climate Change, 6*, 1130–1136. https://doi.org/10.1038/nclimate3115

Liu, B., Asseng, S., Liu, L., Tang, L., Cao, W., & Zhu, Y. (2016b). Testing the responses of four wheat crop models to heat stress at anthesis and grain filling. *Global Change Biology, 22*, 1890–1903. https://doi.org/10.1111/gcb.13212

Liu, B., et al. (2019). Global wheat production with 1.5 and 2.0°C above pre-industrial warming. *Global Change Biology, 25*, 1428–1444. https://doi.org/10.1111/gcb.14542

Lobell, D. B., & Asseng, S. (2017). Comparing estimates of climate change impacts from process-based and statistical crop models. *Environmental Research Letters, 12*, 015001. https://doi.org/10.1088/1748-9326/aa518a

Maiorano, A., et al. (2017). Crop model improvement reduces the uncertainty of the response to temperature of multi-model ensembles. *Field Crop Research, 202*. https://doi.org/10.1016/j.fcr.2016.05.001

Martre, P., et al. (2015). Multimodel ensembles of wheat growth: Many models are better than one. *Global Change Biology, 21*, 911–925. https://doi.org/10.1111/gcb.12768

Mladenova, I. E., Bolten, J. D., Crow, W., Sazib, N., & Reynolds, C. (2020). Agricultural drought monitoring via the assimilation of SMAP soil moisture retrievals into a global soil water balance model. *Frontiers in Big Data, 3*. https://doi.org/10.3389/fdata.2020.00010

Porter, J. R., Xie, L., Challinor, A. J., Cochrane, K., Howden, S. M., Iqbal, M. M., Lobell, D. B., & Travasso, M. I. (2014). Food security and food production systems. In C. B. Field et al. (Eds.), *Climate change 2014: Impacts, adaptation, and vulnerability. Part A: Global and sectoral aspects. Contribution of working group II to the fifth assessment report of the intergovernmental panel on climate change* (pp. 485–533). Cambridge University Press.

Ray, D. K., Gerber, J. S., Macdonald, G. K., & West, P. C. (2015). Climate variation explains a third of global crop yield variability. *Nature Communications, 6*. https://doi.org/10.1038/ncomms6989

Rosenzweig, C., et al. (2013). The agricultural model Intercomparison and improvement project (AgMIP): Protocols and pilot studies. *Agricultural and Forest Meteorology, 170*. https://doi.org/10.1016/j.agrformet.2012.09.011

Rosenzweig, C., et al. (2020). Climate change responses benefit from a global food system approach. *Nature Food, 1*, 94–97. https://doi.org/10.1038/s43016-020-0031-z

Ruane, A. C., et al. (2017). An AgMIP framework for improved agricultural representation in integrated assessment models. *Environmental Research Letters, 12*, 125003. https://doi.org/10.1088/1748-9326/aa8da6

Ruane, A. C., et al. (2021). Strong regional influence of climatic forcing datasets on global crop model ensembles. *Agricultural and Forest Meteorology.*, in press. https://doi.org/10.1016/j.agrformet.2020.108313

Soussana, J.-F., Graux, A.-I., & Tubiello, F. N. (2010). Improving the use of modelling for projections of climate change impacts on crops and pastures. *Journal of Experimental Botany, 61*, 2217–2228. https://doi.org/10.1093/jxb/erq100

Tao, F., et al. (2018). Contribution of crop model structure, parameters and climate projections to uncertainty in climate change impact assessments. *Global Change Biology, 24*, 1291–1307. https://doi.org/10.1111/gcb.14019

Thomey, M. L., Slattery, R. A., Köhler, I. H., Bernacchi, C. J., & Ort, D. R. (2019). Yield response of field-grown soybean exposed to heat waves under current and elevated [CO_2]. *Global Change Biology, gcb.14796*. https://doi.org/10.1111/gcb.14796

Troy, T. J., Kipgen, C., & Pal, I. (2015). The impact of climate extremes and irrigation on US crop yields. *Environmental Research Letters, 10*, 054013. https://doi.org/10.1088/1748-9326/10/5/054013

Urban, D. W., Sheffield, J., & Lobell, D. B. (2015). The impacts of future climate and carbon dioxide changes on the average and variability of US maize yields under two emission scenarios. *Environmental Research Letters, 10*, 045003. https://doi.org/10.1088/1748-9326/10/4/045003

Valdivia, R. O., et al. (2015). Representative agricultural pathways and scenarios for regional integrated assessment of climate change impacts, vulnerability, and adaptation. In Handbook of climate change and agroecosystems (pp.101–145). https://doi.org/10.1142/9781783265640_0005

Wallach, D., & Thorburn, P. J. (2017). Estimating uncertainty in crop model predictions: Current situation and future prospects. *European Journal of Agronomy, 88*. https://doi.org/10.1016/j.eja.2017.06.001

Wallach, D., Mearns, L. O., Ruane, A. C., Rötter, R. P., & Asseng, S. (2016a). Lessons from climate modeling on the design and use of ensembles for crop modeling. *Climatic Change, 139*, 551–564. https://doi.org/10.1007/s10584-016-1803-1

Wallach, D., Thorburn, P., Asseng, S., Challinor, A. J., Ewert, F., Jones, J. W., Rotter, R., & Ruane, A. (2016b). Estimating model prediction error: Should you treat predictions as fixed or random? *Environment Model Software, 84*, 529–539. https://doi.org/10.1016/j.envsoft.2016.07.010

Wallach, D., et al. (2018). Multimodel ensembles improve predictions of crop-environment-management interactions. *Global Change Biology, 24*, 5072–5083. https://doi.org/10.1111/gcb.14411

Yu, Z., Lu, C., Tian, H., & Canadell, J. G. (2019). Largely underestimated carbon emission from land use and land cover change in the conterminous United States. *Global Change Biology, gcb.14768*. https://doi.org/10.1111/gcb.14768

Zhang, S., Tao, F., & Zhang, Z. (2017). Uncertainty from model structure is larger than that from model parameters in simulating rice phenology in China. *European Journal of Agronomy, 87*, 30–39. https://doi.org/10.1016/j.eja.2017.04.004

Zhao, C., et al. (2017). Temperature increase reduces global yields of major crops in four independent estimates. *Proceedings of the National Academy of Sciences of the United States of America, 114*. https://doi.org/10.1073/pnas.1701762114

Open Access This chapter is licensed under the terms of the Creative Commons Attribution 4.0 International License (http://creativecommons.org/licenses/by/4.0/), which permits use, sharing, adaptation, distribution and reproduction in any medium or format, as long as you give appropriate credit to the original author(s) and the source, provide a link to the Creative Commons license and indicate if changes were made.

The images or other third party material in this chapter are included in the chapter's Creative Commons license, unless indicated otherwise in a credit line to the material. If material is not included in the chapter's Creative Commons license and your intended use is not permitted by statutory regulation or exceeds the permitted use, you will need to obtain permission directly from the copyright holder.

Uncertainty in Ecological Models

Dan L. Warren, Lukas Baumbach, Jamie M. Kass, and Alke Voskamp

8.1 Introduction

Unlike other effects of anthropogenic climate change that can potentially be ameliorated over long time scales, biodiversity loss is generally an irreversible process. Species driven to extinction stay extinct and communities altered by climate change may not recover even if anthropogenic disturbances are eliminated. Stakeholders require estimates of how localized climatic changes could impact species and communities so that appropriate steps can be taken to prevent losses (e.g., via changes to conservation threat assessments or land management).

There are many methods for making these estimates at scales ranging from individual species to entire biomes. Here, we focus on one of the most widely used approaches for estimating the effects of climate change on individual species: species distribution models (SDMs), sometimes referred to as ecological niche or habitat suitability models. These models correlate the presence of a species with a set of environmental variables to estimate the species' environmental tolerances. For a typical workflow, see Fig. 8.1.

The utility of SDMs is clear for decision-making, but they are subject to many sources of uncertainty and require careful implementation and interpretation. Time-sensitive decisions may depend on models calibrated with data that are only indirectly related to the phenomenon we are trying to estimate, and which cannot be fully validated. This is problematic when a misinformed decision may have severe financial and opportunity costs. Understanding uncertainties in these models and communicating them effectively is of paramount importance.

This chapter provides an overview of SDM uncertainties and how they interact (summarized in Fig. 8.2), as well as how best to address the issues these uncertainties create. As thousands of SDM manuscripts are published each year, we are only able to present a broad overview of issues that are each backed by their own deep literature. Uncertainties for SDMs are similar to those for other macroecological models, so our discussion should clarify specific issues while offering generalizable insights. Our review will follow a typical SDM workflow: from the collection of occurrence data and selection of environmental variables to model training and evaluation, leading to the generation of potential range maps and forecasting. SDMs are used in both aquatic and terrestrial environments, but for brevity we will concentrate on terrestrial examples in this review.

8.2 Presence Data

Superficially, occurrence data for SDMs are unambiguous: one (species detected in a location) or zero (species not detected). However, those data are products of multiple interacting processes, many of which we would prefer to avoid

D. L. Warren (✉)
Gulbali Institute, Charles Sturt University, Thurgoona, NSW, Australia

Biodiversity and Biocomplexity Unit, Okinawa Institute of Science and Technology, Tancha, Onna-son, Okinawa, Japan

Environmental Science and Informatics, Okinawa Institute of Science and Technology, Tancha, Onna-son, Okinawa, Japan

L. Baumbach
Forestry Economics and Forest Planning, University of Freiburg, Freiburg, Germany

J. M. Kass
Macroecology Laboratory, Graduate School of Life Sciences, Tohoku University, Sendai, Miyagi, Japan

Biodiversity and Biocomplexity Unit, Okinawa Institute of Science and Technology, Tancha, Onna-son, Okinawa, Japan

A. Voskamp
Senckenberg Biodiversity and Climate Research Centre, Frankfurt am Main, Germany

© The Author(s) 2025
L. O. Mearns et al. (eds.), *Uncertainty in Climate Change Research*,
https://doi.org/10.1007/978-3-031-85542-9_8

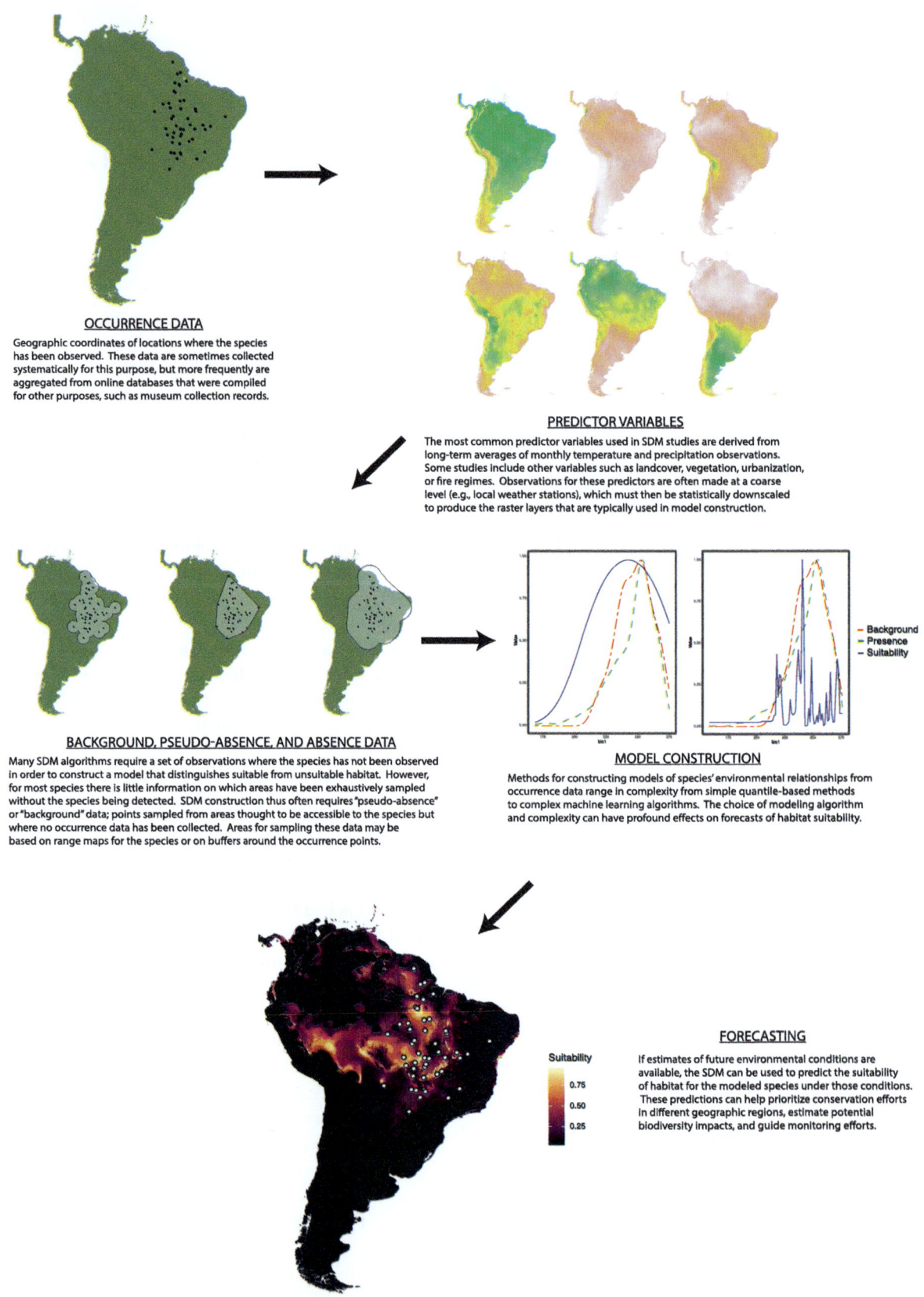

Fig. 8.1 Workflow of a typical SDM analysis. Modeling efforts typically begin with collection and curation of occurrence data (top left). Modelers then select a set of predictor variables (top right) that are thought to be important in limiting the species' distribution, either based on previous biological knowledge or predictive performance of preliminary models. For most modern methods, it is then necessary to select a study area from which to select background points to train the model (middle left). Once data for each predictor have been extracted from the occurrence and background points, modelers construct models to either generalize the environmental distribution of the species (climate envelope methods) or to distinguish environmental conditions at occurrence points from those at background/pseudoabsence points (correlative methods). In this figure (middle right), results of two models are shown for a single predictor.

8 Uncertainty in Ecological Models

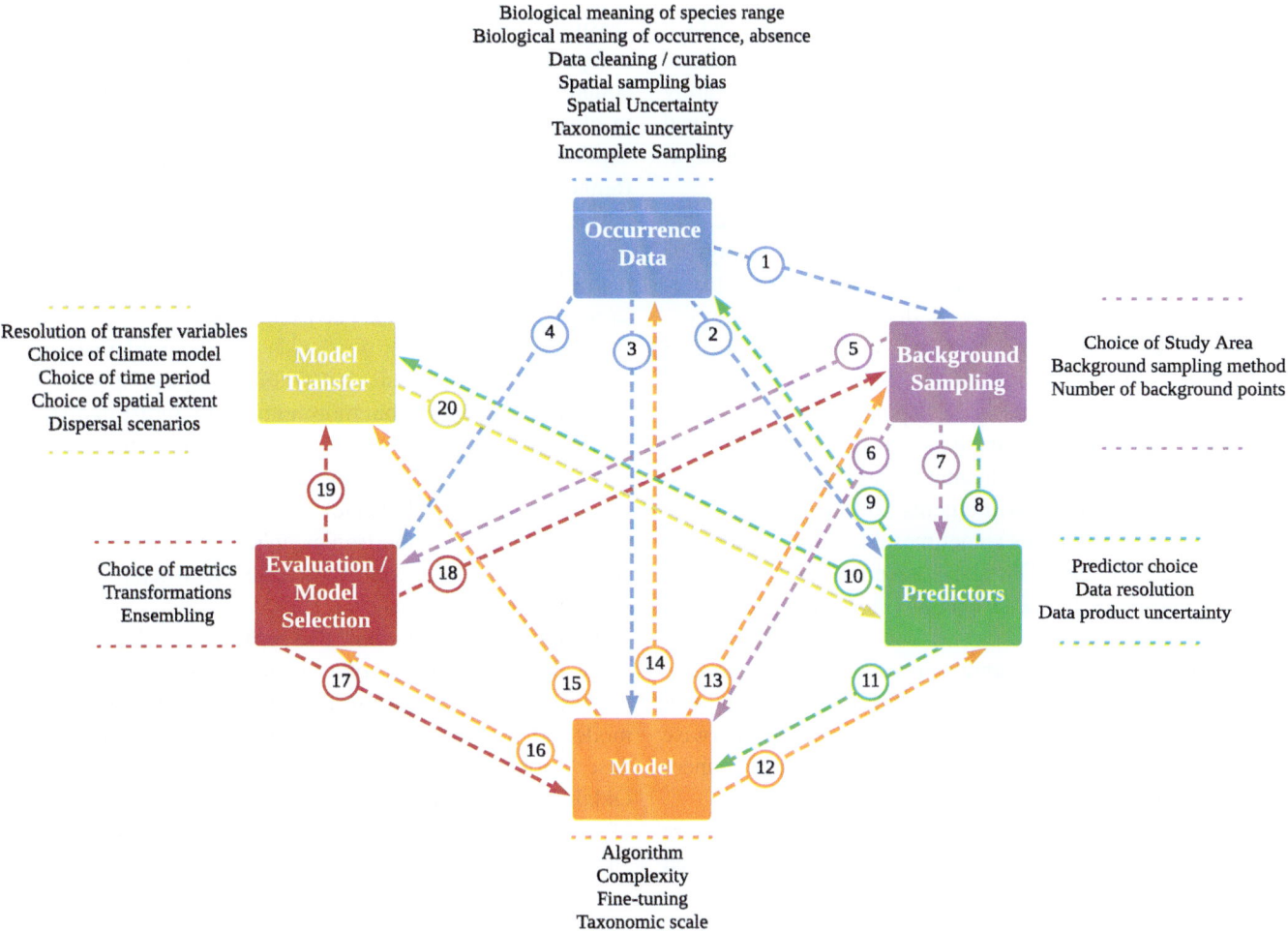

Fig. 8.2 Sources of uncertainty in SDM analyses. Proceeding clockwise from "Occurrence Data", boxes follow a simplified workflow of a typical SDM forecasting study. Sources of uncertainty that are intrinsic to each step of the modeling process are given in adjacent text. In addition to these, there are also many ways in which uncertainty at one step of the modeling process may propagate to other stages, shown here as dashed arrows. Arrows are directional and should be interpreted as "uncertainty at step A generates additional uncertainty at step B". As a result of the iterative nature of many SDM studies, these uncertainties can propagate in both directions. We give brief descriptions of some of the ways this may happen, but given the interconnected nature of these steps, this list should not be considered exhaustive. 1. Uncertainty in occurrence data generates uncertainty in background sampling by affecting the delineation of the study area from which background points are sampled. 2. As occurrence data are integral to a posteriori predictor selection, changes to them may affect predictors selected for modeling. 3. Occurrence data are crucial components of model construction, so uncertainties in occurrence data will propagate to model parameter estimates. 4. Models are typically evaluated and selected based on subsets of the occurrence data, so that uncertainty in occurrence data may result in uncertainty in model selection. 5. Model evaluation often focuses on how well models distinguish presence data from background data, so changes in the background data may affect model selection. In addition, many discrimination metrics are known to be heavily influenced by the size of the background region. 6. Background data are used to represent the

affecting our predictions. For any given data point, there may be profound uncertainties about which processes are driving that observation.

A species' spatial distribution will generally be affected by its environmental niche, but the pattern also reflects geographic barriers, biogeographic history, human land use, species interactions, and stochastic events. Even within this realized distribution, the interpretation of a given occurrence point may be complex. An observation of a species at a locality may be an individual within its home range, or perhaps an individual traveling between resources, midway through a

Fig. 8.1 (continued) Green (simple dashed) and red (compound dashed) lines show the prevalence of different values of the environmental variable at presence and background points, respectively, while solid blue lines indicate the predicted response of the species to that variable from each candidate model. Once models are constructed and one or more models have been selected, a forecast can be made by interpolating or extrapolating from that model onto a set of predicted future conditions for the environmental variables used to build the model (bottom)

seasonal migration, or present due to human-assisted translocation. In most cases, making decisions regarding which observations to include in a model is challenging because information about the circumstances of occurrence is rarely recorded and also because issues with detectability can lead to inconsistent measurements (Everall et al., 2017; Guillera-Arroita, 2017).

It is tempting to conclude from the issues above that species' current environmental distributions will always be a subset of the set of suitable environments, but this is not necessarily true. Many species disperse broadly, for example, via wind or oceanic currents, and propagules often end up in "sink habitat": areas to which species disperse, but which cannot support stable populations (Dias, 1996). If occurrences from sink habitats are used, models erroneously conclude that this habitat is suitable and overestimate the set of suitable environments (Pulliam, 2000). However, marginal populations sometimes adapt to sink habitats (Kawecki, 2008), and SDMs have been used to identify the selective pressures that affect them (Morente-López et al., 2022).

Such conceptual uncertainties are inherent to occurrence data and are largely dealt with during data curation, if at all. This typically involves eliminating outliers or otherwise suspicious points, or points that do not represent permanent habitat according to expert opinion.

There are also many issues that stem from how data are collected and recorded. Spatial sampling bias is widespread in occurrence data due to the logistical challenges of accessing remote areas and variations in sampling intensity and protocols between regions. Sampling bias is usually accounted for by spatial thinning (removing occurrence points in close proximity to others) or by incorporating the bias explicitly into the modeling process (Phillips et al., 2009; Renner et al., 2015). The latter can be achieved by adding a bias estimate (or related predictors) to the model directly or by sampling background/pseudoabsence data (discussed below) with a frequency proportional to the estimated local sampling intensity for the occurrence data.

Taxonomic misidentification is also a problem (Anderson, 2012; Ensing et al., 2013). For instance, species may be genetically distinct but morphologically indistinguishable (i.e., cryptic species, Fig. 8.3), unknown subspecies or variants may be confused for the same species (e.g., Cavers et al., 2013), or several species synonyms may exist. This is particularly true for the tropics, where the sheer number of similar species often makes precise identification difficult (Huettmann, 2015).

Positional accuracy may also be an issue. Modern occurrence data are usually collected using GPS with a spatial resolution much finer than the environmental data used in modeling. However, older records may only contain site descriptions (e.g., "found near reservoir south of Maysville, Oklahoma") and are often geocoded with some estimate of spatial uncertainty, which should be considered when choosing data for analysis.

Fig. 8.2 (continued) set of environments available to the species, and as such have a large effect on the parameterization of correlative models. 7. Choice of study area may determine which predictors are available for model construction. 8. Choice of resolution for predictor data may affect the number of unique background points that can be sampled for model construction. 9. Predictor choice may affect which occurrence data are used for model construction, as some variables may not have complete geographic coverage. Additionally, the resolution of the raster data affects the number of unique combinations of environments that are available at occurrence points, the level of uncertainty in those values due to downscaling, and similar to (8), the number of occurrence points retained for model construction. 10. Choice of predictors for model construction determines which future scenarios may be used for forecasting. 11. Models are fit using the values of the predictor variables that have been extracted using the occurrence and (optionally) background data. Uncertainty in which predictors are instrumental to limiting the species' distribution therefore may have profound impacts on model structure. 12. Predictors are often selected by building one or more candidate models and evaluating their performance with different predictor sets. Uncertainty in optimal model structure may therefore propagate to uncertainty in the selection of ideal predictors. 13. Different algorithms sometimes require different amounts of background/pseudoabsence data, and in some cases different sampling methods for background/pseudoabsence data. 14. Similar to (12), data points are sometimes curated a posteriori based on model performance (e.g., removal of outliers, allowable levels of spatial or taxonomic uncertainty). Uncertainty in optimal model structure may therefore affect a posteriori data curation. 15. The choice of which modeling approaches to employ limits the candidate set of models that may eventually be used in forecasting, and as such uncertainty in selecting appropriate algorithms affects uncertainty in the eventual forecast. 16. Evaluation of models and model selection requires one or more candidate models. The results of model selection are by definition limited by the quality of models in the candidate set. Additionally, some metrics (e.g., information criteria) are only applicable to certain types of models, so that choice of algorithm(s) may limit the set of criteria available for model evaluation. 17. Decisions about which evaluation metrics to calculate and how to weight them when selecting models will often determine which algorithms and settings go into the model(s) used to generate the final forecast. Uncertainty in which evaluation procedures perform best for a given task therefore propagates to decisions about model structure. 18. Decisions about optimal study area and background sampling strategies (e.g., accounting for sampling bias) are sometimes made a posteriori by looking at the performance of models under some set of evaluation metrics or by inspecting spatial predictions or marginal predictor responses of models. Uncertainty in the choice of optimal evaluation metrics may therefore generate uncertainty in optimal choice of background area or sampling. 19. The final step of forecasting is to make predictions from the selected model(s) onto a future scenario for the environmental variables used in model construction. The choice of which models to use is based on how each model performs under the chosen evaluation metric(s), and as such uncertainty in the choice of metrics results in uncertainty in the forecast. 20. Choice of future time period and geographic region may limit which predictor variables are available, so that biologically plausible variables may potentially be unavailable for model construction

Fig. 8.3 Cryptic species are complexes of two or more species that are morphologically indistinguishable. In this example, two species of *Pheidole* ants have minor workers that cannot be distinguished based on any morphological measurements, but major workers of the two species are completely distinct. (Photos courtesy of Alexandre Casadei-Ferreira)

One approach for dealing with uncertainty in presence data is to use Monte Carlo methods. These techniques resample the original set of occurrences either with replacement ("bootstrap") or without ("jackknife"), then build models from the resampled data. By repeating this procedure multiple times, users can estimate variance in estimated habitat suitability and responses to environmental gradients (Efron & Tibshirani, 1994).

An issue that has recently gained attention is taxonomic scale. SDMs are typically built for a single species but occurrences can, in theory, represent any taxonomic unit. Niches or distributions can be modeled for higher taxonomic units (genera or families) in cases where the niche is evolutionarily conserved (Smith et al., 2019), as species within this group would have similar environmental preferences. Conversely, in some cases genetically distinct populations within a species may have differential responses to climate change and should be considered separately in SDMs (Pearman et al., 2010; Zhang et al., 2021). Frameworks have been developed that can jointly model co-occurrence patterns for multiple species (Warton et al., 2015). These models can separate shared environmental preferences from co-occurrence patterns, and can additionally include the effects of phylogeny, traits, or other community data. Due to the relative novelty of this approach, there are no widely accepted methods for estimating uncertainty due to the choice of taxonomic scale.

8.3 Environmental Variables

Often it is not known a priori which environmental variables shape the species' niche or distribution, and thus the selection of variables can be a source of uncertainty. Multicollinearity is common among environmental predictors, which in some cases may involve multiple variables derived from the same source data (e.g., minimum, maximum, and mean temperature). If variables have high multicollinearity, interpretation of variable importance or estimated responses can be highly misleading and model transfer can become problematic (Dormann et al., 2013).

When fitting SDMs, researchers often initially consider multiple predictor variables and take one of two approaches to reduce the number of predictors: (1) for traditional regression or classification models, build candidate models and use model selection to decide which predictors to keep in the final model; or (2) for models that use regularization to penalize complexity (Phillips et al., 2006), build a single model with all variables and have the algorithm decide which variables to retain.

Manual predictor selection based on statistical or biological criteria can be implemented before (a priori) or after (a posteriori) modeling. The statistical a priori approach uses correlation matrices or variance inflation factors (VIF) to exclude variables that surpass a user-defined multicollinearity threshold. The biological a priori approach retains variables based on existing research or other biological justification. In contrast, a posteriori approaches focus on repeating analyses with a reduced predictor variable set after examining a starting model. The statistical a posteriori approach makes such decisions using model performance metrics, whereas the biological approach uses expert opinion of model outputs.

All methods for reducing the set of predictors generate significant uncertainty when multicollinearity is present. Model predictions in the present day may not change much if highly correlated variables are exchanged, but such correlations change over space and time. As a result, the most serious consequences of predictor selection cannot be evaluated based on currently available data. For this reason, predictor selection is often explored with ensemble models, and the resulting model variance used to determine where predictions are most sensitive to these choices.

Selecting a resolution for predictor variables requires consideration of both the data and the intended application. Fine-scale raster datasets can be attractive under the belief that increased resolution will lead to better model accuracy. However, these datasets are typically interpolated from weather station data (Fick & Hijmans, 2017) or downscaled from predictions of global circulation models (GCMs; Karger et al., 2017). Because of uncertainties inherent in the downscaling process, fine-scale estimates for present climate can vary widely between datasets, and those for future climate even more so when comparing different GCMs. Therefore, although coarser resolutions may provide less information, fine-scale data may contribute additional uncertainty. The scale at which stakeholders make decisions can also inform selection of variable resolution. This can be conducive to the application of model products but may not always be the appropriate scale for the data.

8.4 Absence, Pseudoabsence, and Background Data

While many correlative SDM techniques require absence data to compare with presence data, absences are rarely collected in practice. Absence data have the same conceptual issues as presence data, but in addition, demonstrating that an organism is consistently absent from an area is significantly more time-consuming and uncertain than recording its presence. As a result, the most widely used SDM methods attempt to distinguish species occurrences from "background" points (representing available environments) or "pseudoabsence" points (approximating true absences). These points (hereafter, background points) are sampled either randomly or in proportion to some estimate of species dispersal or spatial sampling bias.

The study area defines the geographic region for model training over which environmental values for background points are sampled. Ideally, the study area should include all occurrence records and regions the species has likely been able to reach over long time scales (Barve et al., 2011), but exclude regions that the species is unlikely to have reached. If inaccessible regions are included, they might contain little useful information for the model (Anderson & Raza, 2010; Barbet-Massin et al., 2012) or, at worst, could cause models to parameterize spurious correlations between environmental conditions in those areas and the lack of occurrences of the species. These concerns apply only to the training region; predictions to regions outside the study area are not problematic if they are interpreted as estimates of the species potential distribution (Townsend Peterson et al., 2011).

Data-driven approaches to choosing study areas use buffers or hulls around occurrence data. Alternatively, expert-drawn range maps can be used to delineate study areas. Polygonal data such as administrative or other boundaries without biological meaning should generally not be used to define areas for model training.

Both the number of background points and their proximity to occurrence data can impact model parametrization. The default of 10,000 background points in the Maxent software (a popular implementation of maximum entropy SDMs) has become a standard for SDM studies, but this was originally suggested to avoid computational limitations nearly two decades ago (Phillips et al., 2006)—larger or smaller background samples may be more appropriate depending on study area and resolution. If background points do not adequately sample the available environment, modeled response curves may become truncated, resulting in unrealistic model predictions (Guevara et al., 2018).

Approaches to reducing uncertainty introduced by background selection include weighting background points based on the distance to occurrences and using replicate runs based on randomly sampled sets of background points (Barbet-Massin et al., 2012). Given the large number of background points typically chosen for modeling, variance due to background sampling is typically low compared to the uncertainty introduced by the choice of the study area. In both cases, the effects of uncertainty can be explored by building multiple models with varying study areas and/or repeated random sampling of background points.

8.5 Modeling Methods

SDMs that do not use true absence data are often discussed in two broad classes based on the treatment of the presence data: climate envelope models (presence-only models) and presence/background models. Climate envelope models are simple statistical generalizations of occurrence point distributions in environmental space and require only data on species occurrences (Nix & Busby, 1986; Blonder et al., 2014; Carpenter et al., 1993; Brewer et al., 2016). These models can be based on confidence intervals (e.g., the range of each variable encompassing 95% of the occurrence data) or a distance metric quantifying the difference between a set of environmental conditions and the distribution of occurrences. Climate envelope methods will not extrapolate substantially into climate conditions beyond the current environmental distribution of the species.

Correlative presence/background methods range from more traditional models (e.g., generalized linear models) to machine learning methods (e.g., random forest, maximum entropy, deep neural networks) (Phillips et al., 2006; Liaw & Wiener, 2002; Lek & Guégan, 1999). While these methods are capable of extrapolating to new environmental conditions, and hence can ideally predict the effects of climate change, extrapolation accuracy relies on good model performance under conditions that are not available for model training or validation. Extrapolating to conditions outside those used for model training is not generally seen as good statistical practice because of the high associated uncertainty, but as noted above SDMs are often used in this way because no other option is available.

Managing the extent to which models extrapolate requires users to make decisions regarding model complexity, which requires consideration of how the model will be interpreted and applied (Merow et al., 2014; Warren & Seifert, 2011). Compared to simpler models, complex models are generally more affected by processes unrelated to the species' niche. Model complexity is therefore often desirable when the goal is to estimate the species' current distribution, but results can be misleading for projections to new conditions that may have different correlations between those processes and the environment. Simple models, conversely, may generally be more interpretable and transferable but may make less accurate estimates of species' realized distributions.

Since predictions of species distributions may show significant biases depending on the modeling algorithm used, model ensembling has become a popular approach to detect algorithm-independent patterns (Beaumont et al., 2016). Model ensembles have also been used to explore uncertainty in multiple aspects of SDM analyses, and algorithm choice seems to be associated with particularly high variance (Buisson et al., 2010; Diniz-Filho et al., 2009; Watling et al., 2015; Dormann et al., 2008; Thuiller et al., 2019). Ensemble predictions can be more accurate than individual model predictions, but there is still some disagreement about their utility for SDMs. Ensembling methods are usually applied to model predictions rather than to the models themselves. As a result, if the underlying ecological relationships have been misidentified, or if differences between models are primarily driven by methodological biases instead of statistical noise, ensembling may not be helpful. In some cases, ensembles including poor models may dilute or overrule the predictions of good models. Therefore, construction and selection for ensembled models should be just as rigorous as for single models (Valavi et al., 2021).

Ensembles also add their own sources of uncertainty. First, ensembling requires criteria by which to weigh models; even uniform weighting makes assumptions about models' relative merits. Second, model predictions may differ in nature between algorithms (e.g., probability/suitability values, different presence/absence thresholds, etc.) and thus may not be directly comparable. Often this is addressed simply by normalizing model predictions. In recent years, however, more sophisticated recalibration techniques have been developed, and many approaches are now available (Schwarz & Heider, 2018). However, these may also increase overparameterization and homogenize differences between models (Kindt, 2018).

8.6 Model Evaluation

Model performance is important when deciding how much confidence to place in predictions, but evaluating model fit on training data can lead to overparameterization and poor extrapolation. Cross-validation, an iterative procedure that avoids the need for "independent" datasets for evaluation by building and evaluating models on different subsets of a dataset (Hastie et al., 2001), is often used to evaluate SDMs. Spatial or temporal cross-validation is particularly effective at judging how well models transfer to new times or places (Roberts et al., 2017). However, even evaluations on withheld data can favor models that capture environmental correlates of nontarget processes (e.g., spatial sampling bias). This is particularly true when there are consistent correlations between nontarget processes and one or more environmental predictors (Veloz, 2009).

In practice, model selection in SDM is dominated by the use of discrimination metrics, such as the area under the receiver-operating characteristic curve (AUC) and omission rates (Fielding & Bell, 1997). As commission errors can represent valid predictions outside of sampled areas, they are usually considered a less severe problem than omission errors (Townsend Peterson et al., 2011). Other evaluation metrics have been applied to SDMs (Allouche et al., 2006), but as different discrimination metrics are often highly correlated, metric choice can be relatively arbitrary for model selection (Warren et al., 2019). Other studies have used multiple discrimination metrics to describe different aspects of model performance (Radosavljevic & Anderson, 2014).

In contrast, calibration metrics evaluate how well the observed frequency of species' presence in a set of conditions matches the predicted frequency. However, they require that the prevalence of occurrences in the evaluation data match the true species' prevalence. In most cases, this is unknown, but modified calibration metrics can accommodate this issue [e.g., scaled calibration plots (Phillips & Elith, 2010; Hirzel et al., 2006); continuous Boyce index (Phillips & Elith, 2010; Hirzel et al., 2006)]. These metrics are not widely used in SDM studies, but they are frequently more relevant to the intended model application than binary prediction.

As cross-validation does not explicitly penalize models for being too complex, an alternative method is to use information criteria such as Akaike information criterion (AIC), widely used in statistics to manage the tradeoff between model fit and model complexity. Applying AIC to regression-based SDM algorithms is straightforward, but its applicability to more widely used machine learning SDM algorithms is less clear. For example, while simulation studies have shown that AIC performs well for Maxent models, there are both philosophical reasons (e.g., information criteria are typically applied when the focus is on parameterization, not prediction) and practical reasons (e.g., the disconnect between the number of parameters and the effective degrees of freedom of models built with the L1 lasso) to be cautious about this approach (Warren & Seifert, 2011).

Model performance evaluated via discrimination, calibration, and model complexity metrics may be decoupled and, in some cases, even negatively correlated, so metric choice should be given careful consideration. Many R packages for SDMs generate multiple performance metrics that can lead to the selection of different models for downstream analyses—such exercises can help the modeler quantify evaluation uncertainty. Finally, when evaluating models, it is also important to examine spatial predictions and marginal response curves (relationships between individual variables and the model response) for ecological realism (Guevara et al., 2018).

8.7 Forecasting

Once one or more models have been selected to estimate the species' niche, projecting the effects of climate change is superficially simple; the niche estimate is projected onto a future climate scenario within the region of interest. However, this requires users to select climate scenario(s) representing expected changes in the study region, as well as the geographic extent for the projection. Ideally users would make a single optimal choice for each of these aspects of the forecast, informed by existing data or the relative realism of different climate models and knowledge of which geographic regions are relevant to stakeholders. However, in many cases investigators simply bracket their uncertainty with respect to these parameters by projecting niche estimates to a range of climate scenarios over a broad region.

Forecasting future habitat suitability often involves extrapolating to environmental conditions that were not available in the time period and geographic region for model training ("non-analog conditions"). This poses many potential problems, but this is also typically done when few if any forecasting alternatives are available. Non-analog conditions are more prevalent when models are projected to more extreme climate scenarios and broader geographic areas or more distant times; thus, choice of climate scenario and region may profoundly affect model predictions and their associated uncertainties. Environmental similarity metrics can help identify areas where extrapolation is most extreme (Owens et al., 2013).

Uncertainty in the choice of climate models and emissions scenarios is often explored by projecting a preferred model or set of models across a range of future scenarios. In contrast to the use of ensembles to explore uncertainty in model development, this range of projections is typically explored separately rather than averaged; here, the goal is to investigate the effects of different scenarios rather than to average out the differences between them.

Forecasting also assumes that the relationship of habitat suitability to the environmental predictors remains consistent over time and space. This in turn assumes that genetic drift and natural selection are not substantially altering species' tolerances over relevant time scales, but also that phenomena unrelated to the niche that may affect species' relationships to the environment remain unchanged. These assumptions may be particularly problematic with respect to species interactions, for example, if the interacting species is not tracking the environment consistently. This generates a great deal of uncertainty if future change results in a high prevalence of non-analog communities, which may involve the gain and loss of multiple interactions (Blois et al., 2013).

Similarly, many species distributions may be affected by factors that are not typically included in SDMs (e.g., crops, roads). Even for good niche estimates, there may be significant forecasting errors if some habitat patches are otherwise environmentally suitable but rendered unsuitable by land use or other anthropogenic effects. While some studies have explored the interacting effects of climate change and land use (Newbold, 2018; Hof et al., 2018), these are still not frequently considered in SDM studies.

Finally, predicting expected future species distributions requires knowledge of (or assumptions about) the dispersal potential of the species. This varies even between closely related species, and reliable estimates are often not available. Many forecasting studies address this by predicting distributions under a variety of dispersal scenarios, most often simply using a "no dispersal" scenario (i.e., species cannot disperse to any grid cell that is not currently suitable) and a "full dispersal" scenario (i.e., species can disperse to any grid cell predicted to be suitable). If intermediate scenarios are provided, they typically use simple buffers around species' current ranges either of an arbitrarily chosen distance, or based on range extent or species' biological attributes. While simplistic, these approaches allow us to estimate the prediction uncertainty arising from our lack of detailed knowledge about species dispersal. Even when detailed estimates are available, including them in forecasts can be challenging; this requires us to apply a dispersal estimate over a landscape where some habitat is unsuitable, and where suitability of habitat changes over time. As with different climate scenarios, different dispersal scenarios are typically considered separately, rather than averaged to make an ensemble prediction.

8.8 Decision-Making

For SDMs, uncertainty is unavoidable given the data and modeling issues discussed in this chapter, and different approaches can result in greatly contrasting maps for decision-making (Loiselle et al., 2003). Communicating this constructively to stakeholders presents several challenges. Scientists value estimates of uncertainty as they provide information about the reliability of model predictions. In contrast, stakeholders often desire direct answers to management questions rather than lists of possibilities. Villero et al. (2017) call for open communication between researchers and conservation practitioners to help align management goals with research directions, as well as making products for spatial decision-making to promote information exchange. Distilling SDM uncertainties into a portfolio of maps showing several potential scenarios can help create actionable information for stakeholders. Such scenarios could represent the most extreme predictions made under different modeling approaches to help set bounds for future expectations. Another strategy is to reduce the complexity surrounding parameters of uncertainty to a simple categorical scale that facilitates decision-making. For example, Sofaer et al. (2019) developed a rubric for managers consisting of "interpret with caution," "acceptable," and "ideal," and used these to rate SDM inputs and methods.

8.9 Closing Remarks

All the sources of uncertainty outlined above deserve consideration, but it is more productive to consider these in terms of their relative impacts. Many studies have constructed models using a variety of approaches at each step (e.g., choice of study area, predictor variables, algorithm) accompanied by post hoc analyses partitioning the variance in spatial predictions (Diniz-Filho et al., 2009; Watling et al., 2015) or the discrimination accuracy (Dormann et al., 2008; Watling et al.,

2015) among the different sources of uncertainty. The most consistent picture emerging from these studies is that modeling algorithm choice is the primary source of uncertainty—usually by a fairly substantial margin—with GCM (Watling et al., 2015) and data quality (Dormann et al., 2008) also having notable, albeit substantially smaller, effects.

Species distribution models built from presence/background data are popular because they are relatively straightforward to implement and use data that are widely available. However, these models have considerable uncertainty associated with every stage of the analysis, as this chapter highlights. Methodological uncertainty is present in every empirical research study, but how best to communicate, or even leverage, uncertainty for decision-making depends on the particular management goals being addressed. That said, most uncertainty analyses still only cover a few steps in the modeling process. Recent trends toward formal standards for SDM analyses (Araújo et al., 2019) and new tools for organizing and generating SDM metadata (Merow et al., 2019; Zurell et al., 2020) should aid in both reducing and helping communicate methodological uncertainty.

In the case of SDMs, maps of habitat suitability are the product most needed to aid decision-making, and such maps should communicate to practitioners the uncertainties that may affect conservation outcomes. Species distribution modeling will unarguably continue to be a fundamental instrument in our ecological toolbox for the foreseeable future, and thus it is essential that we develop better ways to accurately measure, communicate, and harness uncertainty in these models. Effectively using this information has manifold potential benefits for guiding future management and conservation practice.

References

Allouche, O., Tsoar, A., & Kadmon, R. (2006). Assessing the accuracy of species distribution models: Prevalence, kappa and the true skill statistic (TSS). *Journal of Applied Ecology, 43*, 1223–1232. https://doi.org/10.1111/j.1365-2664.2006.01214.x

Anderson, R. P. (2012). Harnessing the world's biodiversity data: Promise and peril in ecological niche modeling of species distributions. *Annals of the New York Academy of Sciences, 1260*, 66–80. https://doi.org/10.1111/j.1749-6632.2011.06440.x

Anderson, R. P., & Raza, A. (2010). The effect of the extent of the study region on GIS models of species geographic distributions and estimates of niche evolution: Preliminary tests with montane rodents (genus Nephelomys) in Venezuela. *Journal of Biogeography, 37*, 1378–1393. https://doi.org/10.1111/j.1365-2699.2010.02290.x

Araújo, M. B., et al. (2019). Standards for distribution models in biodiversity assessments. *Science Advances, 5*. https://doi.org/10.1126/sciadv.aat4858

Barbet-Massin, M., Jiguet, F., Albert, C. H., & Thuiller, W. (2012). Selecting pseudo-absences for species distribution models: How, where and how many? *Methods in Ecology and Evolution, 3*, 327–338. https://doi.org/10.1111/j.2041-210x.2011.00172.x

Barve, N., Barve, V., Jiménez-Valverde, A., Lira-Noriega, A., Maher, S. P., Peterson, A. T., Soberón, J., & Villalobos, F. (2011). The crucial role of the accessible area in ecological niche modeling and species distribution modeling. *Ecological Modelling, 222*, 1810–1819. https://doi.org/10.1016/j.ecolmodel.2011.02.011

Beaumont, L. J., et al. (2016). Which species distribution models are more (or less) likely to project broad-scale, climate-induced shifts in species ranges? *Ecological Modelling, 342*, 135–146. https://doi.org/10.1016/j.ecolmodel.2016.10.004

Blois, J. L., Zarnetske, P. L., Fitzpatrick, M. C., & Finnegan, S. (2013). Climate change and the past, present, and future of biotic interactions. *Science, 341*, 499–504. https://doi.org/10.1126/science.1237184

Blonder, B., Lamanna, C., Violle, C., & Enquist, B. J. (2014). The n-dimensional hypervolume. *Global Ecology and Biogeography, 23*, 595–609. https://doi.org/10.1111/geb.12146

Brewer, M. J., O'Hara, R. B., Anderson, B. J., & Ohlemüller, R. (2016). Plateau: A new method for ecologically plausible climate envelopes for species distribution modelling. *Methods in Ecology and Evolution, 7*, 1489–1502. https://doi.org/10.1111/2041-210X.12609

Buisson, L., Thuiller, W., Casajus, N., Lek, S., & Grenouillet, G. (2010). Uncertainty in ensemble forecasting of species distribution. *Global Change Biology, 16*, 1145–1157. https://doi.org/10.1111/j.1365-2486.2009.02000.x

Carpenter, G., Gillison, A. N., & Winter, J. (1993). DOMAIN: A flexible modelling procedure for mapping potential distributions of plants and animals. *Biodiversity and Conservation, 2*, 667–680. https://doi.org/10.1007/BF00051966

Cavers, S., et al. (2013). Cryptic species and phylogeographical structure in the tree Cedrela odorata L. throughout the Neotropics. *Journal of Biogeography, 40*, 732–746. https://doi.org/10.1111/jbi.12086

Dias, P. C. (1996). Sources and sinks in population biology. *Trends in Ecology and Evolution 11*, 326–330. https://doi.org/10.1016/0169-5347(96)10037-9

Diniz-Filho, J. A. F., Mauricio Bini, L., Fernando Rangel, T., Loyola, R. D., Hof, C., Nogués-Bravo, D., & Araújo, M. B. (2009). Partitioning and mapping uncertainties in ensembles of forecasts of species turnover under climate change. *Ecography, 32*, 897–906. https://doi.org/10.1111/j.1600-0587.2009.06196.x

Dormann, C. F., Purschke, O., García Márquez, J. R., Lautenbach, S., & Schröder, B. (2008). Components of uncertainty in species distribution analysis: A case study of the Great Grey Shrike. *Ecology, 89*, 3371–3386. https://doi.org/10.1890/07-1772.1

Dormann, C. F., et al. (2013). Collinearity: A review of methods to deal with it and a simulation study evaluating their performance. *Ecography, 36*, 27–46. https://doi.org/10.1111/j.1600-0587.2012.07348.x

Efron, B., & Tibshirani, R. J. (1994). *An introduction to the bootstrap* (p. 456). CRC Press. https://doi.org/10.1201/9780429246593

Ensing, D. J., Moffat, C. E., & Pither, J. (2013). Taxonomic identification errors generate misleading ecological niche model predictions of an invasive hawkweed. *Botany, 91*, 137–147. https://doi.org/10.1139/cjb-2012-0205

Everall, N. C., Johnson, M. F., Wood, P., Farmer, A., Wilby, R. L., & Measham, N. (2017). Comparability of macroinvertebrate biomonitoring indices of river health derived from semi-quantitative and quantitative methodologies. *Ecological Indicators, 78*, 437–448. https://doi.org/10.1016/j.ecolind.2017.03.040

Fick, S. E., & Hijmans, R. J. (2017). WorldClim 2: New 1-km spatial resolution climate surfaces for global land areas. *International Journal of Climatology, 37*, 4302–4315. https://doi.org/10.1002/joc.5086

Fielding, A. H., & Bell, J. F. (1997). A review of methods for the assessment of prediction errors in conservation presence/absence models. *Environmental Conservation, 24*, 38–49. https://doi.org/10.1017/S0376892997000088

Guevara, L., Gerstner, B. E., Kass, J. M., & Anderson, R. P. (2018). Toward ecologically realistic predictions of species distributions:

A cross-time example from tropical montane cloud forests. *Global Change Biology, 24*, 1511–1522. https://doi.org/10.1111/gcb.13992

Guillera-Arroita, G. (2017). Modelling of species distributions, range dynamics and communities under imperfect detection: Advances, challenges and opportunities. *Ecography, 40*, 281–295. https://doi.org/10.1111/ecog.02445

Hastie, T., Friedman, J., & Tibshirani, R. (2001). *The elements of statistical learning: Data mining, inference, and prediction*. Springer. https://doi.org/10.1007/978-0-387-84858-7

Hirzel, A. H., Le Lay, G., Helfer, V., Randin, C., & Guisan, A. (2006). Evaluating the ability of habitat suitability models to predict species presences. *Ecological Modelling, 199*, 142–152. https://doi.org/10.1016/j.ecolmodel.2006.05.017

Hof, C., Voskamp, A., Biber, M. F., Böhning-Gaese, K., Engelhardt, E. K., Niamir, A., Willis, S. G., & Hickler, T. (2018). Bioenergy cropland expansion may offset positive effects of climate change mitigation for global vertebrate diversity. *Proceedings of the National Academy of Sciences of the United States of America, 115*, 13294–13299. https://doi.org/10.1073/pnas.1807745115

Huettmann, F. (2015). On the relevance and moral impediment of digital data management, data sharing, and public open access and open source code in (tropical) research: The Rio convention revisited towards mega science and best professional research practices. In *Central American biodiversity* (Vol. 391–417). Springer. https://doi.org/10.1007/978-1-4939-2208-6_16

Karger, D. N., et al. (2017). Climatologies at high resolution for the earth's land surface areas. *Scientific Data, 4*, 170122. https://doi.org/10.1038/sdata.2017.122

Kawecki, T. J. (2008). Adaptation to marginal habitats. *Annual Review of Ecology, Evolution, and Systematics, 39*, 321–342. https://doi.org/10.1146/annurev.ecolsys.38.091206.095622

Kindt, R. (2018). Ensemble species distribution modelling with transformed suitability values. *Environmental Modelling & Software, 100*, 136–145. https://doi.org/10.1016/j.envsoft.2017.11.009

Lek, S., & Guégan, J. F. (1999). Artificial neural networks as a tool in ecological modelling, an introduction. *Ecological Modelling, 120*, 65–73. https://doi.org/10.1016/S0304-3800(99)00092-7

Liaw, A., Wiener, M., et al. (2002). Classification and regression by randomForest. *R news, 2*, 18–22.

Loiselle, B. A., Howell, C. A., Graham, C. H., Goerck, J. M., Brooks, T., Smith, K. G., & Williams, P. H. (2003). Avoiding pitfalls of using species distribution models in conservation planning. *Conservation Biology, 17*, 1591–1600. https://doi.org/10.1111/j.1523-1739.2003.00233.x

Merow, C., et al. (2014). What do we gain from simplicity versus complexity in species distribution models? *Ecography, 37*, 1267–1281. https://doi.org/10.1111/ecog.00845

Merow, C., Maitner, B. S., Owens, H. L., Kass, J. M., Enquist, B. J., Jetz, W., & Guralnick, R. (2019). Species' range model metadata standards: RMMS. *Global Ecology and Biogeography, 28*, 1912–1924. https://doi.org/10.1111/geb.12993

Morente-López, J., Kass, J. M., Lara-Romero, C., Serra-Diaz, J. M., Soto-Correa, J. C., Anderson, R. P., & Iriondo, J. M. (2022). Linking ecological niche models and common garden experiments to predict phenotypic differentiation in stressful environments: Assessing the adaptive value of marginal populations in an alpine plant. *Global Change Biology, 28*, 4143–4162. https://doi.org/10.1111/gcb.16181

Newbold, T. (2018). Future effects of climate and land-use change on terrestrial vertebrate community diversity under different scenarios. *Proceedings of the Biological Sciences, 285*. https://doi.org/10.1098/rspb.2018.0792

Nix, H. A., and Busby, J. (1986). BIOCLIM, a bioclimatic analysis and prediction system. *Annual report CSIRO*. CSIRO Division of Water and Land Resources, Canberra, https://doi.org/10.1071/BT13017.

Owens, H. L., et al. (2013). Constraints on interpretation of ecological niche models by limited environmental ranges on calibration areas. *Ecological Modelling, 263*, 10–18. https://doi.org/10.1016/j.ecolmodel.2013.04.011

Pearman, P. B., D'Amen, M., Graham, C. H., Thuiller, W., & Zimmermann, N. E. (2010). Within-taxon niche structure: Niche conservatism, divergence and predicted effects of climate change. *Ecography, 33*, 990–1003. https://doi.org/10.1111/j.1600-0587.2010.06443.x

Phillips, S. J., & Elith, J. (2010). POC plots: Calibrating species distribution models with presence-only data. *Ecology, 91*, 2476–2484. https://doi.org/10.1890/09-0760.1

Phillips, S. J., Anderson, R. P., & Schapire, R. E. (2006). Maximum entropy modeling of species geographic distributions. *Ecological Modelling, 190*, 231–259. https://doi.org/10.1016/j.ecolmodel.2005.03.026

Phillips, S. J., Dudík, M., Elith, J., Graham, C. H., Lehmann, A., Leathwick, J., & Ferrier, S. (2009). Sample selection bias and presence-only distribution models: Implications for background and pseudo-absence data. *Ecological Applications, 19*, 181–197. https://doi.org/10.1890/07-2153.1

Pulliam, H. R. (2000). On the relationship between niche and distribution. *Ecology Letters, 3*, 349–361. https://doi.org/10.1046/j.1461-0248.2000.00143.x

Radosavljevic, A., & Anderson, R. P. (2014). Making better Maxent models of species distributions: complexity, overfitting and evaluation. *Journal of biogeography, 41*(4), 629–643. https://doi.org/10.1111/jbi.12227

Renner, I. W., Elith, J., Baddeley, A., Fithian, W., Hastie, T., Phillips, S. J., Popovic, G., & Warton, D. I. (2015). Point process models for presence-only analysis. *Methods in Ecology and Evolution, 6*, 366–379. https://doi.org/10.1111/2041-210X.12352

Roberts, D. R., et al. (2017). Cross-validation strategies for data with temporal, spatial, hierarchical, or phylogenetic structure. *Ecography, 40*, 913–929. https://doi.org/10.1111/ecog.02881

Schwarz, J., & Heider, D. (2018). GUESS: Projecting machine learning scores to well-calibrated probability estimates for clinical decision making. *Bioinformatics, 35*, 2458–2465. https://doi.org/10.1093/bioinformatics/bty984

Smith, A. B., Godsoe, W., Rodríguez-Sánchez, F., Wang, H.-H., & Warren, D. (2019). Niche estimation above and below the species level. *Trends in Ecology & Evolution, 34*, 260–273. https://doi.org/10.1016/j.tree.2018.10.012

Sofaer, H. R., et al. (2019). Development and delivery of species distribution models to inform decision-making. *Bioscience, 69*, 544–557. https://doi.org/10.1093/biosci/biz045

Thuiller, W., Guéguen, M., Renaud, J., Karger, D. N., & Zimmermann, N. E. (2019). Uncertainty in ensembles of global biodiversity scenarios. *Nature Communications, 10*, 1446. https://doi.org/10.1038/s41467-019-09519-w

Townsend Peterson, A., Soberón, J., Pearson, R. G., Anderson, R. P., Martínez-Meyer, E., Nakamura, M., & Araújo, M. B. (2011). *Ecological niches and geographic distributions (MPB-49)* (p. 328). Princeton University Press. https://doi.org/10.1515/9781400840670

Valavi, R., Guillera-Arroita, G., Lahoz-Monfort, J. J., & Elith, J. (2021). Predictive performance of presence-only species distribution models: A benchmark study with reproducible code. *Ecological Monographs, 1*. https://doi.org/10.1002/ecm.1486

Veloz, S. D. (2009). Spatially autocorrelated sampling falsely inflates measures of accuracy for presence-only niche models. *Journal of Biogeography, 36*, 2290–2299. https://doi.org/10.1111/j.1365-2699.2009.02174.x

Villero, D., Pla, M., Camps, D., Ruiz-Olmo, J., & Brotons, L. (2017). Integrating species distribution modelling into decision-making to inform conservation actions. *Biodiversity and Conservation, 26*, 251–271. https://doi.org/10.1007/s10531-016-1243-2

Warren, D. L., & Seifert, S. N. (2011). Ecological niche modeling in Maxent: The importance of model complexity and the performance of model selection criteria. *Ecological Applications, 21*, 335–342. https://doi.org/10.1890/10-1171.1

Warren, D. L., Matzke, N. J., & Iglesias, T. L. (2019). Evaluating species distribution models with discrimination accuracy is uninformative for many applications. *Journal of Biogeography, 47*, 167–180. https://doi.org/10.1111/jbi.13705

Warton, D. I., Blanchet, F. G., O'Hara, R. B., Ovaskainen, O., Taskinen, S., Walker, S. C., & Hui, F. K. C. (2015). So many variables: Joint modeling in community ecology. *Trends in Ecology & Evolution, 30*, 766–779. https://doi.org/10.1016/j.tree.2015.09.007

Watling, J. I., Brandt, L. A., Bucklin, D. N., Fujisaki, I., Mazzotti, F. J., Romañach, S. S., & Speroterra, C. (2015). Performance metrics and variance partitioning reveal sources of uncertainty in species distribution models. *Ecological Modelling, 309-310*, 48–59. https://doi.org/10.1016/j.ecolmodel.2015.03.017

Zhang, Z., et al. (2021). Lineage-level distribution models lead to more realistic climate change predictions for a threatened crayfish. *Diversity and Distributions, 27*, 684–695. https://doi.org/10.1111/ddi.13225

Zurell, D., et al. (2020). A standard protocol for reporting species distribution models. *Ecography, 43*, 1261–1277. https://doi.org/10.1111/ecog.04960

Open Access This chapter is licensed under the terms of the Creative Commons Attribution 4.0 International License (http://creativecommons.org/licenses/by/4.0/), which permits use, sharing, adaptation, distribution and reproduction in any medium or format, as long as you give appropriate credit to the original author(s) and the source, provide a link to the Creative Commons license and indicate if changes were made.

The images or other third party material in this chapter are included in the chapter's Creative Commons license, unless indicated otherwise in a credit line to the material. If material is not included in the chapter's Creative Commons license and your intended use is not permitted by statutory regulation or exceeds the permitted use, you will need to obtain permission directly from the copyright holder.

9. Uncertainty in Hydrologic and Water Resources Modelling

Robert L. Wilby and Geoff Darch

9.1 Introduction

Uncertainty is the 'new normal' (Deloitte, 2017). Prescient words indeed, yet the authors could hardly have imagined the unprecedented global consequences of the 2020 pandemic. Climate scientists are accustomed to thinking about 'bad stuff'—such as more frequent and/or severe weather extremes and gradual disruptions to Earth system processes, perhaps leading eventually to 'tipping' points. But Covid-19 raised difficult questions around foresight and preparedness for extraordinary shocks. Pandemic flu, emerging infectious diseases and climate-related impacts all featured as significant risks in the UK Government's National Risk Register.[1] But it is unclear whether indirect consequences are fully considered, that combinations of hazards are evaluated or that the risk assessment leads to adequate preparation. Fundamental questions remain unanswered: what, where, when and how bad?

This chapter has, by comparison, set the rather modest goal of explaining how managers can develop more secure water infrastructure and resource plans, despite deep uncertainty about future climate, water supplies and demand. Nonetheless, given that the United Nations Sustainable Development Goals for 2030 are intrinsically water-related, climate threats to the water sector have significant ramifications for human welfare and the environment over coming decades. The interaction of water and Covid-19 has illustrated this, with water being the primary medium for cleaning. For UK water resources managers, the 'lockdown' in response to Covid-19 caused significant changes to patterns of water use in space and time, whilst a concurrent hot dry spring led to record levels of demand in some areas.

We already know a lot about the relative importance and characteristics of hydrologic uncertainty arising from different socioeconomic pathways, associated greenhouse gas emissions, climate system feedbacks, regional climate change and water sector impacts (Clark et al., 2016). We recognise that the data used to calibrate our hydrological models are imperfect and can be hard to obtain in some regions (Wilby et al., 2017). Furthermore, we know that hydrological model structures and parameters are uncertain and non-stationary (Broderick et al. 2016; Wilby, 2005); fundamental water balance terms like evapotranspiration are sensitive to carbon dioxide concentrations, yet this is seldom reflected in model experiments (Prudhomme et al., 2014); changing land-surface properties and hydrological conditions can affect future floods and droughts too (Hutchins et al., 2018; Fowler et al., 2022). Unsurprisingly, indices of hydrologic change are sensitive to all these components and can accrue large uncertainty in their estimates (Ekström et al., 2018).

Faced with such a daunting list of uncertainties, it is not surprising that there are diverse reactions (Curry & Webster, 2011). 'Wait and see' or 'hiding' from the issue may be tempting but rarely permissible—especially in situations where action is mandatory. For instance, UK water companies are required by law to revisit strategic water plans every 5 years and, where necessary, make investments to improve their asset and network resilience to climate hazards.[2] A

---

[1] https://www.gov.uk/government/publications/national-risk-register-2020

R. L. Wilby (✉)
Geography and Environment, Loughborough University, Loughborough, UK
e-mail: r.l.wilby@lboro.ac.uk

G. Darch
Anglian Water Ltd., Huntingdon, Cambridgeshire, UK
e-mail: gDarch@anglianwater.co.uk

[2] https://webarchive.nationalarchives.gov.uk/ukgwa/20170602144659/http://www.ofwat.gov.uk/regulated-companies/resilience-2/climate-change/

'wait and see' approach is also inadvisable for two reasons. First, water resources management has a significant focus on drought risk management and in situations, where drought resilience standards are high[3], the chance of severe restrictions (e.g. cuts to supplies or standpipes) are very low. Therefore, it is difficult to determine the severity of such an event until it is too late, and this is likely to be exacerbated by climate change; a water resources manager will not know in advance how much more severe climate change will make the next drought event! Second, with the exception of some demand management measures, it is generally impossible to 'build your way out' of drought events due to the time taken to develop new supplies (especially more complicated options such as desalination or reuse).

'Reducing uncertainty' is an appealing idea, but experience tells us that this is unlikely. After decades of research, we can better characterise uncertainty, but there are few cases where ambiguity has actually been reduced (e.g. Rowell, 2019). 'Taking no chances' with the uncertainty may manage the risk of regrettable outcomes, but there could be large opportunity costs. A precautionary approach is justifiable, where the risks to people and/or the environment from climate-related hazards are too great to countenance, such as large safety margins in engineering designs to avoid dam collapse or damage to critical infrastructure (e.g. Wilby et al., 2011). 'Adaptively managing' uncertainty is about enhancing flexibility and options to allow timely responses to emerging climate threats and opportunities. The Thames Estuary 2100 Project is widely cited as a classic example of this approach (see: Ranger et al., 2013). But this strategy is based on a single, slowly evolving variable (sea level); assumes longevity of monitoring; and requires governance systems to oversee phased adaptations that could span decades.

Given the above context, this chapter focuses on two key challenges for water resource management: climate-adjusted infrastructure (projects) and climate resilient water systems (plans). Other pressing needs around filling data sparse places, predicting peak water, understanding the physical drivers of mega floods and droughts, evaluating hyper-resolution hydrological models or managing compound hazards have been addressed elsewhere (Wilby, 2019). By concentrating on 'projects and plans', the tone is solution-orientated; accepting uncertainty as a given, there is still much that can be done in practice to adapt water resource systems to climate change. The chapter begins with a synopsis of modelling frameworks for evaluating hydrological change. This is followed by a critique of techniques for incorporating allowances for climate change in detailed engineering designs. Next, water system resilience to credible 'storylines' of change and climate extremes are considered. Finally, we discuss outstanding knowledge gaps and research opportunities for improving water security *despite* uncertainty about climate change.

9.2 Hydrologic and Water Resource Modelling Frameworks

There is already much literature on hydrological modelling and decision-making under uncertainty (see: Wilby, 2016; Wilby & Murphy, 2018; Slater et al., 2021; Chan et al., 2022a; plus references therein). It is helpful to start with a reprise of modelling frameworks because they reflect the overall rationale for uncertainty analysis as well as the practicalities involved. Smith et al. (2018) assert that there are three analytical approaches to uncertainty—conveniently labelled A, B and C (Fig. 9.1).

'A' is for 'Analyse'. This is the conventional approach, sometimes called a 'top down' assessment. The modelling team strives to capture all dimensions of uncertainty in their analysis, such as emissions scenario, climate model ensemble, regional climate downscaling techniques, hydrological model structures and parameters. Uncertainty tends to cascade as outputs from one part of the modelling chain are handed to the next. Early studies explored relatively few dimensions and it was recognised that the sampling captured only a fraction of the true uncertainty domain. For example, Wilby and Harris (2006) investigated the uncertainty in future low flows in the River Thames using two emissions scenarios, four climate models, one statistical downscaling technique, two hydrological model structures and two hydrological model parameter sets (i.e. just[!] 32 dimensions of uncertainty in total). Even so, there was a wide range of projected uncertainty in low flows, which expanded over time: from −20% to +70% (by the 2020s), then −40% to +60% (by the 2050s) and −50% to +80% (by the 2080s) (Fig. 9.2).

Subsequent studies implemented more elaborate modelling of the same river basin. For instance, Borgomeo et al. (2016) applied an ensemble of 10,000 climate model projections to a water resource system model with an algorithm to identify an optimal set of water resource management plans across a 25-year planning horizon. The results showed adaptation costs spanning an order of magnitude depending on attitude to risk. Others have similarly performed super-ensemble/hyper-matrix experiments that show simulated hydrological impacts are sensitive to sources of uncertainty analysed and models used (New et al., 2007; Manning et al., 2009). More recently, UNSEEN (UNprecedented Simulated Extremes using Ensembles) methods apply very large ensembles of retrospective forecasts to evaluate the chance

---

[3] In England and Wales, water companies are moving towards a one in 200-year drought resilience standard in the short-term, and a one in 500-year drought resilience standard by 2039; this standard means that severe restrictions (rota cuts and standpipes) would only be required in events more extreme than the respective return period.

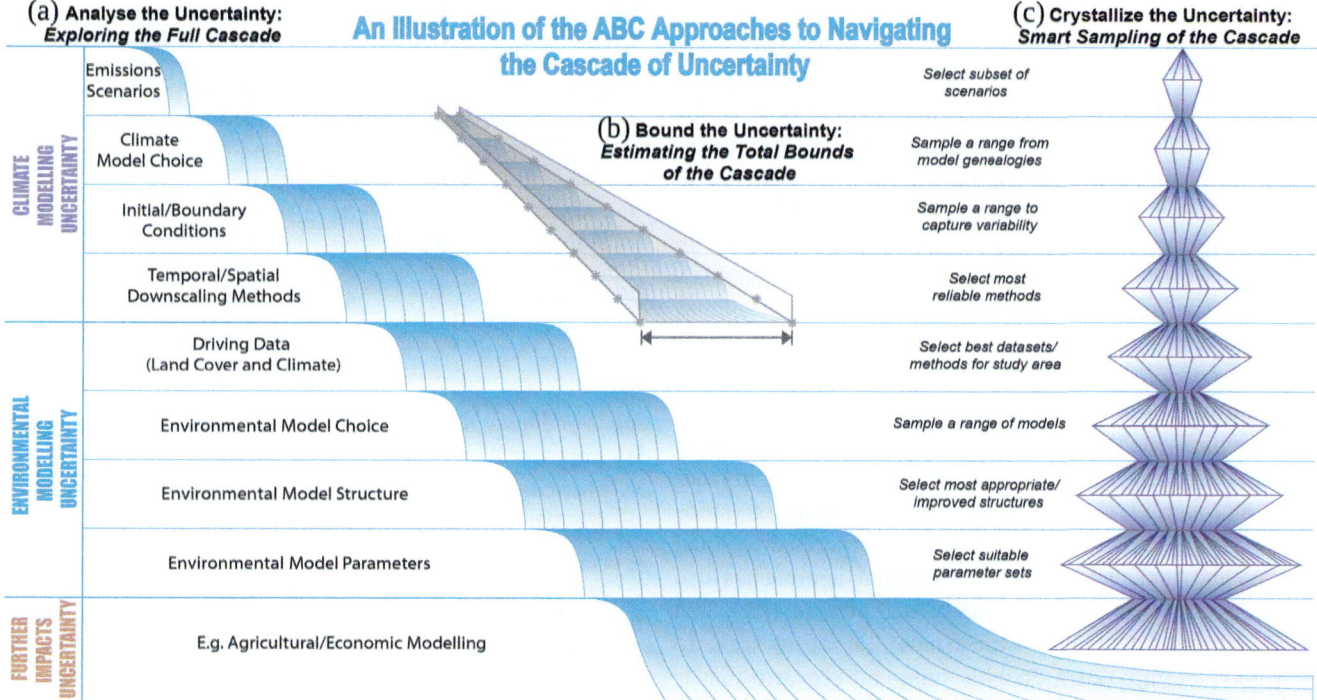

Fig. 9.1 An illustration of the "Analyse", "Bound" and "Crystallise" approaches to tackling the cascade of uncertainty in environmental impacts of climate change. Source: Smith et al. (2018)

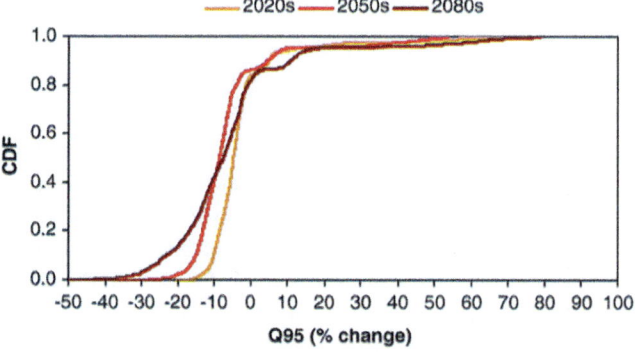

Fig. 9.2 Example cumulative distribution functions (CDFs) for changes (%) in low flows (95th percentile of daily flows) in the River Thames by the 2020s, 2050s and 2080s reflecting emissions, climate model, empirical downscaling, hydrological model and parameter uncertainty. Source: Wilby et al. (2006)

of exceeding extreme weather records even under current climate variability (e.g. Thompson et al., 2017; Kelder et al., 2022).

Returning to Fig. 9.1, 'B' is for 'Bound' the uncertainty (others might refer to the category as 'bottom up'). This framework involves setting 'credible' upper and lower limits to the values used in a modelling chain, such that the full range of system uncertainty is explored. The main advantage is that the computational burden and resources required are considerably reduced compared with the analyse-all strategy, although it can be significant if wide uncertainty bounds are initially used. Bounded sensitivity testing of hydrologic systems has been undertaken in a range of contexts and is an essential feature of decision-scaling (Brown et al., 2012) and scenario-neutral (Prudhomme et al., 2010) methods. These create response surfaces showing changes in meaningful impact metrics (e.g. water in storage, flood areas, annual costs) linked to prescribed changes in climate drivers (e.g. precipitation variability) within the specified bounds (e.g. Poff et al., 2016).

The Bound framework is particularly useful for discerning trade-offs, critical vulnerabilities and thresholds in system behaviour, with and without adaptations; it can also be used to guide the further application of scenarios or 'storylines'. For example, Broderick et al. (2019) demonstrated that a 20% allowance for changes in future flood magnitude in Ireland could afford protection against 48% to 98% of the uncertainty in the CMIP5 ensemble range. However, defining credible maxima and minima for sensitivity testing is not always straightforward. Some recommend perturbation ranges beyond even those of climate model projections to identify system limits/breaking points (e.g. Culley et al., 2016); others have developed stochastic weather generators with vulnerability assessments in mind (e.g. Kilsby et al., 2007; Steinschneider et al., 2019; Wilby et al., 2014). This chapter will return to some of these tools later.

Finally, framework 'C' is for 'Crystallise'. Now the emphasis is on distilling considerable amounts of climate model information into a much smaller set of scenarios or 'storylines' whilst capturing most of the variability within the ensemble and related climate model experiments (Shepherd et al., 2018; Chan et al., 2022b). This not only reduces the amount of work that needs to be done but is also a very powerful communication device, applied most notably in the 2014 climate scenarios for the Netherlands.[4] Moreover, contrasting, yet plausible futures can be co-developed with stakeholders, then used to test water system resilience. For instance, Yates et al. (2015) investigated various storylines of future warming and drying in Colorado with associated narratives of forest dieback (due to fire and/or beetle attack) and more dust on snow events (shifting the timing and volume of snowpack melt). Hydrological model parameters were adjusted accordingly to represent attendant changes in land-surface properties, thereby creating water resource scenarios that were internally consistent with the prescribed regional climate and land surface changes. Storyline approaches also make space to imagine the barely unthinkable (so-called black, grey, and now green swan events). The H++ scenario is a good example of a credible high-end scenario that is now firmly embedded in UK climate risk assessment and sensitivity testing (Ranger et al., 2013).

Although it has been convenient to categorise uncertainty frameworks as A, B or C, in practice, the water sector is not afraid to adopt a mixed methods approach—to draw on the strengths of all. The next two sections show how hydrologists have been applying climate model information in practice to ensure that assets and long-term plans continue to deliver intended benefits regardless of the climate outlook.

9.3 Allowances for Climate Change

Most water infrastructure is long-lived. Hence, there are growing risks of failure or lower levels of service if assets are not designed with climate variability and change in mind. Unfortunately, there is limited guidance on how allowances for climate change should be derived and then incorporated within detailed engineering designs. The UK Government has developed advice with regional allowances for design variables such as peak river flows, intense rainfall and sea level[5]; the US Army Corps of Engineers created an online *Sea-Level Change Curve Calculator*[6] to help with vertical construction infrastructure and flood proofing in the coastal zone. Other rare examples include the Netherlands Ministry of Infrastructure and *Environment Guideline for Stress Test-*

Table 9.1 Climate change adjustment factors (%) for annual 1-day maximum rainfall (Rx1day) in Viet Nam. These values were based on CMIP5, under RCP8.5. Return-period estimates are from the Gumbel distribution. All changes are relative to 1986–2005, for the 97.5th percentile of the credible ensemble, rounded up to the nearest 5%. Source: ADB (2020)

Future period	Return period (years)				
	2	5	10	20	25
2016–2035	15	20	25	25	25
2036–2055	35	25	30	30	35
2056–2075	50	45	45	45	45
2076–2095	80	75	75	70	70

ing the Climate Resilience of Urban Areas,[7] the International Hydropower Association's *Hydropower Sector Climate Resilience Guide,*[8] and the Asian Development Bank's *Manual on Climate Change Adjustments for Detailed Engineering Design of Roads.*[9]

No doubt such guidance is in short supply because the processes involved in creating it are technically demanding and protracted. Any recommendations have to be aligned with national standards and procedures. Allowances must also be provided for typical design variables, but these quantities are not always directly available or archived by climate modelling centres. Physically meaningful performance metrics are then needed to select climate models that are most skilful at representing (sub-daily) extremes (e.g. Cortés-Hernández et al. 2016). Next, climate model information has to be processed and presented in forms that are familiar to practitioners. Hence, 'look up tables' of allowances are preferred (e.g. Table 9.1). This involves making decisions about what models to include or exclude, what extreme value distribution to fit, what baseline period to use as the reference point, and what emissions scenarios and future periods are relevant. How climate information features in decision-making is important too. This depends on how precautionary the end-use will need to be, or whether, and how other uncertainties are included. Above all, working assumptions and methodological details should be transparent, with governance structures in place for periodic review and updating of allowances as their scientific basis evolves.

Figure 9.3 shows what was involved in the creation of climate change adjustments for Viet Nam (Table 9.1). ADB (2018) developed a preliminary set of adjustment factors for each province using a single regional climate model driven by three Global Climate Models (GCMs) under RCP8.5. ADB (2020) followed the same procedure but used an ensemble of 16 GCMs from CMIP5, also for RCP8.5, but without any

---

[4] http://www.climatescenarios.nl/images/Brochure_KNMI14_EN_2015.pdf

[5] https://www.gov.uk/guidance/flood-and-coastal-risk-projects-schemes-and-strategies-climate-change-allowances

[6] https://cwbi-app.sec.usace.army.mil/rccslc/slcc_calc.html

[7] https://climate-adapt.eea.europa.eu/metadata/guidances/guideline-for-stress-testing-the-climate-resilience-of-urban-areas/11258895

[8] https://www.hydropower.org/publications/hydropower-sector-climate-resilience-guide

[9] https://doi.org/10.22617/TIM200147-2

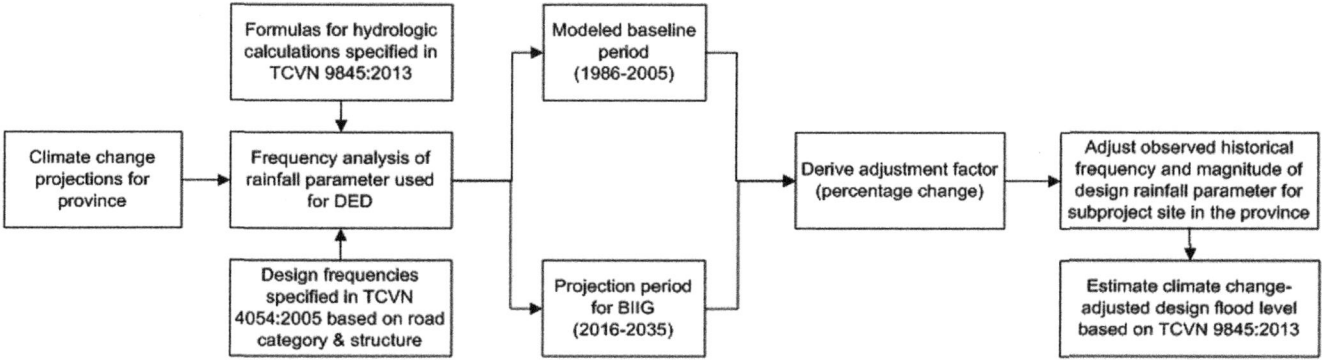

Fig. 9.3 Procedure for incorporating climate change allowances in detailed engineering design (DED). Key: BIIG = Basic Infrastructure for Inclusive Growth (projects), TCVN = national standards of Viet Nam. Source: ADB (2018)

downscaling. Given the low confidence in regional projections, average adjustment factors were derived for the whole of Vietnam rather than at the provincial level. The choice of climate variables (of specified durations and return periods) was mandated by Vietnam national standards for roads (TCVN 4054:2005) and bridges (TCVN 9845:2013). These require a design flood level of 1 in 100 years for expressways and 1 in 25 years for category 3, 4 or 5 rural roads (Fig. 9.3). The 1-day annual maximum rainfall total (Rx1day) is a critical element in their hydrologic formulae, including for design water discharge, flood level and discharge velocity, so this variable was given special attention.

Allowances like those shown in Table 9.1 are implemented in four steps. First, the required design standard is determined by the type of structure. For example, this might be Rx1day with 25-year return period for a culvert draining a rural road with expected service life of 20 years. Second, a baseline value for Rx1day, with same return period, is obtained from historic records for the site (or nearest representative meteorological station). This might be Rx1day = 320 mm. Third, the appropriate allowance is taken from Table 9.1. For Rx1day with 25-year return period and project life extending into 2036–2055, the allowance is +35%. Finally, the uplifted design value of Rx1day is 432 mm (i.e. 320 mm plus 35%). This quantity is used in subsequent calculations of river discharge, water levels and velocities to establish the dimension of the culvert. Headroom is built into Table 9.1 by using the 97.5th percentiles of the credible ensemble range, rounded up to the nearest 5%.

Although climate change allowances depend on location and application, some aspects are transferrable. Four guiding principles sat behind the development of ADB (2020:xii). These were that adjustment factors should (1) draw on strong scientific evidence yet be pragmatic, proportionate in terms of the effort involved, and reflective of key uncertainties; (2) adopt national engineering design standards and procedures (exemplified above for Vietnam); (3) require modest amounts of data (in this case for creating scenarios of changes in extreme rainfall, regional sea level rise and high-end water levels); and (4) apply fully transparent calculations for common design parameters, such as channel discharge, flow depth and velocity, mean sea level, storm surge and wave height and coastal erosion.

Similar workflows have been followed by other national agencies and professional bodies (e.g. Dale et al., 2018). These tend to be pragmatic, but there remains a wider need for revamping upstream decision rules and evaluation principles for project justification (Stakhiv, 2011). In particular, conventional approaches to discounting can present obstacles when the costs for higher standards of design (i.e. adaptation) are borne up front, whereas the benefits from these adjustments to the project may not be realised for decades to come. Indeed, there is a danger that major investments are continuously postponed, leading to increasing risk.

One possible approach to managing this would be to develop adaptive designs for the infrastructure itself (or making the infrastructure system adaptable). Some assets are sufficiently short-lived that they will be replaced in time (e.g. access road resurfacing), but others (e.g. major pipelines, dams, fixed parts of treatment works) are expected to be operational in the late twenty-first century and beyond. As discussed above—where safety is concerned—a precautionary approach can be justified, but in other cases, costs may be prohibitive. However, it is possible to consider how assets might be extended in the future or whether they could be designed to be quickly adapted. For instance, additional mechanical equipment could be brought in during a drought. Use of real options analysis or least worse regret analysis can help end-users balance opportunity cost with regret too.

The following section steps back from the detail of individual projects to consider how plans for entire water supply systems can be tested and bolstered against climate threats.

9.4 Climate Resilient Water Supply Systems

Worldwide, water suppliers face acute and chronic threats from climate variability and change. Short-duration, hydroclimatic extremes can disrupt operations or even threaten the safety of critical assets, such as the near collapse of the emergency spillway at Oroville Dam in 2017 following record-breaking winter rainfall (Vano et al., 2019). At the other extreme, the 2012–2014 California drought may have been the most severe in the last 1200 years (Griffin & Anchukaitis, 2014). In addition, long-term changes in the climatic-drivers of water supply and demand present systemic risks to the industry. These may be countered by a raft of measures, including investments in new infrastructure, water reuse, source protection, improved efficiency and behaviour change, regional water transfer and reallocation of water between uses. However, planners need future climate scenarios to test the robustness of existing water systems to both acute and chronic climate risks; scenarios are also used to explore the efficacy of adaptation measures and trade-offs between options. These scenarios are typically created from ensembles of regional climate simulations ('Analyse' framework), via weather generators ('Bound' framework), but less so for storylines ('Crystallise' framework). Following some contextual matter, the use of these frameworks is illustrated below with reference to the UK water sector.

The UK water regulator (Ofwat) has a 'resilience duty'[10] under the 2014 Water Act. Resilience has been defined by the UK Government as 'the ability of assets, networks and systems to anticipate, absorb, adapt to and/or rapidly recover from a disruptive event'. Improved resilience may be delivered by (1) avoiding dependence on single assets (i.e. by increasing redundancy), (2) enhancing the capacity of systems to withstand disruptions such as floods (i.e. by increasing resistance), (3) adopting design standards to ensure the system functions irrespective of risks (i.e. by increasing reliability) and (4) testing procedures to ensure quick return of disrupted services (i.e. by increasing rates of recovery). The National Infrastructure Commission (2018) placed particular emphasis on delivering resilience to water shortages as a consequence of climate change and population growth whilst protecting the environment.[11] In their Water Resource Management Plan 2019, water companies set out measures over the period 2025 to 2050 for achieving resilience to 'worst historic', 'severe' and 'extreme' droughts (with respectively 1%, 0.5% and 0.2% annual probability). At that time, the cost of proactive long-term resilience improvements was estimated to be £18–21 billion, compared with costs of relying on emergency measures for severe droughts of £25–40 billion (Fig. 9.4). Hence, the costs of achieving resilience to the combined uncertainty in the assessed population and climate scenarios by 2050 were in the range of £3–15 billion.

The above requirements presented significant technical challenges, not least because there are very few river-flow records longer than 50 years and, for rainfall, more than 150 years *anywhere*. How then did the Commission and water companies generate information about a 500-year drought that is credible for the present and climate by 2050? The Commission calculated future water balances using two climate change scenarios taken from the *Future Flows Hydrology* archive[12]: central, medium emissions, average water balance scenario, and dry, medium emissions with less water in South East England. These were based on an ensemble of 11 variants of the Hadley Centre Regional Climate Model HadRM3-PPE under SRES A1B emissions, for 282 river sites and 24 boreholes (Prudhomme et al., 2013). Bias-correction was applied to the HadRM3-PPE precipitation and temperature series, which were then re-gridded to 1 km resolution and input to three hydrological models. Hence, the resilience projections were developed using a Type A framework (see Sect. 9.2) with partial representation of uncertainties.

All water companies in England and Wales are required to publish a water resources management plan every 5 years that looks forward at least 25 years. These plans include key information to support financing for water resources, demand management and leakage services. Anglian Water provides water supply and sewage services to more than six million customers in the East of England. The company covers the largest geographical area with some of the driest locations and regions of most rapid population growth in the UK. Hence, this company faces some of the most pressing climate-supply-demand challenges in the country.

Anglian Water's £6.5 billion investment plan for 2020 to 2025 was submitted to the economic regulator Ofwat in September 2018; of this, approximately £300 M was for adaptation of the water resources system to climate change impacts. The Water Resource Management Plan 2019 (WRMP19) supply forecast[13] incorporated analysis of historical and stochastically generated droughts, including for an 'extreme' event with approximately 1 in 500-year return period (judged to have a 5% chance of occurring during the 25-year planning period). Synthetic droughts were based on output from a spatially coherent weather generator (Serinaldi & Kilsby, 2012, as updated by Met Office, 2016) followed by stringent selection of a subset of severe events with different rainfall deficits, durations and geographic properties. Short-listed 200-year return period droughts and one 500-year drought were analysed using the Aquator water

---

[10] https://www.ofwat.gov.uk/regulated-companies/resilience-2/
[11] https://nic.org.uk/app/uploads/NIC-Preparing-for-a-Drier-Future-26-April-2018.pdf
[12] https://www.ceh.ac.uk/services/future-flows-maps-and-datasets
[13] https://www.anglianwater.co.uk/siteassets/household/about-us/supply-forecast.pdf

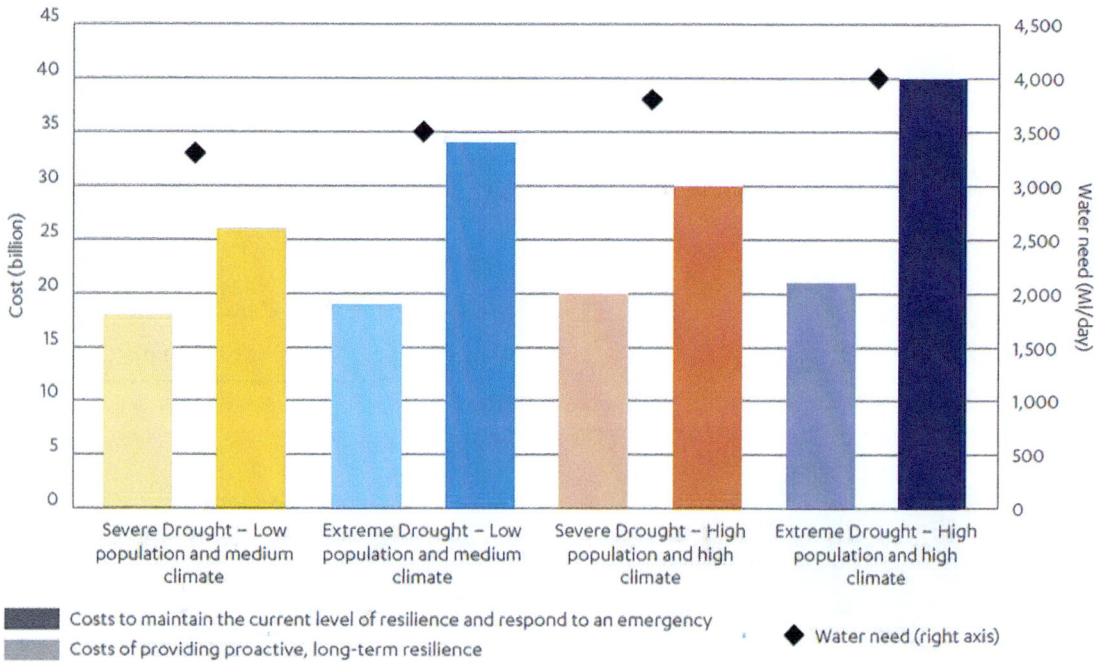

Fig. 9.4 Costs (£ billion) of providing proactive, long-term resilience versus relying on emergency response for droughts beyond current resilience levels. Costs are expected present values to 2050 and include maintaining current levels of resilience. Source: National Infrastructure Commission (2018)

resource zone (WRZ) model to assess the impacts on water yield and deployable output.

Future climate change impacts were evaluated using the UK Climate Projections 2009 (UKCP09) following guidance issued to water companies by the Environment Agency. All 33 of the available HadRM3-PPE projections were run through rainfall-runoff and recharge models to evaluate impacts on surface water and groundwater recharge. Then, a representative subset of scenarios were fed into groundwater yield models and the Aquator model to simulate deployable output. Finally, the ensemble median (in terms of surface water and groundwater yields) under the medium emissions was used to calculate the reduction in deployable output to reveal the most vulnerable WRZs (Fig. 9.5). This central scenario shaped the 'Preferred Plan' in the WRMP19, alongside 'representative' high and low scenarios to capture uncertainty, which were included in 'headroom', a planning margin between supply and demand. These investigations revealed a potential 57.7 Ml/d reduction in deployable output by 2045—the basis for then assessing investment options. In summary, Anglian Water implemented a mix of the 'Bound' and 'Crystallise' frameworks in their WRMP19.

9.5 Discussion and Concluding Remarks

The previous examples show that considerable effort is required to translate climate model output into decision-relevant information for project justification and resilience testing. Time and resource constraints mean that exhaustive assessment of uncertainties (i.e. the 'Analyse' framework) is seldom feasible in the commercial world. Evident in all water-sector literature surveyed, there is an appetite to embrace uncertainty through judicious selection of representative or 'marker' scenarios. Perhaps there is scope within the research community to support this readiness through development of guiding principles to 'Bound' assessments in objective and consistent ways, both in relation to climate change projections and hydrological model uncertainty.

As has been shown, weather generators (and more recently their dynamical modelling equivalent—UNSEEN techniques) are already widely used for climate risk assessment and options appraisal. Rather than a confounding factor, it appears that uncertainty analysis is contributing to more rigorous and systematic appraisal of water supply systems and plans. By stochastically generating extreme events (such as a 500-year drought) then identifying water system vulnerabilities, it is possible to focus resources on improving resilience to almost unimaginable events. However, confidence in the credibility of these stress tests depends on thorough validation of weather generator skill at reproducing physically plausible extremes (high-intensity sub-daily rainfalls through to multi-season or even multi-year droughts) (Kilsby et al., 2007; Steinschneider et al., 2019). Sensitivity testing should also look beyond perturbations to primary variables such as seasonal rainfall

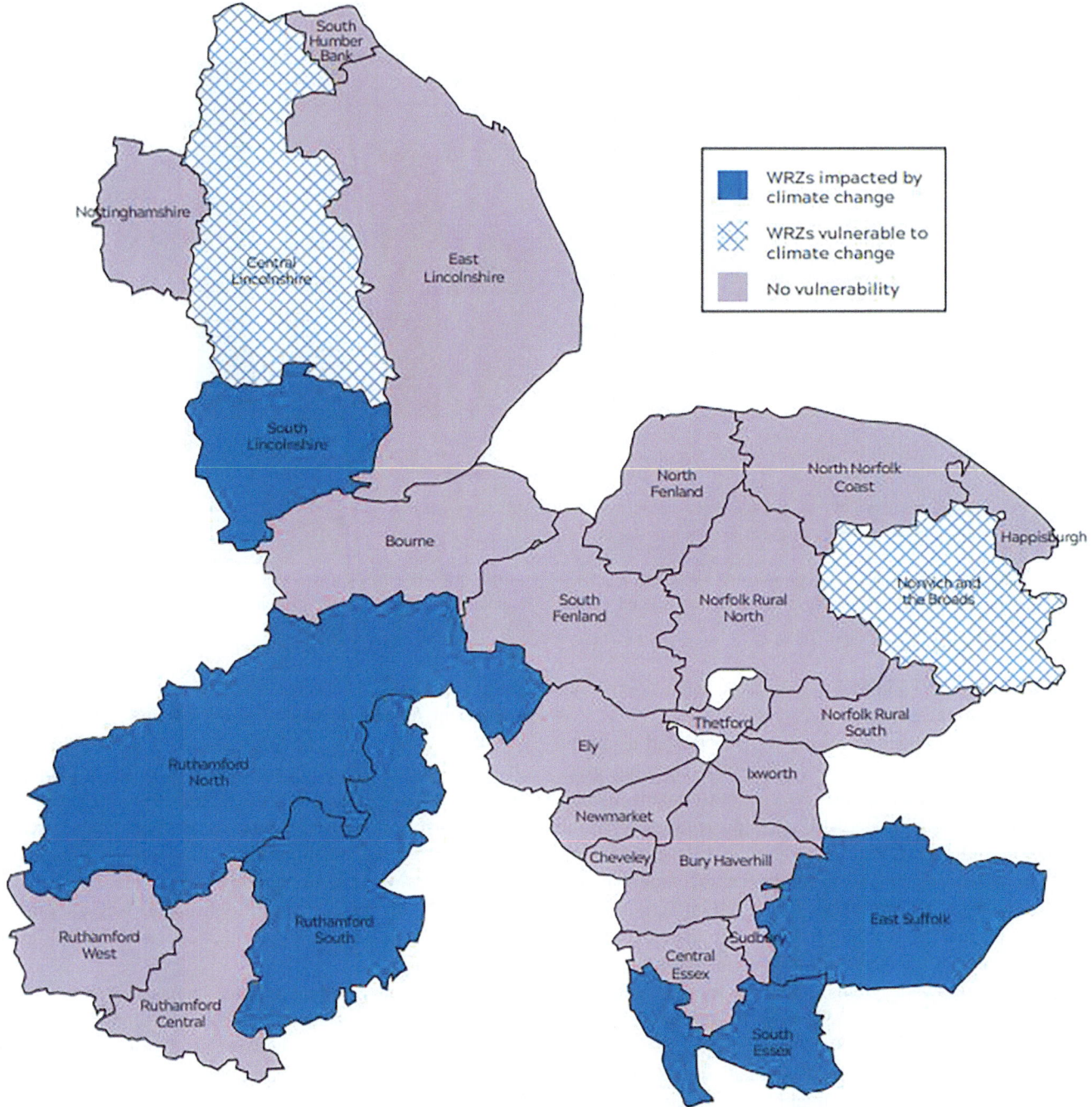

Fig. 9.5 Water resource zones (WRZs) affected by climate change by 2045. Source: Anglian Water Ltd. (2019) *Water Resources Management Plan 2019*

and mean temperature to more elaborate adjustments covering variability, intermittency, extremes, seasonality and/or inter-annual persistence (Culley et al., 2019: 121).

There are other opportunities. So far, storyline applications appear to be under-represented. Yet, they offer a way of rationalising very large numbers of scenarios whilst providing a richer, more coherent and internally consistent set of drivers for exploring system vulnerabilities and adaptation outcomes (Shepherd et al., 2018). Storylines also enable contemplation of rare but high-risk compound events—such as an H++ sea level rise combined with storm surge (Wilby et al., 2011) or deadly heat following a powerful cyclone (Matthews et al., 2019). As seen with Covid-19, previously unanticipated combinations of circumstances and impacts have proved significant. Greater attention to scenario generation and uncertainty analysis is needed to support adaptation efforts in data-scarce regions too. Even where ground-based hydroclimatic information is sparse, it is possible to blend this with remotely sensed data to parameterise weather generators and create atlases of climate hazards (e.g. Wilby & Yu, 2013).

In conclusion, this chapter focused on a narrow set of issues drawn from the very broad topic of uncertainty in hydrologic and water resource modelling. Emphasis was placed on how uncertainty frameworks have facilitated the application of climate science within the water sector by creating workflows for new design standards and testing the resilience of water supply systems. Unfortunately, there was not space to cover other important topics like uncertainties in global drought and flood hotspots, in the integrated assessment of the climate-water-energy-food nexus or in the outlook for global 'water towers' and transboundary water flows or for water-related health hazards or climate impacts on freshwater ecosystems.

However, recent experience of the water sector demonstrates that water managers and planners are learning to live with uncertainty, using climate change allowances, robust decision-making, stress testing of systems and options appraisal, adaptive management and decision-scaling techniques. Nonetheless, more can still be done to strengthen the enabling environment for practitioners, not least, by providing easier access to climate model products and developing guidance for embedding climate adjustments in asset design, operating rules and water resource plans. Researchers and practitioners can also work together on emerging and potential impacts beyond those conventionally modelled, including abrupt shifts in hydrological behaviour under climate change, water quality impacts and future environmental flow requirements. These all offer exciting frontiers for future research and development.

Acknowledgements The authors thank Hayley Fowler for her constructive feedback on an earlier version of this chapter.

References

Anglian Water Ltd. (2019). *Water resources management plan 2019*. Anglian Water Ltd.. https://www.anglianwater.co.uk/corporate/strategies-and-plans/water-resources-management-plan/wrmp19/

Asian Development Bank (ADB) (2018). *Adjusting hydrological inputs to road design for climate change risk based on extreme value analysis*. Report ADB PPTA 8957-VIE. Manila.

Asian Development Bank (ADB) (2020). *Manual on climate change adjustments for detailed engineering design of roads using examples from Viet Nam*. Asian Development Bank, Manila, Philippines. https://doi.org/10.22617/TIM200147-2.

Borgomeo, E., Mortazavi-Naeini, M., Hall, J. W., O'Sullivan, M. J., & Watson, T. (2016). Trading-off tolerable risk with climate change adaptation costs in water supply systems. *Water Resources Research, 52*, 622–643. https://doi.org/10.1002/2015WR018164

Broderick, C., Matthews, T., Wilby, R. L., Bastola, S., & Murphy, C. (2016). Transferability of hydrological models and ensemble averaging methods between contrasting climatic periods. *Water Resources Research, 52*, 8343–8373. https://doi.org/10.1002/2016WR018850

Broderick, C., Murphy, C., Wilby, R. L., Matthews, T., Prudhomme, C., & Adamson, M. (2019). Using a scenario-neutral framework to avoid potential maladaptation to future flood risk. *Water Resources Research, 55*, 1079–1104. https://doi.org/10.1029/2018WR023623

Brown, C., Ghile, Y., Laverty, M., & Li, K. (2012). Decision scaling: Linking bottom-up vulnerability analysis with climate projections in the water sector. *Water Resources Research, 48*, W09537. https://doi.org/10.1029/2011WR011212

Chan, W. C. H., Shepherd, T. G., Facer-Childs, K., Darch, G., & Arnell, N. W. (2022a). Tracking the methodological development of climate change projections for UK river flows. *Progress in Physical Geography, 46*, 589–612. https://doi.org/10.1177/03091333221079201

Chan, W. C. H., Shepherd, T. G., Facer-Childs, K., Darch, G., & Arnell, N. W. (2022b). Storylines of UK drought based on the 2010–2012 event. *Hydrology and Earth System Sciences, 26*, 1755–1777. https://doi.org/10.5194/hess-26-1755-2022

Clark, M. P., et al. (2016). Characterizing uncertainty of the hydrologic impacts of climate change. *Current Climate Change Reports, 2*, 55–64. https://doi.org/10.1007/s40641-016-0034-x

Cortés-Hernández, V. E., Zheng, F., Evans, J., Lambert, M., Sharma, A., & Westra, S. (2016). Evaluating regional climate models for simulating sub-daily rainfall extremes. *Climate Dynamics, 47*, 1613–1628. https://doi.org/10.1007/s00382-015-2923-4

Culley, S., Noble, S., Yates, A., Timbs, M., Westra, S., Maier, H. R., Giuliani, M., & Castelletti, A. (2016). A bottom-up approach to identifying the maximum operational adaptive capacity of water resource systems to a changing climate. *Water Resources Research, 52*, 6751–6768. https://doi.org/10.1002/2015WR018253

Culley, S., Bennett, B., Westra, S., & Maier, H. R. (2019). Generating realistic perturbed hydrometeorological time series to inform scenario-neutral climate impact assessments. *Journal of Hydrology, 576*, 111–122. https://doi.org/10.1016/j.jhydrol.2019.06.005

Curry, J. A., & Webster, P. J. (2011). Climate science and the uncertainty monster. *Bulletin of the American Meteorological Society, 92*, 1667–1682. https://doi.org/10.1175/2011BAMS3139.1

Dale, M., Hosking, A., Gill, E., Kendon, E., Fowler, H. J., Blenkinsop, S., & Chan, S. (2018). Understanding how changing rainfall may impact on urban drainage systems; lessons from projects in the UK and USA. *Water Practice & Technology, 13*, 654–661. https://doi.org/10.2166/wpt.2018.069

Deloitte Mergers and Acquisitions Index (2017). *Uncertainty is the "new normal"*. Outlook for 2017. Deloitte, London, https://www2.deloitte.com/content/dam/Deloitte/uk/Documents/corporate-finance/deloitte-uk-ma-uncertainty.pdf

Ekström, M., Gutmann, E. D., Wilby, R. L., Tye, M. R., & Kirono, D. G. C. (2018). Robustness of hydroclimate metrics for climate change impact research. *WIRES Water, e1288*. https://doi.org/10.1002/wat2.1288

Fowler, K., Peel, M., Saft, M., Nathan, R., Horne, A., Wilby, R. L., McCutcheon, C., & Peterson, T. (2022). Hydrological shifts threaten water resources. *Water Resources Research, 58*, e2021WR031210. https://doi.org/10.1029/2021WR031210

Griffin, D., & Anchukaitis, K. J. (2014). How unusual is the 2012–2014 California drought? *Geophysical Research Letters, 41*, 9017–9023. https://doi.org/10.1002/2014GL062433

Hutchins, M. G., Abesser, C., Prudhomme, C., Elliott, J. A., Bloomfield, J. P., Mansour, M. M., & Hitt, O. E. (2018). Combined impacts of future land-use and climate stressors on water resources and quality in groundwater and surface waterbodies of the upper Thames river basin, UK. *Science of the Total Environment, 631*, 962–986. https://doi.org/10.1016/j.scitotenv.2018.03.052

Kelder, T., Wanders, N., Van der Wiel, K., Marjoribanks, T. I., Slater, L. J., Wilby, R. L., & Prudhomme, C. (2022). Interpreting extreme climate impacts from large ensemble simulations—Are they unseen or unrealistic? *Environmental Research Letters, 17,* 044052. https://doi.org/10.1088/1748-9326/ac5cf4

Kilsby, C. G., et al. (2007). A daily weather generator for use in climate change studies. *Environmental Modelling and Software, 22,* 1705–1719. https://doi.org/10.1016/j.envsoft.2007.02.005

Manning, L. J., Hall, J. W., Fowler, H. J., Kilsby, C. G., & Tebaldi, C. (2009). Using probabilistic climate change information from a multimodel ensemble for water resources assessment. *Water Resources Research, 45,* W11411. https://doi.org/10.1029/2007WR006674

Matthews, T., Wilby, R. L., & Murphy, C. (2019). An emerging tropical cyclone-deadly heat compound hazard. *Nature Climate Change, 9,* 602–606. https://doi.org/10.1038/s41558-019-0525-6

Met Office (2016). *The water resources East Anglia rainfall generator: Technical Report on model set up and quality assurance of stochastic outputs*, March 2016. Met Office, Exeter.

National Infrastructure Commission. (2018). *Preparing for a drier future: England's water infrastructure needs*. Her Majesty's Stationery Office. https://nic.org.uk/app/uploads/NIC-Preparing-for-a-Drier-Future-26-April-2018.pdf

New, M., Dessai, S., Lopez, A., & Wilby, R. L. (2007). Challenges in using probabilistic climate change information for impact assessment: An example from the water sector. *Philosophical Transactions of the Royal Society, 365,* 2117–2131. https://doi.org/10.1098/rsta.2007.2080

Poff, N. L., et al. (2016). Sustainable water management under future uncertainty with eco-engineering decision scaling. *Nature Climate Change, 6,* 25–34. https://doi.org/10.1038/nclimate2765

Prudhomme, C., Wilby, R. L., Crooks, S., Kay, A. L., & Reynard, N. S. (2010). Scenario-neutral approach to climate change impact studies: Application to flood risk. *Journal of Hydrology, 390,* 198–209. https://doi.org/10.1016/j.jhydrol.2010.06.043

Prudhomme, C., et al. (2013). Future flows hydrology: An ensemble of daily river flow and monthly groundwater levels for use for climate change impact assessment across Great Britain. *Earth System Science Data, 5,* 101–107. https://doi.org/10.5194/essd-5-101-2013

Prudhomme, C., et al. (2014). Hydrological droughts in the 21st century, hotspots and uncertainties from a global multi model ensemble experiment. *Proceedings of the National Academy of Sciences of the United States of America, 111,* 3262–3267. https://doi.org/10.1073/pnas.1222473110

Ranger, N., Reeder, T., & Lowe, J. (2013). Addressing 'deep' uncertainty over long-term climate in major infrastructure projects: Four innovations of the Thames estuary 2100 project. *EURO Journal on Decision Processes, 1,* 233–262. https://doi.org/10.1007/s40070-013-0014-5

Rowell, D. P. (2019). An observational constraint on CMIP5 projections of the east African long rains and southern Indian Ocean warming. *Geophysical Research Letters, 46,* 6050–6058. https://doi.org/10.1029/2019GL082847

Serinaldi, F., & Kilsby, C. G. (2012). A modular class of multi-site monthly rainfall generators for water resource management and impact studies. *Journal of Hydrology, 464-465,* 528–540. https://doi.org/10.1016/j.jhydrol.2012.07.043

Shepherd, T. G., et al. (2018). Storylines: An alternative approach to representing uncertainty in climate change. *Climatic Change, 151,* 555–571. https://doi.org/10.1007/s10584-018-2317-9

Slater, L. J., et al. (2021). Nonstationary weather and water extremes: A review of methods for their detection, attribution, and management. *Hydrology and Earth System Sciences, 25,* 3897–3935. https://doi.org/10.5194/hess-25-3897-2021

Smith, K. A., Wilby, R. L., Broderick, C., Prudhomme, C., Matthews, T., Harrigan, S., & Murphy, C. (2018). Navigating cascades of uncertainty—As easy as ABC? Not quite.... *Journal of Extreme Events, 1850007.* https://doi.org/10.1142/S2345737618500070

Stakhiv, E. Z. (2011). Pragmatic approaches for water management under climate change uncertainty. *Journal of the American Water Resources Association, 47,* 1183–1196. https://doi.org/10.1111/j.1752-1688.2011.00589.x

Steinschneider, S., Ray, P., Rahat, S. H., & Kucharski, J. (2019). A weather-regime-based stochastic weather generator for climate vulnerability assessments of water systems in the Western United States. *Water Resources Research, 55,* 6923–6945. https://doi.org/10.1029/2018WR024446

Thompson, V., Dunstone, N. J., Scaife, A. A., Smith, D. M., Slingo, J. M., Brown, S., & Belcher, S. E. (2017). High risk of unprecedented UK rainfall in the current climate. *Nature Communications, 8,* 107. https://doi.org/10.1038/s41467-017-00275-3

Vano, J. A., Miller, K., Dettinger, M. D., Cifelli, R., Curtis, D., Dufour, A., Olsen, J. R., & Wilson, A. M. (2019). Hydroclimatic extremes as challenges for the water management community: Lessons from Oroville dam and hurricane Harvey. *Bulletin of the American Meteorological Society, 100,* S9–S14. https://doi.org/10.1175/BAMS-D-18-0135.1

Wilby, R. L. (2005). Uncertainty in water resource model parameters used for climate change impact assessment. *Hydrological Processes, 19,* 3201–3219. https://doi.org/10.1002/hyp.5819

Wilby, R. L. (2016). Managing rivers in a changing climate. In D. M. Gilvear, M. Greenwood, M. Thoms, & P. Wood (Eds.), *River science: Research and applications for the 21st century*. Wiley & Sons Ltd.. Chapter 25.

Wilby, R. L. (2019). A global hydrology research agenda fit for the 2030s. *Hydrology Research, 50,* 1464–1480. https://doi.org/10.2166/nh.2019.100

Wilby, R. L., & Harris, I. (2006). A framework for assessing uncertainties in climate change impacts: Low flow scenarios for the River Thames, UK. *Water Resources Research, 42,* W02419. https://doi.org/10.1029/2005WR004065

Wilby, R. L., & Murphy, C. (2018). Decision making by water managers despite climate uncertainty. In W. T. Pfeffer, J. B. Smith, & K. L. Ebi (Eds.), *Oxford handbook of planning for climate change hazards*. Oxford University Press. https://doi.org/10.1093/oxfordhb/9780190455811.013.52

Wilby, R. L., & Yu, D. (2013). Rainfall and temperature estimation for a data sparse region. *Hydrology and Earth System Sciences, 17,* 3937–3955. https://doi.org/10.5194/hess-17-3937-2013

Wilby, R. L., Orr, H. G., Hedger, M., Forrow, D., & Blackmore, M. (2006). Risks posed by climate change to the delivery of water framework directive objectives in the UK. *Environment International, 32,* 1043–1055. https://doi.org/10.1016/j.envint.2006.06.017

Wilby, R. L., Nicholls, R. J., Warren, R., Wheater, H. S., Clarke, D., & Dawson, R. J. (2011). Keeping nuclear and other coastal sites safe from climate change. *Proceedings of the Institution of Civil Engineers: Civil Engineering, 164,* 129–136. https://doi.org/10.1680/cien.2011.164.3.129

Wilby, R. L., Dawson, C. W., Murphy, C., O'Connor, P., & Hawkins, E. (2014). The statistical DownScaling model—Decision centric (SDSM-DC): Conceptual basis and applications. *Climate Research, 61,* 259–276. https://doi.org/10.3354/cr01254

Wilby, R. L., et al. (2017). The "dirty dozen" of freshwater science: Detecting then reconciling hydrological data biases and errors. *WIREs Water, 4,* e1209. https://doi.org/10.1002/wat2.1209

Yates, D., Miller, K. A., Wilby, R. L., & Kaatz, L. (2015). Decision-centric adaptation appraisal for water management across Colorado's continental divide. *Climate Risk Management, 10,* 35–50. https://doi.org/10.1016/j.crm.2015.06.001

Open Access This chapter is licensed under the terms of the Creative Commons Attribution 4.0 International License (http://creativecommons.org/licenses/by/4.0/), which permits use, sharing, adaptation, distribution and reproduction in any medium or format, as long as you give appropriate credit to the original author(s) and the source, provide a link to the Creative Commons license and indicate if changes were made.

The images or other third party material in this chapter are included in the chapter's Creative Commons license, unless indicated otherwise in a credit line to the material. If material is not included in the chapter's Creative Commons license and your intended use is not permitted by statutory regulation or exceeds the permitted use, you will need to obtain permission directly from the copyright holder.

Dimensions of Uncertainty in Mitigating Flooding

Sarah Michaels

10.1 Introduction

Climate change exacerbates threats with which we are familiar, such as flooding, requiring us to rethink strategies, in which we have developed confidence if not complacency. At the same time, climate change is one among a number of important sources of uncertainty as we grapple with making decisions about what will be the relationship between people and the environment on which we depend. For example, significant uncertainty surrounds assessing many factors determining flood hazard and flood risk, such as agricultural land expansion, wetland reclamation and deforestation, and projections for them (Kundzewicz et al., 2019).

Defining uncertainty as *"any departure from the unachievable ideal of complete determinism"* (Italics in the original) (Walker et al., 2003) captures the extent and prevalence of what we don't know and suggests uncertainty can range from deviating a little to a lot from complete determinism. In this definition, uncertainty includes risk defined as uncertainty quantified using probabilities (Knight, 1921). The above definition of uncertainty, however, is not sufficient to delineate the different forms uncertainty takes in impinging on decision making. This is what we need to know if we are to consider uncertainty constructively in policymaking.

In the complex, multifaceted, interlinked world of policymaking, those engaged in the process from information providers to decision-makers are rarely confronting a shared, single uncertainty in their different deliberations. Consequently, what is needed is a means of understanding the different dimensions of each uncertainty that create the uncertainty space in which policy relevant choices are made.

While dimensions of uncertainty are applicable regardless of policy sector, policymakers and those who advise them confront uncertainty in making specific policy choices about specific concerns for which they have authority and jurisdiction. In this chapter, we look at flooding, a worldwide devastating phenomenon, specifically flooding in the administratively and culturally valuable vicinity of the National Mall, in the capital of the United States, Washington, DC. To do so, we consider three drivers of the potential to be harmed: hazard, exposure, and vulnerability. The uncertainty associated with each can be understood by considering (1) Where is the uncertainty situated? (2) How much uncertainty is there? (3) What is the nature of the uncertainty (Walker et al., 2003)? The conclusion highlights the value of considering the combined dimensions of salient uncertainties.

10.2 Why Look at Flooding?

Worldwide, flooding is the most prevalent natural disaster (Bailey et al., 2021). A disaster is a situation which exceeds a community's independent ability to rebound fully and in a timely manner. Twenty-nine percent of the world's population, 2.2 billion people, live where some inundation will occur in a one-in-100-year flood event. Nineteen percent of the world population, 1.47 billion people live in high-risk flood zones, exposed to depths of inundation exceeding 0.5 feet (0.15 m) (Rentschler & Salhab, 2020). Flooding can be disastrous in its own right and exacerbate other tragedies (Munich, 2020). Faster than other natural hazards, the number of flood disasters reported has trended upward. Compared to the 1980–1989 decade, the total number of flood disasters in the 2010–2019 decade increased by 181% (Bailey et al., 2021).

There is high confidence in the Intergovernmental Panel on Climate Change (IPCC) Sixth Assessment 2021 Report that extreme weather events can be attributed to climate

S. Michaels (✉)
Department of Political Science and Nebraska Public Policy Center, University of Nebraska-Lincoln, Lincoln, NE, USA
e-mail: sarah.michaels@fulbrightmail.org

change and are aggravating the prospect of floods and droughts (Douville et al., 2021). It is anticipated that extreme precipitation events will intensify with every degree increase in global warming. For every degree of warming, it is anticipated that there will be an additional one or two extreme precipitation events of the extent that, on average, now occur twice per decade (Myhre et al., 2019). The increasing frequency and intensity of flood disasters resulting from more extreme precipitation than occurred in the past is detrimental to ecosystems, societies, and the economy (Tabari, 2020).

Between 1980 and 2019, flood losses worldwide amounted to 40% of all loss-related natural catastrophes and more than US $1 trillion, only 12% of which was insured (Munich, 2020). The deficient investment in flood insurance and underinvestment in flood protection is a function of collectively discounting flood risk even though flooding is pervasive and damaging (Bailey et al., 2021). Yet, compared to other natural hazards, flooding is the one for which precautionary measures are most effective and for which ameliorative actions have the greatest potential (Dieperink et al., 2016). In this context, it is valuable to understand what drives the potential to be harmed.

10.3 Hazard, Exposure, and Vulnerability as Drivers of the Potential to be Harmed

Hazard, exposure, and vulnerability constitute three essential elements of the expected value of a loss. They drive the prospect to be harmed. A hazard constitutes the potential occurrence of a physical event that may adversely impact exposed and vulnerable elements (Cardona et al., 2012). Flood hazard refers to the probability of a flood occurring in a given area with a defined recurrence interval with a particular severity (Santos et al., 2020). Exposure refers to what is in the area in which a hazardous event may happen. It is what is subject to loss should a hazardous event happen (Cardona et al., 2012). People, properties, and infrastructure situated in areas potentially impacted by flooding are flood exposed (Santos et al., 2020). Vulnerability refers to how prone exposed elements—whether it is people, how they make their living, or their assets—are to being adversely impacted by a hazardous event (Cardona et al., 2012, p. 69; Santos et al., 2020). Vulnerability is a function of not having the capacity to ameliorate exposure to a hazardous event or set of circumstances (Cardona et al., 2012). It is determined by the societal capacity to address an extreme event (Koks et al., 2015). While vulnerability is situation specific, it can be exacerbated by underlying conditions. For people, these conditions can involve poverty, isolation, lack of information, and maladaptive practices (Cardona et al., 2012). Flood vulnerability is the propensity to be adversely impacted by the occurrence of flooding, such as through the loss of housing or livelihood or facing flood-related health concerns, such as exposure to mold, waterborne diseases, and mental health issues.

While exposure is a prerequisite to being vulnerable, it is possible to be exposed without being vulnerable. This can happen, for example, if people have sufficient means to mitigate potential harm if exposed to a hazard, such as when living in a floodplain, households can afford to modify the building structure of their homes to withstand a flood event (Cardona et al., 2012).

While disaster is usually triggered by a hazard event, exposure and vulnerability are critical, driving factors (Cardona et al., 2012). Flood risk management is often focused on reducing exposure and vulnerability to flood-related threats (Koks et al., 2015). Increasing exposure and vulnerability to flooding can be attributed to human actions in flood-prone areas that do not implicate climatic mechanisms (Kundzewicz et al., 2020; Mitchell, 2003), such as urbanization, growing populations, and increasing economic activity. These actions do, however, interact with climatic mechanisms. Consequently, to understand composite uncertainty requires a holistic approach to socioeconomic development within the context of climate change, such as employed by Dawson et al. (2011) in assessing how effective nonstructural flood management measures are in the Thames Estuary, England.

Intensifying the prospective consequences of weather events are anthropogenic terrestrial changes, such as regulating rivers, decreasing how much water is retained in catchment, and sealing land surfaces (Mitchell, 2003). To explore further the relationship between hazards, exposure, and vulnerability, it is constructive to consider a specific flood setting.

10.4 Vignette: Flooding and the National Mall, Washington, DC

Washington, District of Columbia (DC), is located in the Mid-Atlantic region of the East Coast of the United States. Washington DC's flood hazard, notably in the vicinity of the National Mall (Fig. 10.1), is considered especially intolerable because of the exposure of buildings housing critical federal activities and highly valued artefacts of national and international significance. The National Mall area because of its downstream location has been impacted acutely by urban development throughout the Potomac watershed. This development has reduced vegetation cover and increased the extent of impervious surfaces (National Capital Planning Commission (NCPC), 2008).

Washington, DC, on the Atlantic Coastal Plain, is flood-prone because of its flat topography, broad floodplains, its location at the confluence of the Potomac and Anacostia

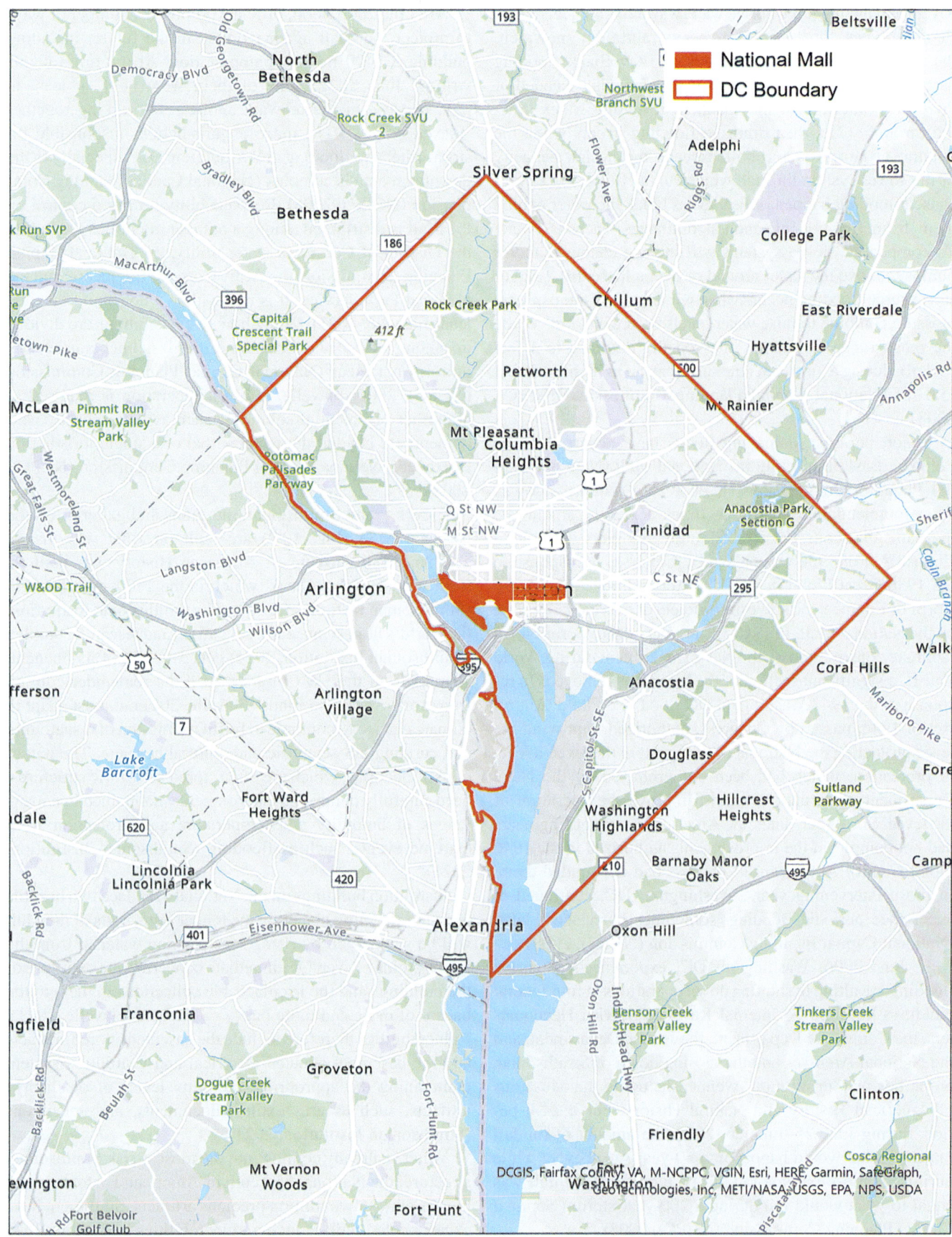

Fig. 10.1 The National Mall in Washington, DC. Source: National Capital Planning Commission (NCPC) (2008)

Rivers and having three buried waterways. It is subject to four flooding types, three caused by heavy rainfall or snowmelt, the fourth by the level of the tide. (1) Overbank flooding happens when the capacity of river channels is exceeded or there is a blockage in the channel preventing water flowing through. (2) Urban drainage flooding occurs when the volume of stormwater runoff exceeds the design capacity of the sewer system through which runoff flows. (3) Levee-caused flooding happens when levees block the water behind them from flowing into natural drainages, such as rivers. To compensate, levee systems will include channels and/or pumps to move the water around or over a levee. (4) Tidal or storm surge flooding occurs when wind and low atmospheric pressure combine to raise water levels before a storm. While the mean range of normal tides on the Potomac is 3 feet (0.9 m), during a hurricane, the surge can be as high as 12 feet (3.6 m) (National Capital Planning Commission (NCPC), 2008).

Major, periodic, damaging floods have occurred in the Potomac Basin impacting the National Mall (National Capital Planning Commission (NCPC), 2008), often followed by recommendations and sometimes action to ameliorate the damaging consequences of future flooding. The Flood Control Act of 1936 was passed by Congress in response to the 1936 Great Flood in DC. The Act authorized the Army Corps of Engineers to solve the problem of overbank flooding on the National Mall. In 1940, the project began operation. To facilitate construction by the Navy Department during World War II, a significant section of the levee was removed. In response to the 1942 Washington DC flood, the Flood Control Act of 1946, passed by Congress, authorized improving the levee protecting the National Mall to restore the levee's level of protection, which had been compromised by the Navy Department construction, and to improve the operation of the levee. Sixty years later, the American Corps of Engineers had not completed the improvements authorized in the 1946 Flood Control Act because Congress had not funded them. In those intervening years, Washington, DC, continued to experience periodic flooding, sometimes hurricane related (National Capital Planning Commission (NCPC), 2008).

In June 2006, Washington, DC, experienced extensive flooding, resulting in shutting down operations in four federal buildings housing the Internal Revenue Service Headquarters, the Commerce Department, the Justice Department, and the National Archives and the closing down of Smithsonian properties on Constitution Avenue, including the Museums of American History and Natural History. For a 24-h period, during June 25–June 26, 2006, the amount of rainfall matched what would happen in a 50-year storm event while during the most extreme rainfall period of 6 h, rainfall was equal to what would fall during a 200-year storm (National Capital Planning Commission (NCPC), 2008).

After the 2006 flood, officials proposed a number of ways to protect the Mall against floods in the future, including building a $400 million pump station. At the time of this writing, June 2022, none of them were built, at least in part, because of the relevant entities not working together (Flavelle, 2021). The maze of jurisdictions responsible for stormwater and flooding in Washington, DC, especially in the downtown area is complex (National Capital Planning Commission (NCPC), 2008). Responsibility for flood control on the Mall is distributed among a number of entities, including the District of Columbia water utility, the National Capital Planning Commission, the Army Corps of Engineers, and the National Park Service. It is not evident who should take the lead (Flavelle, 2021). While how responsibilities are divided up among federal and local authorities is neither consistently uniform nor clear (National Capital Planning Commission (NCPC), 2008; Flavelle, 2021), cooperation is required as a number of potential solutions, such as low impact development, are beyond the wherewithal of a single jurisdiction to execute (National Capital Planning Commission (NCPC), 2008).

Eleven grand Smithsonian museums and galleries on the National Mall (Fig. 10.2) are built on what was marsh. The National Museum of American History was built on what was the Tiber Creek, which in the 1800s was filled in (Smithsonian Institution, 2021). The Smithsonian Institution, the world's largest museum, education, and research complex (Smithsonian Institution, 2021) gets over half of its financial support from the US Congress with the remainder coming from nongovernment funds (Flavelle, 2021). It must adapt to climate change to continue to fulfill its mission of researching and curating US scientific and cultural heritage. The nature of its collections, which are often stored in historic structures, need carefully controlled environments maintained in narrow ranges of humidity and temperature and protection from extreme events, such as flooding (Smithsonian Institution, 2021).

Institution buildings on the National Mall face two hazards from the warming planet. In the longer term, parts of the Mall will be submerged by rising seas pushing water in from the tidal Potomac River. Of immediate concern is the exposure of the buildings and the irreplaceable collections in them to the hazard of more and more heavy rainstorms (Flavelle, 2021). Vulnerabilities therefore include the basement stored artifacts and the basement situated electrical and ventilation system maintaining the appropriate humidity levels of on display artifacts, such as art, textiles, documents, and specimens (Smithsonian Institution, 2021).

Water is already coming into the most at-risk Smithsonian structure, the National Museum of American History, which houses almost two million precious artifacts. Staff there have experimented with reducing vulnerabilities by putting flood barriers outside of windows, installing sensors throughout the

Fig. 10.2 Smithsonian museums and galleries on The National Mall. Source: Smithsonian Institution (2021)

building that are triggered when wet, putting plastic sheeting on a cabinet to guide water leaks into a trash can surrounded by absorbent fabric to catch water that missed the trash can, and situating strategically plastic, wheeled bins that can be deployed where needed to distribute their contents of a variant of cat litter to soak up water (Flavelle, 2021).

In 2021, the Smithsonian was in the early planning stages of a $39 million proposed project to fortify the American History Museum, including building flood gates. In addition, since 2015, the Smithsonian has been requesting federal government funding to begin constructing a $160 million storage site in Suitland, Maryland, to house item from the American History Museum and the National Art Gallery. While relocating collections is time consuming, requiring planning, new facility construction, and meticulous processing of each item (Flavelle, 2021), they would no longer be exposed to the flood hazard in the National Mall Area.

The above vignette illustrates the complexities and nuances of one flood hazard setting with many confounding layers, true of other settings. These include the need for cooperation among entities, dealing with more than one form of the hazard, lack of funding, evolving threats, and coping with the immediate situation while planning for the long term. What is needed is a framework in which to understand the dimensions of uncertainty salient to decision-making in these circumstances.

10.5 The Where, How Much, and Nature of Uncertainty in Decision-Making

The uncertainty space relevant to decision making is created by asking three questions: (1) Where is the uncertainty situated? (2) How much uncertainty is there? (3) What is the nature of the uncertainty (Walker et al., 2003)? The questions are applicable to each of the three drivers of loss: hazard, exposure, and vulnerability.

10.5.1 Where Is the Uncertainty Situated?

The first dimension of uncertainty relevant to decision-making is identifying where uncertainty exists that impinges on how and what policy choices get made. Where uncertainty is situated matters because it determines which people in what roles in the policy process have how much ability, if any, to manipulate the causes of uncertainty. Uncertainty may be situated (Walker et al., 2003):

(a) *Within the policy frame*: In the process of shaping how to interpret the societal problem being confronted, one view gains primacy over others. This results in policy being pushed in a particular direction. For example, in the vignette above, the dominant view is to defend the National Mall area from the flood hazard. That is to reduce the vulnerability of what is exposed to flooding rather than to reduce exposure. While planning has begun on developing a storage site outside of the area for the Museum of American History and the National Art Gallery, management and planning are focused on protecting the collections of artefacts *in situ* and flood-proofing the buildings housing them. Framing influences every stage of the policy process and the concurrent policy analysis.

(b) *The external context*: From a policy perspective, this involves contextual considerations beyond what can be manipulated in a particular policy process. This includes, and is not limited to, the overarching biophysical environment, politics, economics, social issues, and technology. In the vignette above, there are a number of significant contextual considerations. From the biophysical environment, there is the warming planet, sea-level rise, and weather patterns shaping the flood hazard and what is exposed to that hazard. From a political perspective, there is the tension between the District of Columbia and the US Federal Government (National Capital Planning Commission (NCPC), 2008; Flavelle, 2021). From an economics perspective, considerable funding is required from the federal government to reduce vulnerability within the District of Columbia. As illustrated in the vignette, authorizing legislation is no guarantee funding is forthcoming. Funding to reduce the vulnerability of buildings and what they house in the National Mall competes with other pressing social issues in DC, across the US, and US interests outside the country. For example, in a world where there is great human poverty, the question is raised whether funds should go to protecting artifacts. Technological advances provide more accurate and responsive monitoring sensors at lower costs than before, making possible the timely detection of exposure, such as water leaks, thus minimizing water damage to vulnerable artefacts. If warnings are effectively disseminated and acted upon, flood forecasting technologies contribute to improved preparedness (Wilby & Keenan, 2012). Technology also makes possible more effective and lower cost in flood defense systems, such as floodgates, to reduce the vulnerability of exposed structures.

(c) *The model of the system*: This refers to the structured, analytical process of creating an abstraction of the system of interest as it exists or as envisioned. That system can involve a hazard, an area exposed to hazard, or what is vulnerable or some combination of the three. A model enables decision support activities to explore how different alternatives effect outcomes of interest under different scenarios and makes possible examining tradeoffs among policies. Models have been used to further understanding of hazards, exposure, and vulnerability. They have been generated to link increases in temperature with flood exposure (Boesch et al., 2018; Masson-Delmotte, 2021) and have been used to envision future flood scenarios impacting the National Mall area (Smithsonian Institution, 2021; National Capital Planning Commission (NCPC), 2008).

(d) *In the outcomes of the policy analysis and the decision process, policy analysis supports*: While those generating policy advice may focus on the uncertainties associated with the results their models produce and how robust the decision support exercise conclusions are, policymakers may focus on the implications of those outcomes for the goals of their organizations (Walker et al., 2003). For example, the independent consultants engaged to determine the cause of the 2006 DC flood were concerned not to present their findings as more definitive than they were, resulting in their concluding that the rapid, intense rainfall occurring in a short time could not be definitively declared as the cause of the flood (National Capital Planning Commission (NCPC), 2008). In recognition of the assessments of the vulnerability of particular collections housed on the National Mall, Smithsonian decision-makers have proceeded with creating storage space for some of the Smithsonian collections outside the area of the National Mall (Bechtol, 2021).

(e) *In the weights different people assign to those outcomes*: A key factor in decision-making under uncertainty is how individuals perceive uncertainty (Slovic, 1987 uses the term risk rather than uncertainty). Uncertainty judgments vary among individuals because of differences in information about the uncertainty, different levels of uncertainty, relationships to power, or specific concerns (Slovic, 1987). That Congressional decision-makers have broader responsibilities and balance wider ranging interests than do Smithsonian decision-makers means they are unlikely to weigh the outcomes of analyses and decision processes related to the flood hazard and the exposure and vulnerability it creates for Smithsonian buildings and collections on the National Mall the same.

10.5.2 How Much Uncertainty Is There?

The second dimension of uncertainty relevant to decision-making is where the uncertainty is on the spectrum between absolute certainty and complete ignorance (Walker et al., 2003). On the low end of the uncertainty continuum, there are clear and strong indicators about the outcomes in a given policy sphere (Jakobsen, 2016). At the high end of the uncertainty continuum, in novel situations, there are not clear and strong indicators about what the future conditions or the outcomes will be in a given policy sphere or when changes may happen (Walker et al., 2003). Depending on the extent of uncertainty, different analyses are undertaken to form projections. For example, the Maryland Expert Group on Sea-level Rise has determined sea-level rise probabilities salient to Washington DC's flood hazard. DC is on land the State of Maryland ceded to be the nation's seat of government. There is a 66% probability by 2050 in Maryland there will be a 0.8–1.6 feet (0.2–0.4 m) relative mean sea-level rise. There is a 5% chance the mean sea-level rise will be greater than 2.0 feet (0.6 m) and a 1% chance it will be greater than 2.3 feet (0.7 m). These 2018 projections consider Mid-Atlantic US regional factors, such as ocean currents, subsidence, and how far the region is from melting glaciers and polar ice sheets. They were also tied to the larger-scale 2014 Intergovernmental Panel on Climate Change Fifth Assessment Report projections of global sea-level rise (Boesch et al., 2018). The Maryland Expert Group Sea-level Rise projections are of value in ascertaining the exposure of the Smithsonian properties on the National Mall in Washington, DC. Compared to the uncertainty associated with the hazard of sea-level rise noted above, there is less uncertainty about the exposure to flooding in the vicinity of the National Mall from precipitation runoff and from the nearby Potomac River. The National Museum of American History and the National Museum of Natural History on the National Mall have had well-documented experience of water getting into where artifacts are stored. Both have valuable collections on their lower floors. There is the least uncertainty about how to address the vulnerabilities of the structures and collections as the Smithsonian has already undertaken vulnerability assessments and developed plans to address them (Bechtol, 2021). The relative differences in the extent of uncertainty between hazard, exposure and vulnerability are worth noting. This appreciation explains why it is appropriate for the American History Museum to reduce the vulnerability of its collection even while there is a considerable range in the climate change related flood hazard to which the Museum will be exposed.

10.5.3 What Is the Nature of the Uncertainty?

The nature of uncertainty is the third dimension to identify (Walker et al., 2003). Is it because of what we don't know, imperfections in knowledge, epistemic uncertainty (van Dorsser et al., 2018), which has the potential to be reduced by increasing what we know? Or is it because of the inherent variability of the phenomena under scrutiny, which creates irreducible or aleatory uncertainty (Michaels & Tyre, 2012; van Dorsser et al., 2018)? With climate change, variability may be a function of different possibilities for human intervention. For example, the Maryland Expert Group on Sea-level Rise calculated, after 2050, the probability distribution estimates of relative sea-level rise for three different pathways for greenhouse gas emissions: meeting the Paris agreement, stabilized emissions, and growing emissions (Boesch et al., 2018).

Asking what the nature of uncertainty is creates two different paths for how to address the uncertainty being confronted. This is because research, which by definition cannot reduce irreducible uncertainty, can reduce epistemic uncertainty. Research can reduce epistemic uncertainty while acknowledging aleatory uncertainty. Such is the case when the Maryland Expert Committee on sea-level rise determined three different sets of probability-based projections for sea-level rise for the latter half of this century using 2000 as the base year. The Committee acknowledged the future pathway of global emissions was increasingly going to determine sea-level rise rates. If the Paris Climate Agreement goals are met, resulting in emissions being limited, so the global mean temperature is increased to less than 2 °C above preindustrial levels, there is a 66% likelihood sea-level rise in 2100 will be between 1.2 and 3 feet (0.4–0.9 m) with a 5% probability it will be greater than 3.7 feet (1.1 m). In contrast, if greenhouse gas emissions continue to grow after 2050, there is a 66% probability sea-level rise in Maryland by 2100 will be between 2.0 and 4.2 feet (0.6–1.3 m). This is twice to four times what sea-level rise was in the last century. These ranges of estimates for the sea-level rise hazard can be used as inputs into projecting tidal range and storm surge changes to estimate future exposure (Boesch et al., 2018).

The predominance of epistemic over aleatory uncertainty in understanding the vulnerability of the Smithsonian collections and buildings on the National Mall has made possible effective action to mitigate the flood hazard. Enough research has been done for the conservators in the American History Museum to take ameliorative action or at least know what needs to be done, if there are funds to do so (Smithsonian Institution, 2021). The archival and library collections from the basement of the National Air and Space Museum, on the National Mall, have been moved 30 miles (48 km) away (Bechtol, 2021). With the benefit of engineering studies, the latest addition to the Mall, the Smithsonian's National Museum of African American History and Culture was built with pumps sufficiently powerful to keep ground water from permeating its lower floors (Flavelle, 2021).

In sum, to identify uncertainty that can be addressed in policymaking three questions warrant responses: (1) Where is the uncertainty situated? (2) What is uncertain? (3) How much uncertainty is there? All three dimensions of uncertainty matter individually and in combination because they shape what policy responses are appropriate. In turn, the policies implemented shape the future uncertainty space and with that future policy options.

10.6 Concluding Remarks

Uncertainty needs to be considered now more than ever as climate change reveals the extent the past is not a predictor of future flooding scenarios (Milly et al., 2008). Yet, we have still to develop fully operational guidance to think through uncertainty, systematically, effectively, and appropriately as a parameter throughout deliberations shaping flood mitigation policy.

As one step toward doing so, this chapter has brought to the fore the necessity of recognizing three critical dimensions of uncertainty in decision-making. We have done so in the context of flooding, worldwide the most devastating natural hazard, using the specific example of the flood hazard on Washington, DC's National Mall. It is important to recognize the relationship between the appreciation of the drivers of harm and their associated uncertainties. Indeed, the Maryland Sea-level Rise Expert Group made clear their probabilistic projections of the hazard of sea-level rise were intended to be used in determining exposure. This includes estimating tidal range and storm surge changes and developing tools for mapping inundation and in reducing vulnerability, such as developing high-tide flooding adaptation strategies, in planning and regulating, and siting and designing infrastructure (Boesch et al., 2018).

With no expectation of being able to eliminate surprise, the most constructive strategy is to face it, however imperfectly or incompletely and to hedge against it (Schlesinger, 1967). This chapter has argued this needs to be done not for an omnibus uncertainty but rather for the compilation of the sets of uncertainties actors confront in executing their individual and collective responsibilities in the policy process.

Acknowledgments Thanks to Linda Mearns and Robert Wilby for helpful remarks on an earlier draft. Support for writing this chapter was provided by a Fulbright Canada Distinguished Research Chair in Environmental Science, Carleton University, and a University of Nebraska-Lincoln Faculty Development Fellowship.

References

Bailey, R., Saffioti, C., & Drall, S. (2021). *Sunk costs: The socioeconomic impacts of flooding, rethinking flood series, report 1* (p. 32). Marsh McLennan.

Bechtol, N. J. (2021). *Oversight of the Smithsonian Institution: Protecting Smithsonian facilities and collections against climate change hearing* (p. 4). U.S. House of Representatives Committee on House Administration.

Boesch, D. F., et al. (2018). *Sea-level rise: Projections for Maryland* (p. 27). University of Maryland Center for Environmental Science.

Cardona, O. D., et al. (2012). Determinants of risk: Exposure and vulnerability. In C. B. Field (Ed.), *Managing the risks of extreme events and disasters to advance climate change adaptation* (pp. 65–108). Cambridge University Press.

Dawson, R. J., Ball, T., Werritty, J., Werritty, A., Hall, J. W., & Roche, N. (2011). Assessing the effectiveness of non-structural flood management measures in the Thames estuary under conditions of socioeconomic and environmental change. *Global Environmental Change, 21*, 628–646. https://doi.org/10.1016/j.gloenvcha.2011.01.013

Dieperink, C., Hegger, D. L. T., Bakker, M. H. N., Kundzewicz, Z. W., Green, C., & Driessen, P. P. J. (2016). Recurrent governance challenges in the implementation and alignment of flood risk management strategies: A review. *Water Resources Management, 30*, 4467–4481. https://doi.org/10.1007/s11269-016-1491-7

Douville, H., et al. (2021). Water cycle changes. In V. Masson-Delmotte (Ed.), *Climate change 2021: The physical science basis* (pp. 1055–1212). Cambridge University Press. https://doi.org/10.1017/9781009157896.011

Flavelle, C. (2021). Saving history with sandbags: Climate change threatens the Smithsonian. *New York Times*. Retrieved from https://www.nytimes.com/2021/11/25/climate/smithsonian-museum-flooding.html.

Jakobsen, M. L. (2016). Uncertainty, ideas, and institutional reform. *Acta Politica, 51*, 102–121. https://doi.org/10.1057/ap.2015.1

Knight, F. H. (1921). *Risk, uncertainty and profit* (p. 381). Houghton Mifflin Company.

Koks, E. E., Jongman, B., Husby, T. G., & Botzen, W. J. W. (2015). Combining hazard, exposure, and social vulnerability to provide lessons for flood risk management. *Environmental Science & Policy, 47*, 42–52. https://doi.org/10.1016/j.envsci.2014.10.013

Kundzewicz, Z. W., Matczak, P., Otto, I. M., & Otto, P. E. (2020). From "atmosfear" to climate action. *Environmental Science & Policy, 105*, 75–83. https://doi.org/10.1016/j.envsci.2019.12.012

Kundzewicz, Z. W., Su, B., Wang, Y., Wang, G., Wang, G., Huang, J., & Tong, J. (2019). Flood risk in a range of spatial perspectives—From global to local scales. *Natural Hazards and Earth System Sciences, 19*, 1319–1328. https://doi.org/10.5194/nhess-19-1319-2019

Masson-Delmotte, V. (2021). Climate change 2021: The physical science basis. In *Contribution of working group I to the sixth assessment report of the intergovernmental panel on climate change* (p. 2391). Cambridge University Press.

Michaels, S., & Tyre, A. J. (2012). How indeterminism shapes ecologists' contributions to managing socio-ecological systems. *Conservation Letters, 5*, 289–295. https://doi.org/10.1111/j.1755-263X.2012.00241.x

Milly, P. C. D., Betancourt, J., Falkenmark, M., Hirsch, R. M., Kundzewicz, Z. W., Lettenmaier, D. P., & Stouffer, R. J. (2008). Stationarity is dead: Whither water management? *Science, 319*, 573–574. https://doi.org/10.1126/science.1151915

Mitchell, J. K. (2003). European river floods in a changing world. *Risk Analysis, 23*, 567–574. https://doi.org/10.1111/1539-6924.00337

Munich, R. E. (2020). *Flood risk: Underestimated natural hazards*. Retrieved from https://www.munichre.com/en/risks/natural-disasters-losses-are-trending-upwards/floods-and-flash-floods-underestimated-natural-hazards.html#-24989000

Myhre, G., et al. (2019). Frequency of extreme precipitation increases extensively with event rareness under global warming. *Scientific Reports, 9*, 10. https://doi.org/10.1038/s41598-019-52277-4

National Capital Planning Commission (NCPC). (2008). *Report on flooding and stormwater in NCPC* (p. 31). National Capital Planning Commission.

Rentschler, J., & Salhab, M. (2020). People in harm's way: Flood exposure and poverty in 189 countries. In *Policy research working paper 9447* (p. 28). The Word Bank. Retrieved from https://documents.worldbank.org/en/publication/documents-reports/documentdetail/669141603288540994/people-in-harms-way-flood-exposure-and-poverty-in-189-countries

Santos, P. P., Pereira, S., Zezere, J. L., Tavares, A. O., Reis, E., Garcia, R. A. C., & Oliveira, S. C. (2020). A comprehensive approach to understanding flood risk drivers at the municipal level. *Journal of Environmental Management, 260*, 110127. https://doi.org/10.1016/j.jenvman.2020.110127

Schlesinger, J. R. (1967). Organizational structures and planning. In R. N. McKean (Ed.), *Issues in defense economics* (pp. 185–216). National Bureau of Economic Research (NBER).

Slovic, P. (1987). Perception of risk. *Science, 236,* 4799. https://doi.org/10.1126/science.3563507

Smithsonian Institution. (2021). Climate change action plan. In *Presented for review by the Federal Chief Sustainability Officer and National Climate Task Force* (p. 29). Smithsonian Institution.

Tabari, H. (2020). Climate change impact on flood and extreme precipitation increases with water availability. *Scientific Reports, 10,* 13768. https://doi.org/10.1038/s41598-020-70816-2

van Dorsser, C., Walker, W. E., Taneja, P., & Marchau, V. A. W. J. (2018). Improving the link between the futures field and policymaking. *Futures, 104,* 75–84. https://doi.org/10.1016/j.futures.2018.05.004

Walker, W. E., Harremoës, P., Rotmans, J., van der Sluijs, J. P., van Asselt, M. B. A., Janssen, P., & Krayer von Krauss, M. P. (2003). Defining uncertainty: A conceptual basis for uncertainty management in model-based decision support. *Integrated Assessment, 4,* 5–17. https://doi.org/10.1076/iaij.4.1.5.16466

Wilby, R. L., & Keenan, R. (2012). Adapting to flood risk under climate change. *Progress in Physical Geography, 36,* 348–378. https://doi.org/10.1177/0309133312438908

Open Access This chapter is licensed under the terms of the Creative Commons Attribution 4.0 International License (http://creativecommons.org/licenses/by/4.0/), which permits use, sharing, adaptation, distribution and reproduction in any medium or format, as long as you give appropriate credit to the original author(s) and the source, provide a link to the Creative Commons license and indicate if changes were made.

The images or other third party material in this chapter are included in the chapter's Creative Commons license, unless indicated otherwise in a credit line to the material. If material is not included in the chapter's Creative Commons license and your intended use is not permitted by statutory regulation or exceeds the permitted use, you will need to obtain permission directly from the copyright holder.

Uncertainty in Transportation Infrastructure

Jennifer M. Jacobs

11.1 Introduction

Transportation is a crucial component underlying (quite literally) almost every aspect of our individual and collective lives. Without planes, trains, automobiles, and ships, our economy would stagnate, our society would be immobile, and our lives, our work, and our recreational activities would be irreversibly changed. Roads and bridges constitute the most widely recognized part of the global infrastructure for the transportation of people and goods. However, the transportation sector consists of a vast, interconnected system of assets and derived services, which also includes railroads, airports, ports, public transportation, and oil and natural gas pipelines as well as the people who build, operate, and maintain that system. In the United States alone, there are over $4.9 trillion (USD) dollars of publicly owned transportation facilities and equipment, including over 4.2 million miles of center lane roads, 600,000 bridges, 136,000 railroad miles, 20,000 airports, 900 ports, and 2.6 million miles of oil and natural gas pipelines (USDOT, 2020).

The transport system has many attributes that make it particularly climate vulnerable. Many of its components have design lives of 10–100 years with the actual service life being even longer. Present-day transportation systems are often built on routes that were established centuries ago before automobiles existed. For example, much of the United Kingdom's road network has its foundations in Roman transport priorities; hence, a network designed 2000+ years ago is still very much in use but exposed to different climate conditions. Globally, many of these roads are increasingly in harm's way due to climate change impacts from flooding, wildfires, and changing winters. In the continental United States, over 19,000 km of coastal roads are already subject to high-tide flood. The frequency and extent of inundation is projected to increase exponentially in the coming decades due to sea-level rise (Fant et al., 2021).

Uncertainty is not a term that is widely used within the transportation engineering practice because it implies that the system or asset is potentially vulnerable to failure, which may undermine the faith of the public who rely on transportation infrastructure to support movement of people, goods, and services. Instead, the treatment of uncertainty is more often framed as "risk and reliability" where risk combines the likelihood that a system or asset failure occurs with the consequences of that event and reliability is the likelihood that the system of interest performs as designed. Regardless, it is understood that transportation assets are designed, built and maintained in a manner in which most of the decisions include some degree of uncertainty and risk regardless of whether the methods and specifications that are used meet design codes, governmental regulations, or client wishes (Faber & Stewart, 2003). An advantage of design codes is that they systematically handle uncertainties that arise from variable material properties and loads. However, beyond the uncertainties addressed by the design code, there are numerous additional uncertainties that may arise due to the potential for design and construction errors, changes in post-construction use and maintenance practices, a mismatch between the specific design and the original model, and the stressors, including travel demand and weather extremes that are applied to the infrastructure over its lifetime (Bulleit, 2008; Chen et al., 2011). The trick is to reduce uncertainty enough to be able to design and build infrastructure within a tolerable risk.

The engineering mindset is that good designs do not fail (Petroski, 1985) and that the infrastructure is built to handle anticipated stresses. However, recent extreme weather events—including the devastating 2021 and 2022 floods in Western Europe, China, Africa, India, and Pakistan; the

J. M. Jacobs (✉)
Department of Civil and Environmental Engineering, University of New Hampshire, Durham, NH, USA
e-mail: Jennifer.Jacobs@usnh.edu

February 2021 North American cold wave, Cyclone Amphan in Bangladesh and India; and droughts, heat waves, and forest fires in Australia, Europe, western United States, and Uttarakhand—highlight the vulnerability of engineered transportation networks and bridges to extreme events. While it is broadly recognized that the global physical infrastructure is at increasing and critical risk from climate-driven stressors and that the situation will likely be exacerbated as projected in future climate scenarios, relatively few infrastructure researchers and practitioners incorporate climate change. Furthermore, rarely is the risk from natural hazards to transportation infrastructure informed by multi-hazard perspectives (Argyroudis et al., 2019; Hillier et al., 2020).

Changes to the transportation system risk portfolio are not only a product of the changing climate. The transportation system is expanding and evolving in response to market demand and innovation. Globally, mobility has been growing rapidly and forming new transportation patterns. Anticipated changes by 2030 include annual passenger traffic increasing by 50% as compared to 2015, global freight volumes growth of 70% compared to 2015, and a doubling of the current 1.2 billion cars on the road (Sustainable Mobility for All, 2017). Transportation innovations, including shared mobility (e.g., car sharing, carpooling, and ride-sourcing), autonomous and electric vehicles, and advanced materials (e.g., light weight design and composite materials, intelligent materials, and hybrid engineering materials), may supplant existing mobility delivery systems. These new demands and innovations as well as deteriorating infrastructure and land-use change will impact a transportation system's ability to perform reliably, safely, and efficiently under a changing climate (Douglas et al., 2017; Lyons & Davidson, 2016; Jacobs et al., 2018). Moreover, the COVID-19 pandemic resulted in unprecedented changes to mobility. Over the short-term, leisure and business travel nearly terminated, international border lockdowns shut down migratory laborers, stranded travelers, and separated families. Pandemic control measures, such as teleworking, reduced passenger traffic volume by over 50% in many urban areas, with even greater reductions in the public transit usage (Abu-Rayash & Dincer, 2020; Falchetta & Noussan, 2020). At the same time, freight transport demand increased due to a shift in the last-mile delivery of goods from customers to freight (Falchetta & Noussan, 2020). It is unclear how travel practices and modality will shift post-pandemic. Effectively managing future risk, including climate change as well as pandemic-proof travel, requires a shift in the transportation infrastructure community's mindset to account for the changing complexity (Chester & Allenby, 2019b).

This chapter provides a synopsis of the main sources of uncertainty viewed through a climate change lens. It first addresses uncertainties facing civil engineers and methods used in practice to address those uncertainties, including initial conditions and observed forcing data/climate model inputs. The discussion then explores why traditional practices are not up to the task of considering climate change in the sector and points to a range of alternative methods that can provide the basis for the needed shift in thinking. The chapter concludes by exploring several topics and open questions for the transportation field as it moves forward in adapting to a changing climate while at the same time reducing carbon emissions from the transportation sector.

11.2 Overview of Engineering Risk-Based Decision-Making

Most significant transportation infrastructure projects involve engineering design. The goal of design is to develop a transportation asset (e.g., bridge, road, port) that meets fixed performance criteria, including safety and serviceability. For example, if a culvert is to be designed to convey water under a roadway, the site conditions and design runoff values are provided to the engineer. Because there are many possible design options that may meet the given performance criteria, analyses are conducted to determine which options will perform as required. Of the viable designs, an optimal option is selected which meets the criteria and often considers the asset's lifetime cost (De Neufville et al., 2004).

Engineers are known for just wanting a number in order to design new infrastructure. That is, traditionally a design analysis is conducted using specific values for material properties and design loads. However, it has always been understood that, in practice, there are many uncertainties in the design process and that any number used for design is uncertain (Galambos & Ravindra, 1981). These uncertainties in transportation infrastructure planning, design, operation, and maintenance arise from many sources, including human errors in design or construction, randomness in material properties, limits of the design model, limited data, and uncertainty in predicting the future conditions or characterizing the past (Bulleit, 2008).

Engineering practice has always managed the risk of bad things happening due to uncertainties that could result in technical failures, which in turn protects the public and the engineers who design, build, and maintain the infrastructure. Over time, this has refined a general approach to analyzing risk for engineered systems (Fig. 11.1) and led to an evolution of quantitative probabilistic methods to account for those uncertainties (e.g., design and construction errors, long-term maintenance, and model adequacy) that are widely adopted by the profession. Present-day transportation infrastructure design standards provide well-defined specifications that account for a range of uncertainties. In some cases, this may be the simple application of a factor of safety (FS) or effectively a multiplier to scale-up the calculated design that just

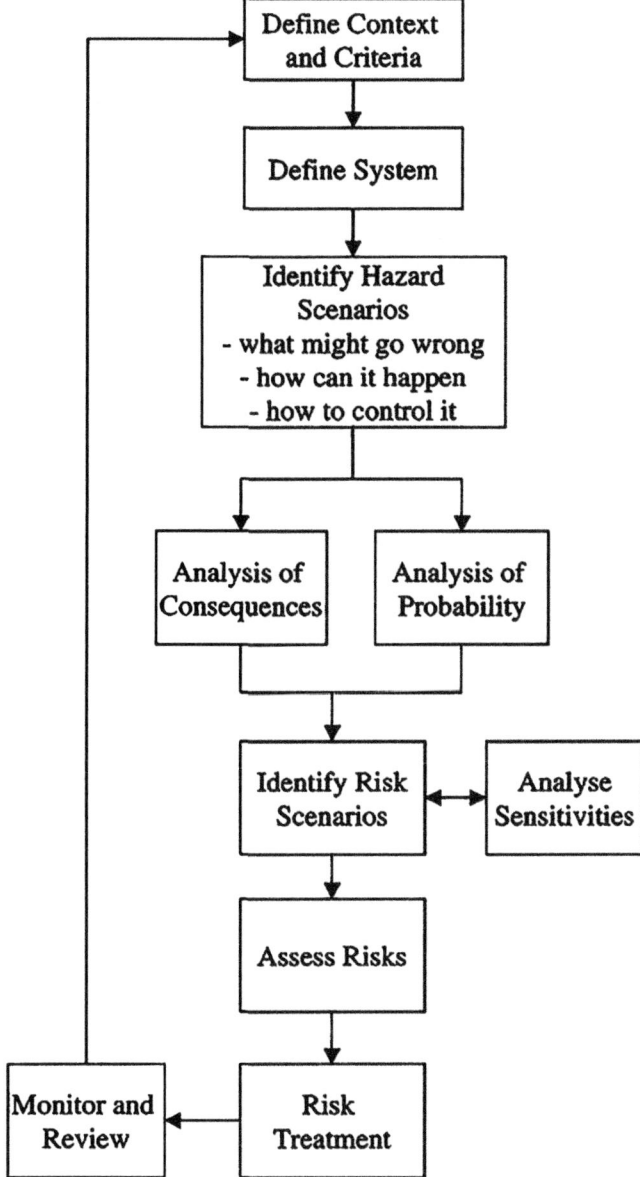

Fig. 11.1 Generic representation of the flow of risk-based decision analysis. Source: Faber and Stewart (2003)

of existing infrastructure requires knowledge about how climate impacts the infrastructure and how those impacts are analyzed. Environmental conditions impact infrastructure in multiple ways. Analytical methods that consider those impacts vary widely. In general, during the design process, an engineer will quantify all the loads, including the environmental loads that a structure will experience over its lifetime and the likelihood of the loads occurring simultaneously. In structures, loads include dead loads that are loads from the structural itself, live loads that are anything that is in or on the structure during its life, and transient loads due to weather, including wind, rain, and snow (Bulleit, 2008). A changing climate will directly modify the individual transient loads and their combinations. In contrast, traffic and its loading are the most important factors that impact pavement design. Changes in environmental conditions such as temperature and precipitation affect the ability of pavement materials to perform as designed. The uncertainty due to changes in temperature or precipitation arises from both the uncertainty about future conditions as well as the uncertainty regarding how those changes affect material properties. Environmental loads are quantified using historical and often outdated observational data, which are recognized as having natural variability, but that variability is assumed to be stationary.

For agencies that manage large portfolios of assets in their transportation system, individual analyses are not practical to understand the relationship between extreme weather events and the system response. At the agency level, the portfolio of risks is beginning to be analyzed in a manner that emphasizes performance, such as a desired state of good repair. Risk management then seeks to achieve performance targets through investments made during a typical planning horizon (e.g., 10 years) to address immediate risks and to take actions to mitigate risks beyond the planning horizon (FHWA, 2017). Many transport agencies are in the early stages of developing risk management practices. The insurance and reinsurance industries have already seen an impact on their business via insurance payouts due to increasing frequency and intensity of extreme events. In response, their risk mapping and catastrophe modelling tools have been updated to include the uncertain future climate (Collier et al., 2021).

At the portfolio scale, risk management for extreme events draws from the risk and disaster resilience community for natural hazard management and planning, which combine the assessment of hazards, asset exposure, and vulnerability (Bernal et al., 2021). These assessments typically use coarser tools such as fragility curves and depth damage functions to identify vulnerable assets and to quantify potential damages to the overall system. Fragility curves provide the likelihood of infrastructure exceeding a particular damage state based on the intensity of the applied hazard. While they are widely known for their application to earthquakes, they have emerged as useful tools for a broader portfolio of hazards, including floods, landslides, and fires. Depth-damage func-

meets requirements to an assumed safer level. More complex approaches such as load and resistance factor design (LRFD) employ multiple considerations. The result is that transportation systems are built stronger than our best estimates of what is needed to exactly meet the specified design conditions (Douglas et al., 2017) in order to account for the uncertainty due to natural variability (aleatory variability).

11.3 Accounting for Environmental Loads in Design and Assessments

Understanding how to consider the changing climate when designing new infrastructure or assessing the vulnerability

tions, which relate inland and coastal floodwater depth versus percent damage, are also used for a variety of infrastructure. The extent and severity of damage to structural components and contents are estimated from the depth of flooding and the application of the assigned depth-damage curve. There are a wide range of methods to use these and other tools within a probabilistic framework that includes the random nature of hazards and vulnerability as well as an uncertain future (Bernal et al., 2021).

11.4 Coping with Climate Change Using Traditional Engineering Approaches

The transportation sector is facing unfamiliar challenges due to climate change and the great uncertainty regarding what the future holds. Extreme weather events are anticipated to become more frequent and more severe than in the past. Sea-level rise is anticipated to cause gradual, long-lasting, and large-scale impacts on roads, rail, ports, and airports. The uncertain nature of climate change includes if, where, when, and how changes to environmental stressors could occur and the infrastructure might respond (Dewar & Wachs, 2008). Because this is new territory for the transportation sector, there is limited reliable guidance regarding how to adapt and even which methods can successfully guide agency practice under climate change uncertainty. Current practice is not well suited to effectively plan and manage under these conditions. However, existing tools can form the basis of traditional approaches that can be employed to "aid in the determination of the relative order of magnitude of environmental forcing uncertainties and associated system sensitivity, risks to existing and planned assets, and to identify critical vulnerability from the decision makers' perspective" (Douglas et al., 2017).

Much of the early work considered the impact of a changing climate in design or vulnerability, leveraged risk-based frameworks, often employing simple top-down approaches or climate analogues. In a top-down approach, one or more climate-change scenario or historical trend is used to project future changes in environmental loadings. These changes are propagated through existing analysis frameworks to identify potential changes to transportation infrastructure performance or to develop climate resiliency through enhanced design, which incorporates future environmental loads (Knott et al., 2019). For example, top-down, risk-based approaches, often informed by an ISO 31000:2009, are at the heart of many national standards including the UK Highways Agency's seven-stage adaptation framework to determine affected infrastructure, risk, and adaptation options (Fig. 11.2). Similar approaches are adopted by the US Federal Highway Administration's (FHWA) Climate Change and Extreme Weather Vulnerability Assessment Framework (Fig. 11.3) (Asam et al., 2015) and the Canadian Climate Lens assessment and the Public Infrastructure Engineering Vulnerability Committee (PIEVC) protocol. Despite their adoption, many agencies noted significant obstacles in implementing the frameworks due to the limits of available climate model output and inadequate treatment of risk for the changing climate (Wall & Meyer, 2013). The evolving climate also requires new infrastructure considerations including the following: (1) the time horizon for the asset's design and operational lifetime and (2) asset criticality classification prior to selecting the analysis approaches and emission scenarios. Many national, regional, and local agencies have found value in first developing procedures to estimate future loadings, to synthesize the results in terms of time horizons and the probability of occurrence of the extreme event, and to translate those results into quantitative values, including adjustment factors for application to design.

Asian Development Bank (2020) has developed top-down treatment procedures for including climate change in detailed engineering design. They provide two detailed case studies from Vietnam, (1) the flooding of the Bao Ninh–Hai Ninh coastal road in Quang Binh province and (2) coastal erosion due to sea-level rise (SLR) at Thinh Long in Nam Dinh province. Their procedures are used to develop adjustment factors for future design precipitation and high-end SLR scenarios, to characterize asset vulnerability from flooding and coastal erosion, and to determine future design needs (e.g., discharge, water level, and streamflow velocity) for an upgraded spillway–culvert. Specific engineering solutions are evaluated including nature-informed solutions as well as performance considerations based on the broader context and area in which the infrastructure is located.

Because results of top-down approaches depend on the chosen climate-change scenarios, a lack of guidance about model and scenario selection plus the inability to assign a probability to the possible futures, among other reasons, has hampered uptake of these methods in practice. Other approaches attempt to bypass these obstacles by eliminating the use of specific climate model output and instead use historical trends, the upper limits of historical data confidence intervals, or climate analogues. Climate analogues identify regions whose current climate is a reasonable analogue for the project's future climate, as a framework for considering adaptation options, as well as to develop an appreciation for the uncertainty and its impact on investment in long-lived assets (Hallegatte et al., 2007).

There are other traditional tools, which are less widely used, that can aid decision-making under uncertainty. These include structured decision analysis techniques, scenario planning, and iterative or cyclical risk-based approaches (Dewar & Wachs, 2008; Wall et al., 2003). A decision tree is a structured decision analysis technique that breaks down a decision into potential outcomes and the likelihood of their occurrence. A decision tree is often represented graphically

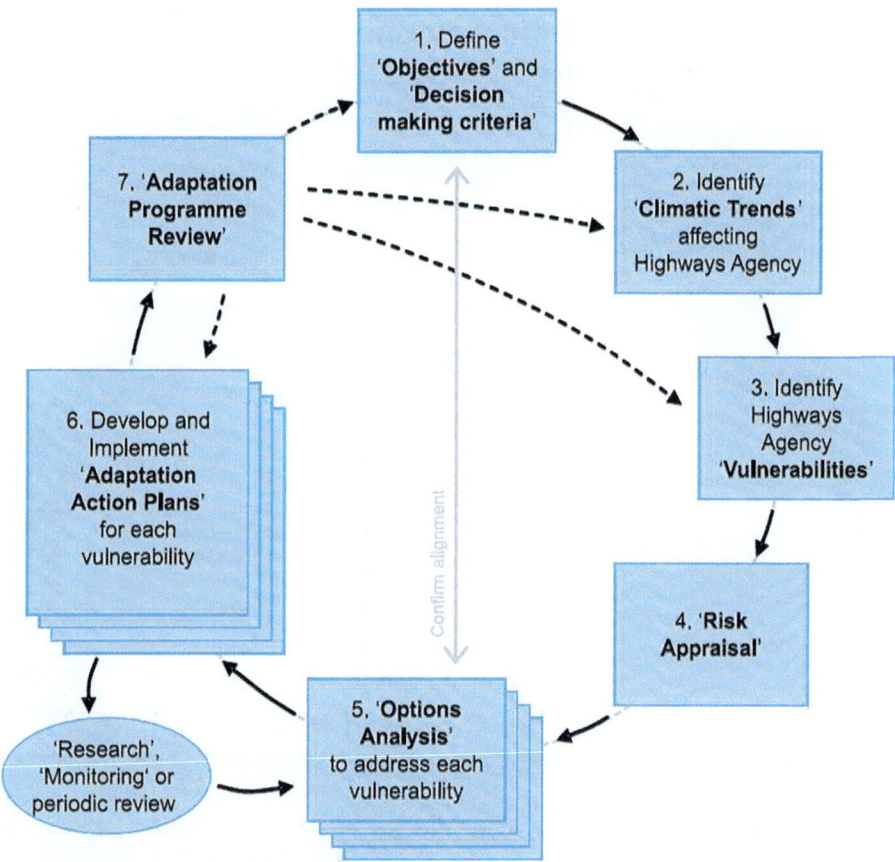

Fig. 11.2 Highways agency adaptation framework model from UK Highways Agency and Parsons Brinckerhoff. Source: Climate Change Adaptation Strategy and Framework. Highways Agency, London, 2009

as a series of options, decisions, and outcome uncertainties from which the expected benefit of each branch's outcome is determined. For example, Qiao et al. (2017) used Bayesian decision trees to determine whether to close a given road after flooding or to keep it open and whether in situ pavement testing should be performed to support the decision. Their approach could be enhanced to support pavement and traffic management under future flooding conditions.

Scenario analysis is another familiar tool in planning which has potential value for the transportation sector. Scenario planning presents multiple potential futures to decision-makers, then asks them to consider how to prepare for each. In some cases, scenario planning has been used to identify plans that are robust across a number of futures. The value of scenario planning is not necessarily the specific plan, but the process of thinking about the problem broadly as well as their possible solutions and the commonalties among those solutions (Chester & Allenby, 2019b). Kirshen et al. (2012) presented a simplified scenario planning approach. Their coastal Maine, USA, case study quantified damages and adaptation costs due to SLR induced flooding by combining multiple SLR scenarios with depth damage curves. The process provides insights to support agency decision-making. Iterative or cyclical risk-based approaches use risk-based adaptation frameworks such as top-down analyses and scenario planning in an iterative manner. As new information becomes available, risks are recalculated and adaptation priorities and approaches may be adjusted (Wall et al., 2003).

11.5 Recent Advances in Uncertainty Approaches to Cope with Climate Change

While the transportation sector has always operated under uncertain conditions, climate change is more challenging because of the "deep uncertainty" posed by an evolving technology, climate, and social systems nexus. As the sector struggles to use tried and tested approaches, it is becoming increasingly clear that the changing complexity requires a shift in thinking about how to provide more agile and flexible infrastructure in the face of uncertainty (Chester & Allenby, 2019b). The following sections outline a range of thinking about the challenges and approaches. The first covers real options and systems dynamics, which incrementally change the sector's way of doing business. The second section presents

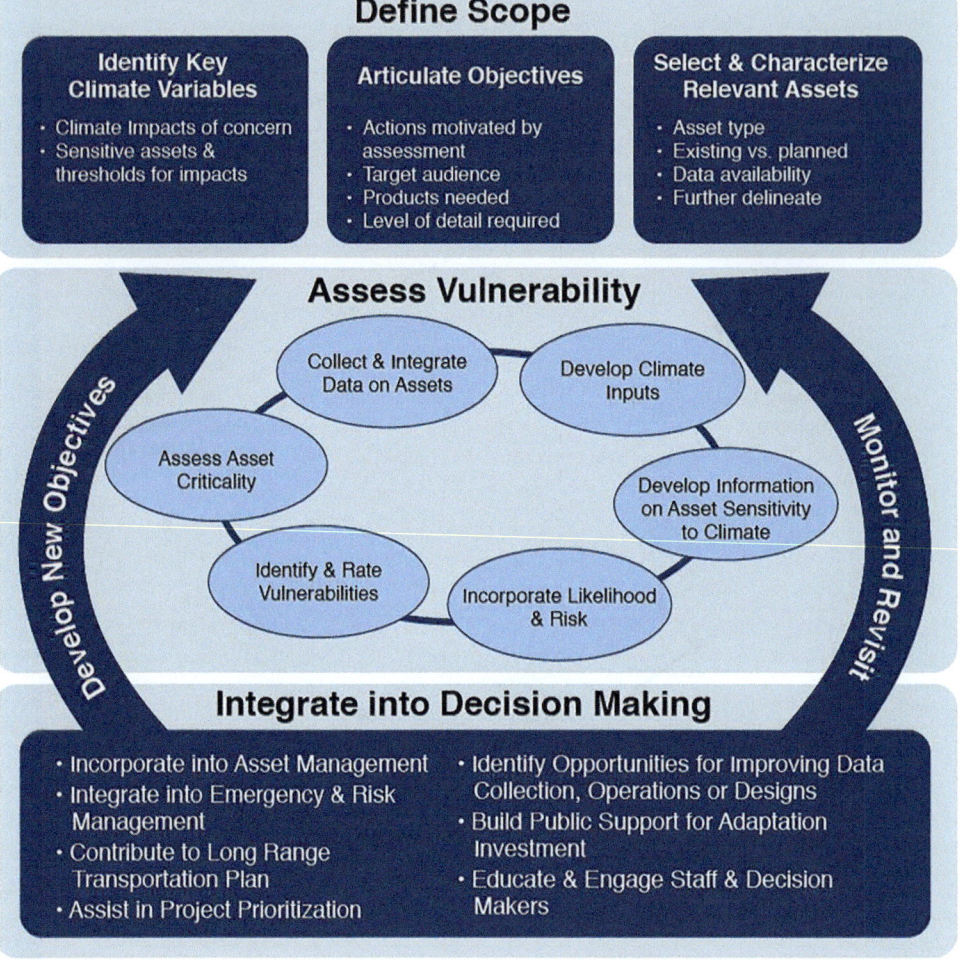

Fig. 11.3 Federal highway administration (FHWA) climate change and extreme weather vulnerability assessment framework (Asam et al., 2015). Source: US FHWA. Permission requested

methods (such as decision-making under deep uncertainty, dynamic adaptive planning in operation and decision-making options, and socioecological and technological systems) that embrace the complexity of the current age but would require a paradigm shift in the transportation sector.

11.5.1 Approach 1: Repurposing Existing Risk-Based Decision-Making

Real options address the challenge that arises when attempting to value investments in long-lived infrastructure given that climate change may radically change the stressors on and demands for that infrastructure. Historically, net present value (NPV) and life-cycle cost assessments (LCAs) have been used to identify efficient investments and costs. However, NPV and LCA are not designed to handle an unknowable future. Introduced for the financial markets, real options are the rights without obligations to delay a decision until more is known about the uncertainty. In the context of civil engineering, this provides a means to build in flexibility to change the infrastructure in the future. Real options analysis techniques extend NPV to include the value of flexible design and management of infrastructure. "The development of options analysis holds the promise of enabling the engineering profession to calculate the value of flexibility" (De Neufville et al., 2004). The decision tree discussed previously is a simplified approach to a real options analysis, but it does not perform as well as real options when faced with the "wicked" uncertainty of future environmental stressors (Van den Boomen et al., 2019). For instance, Kim et al. (2017) present a case study in which real options are used to determine the value of a flexible adaptation strategy for managing current and future floods in Seoul, Korea, under a range of Representative Concentration Pathways. Their work provides a replicable decision approach for including the uncertainty of climate scenarios in a complex urban area. While real options are sometimes criticized for their limited applicability to infrastructure, at its core the method attempts to encourage flexible design that recognizes that the future is uncertain, hence supports infrastructure design that is adaptable in the future. Its strengths are recognized and used to support flexibility in dynamic strategic planning within an adaptive planning framework.

System dynamics can be used to understand complex systems that change over time. It is a mature approach that has been widely used to understand sustainability and more recently climate change and infrastructure (Mallick et al., 2014). This simulation approach helps to provide a strategic view of a "system." This could be a road network, a railway, an airport or even a pavement system, with independencies, feedback loops, and dynamic interactions, among the various components. Because the approach tracks the dynamic nature of the system as opposed to a single snapshot in time, it is well suited to represent nonstationary climate impacts on transportation systems. Mallick et al. (2014) used system dynamics to understand the impact on pavements from multiple climate stressors under climate change in coastal and inland environments. Their system dynamics model shows the complex linkages and feedbacks of the pavement system under a changing climatic and includes those pavement parameters that are affected by climate change that lead to the resulting pavement performance, maintenance requirements, and costs. When used in combination with Monte Carlo simulation, the uncertainty in climate change projections can be used to develop a probabilistic understanding of an outcome (e.g., pavement lifetime or condition) that is relevant to decision-makers.

11.5.2 Approach 2: Embracing Complexity in Engineering Risk-Based Decision-Making

There is increasing evidence that conventional ways of handling uncertainty in engineering practice are antiquated. The rigid infrastructure systems resulting from fail-safe designs are locking-in investments and constraining options for delivering reliable transportation for people, goods, and services in the future. Yet, even with massive investments, the systems are ill-prepared to handle unknowable changes to the climate and unprecedented extreme events as well as changes in demand and technology. When dealing with "deep uncertainty" (Lempert, 2003) from climate change, the transportation sector's risk-based approaches (that rely on probability distributions) are not able to handle the complexity of an unknowable future. Instead, adaptive methods such as real options within dynamic strategic planning, robust decision-making, info-gap analysis, adaptive policymaking, and adaption pathways are better suited to characterize and manage today's climate change challenges (Helmrich & Chester, 2020). The methods outlined below embrace the complexity of the current age but would require a paradigm shift in the transportation sector.

Adaptive management of infrastructure is a broad term that encompasses approaches that manage a system that is likely to change. The underlying assumption is that the current worldview is incorrect and that over time additional information needed to manage the system will come to light. A critical aspect of adaptive management is gathering information to inform the management cycle of assess, design, implement, monitor, evaluate, and adjust (Dewar & Wachs, 2008). The ability to adjust requires "both agility and flexibility, respectively, the ability to maintain function in both physical structure and institutional rules despite a nonstationary future, and the ability to respond to changes in demand beyond regular or incremental changes" (Chester & Allenby, 2019a; Gilrein et al., 2019).

Two types of adaptive management approach are assumption-based planning (ABP) and dynamic adaptive planning (DAP). ABP, developed by the RAND Corporation, spells out all assumptions and uncertainties in a plan then identifies which assumptions are critical to the success of the plans as well as which are likely to change in the future (Wall et al., 2003). The reduced suite of assumptions is monitored using "signposts" or warning signs to detect errors in them. ABP also recommends taking "shaping actions" to coerce the assumptions to play out favorably and "hedging actions" to reduce the impacts of incorrect assumptions. ABP can broadly address many aspects of climate uncertainty, including direct and indirect impacts on traffic patterns, condition of infrastructure, and land-use change (Dewar & Wachs, 2008). DAP builds on ABP by developing a basic plan, identifying the vulnerabilities of the plan, implementing a monitoring program, and predetermined adaptations as needed (Wall et al., 2003, 2015).

Given the complexity of these methods and the nuances in their application, we point to three case studies that illustrate how they can support agility and flexibility in the face of deep uncertainty. In the first case study, Wall et al. (2015) demonstrate the use of DAP for a transportation link between San Francisco, CA, and an adjacent community that is vulnerable to SLR plus storm surge, where uncertainties included future SLR, travel patterns, and land use, as well as social uncertainties. They provide details of the adaptive planning steps, including the initial stage-setting work of defining objectives, constraints, and options; selecting a basic plan from the options; and understanding the conditions under which it will be successful (assumptions). These increase the plan's robustness and a monitoring program, with signposts and trigger levels, enables corrective actions (Kwakkel et al., 2010).

The second example applies a dynamic adaptive policy pathways approach to water management (i.e., flood protection and freshwater supply) for the Rhine Delta in the Netherlands (Haasnoot et al., 2013). This approach also

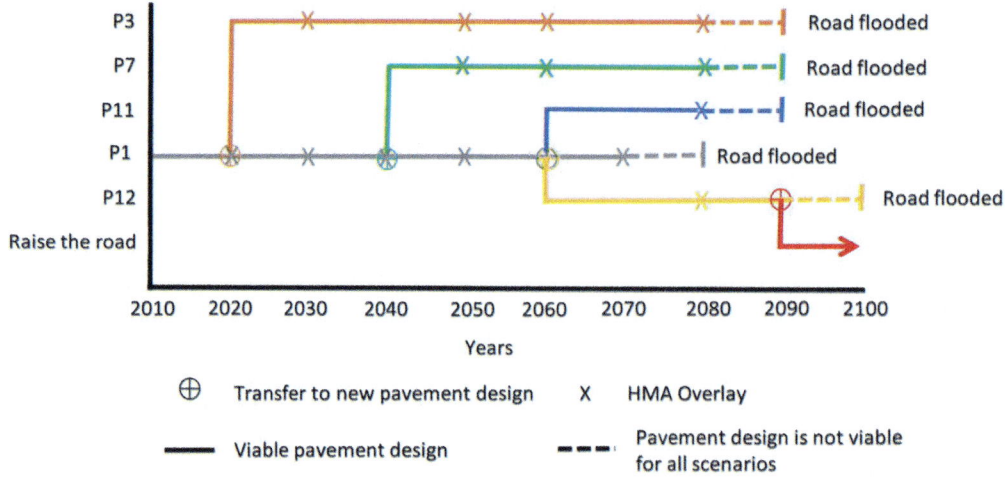

Fig. 11.4 Map showing the most cost-effective adaptation pathways and the cost scorecard. The timing of Hot Mix Asphalt (HMA) overlays is illustrated with X and tipping points, where a transfer to a different pathway may occur, are shown with a bull's eye symbol. Solid lines represent viable pavement structures for all scenarios and dashed lines represent viable pavement structures for some but not all scenarios. PV is present value. Source: Knott et al. (2019)

uses the adaptive planning steps of (Kwakkel et al., 2010) but combines them with an adaptation pathways approach that includes dynamic decisions to address an ensemble of scenarios. Drawing from preexisting water-related scenarios, they determine that there could be stresses on freshwater supply and increased flood risk due to direct and indirect impacts from climate change. The impacts can be managed through a range of actions, each with their own efficacy and effects on other aspects of the system. Adaptation pathways are mapped over time where early actions may transition to other actions in the future. Similar to Wall et al. (2015), a specific monitoring program with signposts, a plan for how to keep future options open, and adaptation opportunities is presented.

The final case study demonstrates a stepwise and flexible pavement adaptation plan using hybrid bottom-up (asset-based)/top-down (scenario-based) approach to address climate-change-induced temperature and groundwater rise at a coastal site in New Hampshire (Knott et al., 2019). In partnership with stakeholders, they identify climate-change-induced temperature and groundwater rise as critical stressors of a coastal roadway, then demonstrate how to determine impacts to pavement life using a scenario-based analysis. They use the adaptation pathways framework (Haasnoot et al., 2013) to present a specific example of how an agency could use the pathways maps with supporting information (e.g., thickness of overlays or rehabilitation of base-layers) to make decisions. The potential routes through time encompass future uncertainty and then translate pathways to present values dollars over a 60-year time horizon (Fig. 11.4).

11.6 Discussion and Conclusions

This chapter has focused on the adaptation of infrastructure or transportation systems to uncertain climate and non-climatic factors. However, these measures will not operate in a vacuum. Within the transportation sector, there is a strong mitigation-adaptation nexus that is largely underserved despite the potential for co-benefits and synergies (Sharifi, 2021). In most cases, climate change mitigation and adaptation efforts are conducted independently and in some cases even seen as competitors for the limited resources that are available to address climate change challenges (Jiang et al., 2020). The transportation sector is also connected to and dependent on other critical infrastructure, e.g., climate-transport-health nexus, or climate-transport-energy nexus, or climate-transport-development nexus, or climate-transport-agriculture nexus. Hence, there is considerable scope for transport-related co-benefits and trade-offs with other sectors.

Infrastructure performance depends not only on critical infrastructure assets, systems, and networks but also the ecosystem, institutional, and social landscapes in which they are embedded. There is increasing awareness that technocentric, robustness strategies have limitations, and a more holistic approach is warranted. Approaches that consider these interactions add complexity but can benefit by identifying and avoiding constraining infrastructure ability to change in the future and broadening the suite of potential options to adaptive capacity beyond those that would emerge from traditional approaches (Brunetta et al., 2019; Markolf et al., 2018). It is this latter point that offers optimism for new, creative, and efficient transportation solutions that could emerge from a more inclusive, common vision of climate-related uncertainties.

The engineering community has long recognized that their practice is fraught with uncertainty, so they have included risk analysis in design practice where possible and acknowledged those uncertainties outside of the design code. This chapter identified established methods that the transportation community uses to address uncertainty and provided insights as to why these methods are rapidly becoming inadequate.

A key takeaway from this chapter is that today's transportation sector is faced with increasing complexities in uncertainty due to a changing climate. Climate change must be considered in combination with a rapidly evolving community of practice, the recognition of interdependencies among sectors, and an increasing role of public institutions. In light of these changes, the sector's existing methods fall short. The cited literature is replete with calls to radically change the sectors' means of addressing uncertainty arising from unknown and unknowable futures. What this means in practice remains unclear. However, the wealth of approaches that have recently emerged and were reviewed in this chapter consistently point to the need for more agile and flexible infrastructure in the face of uncertainty.

These adaptive approaches and associated uncertainty challenges are not compartmentalized like the past engineering frameworks. For example, from a practical perspective, the adaptive management strategies not only require designs that are flexible enough to adapt to unpredictable changes in the future but also require agreed metrics for monitoring performance and triggering change (Chester & Allenby, 2019b). As Dewar and Wachs (2008) noted, the cost of information is a significant barrier to adaptive management. Hence, transportation agencies have begun to leverage advances in information technology to support the adaptive management of transport systems. Transportation agencies still need to better understand and track interconnectedness plus move toward approaches capable of reducing uncertainty and delivering performance despite an unknown and changing future.

References

Abu-Rayash, A., & Dincer, I. (2020). Analysis of mobility trends during the COVID-19 coronavirus pandemic: Exploring the impacts on global aviation and travel in selected cities. *Energy Research & Social Science, 68*, 101693. https://doi.org/10.1016/j.erss.2020.101693

Argyroudis, S. A., Mitoulis, S. A., Winter, M. G., & Kaynia, A. M. (2019). Fragility of transport assets exposed to multiple hazards: State-of-the-art review toward infrastructural resilience. *Reliability Engineering & System Safety, 191*, 106567. https://doi.org/10.1016/j.ress.2019.106567

Asam, S., Bhat, C., Dix, B., Bauer, J., & Gopalakrishna, D. (2015). *Climate change adaptation guide for transportation systems management, operations, and maintenance.* Federal Highway Administration.

Asian Development Bank. (2020). *Climate change adjustments for detailed engineering design of roads: Experience from Viet Nam.* Asian Development Bank. https://doi.org/10.22617/TIM200148-2

Bernal, G. A., Cardona, O.-D., Marulanda, M. C., & Carreño, M.-L. (2021). Dealing with uncertainty using fully probabilistic risk assessment for decision-making. In *Handbook of disaster risk reduction for resilience* (pp. 299–340). Springer. https://doi.org/10.1007/978-3-030-61278-8_14

Brunetta, G., et al. (2019). Territorial resilience: Toward a proactive meaning for spatial planning. *Sustainability, 11*, 2286. https://doi.org/10.3390/su11082286

Bulleit, W. M. (2008). Uncertainty in structural engineering. *Practice Periodical on Structural Design and Construction, 13*, 24–30. https://doi.org/10.1061/(asce)1084-0680(2008)13:1(24)

Chen, A., Zhou, Z., Chootinan, P., Ryu, S., Yang, C., & Wong, S. (2011). Transport network design problem under uncertainty: A review and new developments. *Transport Reviews, 31*, 743–768. https://doi.org/10.1080/01441647.2011.589539

Chester, M. V., & Allenby, B. (2019a). Toward adaptive infrastructure: Flexibility and agility in a non-stationarity age. *Sustainable and Resilient Infrastructure, 4*, 173–191. https://doi.org/10.1080/23789689.2017.1416846

Chester, M. V., & Allenby, B. (2019b). Infrastructure as a wicked complex process. *Elementa: Science of the Anthropocene, 7*, 360. https://doi.org/10.1525/elementa.360

Collier, S. J., Elliott, R., & Lehtonen, T.-K. (2021). Climate change and insurance. *Economy and Society, 50*, 158–172. https://doi.org/10.1080/03085147.2021.1903771

De Neufville, R., et al. (2004). Uncertainty management for engineering systems planning and design. In *Engineering systems symposium.* MIT. https://doi.org/10.1061/(asce)is.1943-555x.0000377

Dewar, J. A., & Wachs, M. (2008). *Transportation planning, climate change, and decisionmaking under uncertainty.* Transportation Research Board.

Douglas, E., et al. (2017). Progress and challenges in incorporating climate change information into transportation research and design. *Journal of Infrastructure Systems, 23*, 04017018.

Faber, M. H., & Stewart, M. G. (2003). Risk assessment for civil engineering facilities: Critical overview and discussion. *Reliability Engineering & System Safety, 80*, 173–184. https://doi.org/10.1016/s0951-8320(03)00027-9

Falchetta, G., & Noussan, M. (2020). *The impact of COVID-19 on transport demand, modal choices, and sectoral energy consumption in Europe.* IAEE Energy Forum.

Fant, C., et al. (2021). Mere nuisance or growing threat? The physical and economic impact of high tide flooding on US road networks. *Journal of Infrastructure Systems, 27*, 04021044. https://doi.org/10.1061/(ASCE)IS.1943-555X.0000652

FHWA. (2017). *Incorporating risk management into transportation asset management plans.* Federal Highway Administration.

Galambos, T. V., & Ravindra, M. (1981). Load and resistance factor design. *Engineering Journal, 18*, 78–84.

Gilrein, E. J., Carvalhaes, T. M., Markolf, S. A., Chester, M. V., Allenby, B. R., & Garcia, M. (2019). Concepts and practices for transforming infrastructure from rigid to adaptable. *Sustainable and Resilient Infrastructure, 6*, 213–234. https://doi.org/10.1080/23789689.2019.1599608

Haasnoot, M., Kwakkel, J. H., Walker, W. E., & ter Maat, J. (2013). Dynamic adaptive policy pathways: A method for crafting robust decisions for a deeply uncertain world. *Global Environmental Change, 23*, 485–498. https://doi.org/10.1016/j.gloenvcha.2012.12.006

Hallegatte, S., Hourcade, J.-C., & Ambrosi, P. (2007). Using climate analogues for assessing climate change economic impacts in urban areas. *Climatic Change, 82*, 47–60.

Helmrich, A. M., & Chester, M. V. (2020). Reconciling complexity and deep uncertainty in infrastructure design for climate adaptation. *Sustainable and Resilient Infrastructure, 7*, 83–99. https://doi.org/10.1080/23789689.2019.1708179

Hillier, J. K., Matthews, T., Wilby, R. L., & Murphy, C. (2020). Multi-hazard dependencies can increase or decrease risk. *Nature Climate Change, 10*, 595–598. https://doi.org/10.1038/s41558-020-0832-y

Jacobs, J. M., Culp, M., Cattaneo, L., Chinowsky, P., Choate, A., DesRoches, S., Douglass, S., & Miller, R. (2018). Chapter 12: Transportation. In D. R. Reidmiller, C. W. Avery, D. Easterling, K. Kunkel, K. L. M. Lewis, T. K. Maycock, & B. C. Stewart (Eds.), *USGCRP, 2018: Fourth National Climate Assessment: Volume II - Climate Change Impacts, Risks, and Adaptation in the United States.* U.S. Global Change Research Program, Washington, DC, USA.

Jiang, C., Zheng, S., Ng, A. K., Ge, Y.-E., & Fu, X. (2020). The climate change strategies of seaports: Mitigation vs. adaptation. *Transportation Research Part D: Transport and Environment, 89*, 102603. https://doi.org/10.1016/j.trd.2020.102603

Kim, K., Ha, S., & Kim, H. (2017). Using real options for urban infrastructure adaptation under climate change. *Journal of Cleaner Production, 143*, 40–50. https://doi.org/10.1016/j.jclepro.2016.12.152

Kirshen, P., Merrill, S., Slovinsky, P., & Richardson, N. (2012). Simplified method for scenario-based risk assessment adaptation planning in the coastal zone. *Climatic Change, 113*, 919–931. https://doi.org/10.1007/s10584-011-0379-z

Knott, J. F., Jacobs, J. M., Sias, J. E., Kirshen, P., & Dave, E. V. (2019). A framework for introducing climate-change adaptation in pavement management. *Sustainability, 11*, 4382. https://doi.org/10.3390/su11164382

Kwakkel, J. H., Walker, W. E., & Marchau, V. (2010). Adaptive airport strategic planning. *European Journal of Transport and Infrastructure Research, 10*, 249.

Lempert, R. J. (2003). *Shaping the next one hundred years: New methods for quantitative, long-term policy analysis*. Rand Corporation. https://doi.org/10.7249/mr1626

Lyons, G., & Davidson, C. (2016). Guidance for transport planning and policymaking in the face of an uncertain future. *Transportation Research Part A: Policy and Practice, 88*, 104–116. https://doi.org/10.1016/j.tra.2016.03.012

Mallick, R. B., Radzicki, M. J., Daniel, J. S., & Jacobs, J. M. (2014). Use of system dynamics to understand long-term impact of climate change on pavement performance and maintenance cost. *Transportation Research Record, 2455*, 1–9. https://doi.org/10.3141/2455-01

Markolf, S. A., et al. (2018). Interdependent infrastructure as linked social, ecological, and technological systems (SETSs) to address lock-in and enhance resilience. *Earth's Future, 6*, 1638–1659. https://doi.org/10.1029/2018ef000926

Petroski, H. (1985). *To engineer is human: The role of failure in successful design*. St. Martin's Press.

Qiao, Y., Medina, R. A., McCarthy, L. M., Mallick, R. B., & Daniel, J. S. (2017). Decision tree for postflooding roadway operations. *Transportation Research Record, 2604*, 120–130. https://doi.org/10.3141/2604-15

Sharifi, A. (2021). Co-benefits and synergies between urban climate change mitigation and adaptation measures: A literature review. *Science of the Total Environment, 750*, 141642. https://doi.org/10.1016/j.scitotenv.2020.141642

Sustainable Mobility for All. (2017). *Global mobility report 2017: Tracking sector performance*. Creative Commons Attribution.

U.S. Department of Transportation. (2020). *Bureau of Transportation Statistics, Transportation Statistics Annual Report 2020*. Washington, DC. https://doi.org/10.21949/1520449

Van den Boomen, M., Spaan, M., Schoenmaker, R., & Wolfert, A. (2019). Untangling decision tree and real options analyses: A public infrastructure case study dealing with political decisions, structural integrity and price uncertainty. *Construction Management and Economics, 37*, 24–43. https://doi.org/10.1080/01446193.2018.1486510

Wall, T. A., & Meyer, M. D. (2013). *Risk-based adaptation frameworks for climate change planning in the transportation sector. A synthesis of practice*. Transportation Research Board. https://doi.org/10.17226/22462

Wall, T. A., Walker, W. E., & Marchau, V. A. (2003). Transportation planning methods for coping with climate change uncertainty: An overview. *Integrated Assessment, 4*, 46–55.

Wall, T. A., Walker, W. E., Marchau, V. A., & Bertolini, L. (2015). Dynamic adaptive approach to transportation-infrastructure planning for climate change: San-Francisco-Bay-area case study. *Journal of Infrastructure Systems, 21*, 05015004. https://doi.org/10.1061/(asce)is.1943-555x.0000257

Open Access This chapter is licensed under the terms of the Creative Commons Attribution 4.0 International License (http://creativecommons.org/licenses/by/4.0/), which permits use, sharing, adaptation, distribution and reproduction in any medium or format, as long as you give appropriate credit to the original author(s) and the source, provide a link to the Creative Commons license and indicate if changes were made.

The images or other third party material in this chapter are included in the chapter's Creative Commons license, unless indicated otherwise in a credit line to the material. If material is not included in the chapter's Creative Commons license and your intended use is not permitted by statutory regulation or exceeds the permitted use, you will need to obtain permission directly from the copyright holder.

Science in Coastal Adaptation Decision-Making: Working Effectively with Persistent Uncertainties

Susanne C. Moser

12.1 Introduction

A remarkable paradox characterizes science in coastal adaptation decision-making: sea-level rise (SLR) is one of the most certain and irrefutable consequences of climate warming, and yet exactly how high it will rise over the twenty-first century in different locations is ridden with deep and persistent uncertainties. In the first dispensation of the Intergovernmental Panel on Climate Change's (IPCC) Sixth Assessment trilogy—the Physical Science Basis—the key takeaways for SLR could not be more convincing nor more eyebrow-raising in their implications (Box 12.1; and Chap. 19 on sea-level rise projections, this volume).

> **Box 12.1 High-Level Findings from the IPCC Sixth Assessment Report on Sea-Level Rise**
>
> - Global mean sea level increased by an average of 0.20 m between 1901 and 2018 at a mean rate of 1.3 mm/year in the first 70 years of that period, an accelerated rate of 1.9 mm/year between 1971 and 2006, then further increasing to 3.7 mm/year between 2006 and 2018. Human influence was very likely the main driver of these increases since at least 1971 (IPCC, 2021, p. 6).
> - Global mean sea level has risen faster since 1900 than over any preceding century in at least the last 3000 years, driven by accelerating ocean warming (thermal expansion) and ice loss from land (addition
>
> *(continued)*

> **Box 12.1 (continued)**
>
> of water to the ocean basins) (IPCC, 2021, p. 9). "The rate of ice sheet loss increased by a factor of four between 1992–1999 and 2010–2019. Together, ice sheet and glacier mass loss were the dominant contributors to global mean sea level rise during 2006–2018" (IPCC, 2021, p. 14).
> - It is virtually certain that global mean sea level will continue to rise over the twenty-first century (IPCC, 2021, p. 28) and that the rate of rise will continue to increase.
> - "Many changes due to past and future greenhouse gas emissions are irreversible for centuries to millennia, especially changes in the ocean, ice sheets, and global sea level. ... In the longer term, sea level is committed to rise for centuries to millennia due to continuing deep ocean warming and ice sheet melt, and will remain elevated for thousands of years," rising "over the next 2000 years [...] by about 2–3 m if warming is limited to 1.5 °C, 2–6 m if limited to 2 °C, and 19–22 m with 5 °C of warming, and [continuing] to rise over subsequent millennia" (IPCC, 2021, p. 28).
> - While far out, difficult to imagine and in many decision-makers' minds irrelevant to today's decisions, there is remarkable confidence in where modern-day SLR is headed, given that "projections of multi-millennial global mean sea level rise are consistent with reconstructed levels during past warm climate periods" (IPCC, 2021, p. 29).

Despite these stark and mostly high-confidence findings in the IPCC Sixth Assessment Report, a look at the detailed SLR projections over the twenty-first century reveals only medium and low confidence, confounded by regional di-

S. C. Moser (✉)
Susanne Moser Research and Consulting, Hadley, MA, USA

Antioch University, New England, Keene, NH, USA
e-mail: promundi@susannemoser.com

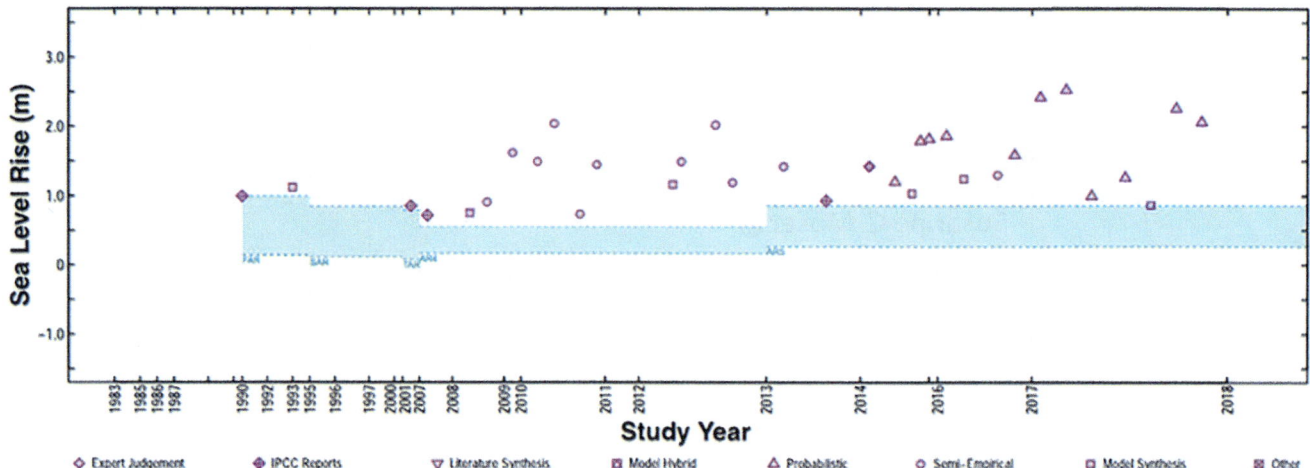

Fig. 12.1 Comparison of upper-range SLR projections illustrates how conservative the IPCC projections (light blue) have been compared to the wider scientific literature (pink symbols). Source: Garner et al. (2018), figure reproduced under the CC BY-NC-ND 4.0 license terms. See: https://creativecommons.org/licenses/by-nc-nd/4.0/

vergence from these global estimates. Thus, the challenge that coastal decision-makers must confront. Relative to the baseline established between 1995 and 2014, global mean sea level in 2100 has at least a 2/3 chance to rise by as much as

- 0.28–0.55 m under the very low greenhouse gas (GHG) emissions scenario
- 0.32–0.62 m under the low GHG emissions scenario
- 0.44–0.76 m under the intermediate GHG emissions scenario
- 0.63–1.01 m under the very high GHG emissions scenario
- A small chance, it could even go up to 2 m if there is catastrophic ice loss.

Scientific projections of SLR over the past four decades have varied notably, reflecting remarkable advances in observation, modeling and methodology, but also changes in the underlying emissions scenarios and persistent deep uncertainties in scientists' understanding of the fundamental processes driving global sea-level rise (Garner et al., 2018). As a systematic review of the literature reveals, "Results show a reduction in the range of SLR projections from the first studies through the mid-2000s that has since reversed. In addition, [the analysis reveals] a tendency for [IPCC] reports to *err on the side of least drama* [a term coined by (Brysse et al., 2013)]—a conservative bias that could potentially impede risk management" (Garner et al., 2018, p. 1603). That look at the broader literature illustrated that the greatest change and variation has been on the more dangerous *upper* end of projections (Fig. 12.1).

For coastal decision-makers, wide, widening, and changing ranges of scientific projections are difficult to deal with given the high-stakes decisions they face. How to decide when faced with questions such as the following: Should further development along vulnerable shorelines be permitted, given the sea-level rise outlook over the long-term even if there is great near-term economic benefit to such development? How can coastal erosion and flooding—some of the greatest risks from sea-level rise and coastal storms—be most effectively managed? Should storm-damaged homes be restored and repaired in place, and if so, to what level of protection? How much time can be gained by building nature-based buffers between the sea and coastal structures, and is that worth the investment? What are the costs and benefits of different adaptation strategies? Is relocation from the shoreline necessary and how soon? Different rates of SLR would result in different answers. Some of these responses may later turn out to be maladaptive, i.e., creating lock-ins and/or greater vulnerabilities, either for at-risk human communities, coastal industries, and natural systems or for adjacent ones to whom risks have been transferred (Schipper, 2020).

Many coastal managers are stymied by these difficult questions, even though coastal environments, economies, and property are already at risk and affected by the impacts of a rising ocean (Fleming et al., 2018; Sayers et al., 2022). Decision-makers tend to plan and call for action, but absent adequate investment often still select to delay adaptation action as long as possible, demanding more reliable, locally relevant data (Fleming et al., 2018; Moser et al., 2014). Others decide to begin adaptation, often selecting a "politically feasible" set of SLR projections, i.e., projections that are most likely to be acceptable to political decision-makers, to advance the process at all. Autonomous market responses and distortions can even encourage further high-intensity development at the shorefront (via climate gentrification), rendering the political economy of adaptation even more challenging and creating—ultimately—even greater vulnerabilities (Keenan et al., 2018).

Taking a step back then from the state of SLR science and coastal adaptation in practice,[1] a mixed picture of certainties and uncertainties in the scientific, observational, and decision-making realms arises. It inevitably leads to several vexing questions:

- Given scientific certainties and the already observed impacts and trends of SLR (observational certainties), why are we not seeing more action?
- Given the range and types of scientific uncertainties, is action likely or possible?
- Given uncertainties in the decision-making arena itself, does scientific (un)certainty matter at all, and if so, how?

This chapter attempts to address these questions and the paradoxical situation they portray. While anchored in SLR science and coastal planning and decision-making, it asks one central question, albeit with wider applicability, namely: what is the relationship between scientific (un)certainty and action? Section 12.2 articulates the theoretical expectations one might hold about this relationship in the context of rational decision-making. Section 12.3 then tests this theory against several brief empirical cases, showing how real-life decision-making often does not respond to certain or uncertain scientific understanding as expected. Rather, as Sect. 12.4 will show, scientific knowledge (whether certain or not) is transformed into a strategic tool in the political process which then attains importance for or against action. If scientists wish to engage and become players in this process, what options do they have to bring science as effectively as possible into the political process? Section 12.5 offers some answers before concluding with a summary and outlook in Sect. 12.6.

12.2 Scientific Uncertainty in the Rational Decision-Making Paradigm: Theoretical Expectations

Rational decision theory has been developed, critiqued, and advanced over the past five decades in a variety of disciplines, including (behavioral) economics, psychology, sociology, law, neuroscience, philosophy, political science, organizational studies, business operations, and planning (e.g., Andrews, 2017; Brown, 2005; Gächter, 2013; Jaeger et al., 2001; Reyna & Rivers, 2008; Wolbring, 2020). It is beyond the scope of this chapter to review this wide-ranging body of work. Suffice it to say, in its simplest form, rational choice theory assumes:

- Individuals act rationally in pursuit of their goals (i.e., aligning means and ends logically) and in a risk-averse manner (i.e., maximizing utility, satisfaction or gains while minimizing losses).
- Individuals have sufficient and unambiguous information to establish their preferences.
- Those choices can be influenced by incentives.

Because decision-makers often appear to act seemingly against their own stated goals or self-interest, respond counterintuitively to incentives, and/or the information available to them is *not* unambiguous or sufficient (much less complete and certain), much research has gone into understanding the environmental (e.g., organizational, contextual) and internal (cognitive and affective) factors that could explain these empirically observed deviations from theory. The result has been a plethora of alternative models of decision-making in the face of risk and uncertainty.

One logical implication of the basic tenets of rational choice theory, however, is that when decision-makers face a reasonably certain future, they will be in a better position to strategically align means and ends and determine a course of action in ways that maximize gains while minimizing potential losses. In other words, certainty about the future should enable swift and rational action, while uncertainty about future outcomes should stymie it. This does not mean, decision-makers have no tools available to act in the face of uncertainty, but the theory implies that uncertainty makes it more difficult to act, and it would not be unreasonable to see delays in action, particularly if the gains are unclear and/or the stakes (the potential losses) are high (Kasperson, 2009).

This leads to a simple matrix which relates scientific uncertainty to action (Fig. 12.2). While this caricature does not do justice to the messiness of real-life situations, it helps to structure the discussion below. It suggests that in cases of low levels of uncertainty action is more likely (rational action; dark green), whereas in cases of high uncertainty, no action should be expected (rational inaction; dark red), with the counterintuitive cases being action in the face of high uncertainty (irrational action; light green) and inaction in the face of low uncertainty (irrational inaction; light red).

The next section explores several brief cases from around the USA to illustrate with empirical evidence how decision-makers do or do not act in accordance with the rational decision-making paradigm laid out here.

---

[1] The main focus in this chapter will be on US coastal adaptation. While not directly transferable to other political, socioeconomic, cultural, and legal contexts, many insights gained from this case will resonate elsewhere.

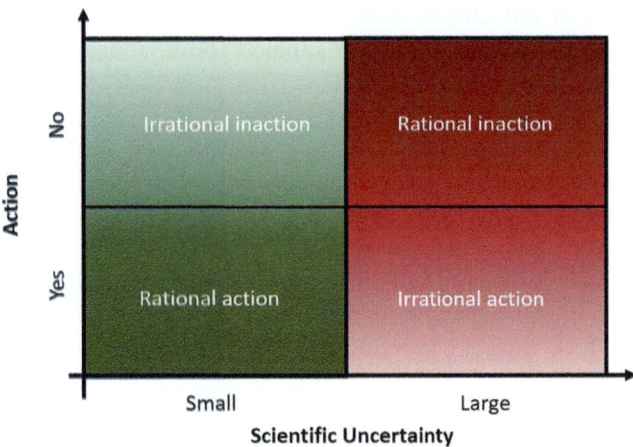

Fig. 12.2 Expected types of action or inaction in the face of different levels of uncertainty

12.3 Testing Theory Against Real-World Cases

12.3.1 Rational Action

The case of greatest certainty is the one where we can look at twentieth-century coastal management action in the face of observed (i.e., well-understood) slowly rising sea level and examine the actions that were taken. Before 1972—the year the federal Coastal Zone Management Act was passed—shoreline development was essentially blind and reactive: development happened, disasters happened, and rebuilding happened. But since the passage of the federal act and its state companions in the decades since, the action seen in the face of low rates of SLR is a patchwork of relatively minor adjustments, including flood insurance, elevation of structures, strengthened zoning or building codes, hard and soft shoreline protections (e.g., seawalls, beach renourishment, protections of natural buffers), setbacks, and often no action at all until disaster struck. Rebuilding in the same location to the same building standards has been the oft-repeated history of coastal zone management in the USA (although the US Federal Emergency Management Agency (FEMA) under President Biden in 2022 moved to prevent this by administrative action; see FEMA, 2022). In the past, however, this has left many homeowners more vulnerable to the encroaching sea, natural shorelines severely degraded in many places, and countless coastal communities ill-prepared for current and future storms, floods, and erosion.

In short, under the assumption of slow SLR (i.e., the certainty that comes from extending known trends into the future), many states and communities acted, albeit only minimally, to adapt. The history of coastal disasters makes the unequivocal case that these actions were insufficient. We have never been over-adapted. Spending too much money for coastal protections that would later turn out to be too much has never been our problem.

By implication, this past experience suggests that there is one uncertainty that coastal decision-makers can safely ignore: the low end of future SLR projections. Decision-makers needn't worry about the most conservative end of SLR projections. Given the physics of ocean warming and accelerating ice loss from land, resulting in the increasing rates of SLR already being observed, there is no reasonable case that can be made that twenty-first-century SLR will be merely an extension of the past. Nor can a good case be made that coastal storms will lessen in terms of intensity or frequency, that El Niño will cease to play a major role along the West Coast of the USA, that erosion will slow down, or that wetland loss come to a halt. Thus, with US coastal communities already not well adapted to current rates of SLR and coastal hazards, more *must* be done. The only question is how much more, which points toward the higher ends of projections, i.e., to the far more uncertain end of the spectrum.

12.3.2 Irrational Action

The earliest known state-level policy in the USA regulating shoreline development in the face of uncertain SLR was put in place in Maine more than 30 years ago, in 1988. The state passed its so-called Sand Dune Rules, which later were slightly weakened due to property takings concerns but were essentially upheld against these challenges. The law explicitly acknowledges uncertainties in the science and the existence of divergent SLR projections but drew on expert judgment of the state of science at that time to defend its rule-making and specific choice of one SLR scenario (3 ft. by 2100; later revised under legal pressure to 2 ft.) (Moser, 2005). An in-depth examination of that case revealed that the motivation behind the policy had little to do with SLR per se, but with a perceived defacement of the coast from increasing high-density coastal development experienced at that time. Nevertheless, SLR science was used to embolden the case against such development and stands as the earliest example in US coastal policy to put in place a SLR regulation despite uncertain future projections.

A second case comes from California, namely, the 2011 amendment to the San Francisco Bay Plan (San Francisco Bay Conservation and Development Commission (BCDC), 2020), with further, complementary updates in 2019. The policy directing shorefront development along the San Francisco Bay fully acknowledges uncertain SLR projections, particularly beyond mid-century, but does not prohibit new development. Instead, it requires developers to demonstrate resilience until 2050 under all SLR projections and demands that they present a feasible adaptation plan thereafter, i.e., to make clear how their development will be protected under different SL scenarios. It also places the economic burden of that adaptation on the developer.

12.3.3 Rational Inaction

The now infamous case of the State of North Carolina "legislating away" SLR might count as a case of rational inaction in the face of significant uncertainty. The state legislature chose to ignore long-term SLR trends, nominally because the "science was bad" and the uncertainties too great to support regulation (Opt & Low, 2017). In 2009, the state Coastal Resources Commission had directed its Science Panel to produce state-level SLR projections to guide coastal regulation, but when those scientific recommendations were delivered a year later, pointing to potentially very high sea levels, the state legislature stipulated that the projections to 2100 could not be applied for such purposes. While well understood as having been politically motivated by development interests, the law did not lay out an adaptation path forward in the face of uncertain albeit potentially very high risk but instead shaped a path—if only temporarily—that enabled continued coastal development with minimal adaptation in the face of a false sense of certainty (i.e., ignoring the significant rise projected in the latter half of the century).

The only concession made was to require updates on the SLR science every 5 years. As a result, the Science Panel has updated its projections twice (still only projecting to 2050, but acknowledging that SLR is accelerating). The state's coastal zone management program is providing technical assistance to local communities wishing to adapt, and the state has advanced comprehensive resilience planning. But the science has still not been deemed certain enough to be used in regulatory and permitting decisions (Allen, 2020).

12.3.4 Irrational Inaction

A case of irrational inaction (i.e., not adapting in the face of evident risk and agreed-upon science) is the case of Florida permitting the expansion of the Turkey Point nuclear power plant to twice its generating capacity. The power plant, located near Homestead, Florida, some 25 miles south of Miami, is situated at sea level in a high-hazard flood zone. The Southeast Florida Regional Climate Change Compact—a four-county partnership—had previously agreed on SLR scenarios to be used for such decisions, namely, 23–61 cm of SLR by 2060 (Moser et al., 2014). While a politically significant step to have a common set of projections across the four-county region, the figures are remarkably low, i.e., not particularly risk-averse or precautionary, when compared to the numbers used in other regions of the world (e.g., the highly precautionary figures [+4.3 m for extreme sea levels by 2100] used for the Sizewell nuclear power site in the UK; see Wilby et al., 2011). Given the already-existent and clearly growing flood risk under any SLR scenarios to this high-risk infrastructure, relocation to higher ground might be considered a rational choice. However, plant expansion plans were only slightly modified to account for potential flooding of access roads, while the plant itself was not fortified any further in place. Relocation was dismissed as it was deemed too expensive and electricity rate payers were thought to not accept that added cost.

12.3.5 Implications

A first lesson from these short vignettes is that even with perfect knowledge or well-supported scientific evidence, action is not guaranteed because science does not compel action. Decision-makers' goals, intentions, and underlying value commitments (e.g., to growth, development, profit) do. And even when action is being taken, that adaptive action has been generally insufficient—a "fig leaf" that allows decision-makers to say they have acted but one that allowed them to side-step political backlash or attacks due to unpopular choices. One critical implication is that focusing only on reducing scientific uncertainties by way of additional research is inadequate at best and will not guarantee that adaptation decisions will in any way be scientifically informed. Differently put, further reduction in scientific uncertainties will not guarantee appropriate, sufficient, or timely action to prevent significant losses.

A parallel, second insight from these cases is that uncertainty is not a show-stopper to action. In fact, both certain and uncertain science have been used to justify action. The question, rather, is what is at stake and what motivates people to action, and how is science used to bolster the case for or against action in the face of uncertainty?

Finally, and at first seemingly in contradiction to these observations, is that in each of these cases, decision-makers did create or depended on some kind of certainty that allowed them to move forward on the path of action (adaptive or maladaptive) they decided to pursue—just not necessarily scientific certainty. Rather, each describes a case in which decision-makers arrived at a kind of political or psychological certainty by accepting or ignoring, curtailing, or selectively using the science. In turn, that certainty-imbued science became a strategic tool in the political process. This violates the rational decision-making paradigm but does not mean science is not relevant to decision-making. It simply tells the wrong story of how that is so.

12.4 Scientific Knowledge as Strategic Tool in the Political Process

How then does scientific knowledge—uncertain or not—come to matter in the political process? In each of the stories relayed above, more or less uncertain scientific knowledge got transformed into a psychologically "certain" and politically persuasive argument for or against action:

- In the case of rational action in the face of scientifically well-established knowledge of SLR trends, the risks of a rising sea level seem to have been ignored or downplayed (a seeming reduction of uncertainty) to justify continued, if minimally hazard-cognizant coastal development.
- In the case of irrational action, two pathways to action emerged: in one, expert judgment was used to legitimize the selection of one out of several SLR scenarios (again, a seeming reduction of scientific uncertainty to establish regulatory certainty). In the other example, the relatively small uncertainty in SLR in the next few decades was embraced and adaptive action by coastal developers required, with the added stipulation to illustrate both financial capability and flexible yet feasible adaptive pathways over the long-term should sea level be higher than expected over the lifetime of the structure (regulatory and financial certainty).
- In the case of rational inaction, the available science was termed "bad" and unreliable and thus supported legislators in their ideological desire to ignore its implications (creating a psycho-political certainty). However, the door for future political consideration was left open by demanding periodic science updates.
- And finally, in the case of irrational inaction, an economic and associated sociopolitical certainty (high costs of relocation and the perceived lack of the public's willingness to foot that bill) won out over another certainty, namely, the already evident and growing risks and clearly established scientific guidance on which science to use in planning and decision-making.

Not only do scientific uncertainties become transmuted into political certainties. Scientific certainties can also be transfigured into political uncertainties to argue against action. Figure 12.3 illustrates some of the common motivations that drive this transmutation, including personal or political motivation as well as actual or perceived economic benefits from acting or postponing it but also reputational, economic or legal liabilities and policy requirements (see also Curry & Webster, 2011).

These observations clearly help to better understand why actions and inactions are being observed against the spectrum of scientific (un)certainties. They tell an empirically more

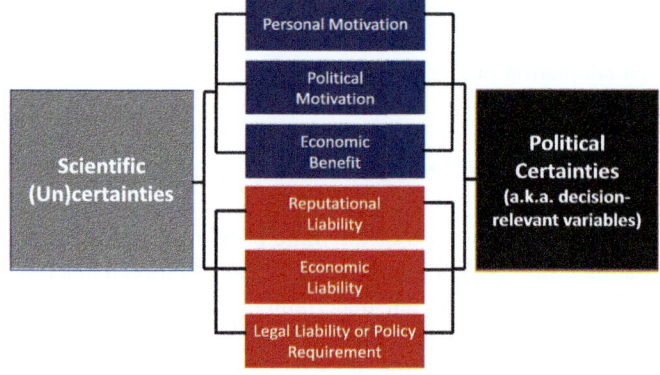

Fig. 12.3 The transmutation of scientific (un)certainties into political certainties

accurate, more nuanced, and thus more persuasive story. This alone, however, would paint a picture of science essentially being no more than a political football (incidentally, a bias some scientists hold and actively use to stay a long way away from political engagement). Clearly, such a story would still be inadequate because it diminishes scientists' agency to simply being servants to a process they cannot influence and which reduces them to one-way information-delivery "automatons," rather than actors who themselves can act strategically and ethically in their engagement with information users.

Differently put, for scientists to ignore the political needs of decision-makers and the range of motivations that underlie their choices (as depicted in Fig. 12.3) is no better than politicians ignoring scientific facts just because they are inconvenient. The task for scientists instead is to become smart and practiced in learning about these political (and underlying social and psychological) motivations and learning to work with them. As they learn about these underlying decision drivers, scientists can play more effective roles as issue advocates (more activist) or honest brokers (less activist) (Pielke Jr., 2007). In the former case, they may help advocate for flexible adaptation actions that reduce risk, minimize losses, benefit at-risk communities and natural habitats, and preserve positive choices for future generations while protecting and restoring the life support systems on which humans depend. In the latter case, they may simply choose to lay out the choices more clearly without taking any side as to what decision-makers should choose.

Incidentally, doing so involves learning some of the very lessons that have led to modifications of the all-too-simplistic rational decision-making paradigm (Samson, 2014) (Table 12.1).

Which of these influences matter most at any one time, and with different decision-makers, cannot be predicted. The important takeaway is that these psychological influences *always* come into play. The purely "rational" (emotion-free)

Table 12.1 Social-science insights that have modified the rational decision-making paradigm

Biases and heuristics	People employ biases and heuristics in forming judgments in the face of uncertainty (Tversky & Kahneman, 1974)
Mental models and confirmation bias	People process information through preexisting mental models and exhibit confirmation bias, which makes uptake of new information that challenges people's beliefs and values more difficult (Johnson-Laird, 2010; Plous, 1993)
Slow vs. fast information processing	People process information not just carefully, analytically, and systematically ("slowly") but also affectively and intuitively ("fast") (Kahneman, 2013)
Choice framing	People respond differently to different kinds of uncertainties and framings of choices (Ho & Budescu, 2019; Levin et al., 1998)
Trust in information sources	People do not process all information equally but tend to pay more attention to information from trusted sources (Sarathchandra & Haltinner, 2020)
Denial of existential threats	People avoid and deny information that signals existential threats (Wullenkord & Reese, 2021)
Context-dependence of decision-making	People make decisions that don't just depend on the information they receive but the environment and context in which they make them (Gigerenzer & Goldstein, 1996)
Context-dependent valence of values	People hold multiple values with different, often context-dependent valence as they choose among options and those preferences don't stay the same over time (Thaler, 2015)
Loss aversion	People care more about losing something they have than gaining something they don't yet have (Kahneman & Tversky, 1979)
Social norms	People are often more influenced by social norms than economic norms (Kinzig et al., 2013)
Emotions and values	People make better choices when factual information is linked to emotions and values (Reyna, 2021)

decision does not exist and is, in fact, not desirable (Reyna, 2021). Drawing on these insights, however, suggests scientists have significant power in shaping how information is delivered and heard, how decision options are framed, and how scientists can help decision-makers make adaptation decisions without risking their political survival. It changes scientists from being deliverers of information to being sophisticated partners in a transaction, which makes the uptake of (uncertain) science in adaptive decision-making more likely.

12.5 Opportunities for Science to Inform Coastal Adaptation Decision-Making

Returning to the practical, what are some concrete ways then for scientists to help inform and shape coastal adaptation decision-making in the face of scientific uncertainty? The proposed entry points discussed below illustrate not only ways in which scientists can help inform decision-makers' understanding of risks, uncertainties and the options, values, and preferences they have but also the necessity of relationship building, continuity of engagement, and reflexivity (on all sides) to grapple with the inherently nonscientific, normative dimensions of decision-making.

12.5.1 Help in Prioritizing Risks

With climate change impacts emerging faster and faster, and many of them creating a sense of foreboding and overwhelm, busy decision-makers first need help with not just identifying the entire universe of possible risks, but which ones to focus on first. Breaking down overwhelmingly large (and profoundly threatening) problems into a series of tractable ones makes it more likely to get publics' and decision-makers' attention. Pragmatically, often this means tying climate change and adaptation to an existing problem that already has their attention. For example, as coastal managers update existing road or water infrastructure in coastal areas, careful consideration of sea-level rise (and other climate change impacts) projections and an appropriate economic analysis of adaptation options (e.g., robust decision-making under deep uncertainty) can help meet multiple objectives while minimizing risks for decades. Linking to such immediate needs helps planners, stakeholders, and elected officials with limited attention spans to focus on adaptation; it is also often where there are resources to address the growing risks from climate change.

This may still not reduce the universe of possible climate change impacts enough to help with focus. Where then are the most meaningful points of intervention? Walker et al. (2010) proposed a triage approach distinguishing three categories of situations:

1. No matter the climate scenario, impacts will be minimal.
2. No matter what we do, losses will be severe and irreversible.
3. Adaptation promises to make the greatest difference on impacts.

It is only in that third category where questions of risks, benefits, and harms (and associated uncertainties), the relevant actors/affected parties to involve, scope, scale, urgency/timeframe, and the feasibility of adaptation actions need to be addressed. Thus, rather than delivering all-encompassing climate impacts or risks assessments, scientists can be most helpful to decision-makers if they go a step further and identify this third space where adaptation could make a real difference. This opens the door to being decision-relevant.

12.5.2 Address What Is Unique about Adaptation

In the next instance, scientists can be helpful in elucidating aspects that are unique about adaptation. While many decisions entail uncertainty, many adaptation decisions must contend with *deep uncertainty*, *indirect costs and benefits*, and with *long time horizons* at once.

Importantly, there are not only scientific or predictive uncertainties to worry about but also values uncertainties (Bammer & Smithson, 2008; Moser, 2005). Regarding values uncertainties, a unique challenge in adaptation is that what and how much society values something currently vs. in the future is not necessarily the same. This may affect what gets protected now vs. what is still seen as worthy of protection later, how much society is willing to pay for something now vs. later, and so forth. Often, there is ambiguity in individuals' values and the sum of individual values is not necessarily the same as collective values due to competition among the values we hold. Typically, these value uncertainties are hidden in model assumptions and not made visible or understandable to decision-makers. Thus, rather than glossing over or trying to resolve all these uncertainties, it is often more important for decision-makers to understand the role and implications of different values so they can arrive at their own judgment. Scientists can further help decision-makers and stakeholders with different values identify their own implicit assumptions and facilitate deliberate reflection and deliberation on these values.

Another (not-entirely unique but nonetheless crucial) feature of adaptation is that the cost of anticipatory adaptation is born now, even if its ultimate benefits are only reaped in the future. As with greenhouse gas emissions reduction, decision-makers will be keenly interested in a) distributing the cost not just across societal groups in the present but also over time and b) favor adaptation options that have near-immediate (co-)benefits. In many ways, however, this uneven distribution of costs and benefits gives adaptation often the character of public goods, provided best by institutions with a responsibility toward the collective, rather than by private actors, seeking profit maximization. Given limited public funds to date for adaptation, there has been a growing interest in involving the private sector in financing adaptation. In some instances, this may not only open up greater pools of resources but may also require improved support from the scientific community in placing economic value on non-monetized risks and benefits and help with comparing costs and benefits that are often separated by disciplinary and governance silos.

Finally, adaptation must contend with the necessarily long-term planning horizons for future climate risks versus short-term planning cycles. In many instances, existing institutions (much less current decision-makers) may not have the longevity required to sustain and/or repeat adaptation efforts into that long-term future. Scientists must help facilitate conversations about how to chart that path, help identify ways for decision-makers to feel comfortable and able to make commitments over time, and help them find feasible near-term and interim steps that maintain long-term flexibility. The important work on adaptation pathways and robust decision-making in the face of deep uncertainty (see Chaps. 3 and 4, this volume) provide tools to do so.

12.5.3 Help in Choosing Adaptation Options

Having identified risks where adaptation can make a real difference and having recognized the unique challenges decision-makers face, scientists can then support the adaptation process by making it possible for stakeholders and decision-makers to choose among different adaptation options. Not only do people need to understand the pros and cons and costs and benefits of different adaptation options, they also need guidance in working through these decisions.

A useful step here is to help assess the adaptation options under consideration as to whether and how they address the unique aspects of adaptation (Table 12.2).

In addition, adaptation decisions must also consider issues like safety of operation, ease of implementation, and other implementation issues. Clearly, the decision for or against different adaptation options involves profound values choices. Scientists—using structured decision-making processes—can help planners, decision-makers, and stakeholders make the criteria transparent and then deliberate the values-side of their choices (Gregory et al., 2012). This may entail explicitly exploring visions of the future, including the desirable, plausible/constrained, and possible futures. This also means listening carefully and helping to surface implicit values in how people discuss these futures and choices. It can empower all involved in the decision-making process and enable them to participate more effectively. Finally, it can help to improve the quality of decisions itself, i.e., by helping to refine the problem definition, clearly elicit and discern the objectives, define a range of alternatives that are linked to those objectives, and then help assess the consequences of pursuing any one of the alternatives and confronting trade-offs (and possible synergies).

Notably, in none of these instances are scientists unduly influencing or making any of these choices for the decision-

Table 12.2 Preferable adaptation options that address the unique features of adaptation

Adaptation options that address deep uncertainty issues	Adaptation options that address indirect benefit issues	Adaptation options that address long time horizons
• Have net benefits, regardless of future climate • Include the possibility of low-cost safety margins • Can easily be changed, avoid lock-in • Fit with short-lived planning horizons, allowing repeats • Simplify or streamline decision-making complexity • Build toward wider range of variance, not average • Include inbuilt mechanisms for routine, periodic review	• Involve mechanisms that lower direct costs to actors now, spread to, or share costs with future actors • Involve mechanisms that provide near-term benefits and address trade-offs • Facilitate cross-sector alignment and thus enable sharing of costs and benefits • Lower transaction costs now, e.g., by allowing more frequent, smaller decisions resulting in learning and familiarity or through "mainstreaming"	• Identify options to fill the institutional gap • Involve mechanisms that bridge short-term horizons • Require periodic revisiting of decisions • Build in monitoring and evaluation, and establish agreed thresholds which—when reached—trigger subsequent adaptive action

Source: Adapted from Walker et al. (2010)

makers, i.e., they remain objective in the sense of not imposing their values on a decision. Scientists—as citizens—have no more a voice than other citizens. Rather, to be useful and decision-relevant, scientists assist better decision-making. They do so not only by conducting research to answer decision-relevant questions but by helping to facilitate a process in which all voices are heard and given appropriate consideration so that the decision and its consequences become clearer to all involved. They support joint fact-finding and knowledge coproduction and assist in making risks, uncertainties, and decision consequences meaningful (Hilger et al., 2021).

12.6 Summary and Outlook

This chapter examined the ways in which uncertainties in and beyond sea-level rise science can, but does not necessarily, delay coastal adaptation action and how scientists can work more effectively with coastal practitioners and communities to make uncertainties intelligible and decision-relevant while still facilitating action. The argument launched from a review of an outdated way of thinking wherein scientific uncertainty is thought to be "the problem," i.e., the reason for delayed decision-making, resulting in an assumption that uncertainty needs to be reduced in order to see "right action." With limited empirical evidence to support this simplistic assumption, the chapter then proposed an alternative paradigm that more adequately captures the role of (uncertain) science in decision-making. It showed how scientific uncertainty is transmuted in the political process into a "political certainty" so that it can bolster the case for action or inaction, as the case may be. In this sense, uncertainty becomes "politically constructed." Importantly, however, scientists are not just powerless bystanders to this process but can actively "co-construct" the meaning, importance, and interpretation of uncertainty. While more science may be useful and some scientists are better at advancing the knowledge frontier than public deliberation of adaptation options, this recommendation shifts the attention from "doing more science" to "working effectively at the science-policy interface."

The chapter argued that there are not only uncertainties in all dimensions of climate risk assessment and coastal impacts research but also in all aspects of coastal adaptation decision-making and risk governance (Moser, 2005; Renn, 2008). Scientists must learn to navigate this complex territory with greater sophistication, drawing on what is understood about how people process information, form judgments, and make decisions in the real world. With that understanding, scientists can more usefully support coastal decision-making by helping to identify coastal adaptation priorities, address the unique aspects of adaptation, and identify and assess possible adaptation options suitable in different coastal contexts. At minimum, the shift that the chapter proposes is thus one from a "scientifically rational" decision-making paradigm to a "politically rational" one. Maybe more important even is the move away from seeing (or drawing) a sharp line between the scientific and the decision-making processes to seeing the two as transactionally and relationally intertwined. It asks that both scientists and decision-makers get better at working with each other. Then, uncertainty can no longer be seen as being inherently important to decision-making but as a condition that attains cocreated political significance. In short, no knowledge is inherently valuable; no knowledge is inherently "certain enough" for action; and no uncertainty is inherently decision-relevant or decision-limiting. Instead, all forms of knowledge can attain value in someone's eyes, in some contexts; all knowledge can be "good enough" to act on; and all certainties and uncertainties can be made decision-relevant.

This implies—for both the coast and for other sectoral contexts—not only a different kind of training of scientists and decision-makers to build the necessary skills. It also demands doubled efforts in strengthening, normalizing, and institutionalizing science-policy interactions. In the face of accelerating climate changes, there is no time to lose in making sophisticated science-policy interactions commonplace so that trust and familiarity are established as the foundational conditions for the difficult choices coastal communities now face in an always-uncertain and increasingly high-stakes environment.

References

Allen, J. (2020). NC's first sea level rise report, 10 years on. *Coastal Review*. Retrieved June 24, 2020, from https://coastalreview.org/2020/06/ncs-first-sea-level-rise-report-10-years-on/

Andrews, C. J. (2017). Rationality in policy decision making. In F. Fischer et al. (Eds.), *Handbook of public policy analysis: Theory, politics, and methods* (pp. 161–172). Routledge.

Bammer, G., & Smithson, M. (2008). *Uncertainty and risk: Multidisciplinary perspectives* (p. 400). Earthscan.

Brown, R. (2005). *Rational choice and judgment: Decision analysis for the decider* (p. 280). Wiley.

Brysse, K., Oreskes, N., O'Reilly, J., & Oppenheimer, M. (2013). Climate change prediction: Erring on the side of least drama? *Global Environmental Change, 23*, 327–337. https://doi.org/10.1016/j.gloenvcha.2012.10.008

Curry, J. A., & Webster, P. J. (2011). Climate science and the uncertainty monster. *Bulletin of the American Meteorological Society, 92*, 1667–1682. https://doi.org/10.1175/2011BAMS3139.1

FEMA. (2022). *Partial implementation of the federal flood risk management standard for public assistance (interim). FEMA policy 104-22-0003*. FEMA. Retrieved from https://www.fema.gov/sites/default/files/documents/fema_fp-104-22-0003-partial-implemetnation-ffrms-pa-interim.pdf

Fleming, E., et al. (2018). Coastal effects. In D. R. Reidmiller et al. (Eds.), *Impacts, risks, and adaptation in the United States: Fourth national climate assessment* (Vol. 2, pp. 322–352). U.S. Global Change Research Program.

Gächter, S. (2013). The handbook of rational choice social research. In W. Rafael et al. (Eds.), *Rationality, social preferences, and strategic decision-making from a behavioral economics perspective* (pp. 33–71). Stanford University Press.

Garner, A. J., et al. (2018). Evolution of 21st century sea level rise projections. *Earth's Future, 6*, 1603–1615. https://doi.org/10.1029/2018EF000991

Gigerenzer, G., & Goldstein, D. G. (1996). Reasoning the fast and frugal way: Models of bounded rationality. *Psychological Review, 103*, 650–669. https://doi.org/10.1037/0033-295X.103.4.650

Gregory, R., Falling, L., Harstone, M., Long, G., McDaniels, T., & Ohlson, D. (2012). *Structured decision-making: A practical guide to environmental management choices* (p. 312). Wiley-Blackwell.

Hilger, A., Rose, M., & Keil, A. (2021). Beyond practitioner and researcher: 15 roles adopted by actors in transdisciplinary and transformative research processes. *Sustainability Science, 16*, 2049–2068. https://doi.org/10.1007/s11625-021-01028-4

Ho, E. H., & Budescu, D. V. (2019). Climate uncertainty communication. *Nature Climate Change, 9*, 802–803. https://doi.org/10.1038/s41558-019-0606-6

IPCC. (2021). Summary for policy makers. In V. Masson-Delmotte et al. (Eds.), *Climate change 2021: The physical science basis* (p. 41). Cambridge University Press.

Jaeger, C. C., Renn, O., Rosa, E. A., & Webler, T. (2001). *Risk, uncertainty, and rational action* (p. 320). Earthscan.

Johnson-Laird, P. N. (2010). Mental models and human reasoning. *Proceedings of the National Academy of Sciences, 107*, 18243–18250. https://doi.org/10.1073/pnas.1012933107

Kahneman, D. (2013). *Thinking, fast and slow* (p. 512). Farrar, Straus and Giroux.

Kahneman, D., & Tversky, A. (1979). Prospect theory: An analysis of decision under risk. *Econometrica, 47*, 263–291.

Kasperson, R. E. (2009). Coping with deep uncertainty: Challenges for environmental assessment and decision-making. In G. Bammer & M. Smithson (Eds.), *Uncertainty and risk: Multidisciplinary perspectives* (pp. 337–347). Earthscan.

Keenan, J. M., Hill, T., & Gumber, A. (2018). Climate gentrification: From theory to empiricism in Miami-Dade County, Florida. *Environmental Research Letters, 13*, 054001. https://doi.org/10.1088/1748-9326/aabb32

Kinzig, A. P., et al. (2013). Social norms and global environmental challenges: The complex interaction of behaviors, values, and policy. *Bioscience, 63*, 164–175. https://doi.org/10.1525/bio.2013.63.3.5

Levin, I. P., Schneider, S. L., & Gaeth, G. J. (1998). All frames are not created equal: A typology and critical analysis of framing effects. *Organizational Behavior and Human Decision Processes, 76*, 149–188. https://doi.org/10.1006/obhd.1998.2804

Moser, S. C. (2005). Impact assessments and policy responses to sea-level rise in three US states: An exploration of human-dimension uncertainties. *Global Environmental Change, 15*, 353–369. https://doi.org/10.1016/j.gloenvcha.2005.08.002

Moser, S. C., et al. (2014). Coastal zone development and ecosystems. In J. M. Melillo, T. T. C. Richmond, & G. W. Yohe (Eds.), *Climate change impacts in the United States: The third national climate assessment* (pp. 579–618). U.S. Global Change Research Program.

Opt, S., & Low, R. (2017). Dividing and uniting through naming: The case of North Carolina's sea-level rise policy. *Environmental Communication, 11*, 218–230. https://doi.org/10.1080/17524032.2016.1157507

Pielke, R. A., Jr. (2007). *The honest broker: Making sense of science in policy and politics* (p. 198). Cambridge University Press.

Plous, S. (1993). *The psychology of judgment and decision making* (p. 302). McGraw-Hill.

Renn, O. (2008). *Risk governance: Coping with uncertainty in a complex world* (p. 476). Earthscan.

Reyna, V. F. (2021). A scientific theory of gist communication and misinformation resistance, with implications for health, education, and policy. *Proceedings of the National Academy of Sciences, 118*, e1912441117. https://doi.org/10.1073/pnas.1912441117

Reyna, V. F., & Rivers, S. E. (2008). Current theories of risk and rational decision making. *Developmental Review, 28*, 1–11. https://doi.org/10.1016/j.dr.2008.01.002

Samson, A. (2014). *The behavioral economics guide 2014: Introduction to behavioral economics*. Behavioral Science Solutions Ltd. Retrieved from https://www.behavioraleconomics.com/be-guide/the-behavioral-economics-guide-2014/

San Francisco Bay Conservation and Development Commission (BCDC). (2020). *San Francisco bay plan*. BCDC. Retrieved from https://bcdc.ca.gov/plans/sfbay_plan.html

Sarathchandra, D., & Haltinner, K. (2020). Trust/distrust judgments and perceptions of climate science: A research note on skeptics' rationalizations. *Public Understanding of Science, 29*, 53–60. https://doi.org/10.1177/0963662519886089

Sayers, P., Moss, C., Carr, S., & Payo, A. (2022). Responding to climate change around England's coast-the scale of the transformational challenge. *Ocean and Coastal Management, 225*, 106187. https://doi.org/10.1016/j.ocecoaman.2022.106187

Schipper, E. L. F. (2020). Maladaptation: When adaptation to climate change goes very wrong. *One Earth, 3*, 409–414. https://doi.org/10.1016/j.oneear.2020.09.014

Thaler, R. H. (2015). *Misbehaving: The making of behavioral economics* (p. 432). W. W. Norton & Company.

Tversky, A., & Kahneman, D. (1974). Judgments under uncertainty: Heuristics and biases. *Science, 185*, 1124–1131.

Walker, W. D., Liebl, D. S., Gilbert, L., LaGro, J., Nowak, P., & Sullivan, J. (2010). *Adapting to climate change: Why adaptation policy is more difficult than we think (and what to do about it)*. Wisconsin Initiative on Climate Change Impacts (WICCI).

Wilby, R. L., Nicholls, R. J., Warren, R., Wheater, H. S., Clarke, D., & Dawson, R. J. (2011). Keeping nuclear and other coastal sites safe from climate change. *Proceedings of the Institution of Civil Engineers: Civil Engineering, 164*, 129–136. https://doi.org/10.1680/cien.2011.164.3.129

Wolbring, T. (2020). Rationality, strategic interactions, and theories of middle range. *Social Science Information, 59*, 569–574. https://doi.org/10.1177/0539018420964390

Wullenkord, M. C., & Reese, G. (2021). Avoidance, rationalization, and denial: Defensive self-protection in the face of climate change negatively predicts pro-environmental behavior. *Journal of Environmental Psychology, 77*, 101683. https://doi.org/10.1016/j.jenvp.2021.101683

Open Access This chapter is licensed under the terms of the Creative Commons Attribution 4.0 International License (http://creativecommons.org/licenses/by/4.0/), which permits use, sharing, adaptation, distribution and reproduction in any medium or format, as long as you give appropriate credit to the original author(s) and the source, provide a link to the Creative Commons license and indicate if changes were made.

The images or other third party material in this chapter are included in the chapter's Creative Commons license, unless indicated otherwise in a credit line to the material. If material is not included in the chapter's Creative Commons license and your intended use is not permitted by statutory regulation or exceeds the permitted use, you will need to obtain permission directly from the copyright holder.

Uncertainty in Determining Impacts of Climate Change on Human Health

Kristie L. Ebi, Mary H. Hayden, Morgan E. Gorris, Christopher K. Uejio, and Jennifer Vanos

13.1 Introduction

Changing weather patterns are altering the population burden of climate-sensitive injuries, illnesses, and deaths and impacting health systems and access to healthcare. The health impacts can be direct (heat, flood, wildfire, storms, and drought), ecosystem-mediated (vector-borne diseases, food- and water-borne infections; air quality), or involve human agency (e.g., undernutrition, migration, and conflict) (Cisse et al., 2022). Climate change can also reduce access to and jeopardize critical infrastructure (e.g., water and sanitation services, hospitals, healthcare facilities) and destabilize systems that maintain population health (e.g., flooding events that reduce food security, or sea-level rise in coastal regions that limits access to healthcare). Risks to population health and health systems arise from the interactions of climate-related hazards with the vulnerability of individuals, communities, facilities, and geographic regions exposed to those hazards; and the capacity to prepare for and effectively manage the associated risks. For each health outcome, a mix of interventions can be implemented along the pathways from exposure to health impact, which are affected by vulnerability and capacity to manage. Inherent uncertainties in data, analyses, models, and health system responses, as well as projections of health risks of climate change, should be factored into health decision-making to increase resilience and health equity.

The chapter first outlines a framework for understanding the health risks of climate change and then discusses sources of uncertainty in projecting health risks and of incorporating uncertainties into health decision-making. The chapter ends with a brief discussion section highlighting the importance of proactive, timely, and sufficient investments to reduce health risks in a changing climate.

13.2 Framework for Understanding the Health Risks of Climate Change

Any health outcome that is climate-sensitive could be affected by a changing climate. The extent to which climate-related hazards affect population health and health systems depends on various interacting environmental, socioeconomic, and health factors and trends. Climate change is a risk multiplier that interacts with upstream determinants of exposure and vulnerability, including environmental factors (e.g., air pollution, land use change, biodiversity loss, and desertification), socioeconomic factors (e.g., demographic change, economic growth, investments in science and technology development, urbanization, and inequities), and vulnerability and susceptibility (e.g., population health status, social infrastructure, and political commitment) (Haines & Ebi, 2019). Each of these factors has multiple components that may affect the burden of climate-sensitive health outcomes and the effectiveness of health

K. L. Ebi
Center for Health and the Global Environment, University of Washington, Seattle, WA, USA
e-mail: krisebi@uw.edu

M. H. Hayden
Lyda Hill Institute for Human Resilience, University of Colorado, Colorado Springs, CO, USA
e-mail: mhayden@uccs.edu

M. E. Gorris
Information Systems and Modeling Group, Los Alamos National Laboratory, Los Alamos, NM, USA
e-mail: mgorris@lanl.gov

C. K. Uejio
Department of Geography and Program in Public Health, Florida State University, Tallahassee, FL, USA
e-mail: cuejio@fsu.edu

J. Vanos
School of Sustainability, Arizona State University, Tempe, AZ, USA
e-mail: jvanos@asu.edu

Table 13.1 Exposure pathways with examples of health outcomes

Exposure pathway	Examples of health outcomes
Extreme weather and climate events	Injuries, fatalities, mental health effects, reduced functioning of or access to healthcare facilities
Heat stress	Heat-related illnesses and death; exacerbation of existing chronic diseases; injuries; reduced labor capacity, income, and nutrition
Air quality, including air pollution and aeroallergens	Exacerbations of asthma and other respiratory diseases, respiratory allergies, cardiovascular disease
Water quality and quantity	Campylobacter infection, cholera, cryptosporidiosis, harmful algal blooms, leptospirosis, viruses, salinity and heavy metal (e.g., arsenic) poisoning
Food safety and security	Undernutrition, *salmonella* food poisoning, *campylobacter infection*
Vector distribution and ecology	Dengue; malaria; Zika, Lyme disease; chikungunya; rift valley fever; hantavirus infection; West Nile infection; outbreaks affecting the functioning of healthcare facilities
Social factors	Physical and mental health effects of violent conflict, civil unrest and forced migration

systems, such as the extent to which urban growth plans address the quality of housing stock. Together, these factors interact to affect exposure pathways that determine the magnitude and pattern of climate-sensitive health outcomes at various scales, including extreme weather and climate impacts, heat stress, air quality, water quality and quantity, food security and safety, vector distribution and ecology, and social factors that can exacerbate or ameliorate conflict and migration (see Table 13.1). The level of climate resilience of health systems then determines the extent to which climate-related shocks and stresses impact population health and access to and functioning of healthcare infrastructure.

Natural and human systems are deeply integrated, requiring systems-based approaches to understand the scale of the challenges and the options to manage. Humans affect natural systems that, in turn, can alter population health. For example, essential nutrients, including carbohydrates, proteins, fats, vitamins, and minerals, are required for human health and development (Ebi, Anderson, et al., 2021). Inadequate intake can negatively affect development and result in a wide range of adverse health outcomes. At the same time, human land use directly affects more than 70% of the global, ice-free land surface that supplies food, freshwater, and other nature services (IPCC, 2019). Global population growth and changes in per capita consumption of food, feed, fiber, timber, and energy have caused unprecedented rates of land and freshwater use. About 23% of total anthropogenic greenhouse gas emissions derive from agriculture, forestry, and other land use (IPCC, 2019). Increased carbon dioxide emissions are reducing the nutritional quality of major cereal crops, including wheat and rice, that are projected to affect millions of people later this century (Ebi, Anderson, et al., 2021). Changing temperature and precipitation patterns, particularly in tropical regions, are decreasing crop yields, leading to increasing rates of food insecurity in vulnerable countries and regions. Climate change also impacts food security through loss of livelihoods. A better understanding of this complex food-health-climate system requires addressing data and knowledge gaps surrounding plant biology and policy responses. There are limited data on nutrients other than protein, iron, and zinc, leading to a limited mechanistic understanding of the response of plants to elevated carbon dioxide concentrations. Further, data are largely missing on nutritional status and food safety in low- and middle-income countries, limiting assessments of the magnitude of the risks. Transdisciplinary research involving at least ecologists, plant physiologists, economists, and experts in human nutrition is essential for developing a systems-based understanding of the potential impacts on human nutrition and the attendant consequences for achieving food security as part of the sustainable development goals (SDG 2: Zero Hunger; https://www.un.org/sustainabledevelopment/hunger/). This example provides context for the complex nature of understanding the direct and indirect connections between climate change and health outcomes in diverse settings.

13.3 Sources of Uncertainty in Projecting Health Risks of Climate Change

There are multiple sources of uncertainty when projecting how the magnitude and pattern of health risks and impacts on healthcare infrastructure would change with additional climate change, including:

- Health data access and quality.
- Quantifying relationships between weather/climate data and health outcomes.
- Modeling weather/climate and health.
- Future development (e.g., infrastructure) and population growth/dynamics.
- Spatial scales of data and impact on decision-making at differing levels/scales.

Figure 13.1 illustrates the compounding uncertainties in projecting health risks of climate change.

13.3.1 Health Data

As with all data, there are uncertainties in the accuracy and completeness of health data. An outbreak of an infectious

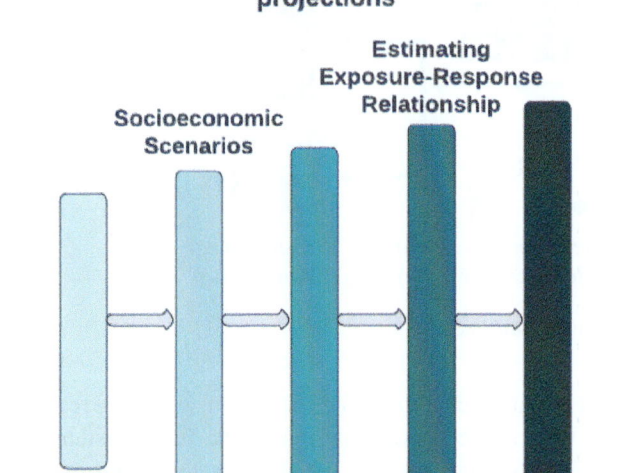

Fig. 13.1 Compounding uncertainties in projecting the health risks of climate change

disease following a flooding event is an example wherein completeness and accuracy of epidemiological data are often lacking: it is likely that not all infectious disease cases were identified and accurately diagnosed because (1) individuals may have chosen not to visit a healthcare provider, (2) the disease was misdiagnosed, or (3) health systems were too overwhelmed to accurately record the information.

Mortality associated with extreme heat is another example. The official statistics from the Centers for Disease Control and Prevention (CDC) state that more than 600 people in the USA die annually from extreme heat (https://www.cdc.gov/disasters/extremeheat/index.html), yet the National Weather Service estimates merely 158 deaths annually from extreme heat (https://www.weather.gov/hazstat/), and Maricopa County in Arizona alone reports 229 deaths annually from extreme heat since 2016 (https://www.maricopa.gov/ArchiveCenter/ViewFile/Item/5404). However, epidemiological modeling used to estimate excess deaths from high ambient temperature indicates that about 12,000 heat-related mortality deaths occur annually in the USA (Shindell et al., 2020). These varying numbers underline the inherent uncertainty in quantifying current (and thus future) impacts of extreme heat on human health. There is an immediate need to align these numbers using a systematic method for identifying heat-caused and heat-associated deaths across counties, which involves medical examiner or coroner decision-making/approaches and accounting for underlying illnesses and contextual factors of death (location, timing, activities, etc.).

Human disease case data are collected through disease surveillance systems, wherein human case reports are entered into a system and shared among levels of public health (local, state, territorial, federal, and international). These systems aim to monitor, control, and prevent the occurrence and spread of state-reportable and nationally notifiable infectious (and some noninfectious) diseases and conditions. In the USA, common surveillance systems include the CDC, National Notifiable Diseases Surveillance System (NNDSS), Environmental Public Health Tracking Network, and the National Syndromic Surveillance Program (Centers for Disease Control and Prevention, 2021). Recent federal government efforts have focused on providing near-real-time data and proactively forecasting influenza and COVID-19 (CDC; https://ephtracking.cdc.gov/Applications/heatTracker/).

Data surveillance programs are either passive or active (Murray & Cohen, 2017). Passive disease surveillance occurs when cases are reported to a public health agency through a reporting system upon patient diagnosis. Active data surveillance occurs when public health staff seek out and receive reports of disease cases, including calling or visiting health facilities or interviewing patients. Combining these two data surveillance programs into the same analysis may result in spatial and temporal statistical bias in the human disease data.

Disease surveillance systems are subject to two major uncertainties: under-ascertainment and underreporting. First, at the community level, not all cases seek healthcare, leading to under-ascertainment (Gibbons et al., 2014). Often, a person may not be symptomatic or have mild symptoms and can recover without seeking medical care. As a result, reported disease cases tend to be more serious. Second, at the healthcare level, there is a failure to adequately report symptomatic cases that have sought medical advice, causing underreporting (Gibbons et al., 2014). Underreporting may be due to human error in entering disease billing codes into the system, physician awareness, or differences in diagnostic testing or uncertainties (false-positives or false-negatives) in the diagnostics. These issues could result in spatial heterogeneity in disease reporting. Disease cases may be labeled as suspected, probable, or confirmed to address differences in diagnostic practices.

Because not all states in the USA are required to report each disease, there is spatial heterogeneity in human disease case data. Furthermore, the location where a case was contracted and where the official case report is submitted does not always match and could be based on where the case resides; hence, spatial noise exists in the data. Such spatial heterogeneity and noise in the data can result in incongruent decision-making to mitigate current or future health impacts in a given location (Solís et al., 2017). Further, the diseases that are nationally notifiable are subject to change and may cause a temporal gap in reporting. Federally funded facilities and tribal governments are not required to report cases to lo-

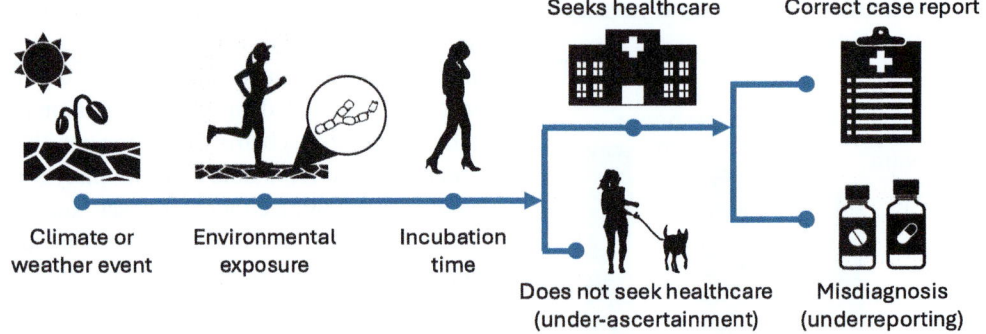

Fig. 13.2 Pathway from weather or climate event to reported climate-sensitive health outcomes

cal health agencies (McCotter et al., 2019), so disease counts of particular populations, including Indigenous people, may not be accurately reflected within US surveillance data.

Disease vector data and vector surveillance programs are other common forms of health data used for climate change and human health analyses, especially because climate change is projected to change the distribution, abundance, and seasonality of disease vectors and pathogens. Disease vector taxonomy, especially before advanced genomic sequencing, was frequently identified by eye and subject to incorrect assignment. Where disease vectors are surveyed and collected often coincides in urban and suburban areas where people are; hence, there is often a spatial bias in vector sampling. Additionally, different vector surveillance techniques are used for data collection, which can bias their results. For example, some types of mosquito traps yield a higher abundance and diversity than others (Giordano et al., 2020).

A more comprehensive and strategic approach is needed to build the public health surveillance capacity to climate change and its compounding impacts on human health (Moulton & Schramm, 2017).

Figure 13.2 illustrates the pathway from a weather or climate event to reported health outcomes.

13.3.2 Quantifying Relationships between Weather/Climate and Health

The limited number of published studies reporting climate and health exposure-response (e.g., damage) functions is one of the largest challenges in linking health outcomes to climate-related factors, thus resulting in modeling uncertainties. Even for relatively well-studied systems, such as air pollution (extreme heat-related morbidity and mortality is another example), there are still critical knowledge gaps related to emission factors, fuel load, and combustion completeness (e.g., McCarty, 2011) and the specific impacts on health. Intuitively, exposure-response functions are best suited for the place in which they were derived and should only be judiciously extrapolated to other locations.

There is an immediate need for multidisciplinary research teams with a deep understanding of the Earth system processes and climate metrics to estimate exposure-response functions more accurately. For example, public health researchers may not be aware that ground-based rain gauges can contain 10–40% errors in temperate regions (Aguado & Burt, 2000). Some guidance is available on appropriate environmental metrics for human health applications, but more cross-disciplinary fertilization would be beneficial (e.g., Davis et al., 2016). For example, the misapplication of gridded climate datasets obscured the positive climate warming and malaria transmission association in Kenya (Omumbo et al., 2011). This association was only revealed by cocreating local climate at the same analysis level as the health outcomes.

Generally, common statistical methods (e.g., time series analysis, case-crossover study designs) for developing exposure-response functions produce similar results (Lu & Zeger, 2007). Nonetheless, results are strongly influenced by key statistical model choices such as the exposure-response functional form (e.g., linear, nonlinear, threshold-based) and lag between exposure and onset of symptoms (e.g., Krow-Lucal et al., 2017). Theoretical, experimental, and observational studies indicate that environmental exposures and/or infectious disease risks are frequently nonlinear. Simplifying nonlinear relationships as linear or threshold-based introduces statistical bias into the analysis (e.g., Vicedo-Cabrera et al., 2019). Time series methods such as distributed lag nonlinear models or generalized additive mixed models are popular for considering different exposure-response functional forms and temporal lags (e.g., Gasparrini et al., 2010).

13.3.3 Modeling Climate Change and Health

Climate models project various plausible climatic conditions based on greenhouse gas emission growth trajectories (representative concentration pathways). Projecting the future is inherently uncertain due to natural climate variability (e.g., El Niño Southern Oscillation), scientific uncertainty (e.g., climate model sensitivity to greenhouse gases, feedbacks), and societal choices (e.g., population growth, consumption, degrowth) (see chapters in this volume by Morris and Monier, Lehner and Deser, and Forest and Collins). To provide an analogy, future COVID-19 cases are related to natural variability, scientific uncertainty (e.g., emergence of new strains or variants, waning immunity, protection against reinfection, etc.), and social choices (e.g., proportion of the population vaccinated and boosted, mask-wearing, social distancing) (see Box 13.1).

> **Box 13.1: COVID-19**
>
> COVID-19 highlighted the importance of recognizing and accommodating uncertainty given the system complexity with multiple interacting components, including nonlinear interactions. The complexity of COVID-19 translated into day-to-day uncertainty about the optimal steps to reduce transmission in a dynamic and unpredictable system whose components adapt readily to interventions (Rutter et al., 2020). Transparent reporting and data surveillance are critical first steps to understanding the scope and breadth of any outbreak. However, from the beginning of the COVID-19 outbreak, uncertainty has been a hallmark of the response because of the inherent limitations of having accurate health surveillance data with a novel pathogen that did not have approved testing available (Koffman et al., 2020). These known limitations require scientists to acknowledge the uncertainty in the data and in the proposed interventions while doing so in a transparent manner that allows for scientific debate as to the next best steps. This transparency in data limitations also allows the public to recognize that COVID-19 (or any novel pathogen) is evolving alongside scientific thought as to the best adaptive behaviors to reduce the risk of transmission. Without the acknowledgment of uncertainty, the scientific community risks rejection of proposed interventions, spread of disinformation, and a tainted view of the scientific process.
>
> The lessons learned from the COVID-19 response can readily be transferred to approaching uncertainty in the climate system (Klenert et al., 2020). There is a great need to educate the public about uncertainty and its place in the scientific process (see Broomell et al. in this volume) and about the need for collective action to reduce future impacts of climate change and novel pathogen introductions.

13.3.4 Future Development

Perhaps the largest source of uncertainty in projecting future health risks involves the extent of action that societies will take to move toward a more resilient and sustainable future, including changing demographic patterns, economic growth, urbanization, investments in research and technology development, and other factors. The shared socioeconomic pathways (SSPs) are used by the climate change community to describe the socioeconomic factors that will alter future burdens of climate-sensitive health outcomes and the status of healthcare infrastructure (Sellers & Ebi, 2017). The SSPs describe potential futures that diverge on axes of increasing challenges to adaptation and mitigation. These potential futures range from very optimistic, where governments can develop sustainable futures that provide the baseline to facilitate mitigation to climate change while adapting to its impacts (SSP1), to very pessimistic, where governments must confront the effects of increasingly severe climate impacts and events while lacking the political will or resources to invest in adaptation or mitigation measures, resulting in poor health outcomes (SSP3) (see also Morris and Monier, this volume). The building blocks of health systems also vary dramatically across the SSPs, from proactive, adaptively managed, interdisciplinary, and strong North-South partnerships under SSP1 to reactive, failure to adapt, and siloed information channels and national governance under SSP3.

Ebi, Boyer, et al. (2021) applied a synthesis approach used in the Intergovernmental Panel on Climate Change (IPCC) assessment reports to illustrate how six health risks could change with further temperature increases under three adaptation scenarios: heat-related morbidity and mortality, ozone-related mortality, malaria incidence rates, incidence rates of dengue and other diseases spread by *Aedes* sp. mosquitos, Lyme disease, and West Nile fever. Adaptation can reduce the magnitude of risks as they increase with additional climate change. Transitions from detectable and attributable risks to severe and widespread risks related to heat could manifest even at warming of less than 1.5 °C above preindustrial temperatures and will continue to develop at warming levels up to about 2.5 °C, depending on the extent to which adaptation is proactive, timely, and effective (Ebi, Boyer, et al., 2021). The high adaptation scenario that emphasizes international cooperation toward achieving sustainable development has the greatest potential to avoid significant increases in risks under all but the highest warming scenarios; even then, there will be residual risks for health systems to manage. The analysis provides insight into possible limits for adaptation. For example, additional warming may lead to expansion or northern range shifts of tick species capable of transmitting pathogens that cause Lyme disease that, combined with underprepared or overburdened health systems, could lead to communities in certain regions being overwhelmed by disease outbreaks (Cisse et al., 2022).

13.3.5 Spatial Scales of Importance for Decision-Making

The misalignment of climate analysis levels/scales may introduce statistical bias into climate change and health projection studies. Projections may apply temperature/precipitation averaged over a large area (50 km^2) to exposure-response functions developed from a single weather station or city with different statistical properties or found at a different location than the reported health data. Generating higher spatial resolution projections ("downscaling") (see Fowler and Mearns, this volume) can serve the dual purposes of aligning analysis scales and statistically debiasing the climate projections (for statistical downscaling), thus supporting congruent decision-making to reduce climate-related health outcomes. Moreover, low- and middle-income countries (LMICs) often lack spatially representative or accurate health and weather information resulting in minimal data to leverage for decision-making now or within projections. For example, Green et al. (2019) found gaps in the representation of LMICs in heat-impact assessments.

13.4 Incorporating Uncertainties into Decision-Making

Avoidable injuries, illnesses, disabilities, diseases, and deaths caused by climate change show that many individuals, communities, and health systems are not prepared for the current or future health impacts of climate change, demonstrating the presence of an adaptation gap (Cisse et al., 2022; Ebi, 2020; Romanello et al., 2021; United Nations Environment Programme, 2018). Limited human and financial resources and a lack of health adaptation research have constrained needed progress (United Nations Environment Programme, 2018; World Health Organization (WHO), 2021a).

The World Health Organization climate change and health surveys chart progress toward preparing health systems for climate change impacts. The 2021 survey concluded that national planning on health and climate change is advancing, but the comprehensiveness of strategies and plans needs to be strengthened (World Health Organization (WHO), 2021a). For example, only 47 (52%) of 91 countries surveyed had a national health and climate change plan or strategy and thus implementing action on crucial health and climate change priorities remains challenging. The main obstacle is lack of climate finance for health-related adaptation, with less than 0.5% of international adaptation funding going to the health sector, and equally limited research funding in biomedical sciences for climate change and health (Romanello et al., 2021; United Nations Environment Programme, 2018).

Given the current adaptation gap and projected increase in health risks, a rapid scaling up of efforts is needed to increase the resilience of health systems and empower individuals and communities to protect health, using iterative risk management approaches that explicitly consider the range of possible climate futures. To build climate-resilient and environmentally sustainable health and healthcare systems, adaptation approaches need to address the building blocks of health and healthcare systems: (1) leadership and governance; (2) health workforce; (3) health information systems, including vulnerability, capacity, and adaptation assessments, integrated risk monitoring, and early warning and response systems; (4) service delivery, including climate-informed health policies, plans, and programs and management of the environmental determinants of health; (5) the availability and accessibility of complementary social services; (6) climate-resilient and sustainable technologies and infrastructure; and (7) financing and investment (World Health Organization (WHO), 2020).

The first foundational step toward building resilience for most health departments and ministries is conducting a climate change and health vulnerability and adaptation (V&A) assessment (World Health Organization (WHO), 2021a). V&A assessments establish a knowledge base of current and projected health risks for:

- Developing, implementing, monitoring, and evaluating the effectiveness of adaptation options in the context of a changing climate.
- Identifying particularly vulnerable populations, regions, and healthcare infrastructure.
- Detailing the capacity of systems, organizations, and communities to prepare for and manage changes in the magnitude and pattern of climate-related risks.

The process of conducting these assessments must involve partnership building within the health sector to ensure the integration of climate change into relevant policies and programs, such as vector control or maternal and child health programs. The process also builds partnerships across non-health sectors to collaboratively develop decision-support tools such as early warning and response systems and to ensure decisions taken in these sectors promote health and well-being by using a health-in-all-policies approach. An outcome of V&A assessments is prioritized adaptation strategies, policies, and programs, with implementation plans over short to longer terms that incorporate mid-course corrections to reflect changing uncertainties about the magnitude and pattern of climate change and the consequences of development choices (World Health Organization (WHO), 2021b).

13.5 Discussion

The increasing scale and pace of current and projected climate change impacts require determined efforts to rapidly scale-up actions to protect health and well-being, as well as to protect access to and functioning of healthcare infrastructure. Designing effective adaptation options requires explicit consideration of the wide range of uncertainties in data, models, analyses, and actions by individuals, communities, and health systems. Actions are needed to assess risks and vulnerabilities to current and future impacts, to identify populations that suffer disproportionately, and to develop, implement, and evaluate required interventions. The increasing complexity and severity of risks through, for example, compounding or cascading events, requires that health decision-makers work closely with those in other sectors to shape essential determinants of health and develop climate resilient and low carbon health systems. (e.g., Mathews et al., 2019). This process will create new demands on technical knowledge, infrastructure, and capacities and will require enhanced and novel surveillance to strengthen the iterative management of climate change risks to health. Proactive, timely, and sufficient investments in research, data access, and adaptation are critical to ensuring that individuals, communities, and health systems can prepare for a rapidly changing climate.

References

Aguado, E., & Burt, J. E. (2000). *Understanding weather and climate* (2nd ed., p. 528). Prentice Hall.

Centers for Disease Control and Prevention. (2021). *National notifiable diseases surveillance system (NNDSS)*. Retrieved from https://www.cdc.gov/nndss/about/index.html

Cisse, G., McLeman, R., Adams, H., Aldunce, P., Bowen, K., Campbell-Lendrum, D., et al. (2022). Health, wellbeing and the changing structure of communities. In H.-O. Pörtner, R. C. Roberts, M. Tignor, E. S. Poloczanska, K. Mintenbeck, A. Alegría, et al. (Eds.), *Climate change 2022: Impacts, adaptation, and vulnerability* (pp. 1041–1170). Cambridge University Press. https://doi.org/10.1017/9781009325844.009

Davis, R. E., McGregor, G. R., & Enfield, K. B. (2016). Humidity: A review and primer on atmospheric moisture and human health. *Environmental Research, 144*, 106–116.

Ebi, K. L. (2020). Mechanisms, policies, and tools to promote health equity and effective governance of the health risks of climate change. *Journal of Public Health Policy, 41*, 11–13. https://doi.org/10.1057/s41271-019-00212-2

Ebi, K. L., Anderson, C. L., Hess, J. J., Kim, S.-H., Loladze, I., et al. (2021). Nutritional quality of crops in a high CO_2 world; an agenda for research and technology development. *Environmental Research Letters, 16*, 064045.

Ebi, K. L., Boyer, C., Ogden, N., Paz, S., Berry, P., Campbell-Lendrum, D., Hess, J. J., & Woodward, A. (2021). Burning embers: Synthesis of the health risks of climate change. *Environmental Research Letters, 16*, 044042.

Gasparrini, A., Armstrong, B., & Kenward, M. G. (2010). Distributed lag non-linear models. *Statistics in Medicine, 29*(21), 2224–2234.

Gibbons, C. L., Mangen, M. J. J., Plass, D., Havelaar, A. H., Brooke, R. J., Kramarz, P., et al. (2014). Measuring underreporting and under-ascertainment in infectious disease datasets: A comparison of methods. *BMC Public Health, 14*(1), 1–17. https://doi.org/10.1186/1471-2458-14-147

Giordano, B. V., Bartlett, S. K., Falcon, D. A., Lucas, R. P., Tressler, M. J., & Campbell, L. P. (2020). Mosquito community composition, seasonal distributions, and trap bias in northeastern Florida. *Journal of Medical Entomology, 57*(5), 1501–1509. https://doi.org/10.1093/jme/tjaa053

Green, H., Bailey, J., Schwarz, L., Vanos, J., Ebi, K., & Benmarhnia, T. (2019). Impact of heat on mortality and morbidity in low and middle income countries: A review of the epidemiological evidence and considerations for future research. *Environmental Research, 171*, 80–91. https://doi.org/10.1016/j.envres.2019.01.010

Haines, A., & Ebi, K. L. (2019). The imperative for climate action to protect health. *The New England Journal of Medicine, 380*(3), 263–273.

IPCC. (2019). Summary for policymakers. In P. R. Shukla, J. Skea, E. Calvo Buendia, V. Masson-Delmotte, H.-O. Pörtner, D. C. Roberts, et al. (Eds.), *Climate change and land: An IPCC special report on climate change, desertification, land degradation, sustainable land management, food security, and greenhouse gas fluxes in terrestrial ecosystems*. IPCC.

Klenert, D., Funke, F., Mattauch, L., & O'Callaghan, B. (2020). Five lessons from COVID-19 for advancing climate change mitigation. *Environmental and Resource Economics, 76*, 751–778. https://doi.org/10.1007/s10640-020-00453-w

Koffman, J., Gross, J., Etkind, S. N., & Selman, L. (2020). Uncertainty and COVID-19: How are we to respond? *Journal of the Royal Society of Medicine, 113*(6), 211–216. https://doi.org/10.1177/0141076820930665

Krow-Lucal, E. R., Biggerstaff, B. J., & Staples, J. E. (2017). Estimated incubation period for Zika virus disease. *Emerging Infectious Diseases, 23*(5), 841–844. https://doi.org/10.3201/eid2305.161715

Lu, Y., & Zeger, S. L. (2007). On the equivalence of case-crossover and time series methods in environmental epidemiology. *Biostatistics, 8*(2), 337–344.

Mathews, T., Wilby, R. L., & Murphy, C. (2019). An emerging tropical cyclone-deadly heat compound hazard. *Nature Climate Change, 9*(8), 602–606.

McCarty, J. L. (2011). Remote sensing-based estimates of annual and seasonal emissions from crop residue burning in the contiguous United States. *Journal of the Air & Waste Management Association, 61*(1), 22–34. https://doi.org/10.3155/1047-3289.61.1.22

McCotter, O., Kennedy, J., McCollum, J., Bartholomew, M., Iralu, J., Jackson, B. R., et al. (2019). Coccidioidomycosis among American Indians and Alaska natives, 2001–2014. *Open Forum Infectious Diseases, 6*(3), ofz052. https://doi.org/10.1093/ofid/ofz052

Moulton, A. D., & Schramm, P. J. (2017). Climate change and public health surveillance: Toward a comprehensive strategy. *Journal of Public Health Management & Practice, 23*(6), 618. https://doi.org/10.1097/PHH.0000000000000550

Murray, J., & Cohen, A. L. (2017). Infectious disease surveillance. In *International encyclopedia of public health* (pp. 222–229). Elsevier. https://doi.org/10.1016/B978-0-12-803678-5.00517-8

Omumbo, J. A., Platzer, B. L., Girma, A., & Connor, S. J. (2011, April). *Climate and health in Africa: 10 years on*. Workshop Report, UNECA Conference Center, Addis Ababa. https://doi.org/10.7916/D83J3MV8

Romanello, M., McGushin, A., Di Napoli, C., Drummond, P., Hughes, N., Jamart, L., et al. (2021). The 2021 report of the lancet countdown on health and climate change: Code red for a healthy future. *Lancet, 398*, 1619–1662. https://doi.org/10.1016/S0140-6736(21)01787-6

Rutter, H., Wolpert, M., & Greenhalgh, T. (2020). Managing uncertainty in the COVID-19 era. *BMJ, 370*, 3349. https://doi.org/10.1136/bmj.m3349

Sellers, S., & Ebi, K. L. (2017). Climate change and health under the shared socioeconomic pathway framework. *International Journal of Environmental Research and Public Health, 15*(1), 3.

Shindell, D., Zhang, Y., Scott, M., Ru, M., Stark, K., & Ebi, K. L. (2020). The effects of heat exposure on human mortality throughout the United States. *GeoHealth, 4*, e2019GH000234. https://doi.org/10.1029/2019GH000234

Solís, P., Vanos, J. K., & Forbis, R. E. (2017). The decision-making/accountability spatial incongruence problem for research linking environmental science and policy. *Geographical Review, 107*, 680–704. https://doi.org/10.1111/gere.12240

United Nations Environment Programme. (2018). *The adaptation gap report 2018* (p. 104). United Nations Environment Programme (UNEP).

Vicedo-Cabrera, A. M., Sear, F., & Gasparrini, A. (2019). Hands-on tutorial on a modeling framework for projections of climate change impacts on health. *Epidemiology, 30*, 321–329.

World Health Organization (WHO). (2020). *Guidance for climate-resilient and environmentally sustainable health care facilities*. World Health Organization.

World Health Organization (WHO). (2021a). *Climate change and health vulnerability and adaptation assessment*. World Health Organization.

World Health Organization (WHO). (2021b). *Health and climate change global survey report*. World Health Organization.

Open Access This chapter is licensed under the terms of the Creative Commons Attribution 4.0 International License (http://creativecommons.org/licenses/by/4.0/), which permits use, sharing, adaptation, distribution and reproduction in any medium or format, as long as you give appropriate credit to the original author(s) and the source, provide a link to the Creative Commons license and indicate if changes were made.

The images or other third party material in this chapter are included in the chapter's Creative Commons license, unless indicated otherwise in a credit line to the material. If material is not included in the chapter's Creative Commons license and your intended use is not permitted by statutory regulation or exceeds the permitted use, you will need to obtain permission directly from the copyright holder.

Uncertainty and Socioeconomic Vulnerability to Climate Change

Kirstin Dow, Paty Romero-Lankao, and Olga Wilhelmi

14.1 Introduction

As climate change expresses itself in ways more apparent and compelling to all communities, the pressure to act to reduce risks and vulnerabilities is increasing. Correspondingly, the goal of understanding and potentially reducing or most effectively managing uncertainty associated with vulnerability to climate change is becoming more prominent. Qualitative and quantitative vulnerability analyses are frequently used to better understand the past, present, and future impacts of environmental stressors on groups of people, infrastructure, and ecosystems; to inform decision- and policymakers about how to mitigate these impacts; and to guide interventions seeking to build community resilience. The related disaster risk reduction and climate adaptation debates and decisions are likely to be contentious as they may aggravate existing inequalities and redistribute resources, benefits, and costs under changing conditions. These raise the broader social significance of making uncertainties within vulnerability assessment more transparent.

This chapter examines concepts of uncertainty in relation to socioeconomic vulnerability to climatic hazards. It discusses different types and potential sources of uncertainty in vulnerability assessments. A pair of cases considering vulnerability to flooding examine how uncertainty about vulnerability is reflected in policy processes. The first case focuses on how uncertainty in legal authorities and obligations influences the vulnerability of communities to the impacts on roads of sea level rise-related flooding. This relates to our limited understanding of the interacting factors and actions shaping vulnerability and flood risk. The second case illustrates how two sources of uncertainty can limit stakeholders' actions in building climate resilience. This results from different ways of knowing, or how we come to know the world and our relationship to it, and actions involved in mitigating present climatic risks and adapting to future risks.

14.2 Intersections of Socioeconomic Vulnerability and Uncertainty

Natural hazard and climate change vulnerability analyses are critical steps in disaster risk reduction and climate change adaptation. These analyses may also inform priority setting and identify opportunities. Research in vulnerability, with various schools of thought, includes different framings, the characterization of indicators, and the analytical approaches (Cutter, 2012; Peduzzi et al., 2009; Romero-Lankao et al., 2012). In climate change research, the IPCC (2014) defined vulnerability as "the susceptibility to be harmed... the degree to which a system is susceptible to, and unable to cope with, adverse effects..., including climate variability and extremes. Vulnerability is a function of the character, magnitude, and rate of the threat to which a system is exposed, its sensitivity, and its adaptive capacity."

Following this definition, in this chapter, we approach vulnerability as a multidimensional concept reflecting the dynamic processes that create exposure, sensitivity, and adaptive capacity among groups, sectors, and communities. Vulnerability analyses are an important mechanism to evaluate risk or the potential for consequences from hazardous events

K. Dow (✉)
Department of Geography, University of South Carolina, Columbia, SC, USA
e-mail: KDow@sc.edu

P. Romero-Lankao
Department of Sociology, University of Toronto-Scarborough, Toronto, ON, Canada
e-mail: paty.romerolankao@utoronto.ca

O. Wilhelmi
NSF National Center for Atmospheric Research, Boulder, CO, USA
e-mail: olgaw@ucar.edu

(Thomas et al., 2020; IPCC, 2014) and to inform hazard mitigation and climate adaptation. However, vulnerability is complex and inherently uncertain. While uncertainty in vulnerability analyses and risk modeling has been generally emphasized in the literature, uncertainty in socioeconomic vulnerability has not received as much attention as that given to uncertainty in climate prediction, weather forecasting, or decision-making (Dilling et al., 2018). After several decades of theoretical and empirical work, vulnerability assessments, and the introduction of vulnerability science (Cutter, 2003), some scholars recognized "capturing socioeconomic uncertainty" as one of the main challenges in vulnerability research (Eakin & Luers, 2006).

The discussion about uncertainty in vulnerability and focused efforts to categorize and quantify uncertainty in vulnerability have only gained momentum in the last 15 years (Patt & Dessai, 2005; Eakin & Luers, 2006; Gall, 2007, Vincent, 2007; Eakin & Bojorquez-Tapia, 2008; Aven, 2011; Tate et al., 2011; Tate, 2013). These efforts at conceptualization and categorization of uncertainty differ based on discipline or research focus (e.g., Bijlsma et al., 2011; Di Baldassarre et al., 2016; Lempert et al., 2004; Patt & Dessai, 2005; Pedde et al., 2019). In part, this progress can be attributed to a greater emphasis on a socio-ecological system perspective (Turner et al., 2003), climate change adaptation (Vincent, 2007), and complex and cumulative effects of cascading disasters (Thomas et al., 2020). As emphasized by Adger and Vincent (2005) and recently illustrated by the COVID-19 pandemic, uncertainties associated with economic, social, and political systems and human behavior have much greater range compared to the range of uncertainties associated with the physical (climate) system alone.

14.3 Characterizing the Nature of Uncertainty Related to Vulnerability

Many terms capture aspects of uncertainty (e.g., risk, surprise, ignorance, imprecision, and ambiguity). Based on Doll and Romero-Lankao et al. (2016), we classify the *forms* of uncertainty into three categories:

1. *Epistemic uncertainty* is caused by limited knowledge. Following Bijlsma et al. (2011), we distinguish two types of relevant epistemic uncertainties: *substantive uncertainty* or limited knowledge about the substance (content, subject matter) of the vulnerability problem *and process uncertainty* related to how lack of knowledge about how the actions of participants in a process or those of actors at high levels will influence vulnerability.
2. *Ambiguity* is uncertainty arising from multiple legitimate conceptualizations and framings of vulnerability, such that even in the absence of epistemic uncertainty not everyone should or could agree on what vulnerability is (Brugnach & Ingram, 2012; Kwakkel et al., 2010; Renn, 2008; Renn et al., 2011; Sword-Daniels et al., 2018).
3. *Ontological uncertainty*, also called random, stochastic, or aleatoric uncertainty, refers to the inherent variability of human or natural systems (Ascough et al., 2008). Ontological uncertainty in vulnerability is due to the stochastic nature of weather, climate, and society that makes it difficult to predict the occurrence of both a hazard or stressor and the societal and environmental conditions and human behaviors at the time of the hazard.

To a greater or lesser degree, all understandings of vulnerability, regardless of a framing or a school of thought, are surrounded by these types of uncertainties. High uncertainty in vulnerability stems from a number of factors related to the dynamic nature of exposure, sensitivity, and capacity. As it is related to risks, what is vulnerable also depends on what is valued—health, wealth, and equity—and on how it is framed. Some aspects, such as social capital, can be viewed as "potential energy" or latent traits that contribute to resilience but are not fully activated (Cutter, 2016). Previous studies have attributed uncertainty to varying quantification methods, data availability, characteristics of indicators, research frameworks, multi-scalar issues, or definitions (Bijlsma et al., 2011; Tate et al., 2011).

Numerous methodological techniques attempt to characterize the range of uncertainty. For example, stochastic probability techniques have been developed to simulate climate variability (e.g., Franzke et al., 2015), dynamic commodity prices, and other social environmental phenomena. Multivariate indices attempt to capture the vulnerability of groups, livelihoods, and ecological and infrastructure systems (Cutter, 2012). Analyses of the impact of policy scenarios on environmental risk also consider uncertainties (e.g., Baeza et al., 2019).

The potential to fully overcome analytical challenges is, however, questioned. As Berkes (2007, p. 284) has argued, socio-ecological systems are so complex that "our knowledge of them, and our ability to predict their future dynamics, will never be complete." Some studies assert that vulnerability analyses promise more certainty, and more useful results, than they can deliver (Patt et al., 2005). Discussion in the literature suggests different approaches to analyzing and characterizing uncertainty (Romieu et al., 2010; see also Curry and Webster for other ways to address uncertainty). For example, Romieu et al. (2010) identify a set of frameworks recommending dealing with high uncertainty through a precautionary-based approach related to vulnerability reduction. They also observe that others see high uncertainty as requiring new decision-making tools (Romieu et al., 2010). Dilling et al. (2018) argue that surprise arising from uncertainty, "is an unavoidable component of weather and climate disasters—one that we must acknowledge, learn to anticipate, and incorporate into risk assessment and management efforts. In sum, although it may seem paradoxical, we should be learning how to expect surprise." Hardy and Hauer (2018) point to the need for more research on the most effective planning frameworks to foster adaptation decisions in the face of uncertainty.

The following sections of this chapter introduce the topics contributing to uncertainty (Bijlsma et al., 2011) and point to areas where efforts to reduce uncertainty are advancing, notably in the use of vulnerability indices. Issues related to theoretical frameworks where uncertainty persists are discussed and then two case studies that illustrate the impact of uncertainty on the understanding and management of vulnerability are presented. One case study considers the role of legal systems in shaping vulnerability where rather than aiming to reduce complexity to a level of predictability, new risk management approaches will be needed. The second case illustrates how both limited knowledge and different ways of framing vulnerability can hinder stakeholders' efforts to build resilience.

14.4 Vulnerability Indices and Associated Uncertainties

The complex, multidimensional nature of vulnerability encourages the use of aggregate or composite vulnerability indices—social analogs to the quantitative physical hazard models (Tate, 2012). Various indices aim to quantify many types of vulnerability—such as social, biophysical, and institutional. Typical data in a vulnerability index include demographic characteristics such as number of children, women, elderly, with measures of exposure, infrastructure quality, social and institutional assets, and other site characteristics. Despite widespread use of indices, however, our knowledge about their robustness and reliability is limited (Tate, 2013). The selection, interpretation, and availability of indicators, used in construction of vulnerability indices, also vary, necessitating the use of proxy variables. Indicators of exposure, sensitivity and adaptive capacity may have varying significance in different settings (e.g., poverty carries different implications in places with higher levels of social protection) (e.g., Rohat et al., 2019; Wilby, 2017).

Tate et al. (2011) and Tate (2012, 2013) further explore the partiality and uncertainty in vulnerability indices by expanding the discussion about subjective weighting of indicators to other steps in index construction, namely, indicator selection, scale of analysis, measurement error, data transformation, normalization, weighting, and aggregation. He uses Monte Carlo-based uncertainty analysis to demonstrate that changes in input data and algorithms have the potential to alter the end results. His work shows that uncertainty analyses can improve precision, transparency, and credibility of social vulnerability indices. He argues that "Epistemic uncertainty is associated with all vulnerability models. The lack of its assessment and portrayal does not deny its existence" (Tate, 2013, p. 541). Efforts to assess the empirical validity of social vulnerability indices against Hurricane Sandy impacts show different spatial patterns depending on the vulnerability construct employed and identify the need for more research on the validity of social vulnerability indices (Rufat et al., 2019; Fig. 14.1).

Eakin and Bojorquez-Tapia (2008), Tate (2012, 2013), Romero-Lankao et al. (2016), and others provide further important insights into uncertainty associated with vulnerability indicators and indices. Both Eakin and Bojorquez-Tapia (2008) and Romero-Lankao et al. (2016) discuss the inherent subjectivity and uncertainty of assigning weights and aggregating indicators in vulnerability assessments. They argue that reducing uncertainty (and subjectivity) requires an approach that can handle the diverse types of input variables and can provide a basis for comparing, weighting, and aggregating diverse data into a single index. The multicriteria analysis and fuzzy logic were used to illustrate this process in rural Mexico (Eakin & Bojorquez-Tapia, 2008) and Mumbai India (Romero-Lankao et al., 2016). For addressing weighting-related uncertainty in the assessment of flood risk and vulnerability in the city of Leipzig, Germany, Scheuer et al. (2011) suggested combining systematic weights alteration and subsequent analysis of variation in the resulting index values with the validation by stakeholders, thus combining quantitative and qualitative methods.

Different theoretical underpinnings and methodological approaches to vulnerability and associated uncertainty illuminate contrasting aspects of vulnerability and omit others (see Eakin & Luers, 2006; O'Brien et al., 2007; Romero-Lankao et al., 2012, 2016). Livelihoods and political ecology approaches place greater emphasis on the structural, institutional, cross-scale drivers of differences in populations' vulnerability to hazards. In the natural hazards research community, reliability of vulnerability analyses in part has been addressed by validating the indices with impact data (e.g., de Sherbinin et al., 2019; Gall, 2007; Schmidtlein et al., 2011; Wilhelmi & Morss, 2013). While index validation helps to connect underlying social and environmental drivers to the negative impacts (e.g., mortality, economic losses, damages), the focus on impacts generally does not look explicitly at epistemological process uncertainties, such as inequality or determinants of political power operating across social scales and distance to influence variations in populations vulnerability (Adger, 1999; Ribot, 2010).

The climate change impacts, vulnerability, and adaptation researchers have been tackling similar issues of the difficulties of validating indicators, constructing a single

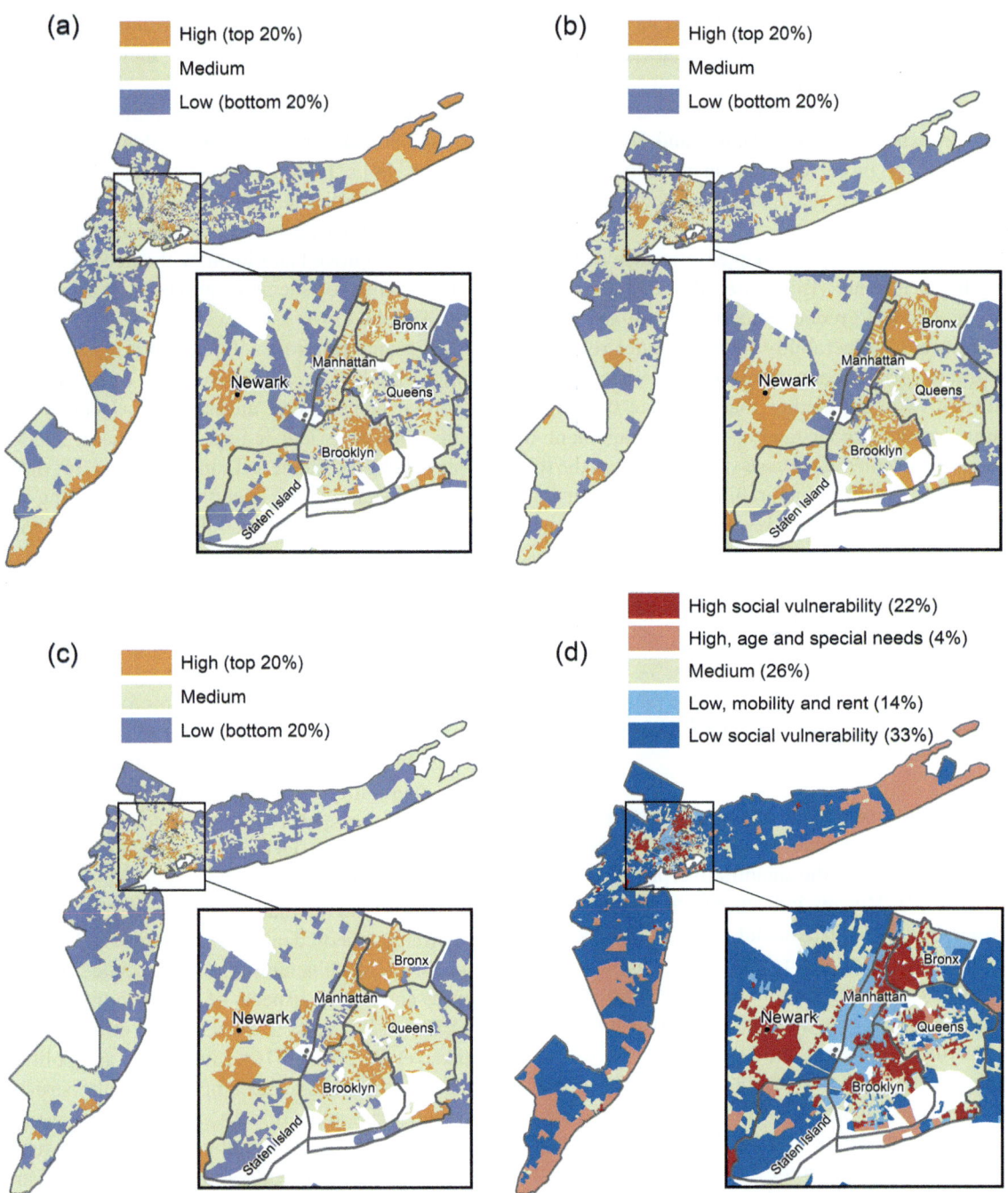

Fig. 14.1 Social vulnerability indexes for New York and New Jersey census tracts affected by Hurricane Sandy. Source: Rufat, S.E. Tate, C. T. Emrich & F. Antolini, 2019: How Valid Are Social Vulnerability Models? Annals of the American Association of Geographers, copyright © 2019 by American Association of Geographers, reprinted by permission of Informa UK Limited, trading as Taylor & Francis Group, www.tandfonline.com on behalf of © 2019 by American Association of Geographers

vulnerability metric, complex human-environment system, and the contextual nature of vulnerability (Adger & Vincent, 2005; Füssel, 2010; Patwardhan et al., 2009; Flanagan et al., 2020; Rohat et al., 2019). Adger and Vincent (2005) and Vincent (2007) point out that the varying contexts of governance among places pose a challenge to developing reliable adaptive capacity indices at the scales required for adaptation decision-making. These issues are in addition to the uncertainty involved in relying on scenarios and socioeconomic projections.

For example, Adger and Vincent (2005) state that uncertainties in adaptive capacity can be significant due to limited knowledge about the direction of change in many of the key input variables. They discuss vulnerability as a dynamic construct that is constantly shifting in response to changing environmental and social conditions. Adger and Vincent (2005, p. 400) write, "it is apparent that uncertainty in the science of adaptation stems more from contested underlying theories of behavior, politics, and risk than of data and observation," suggesting the significance of two categories of uncertainty: epistemic and ambiguity. Patt et al. (2005, p. 412) described how future uncertainties cascading upon each other increase the overall uncertainty, "even more so when translating natural system changes to their effects on an uncertain social system." In one effort to address such uncertainty, Romero-Lankao et al. (2016) turned to a combination of methods for a more nuanced understanding of the multilevel socioeconomic, political, environmental, and infrastructural factors that affect inequality in vulnerability and shed light on uncertainty.

14.5 Uncertainty Arising from Ambiguity

Uncertainty about who and what is vulnerable can also result from the already referred situations of ambiguity. Such ambiguity exists when scientists and stakeholders frame vulnerability differently due to alternative disciplinary traditions, value systems, expectations, experiences and forms of knowledge (Kwakkel et al., 2010; Renn, 2008). Ambiguity remains even if epistemic uncertainty is addressed. Linguistic uncertainty is a form of ambiguous uncertainty arising when vague, equivocal, under-specified, and context-dependent notions of vulnerability, and the meaning of words can change over time (Ascough et al., 2008; Regan & Colyvan, 2000) and made more challenging when people do not share the same native language.

Ambiguity is influential in the development of policy to reduce vulnerability. Policies to address vulnerability are faced with the challenge of moving decision-makers beyond their accustomed ways of framing and managing climate change mitigation and adaptation. For example, some decision-makers focus on projecting climate parameters (e.g., temperature and sea-level) and exploring adaptation options under different scenarios, while others focus on policies (e.g., infrastructure provision, health, and education) that can reduce vulnerability to those hazards. Policy also involves a collective engagement of disparate interests, values, and power relations (Romero-Lankao et al., 2016). For instance, officials within sectors involved in managing climate risk, such as food, energy, water, disaster risk management, and urban planning hold diverse organizational and cultural values. They lack the incentives, rights, financial resources, and responsibilities needed to work across sectors and jurisdictions (Scott et al., 2015). Additionally, decision-makers involved in disaster risk reduction and climate change adaptation policies lack interaction and coordination because of differences in language and political culture (Schipper et al., 2016). An examination of these factors is an essential first step to developing the skill sets, tools, funding, and incentives needed to foster effective risk mitigation and climate adaptation practices under uncertain climate and development dynamics. There is also frequently a mismatch between geographical and topical areas of concern and the jurisdiction or knowledge necessary to manage climate risks. For instance, large-scale hydropower generation, while managed to provide electricity to an urbanizing region, often creates problems of fit among service areas and multiple water basins and externalities, such as risks of floods and water scarcity, for outside regions and downstream communities depending on water basins. Integration across these lines creates the ever-present potential for conflict and misunderstanding.

14.6 Case Study 1: Legal Uncertainty and Barriers to Vulnerability Reduction

The multi-scalar and dynamic traits of vulnerability are two major factors contributing to uncertainty. One source of uncertainty is the interactions of institutional, including legal, processes moving at varied rates, sometimes slowed or accelerated by random physical or societal events, and covering diverse, occasionally poorly defined jurisdictions that can interact with the ability of institutional changes to alter socioeconomic vulnerability. These complex spatial and temporal interactions come to the fore in consideration of what laws enable or disincentivize a local government's willingness to act in different ways to reduce the vulnerability of citizens.

Local governments face a variety of barriers to climate adaptation and vulnerability reduction. Lack of staffing capacity and funding are well documented barriers to adaptation (Nordgren et al., 2016) though the significance and duration of such barriers are uncertain. Smaller and more rural local governments are known to have capacity issues affecting the well-being of citizens (Horney et al., 2017; Jurjonas & Seekamp, 2018), but these differences are not well-explored. A survey of local government practitioners in the USA also identified that regulations and laws at regional, state, and federal levels impede local government efforts to prepare for climate change (Nordgren et al., 2016). Institutional resistance to change (Barnett et al., 2015) or ambiguity and lack of consensus over policy goals can slow the pace of adaptive change. When adaptive change becomes slower than climate change, the pace of institutional responsiveness becomes a limit to adaptation (e.g., the ability of water utilities to shift from a water provision orientation to one of strong water conservation) (Barnett et al., 2015), introducing further uncertainty into the potential for timely implementation of strategies to reduce vulnerability.

The legal and regulatory institutions called out by local governments are slow-to-change institutions, perhaps not as resistant as some discussed by Barnett et al. (2015), but as common law courts, they build from precedent and decisions challenged over time at multiple levels. Adaptation can raise unanticipated or unresolved legal issues requiring engagement with potentially lengthy and expensive legal processes. For example, litigation over one dilapidated coastal home that had been declared a public nuisance took 5 years and resulted in a $1.5 million settlement, a figure that did not include legal costs (Kusnetz, 2017). The institutional uncertainty over rights and jurisdictions coupled with the uncertain pace of resolution are emerging as significant factors in understanding how institutional processes at higher scales influence the dynamics of vulnerability at local and individual scales.

The maintenance of infrastructure is a key government service influencing vulnerability in many forms. For example, even low magnitude flooding reduces the response times of emergency responders in England (Yu et al., 2020). Safe and accessible roads support economic vitality; access to employment, medical care, and education; and, in extreme events, evacuation. Safe wastewater treatment systems protect public health by assuring proper treatment of wastes potentially containing harmful bacteria. Increased extreme rainfall events, more frequent tidal flooding, coastal storms, and sea level rise are likely to place greater stresses on both types of systems at risk to climate-related changes, increasing community vulnerability to a diverse set of economic and other losses.

Jones et al. (2019) conducted a Southeast USA regional analysis of whether tort and local government law can both advance and hinder climate change resilience planning and climate adaptation efforts. In the USA, tort law refers to "A body of rights, obligations, and remedies that is applied by courts in civil proceedings to provide relief for persons who have suffered harm from the wrongful acts of others" (Lehman & Phelps, 2004). Jones et al. (2019, p. 81) find that "Our analysis of how roads are managed in our four-state study leads us to conclude that, even if a governmental entity wanted to make an adaptive choice—say repair, upgrade, or abandon a road—the laws as they currently exist make such choices difficult." For local governments considering adaptive strategies, the laws create a dilemma of having a duty to maintain roads and knowing that the failure to do so could result in tort claims versus the potential to abandon roads that are too expensive, impractical, or impossible to maintain which could result in taking claims.

US Public tort law has been described as "a jerry-built structure, a patchwork, a doctrinal stew" (Schuck, 1983, p. 51; Jones et al., 2019). Ongoing sea-level rise challenges the assumption of unchangeable property rights in regulatory doctrine. In this way, the physical process of rising sea levels highlights the ideal of unchangeable property rights as a clash with very real and drastic coastal transformation (Byrne, 2012). The challenges of both of these areas of law as related to adaptation represent a major barrier with significant potential to delay adaptive actions and, in the process, increase the duration and magnitude of some forms of vulnerability.

14.7 Case Study 2: Epistemic Uncertainty Joins Ambiguity in Boulder CO

In September 2013, Boulder County, Colorado, USA, experienced unprecedented, week-long rainfall corresponding to about 80% of its annual average (MacClune et al., 2014). The resulting flood caused the evacuation of 18,000 people, the destruction of 688 homes, and damages to an additional 9900. The flood unequally affected people and places, hitting mountain communities such as Lyons and Jamestown, CO more and low-income households—e.g., those living in mobile parks—the hardest. The flood serves as an example of the significance of epistemic uncertainty in the ways of framing siloed management. These do not fit with the multidimensional nature of vulnerability and ambiguity manifested in stakeholders' inability to grasp and manage the complex interdependencies between critical infrastructures and actions (Romero-Lankao & Norton, 2018).

Romero-Lankao and Norton (2018) examined the interdependent conditions, during the 2013 flood, that amplified or mitigated cascading negative consequences on people, infrastructure, and places. The researchers combined 17 semistructured qualitative interviews and fuzzy cognitive maps (FCMs), a semiquantitative and participatory method, to analyze the conditions and learning processes (Carvalho, 2013; Kok, 2009). They interviewed representatives from governmental and civil society organizations from most of the key sectors and jurisdictions (e.g., emergency recovery, energy, water, and food) from the local, county, and state levels. The sample also drew from the Cities of Boulder and Longmont on the plains and the towns of Lyons and Jamestown in the mountains.

Six of seven roads following creeks up mountain canyons failed and left the affected populations isolated. The breakdown of energy and transportation networks led to failures in water treatment in Lyons and Longmont, not in Boulder, where backup generators ensured that the water treatment plant stayed online. Electricity, gas, and refrigeration systems in the mountain communities were off and affected food suppliers, restaurants, and markets, which are key within the mountain tourism sector.

Factors such as zoning regulations, greenways, institutional capital, and ability for learning helped to mitigate the impacts of the floods. A key example of institutional capital and ability for learning was a strong preexisting culture of cooperation, which sped response and enabled recovery—e.g., through learning from previous experiences, such as the Four Mile Fire of 2010, to set construction permits to expedite recovery. Community organizations mobilized food for those affected, and gift cards provided residents with a means to offset the costs of electrical bills.

Through cultural brokers, community gathering places and shelters helped Latinos and minority groups. At the same time, channels of communication (e.g., warning in English) were culturally inappropriate to minority groups. Acknowledgment of this barrier led local authorities to suggest strengthening reliance on cultural brokers and community leaders—two trusted means to help disseminate warnings targeting most vulnerable groups.

However, agency was not enough to understand and respond to interdependencies between transportation, energy, food, and other critical sectors, on the one hand, and community responses on the other. For example, the flood triggered impacts in the transportation, energy, food, and water sectors that generated negative cascading effects. These impacts were felt across the economy, livelihoods, and interests that were intricately woven into the social fabric of Boulder, affecting everything from the mountains to the plains, from transport and communication to emergency response, profoundly shaking the experience of everyday life. Interestingly, actions by different sectors and jurisdictions helped to minimize some negative cascading effects. For example, the existence of an integrated Office of Emergency Management structure provided for effective coordination between sectors such as fire and utilities. Through emergency preparedness, different sectors knew what to do during the flood. For instance, they learned that they would need to open a departmental operating center to organize the emergency response.

However, stakeholders were not able to grasp the complexity of both interdependencies and cascading effects. In other words, they were faced with limited and different siloed ways of framing and adapting (ambiguity). It was only through our cognitive FCM exercise already described above that stakeholders were able to grasp some of this complexity, which is a source of epistemic uncertainty.

We also examined changes in respondents' awareness and learning from extreme events in order to do things differently. Almost across the board respondents indicated that their organizations, and many organizations they worked with, implemented a variety of changes following the 2013 floods. For instance, they updated contingency, communication, and recovery plans within each sector, and they introduced updated building and land use codes within sewage and other specific systems.

Notably, most of the learning occurred within organizations with a focus on the systems and infrastructure that each controlled rather than on interdependencies between systems, actions, and interventions that participants were able to grasp during our interviews. Exceptions to this include the creation of the Flood Rebuilding and Permit Information Center, staffed by experts in transportation, septic infrastructure, and floodplain management. We can conclude that even in Boulder County, whose community has historically shown high ability for learning from extreme events, the capacity to grasp the full complexity of vulnerability and the different ways of knowing and action limited the ability to influence deeper, transformative changes in understanding and practices among public and private actors. Examples include actions to address the links between transportation and water. Often roads are parallel to river systems and require enforced banks and setbacks from waterways. Similarly, another option is to move water lines farther away from rivers, as well as burying gas lines 5–6 m underground.

14.8 Closing Reflections

Addressing uncertainty in vulnerability research in general and quantitative assessments, in particular, has broad societal implications for informing priorities in adaptation and increasing resilience. The increasingly visible impacts of climate change have led to greater interest in adaptation coupled with the growing recognition that the scale of the problem will require sometimes controversial priority setting (Dow et al., 2013). Reducing uncertainty is very valuable in this decision context, but the extent to which that is possible is an active area of discussion.

The complex, multidimensional nature of vulnerability is surrounded by multiple forms of uncertainty; some forms (e.g., those related to climate hazards) are more amenable to robust, predictive analysis to inform decision-making than others. Progress is being made in characterizing these forms of uncertainty; understanding uncertainties involved in creating indices; exploring cascades of uncertainty in multilevel processes linking environment, society, and technology; and identifying societal processes that are subject to emerging trends, random events, and enduring ambiguity.

With this progress and despite these challenges, researchers highlight the need for vulnerability analyses, provided the uncertainty is recognized and the implicit and explicit working assumptions made at each stage of constructing vulnerability indices are explained (Hardy & Hauer, 2018; Patt et al., 2005; Vincent, 2007). Hardy and Hauer (2018) contend that despite uncertainty about climate change predictions, effective adaptation planning is feasible (also see Smith et al., 2018 for approaches to incorporate uncertainty into decision-making). However, as suggested by this review, considerable effort will be required to improve the characterizations of the many forms of uncertainty in socioeconomic vulnerability, and then further develop strategies to address those that are more difficult to reduce.

References

Adger, W. (1999). Social vulnerability to climate change and extremes in coastal Vietnam. *World Development, 27*, 249–269.

Adger, W., & Vincent, K. (2005). Uncertainty in adaptive capacity. *Comptes Rendus Geoscience, 337*, 399–410. https://doi.org/10.1016/j.crte.2004.11.004

Ascough, J., Maier, H., Ravalico, J., & Strudley, M. (2008). Future research challenges for incorporation of uncertainty in environmental and ecological decision-making. *Ecological Modelling, 219*, 383–399. https://doi.org/10.1016/j.ecolmodel.2008.07.015

Aven, T. (2011). On some recent definitions and analysis frameworks for risk, vulnerability, and resilience. *Risk Analysis, 31*, 515–522. https://doi.org/10.1111/j.1539-6924.2010.01528.x

Baeza, A., Bojorquez-Tapia, L., Janssen, M., & Eakin, H. (2019). Operationalizing the feedback between institutional decision-making, socio-political infrastructure, and environmental risk in urban vulnerability analysis. *Journal of Environmental Management, 241*, 407–417. https://doi.org/10.1016/j.jenvman.2019.03.138

Barnett, J., Evans, L., Gross, C., Kiem, A., Kingsford, R., Palutikof, J., Pickering, C., & Smithers, S. (2015). From barriers to limits to climate change adaptation: Path dependency and the speed of change. *Ecology and Society, 20*, 200305. https://doi.org/10.5751/ES-07698-200305

Berkes, F. (2007). Understanding uncertainty and reducing vulnerability: Lessons from resilience thinking. *Natural Hazards, 41*, 283–295. https://doi.org/10.1007/s11069-006-9036-7

Bijlsma, R., Bots, P., Wolters, H., & Hoekstra, A. (2011). An empirical analysis of stakeholders' influence on policy development: The role of uncertainty handling. *Ecology and Society, 16*, 16.

Brugnach, M., & Ingram, H. (2012). Ambiguity: The challenge of knowing and deciding together. *Environmental Science & Policy, 15*, 60–71. https://doi.org/10.1016/j.envsci.2011.10.005

Byrne, J. (2012). The cathedral engulfed: Sea-level rise, property rights, and time. *LA Law Review, 73*, 69–118.

Carvalho, J. (2013). On the semantics and the use of fuzzy cognitive maps and dynamic cognitive maps in social sciences. *Fuzzy Sets and Systems, 214*, 6–19. https://doi.org/10.1016/j.fss.2011.12.009

Cutter, S. (2003). The vulnerability of science and the science of vulnerability. *Annals of the Association of American Geographers, 93*, 1–12. https://doi.org/10.1111/1467-8306.93101

Cutter, S. (2012). *Hazards, vulnerability, and environmental justice* (p. 448). Routledge.

Cutter, S. (2016). The landscape of resilience indicators in the USA. *Natural Hazards, 80*, 741–758. https://doi.org/10.1007/s11069-015-1993-2

de Sherbinin, A., et al. (2019). Climate vulnerability mapping: A systematic review and future prospects. *Wiley Interdisciplinary Reviews: Climate Change, 10*, 600. https://doi.org/10.1002/wcc.600

Di Baldassarre, G., Brandimarte, L., & Beven, K. (2016). The seventh facet of uncertainty: Wrong assumptions, unknowns and surprises in the dynamics of human-water systems. *Hydrological Sciences Journal, 61*, 1748–1758. https://doi.org/10.1080/02626667.2015.1091460

Dilling, L., Morss, R., & Wilhelmi, O. (2018). Learning to expect surprise: Hurricanes Harvey, Irma, Maria, and beyond. *Journal of Extreme Events, 4*, 1771001. https://doi.org/10.1142/S2345737617710014

Doll, P., & Romero-Lankao, P. (2016). How to embrace uncertainty in participatory climate change risk management-a roadmap. *Earth's Future, 5*, 18–36. https://doi.org/10.1002/2016EF000411

Dow, K., Berkhout, F., & Preston, B. (2013). Limits to adaptation to climate change: A risk approach. *Current Opinion in Environment Sustainability, 5*, 384–391. https://doi.org/10.1016/j.cosust.2013.07.005

Eakin, H., & Luers, A. (2006). Assessing the vulnerability of social-environmental systems. *Annual Review of Environment and Resources, 31*, 365–394. https://doi.org/10.1146/annurev.energy.30.050504.144352

Eakin, H. & Bojorquez-Tapia, L. (2008). Insights into the composition of household vulnerability from multicriteria decision analysis. *Global Environmental Change, 18*(1), 112–127. https://doi.org/10.1016/j.gloenvcha.2007.09.001

Flanagan, B., Hallisey, E., Sharpe, J., Mertzlufft, C., & Grossman, M. (2020). On the validity of validation: A commentary on Rufat, Tate, Emrich, and Antolini's "how valid are social vulnerability models?". *Annals of the Association of American Geographers, 111*, 4. https://doi.org/10.1080/24694452.2020.1857220

Franzke, C., O'Kane, T., Berner, J., Williams, P., & Lucarini, V. (2015). Stochastic climate theory and modeling. *Wiley Interdisciplinary Reviews: Climate Change, 6*, 63–78. https://doi.org/10.1002/wcc.318

Füssel, H.-M. (2010). Review and quantitative analysis of indices of climate change exposure. In *Adaptive capacity, sensitivity, and impacts*. World Bank.

Gall, M. (2007). *Indices of social vulnerability to natural hazards: A comparative evaluation* (p. 231). University of South Carolina.

Hardy, R., & Hauer, M. (2018). Social vulnerability projections improve sea-level rise risk assessments. *Applied Geography, 91*, 10–20. https://doi.org/10.1016/j.apgeog.2017.12.019

Horney, J., Nguyen, M., Salvesen, D., Dwyer, C., Cooper, J., & Berke, P. (2017). Assessing the quality of rural hazard mitigation plans in the southeastern United States. *Journal of Planning Education and Research, 37*, 56–65. https://doi.org/10.1177/0739456X16628605

IPCC. (2014). Summary for policymakers. Climate change 2014: Impacts, adaptation, and vulnerability. In C. B. Field et al. (Eds.), *Part A: Global and sectoral aspects*. Cambridge University Press.

Jones, S., Ruppert, T., Deadly, E., Payne, H., Pippin, J.S., & Huang, L.-Y. (2019). Roads to nowhere in four states: State and local governments in the Atlantic southeast facing sea-level rise. *Columbia Journal of Environmental Law, 44*, 67–136. https://doi.org/10.7916/cjel.v44i1.806

Jurjonas, M., & Seekamp, E. (2018). Rural coastal community resilience: Assessing a framework in eastern North Carolina. *Ocean and Coastal Management, 162*, 137–150. https://doi.org/10.1016/j.ocecoaman.2017.10.010

Kok, K. (2009). The potential of fuzzy cognitive maps for semi-quantitative scenario development, with an example from Brazil. *Global Environmental Change, 19*, 122–133. https://doi.org/10.1016/j.gloenvcha.2008.08.003

Kusnetz, N. (2017). In the Outer Banks, officials and property owners battle to keep the ocean at bay. *Inside Climate News*. 28 November, 2017. https://insideclimatenews.org/news/28112017/nags-head-north-carolina-beach-erosion-climate-change-sea-level-rise/#::text=Their0lawsuits0dragged0for0years,are0now0in0a0stalemate

Kwakkel, J. H., Walker, W. E., & Marchau, V. (2010). Classifying and communicating uncertainties in model-based policy analysis. *International Journal of Technology Management, 10*, 299–315. https://doi.org/10.1504/IJTPM.2010.036918

Lehman, J., & Phelps, S. (2004). Tort law. In *West's encyclopedia of American law* (2nd ed.). Thomson/Gale.

Lempert, R., Nakicenovic, N., Sarewitz, D., & Schlesinger, M. (2004). Characterizing climate-change uncertainties for decision-makers - An editorial essay. *Climate Change, 65*, 1–9. https://doi.org/10.1023/B:CLIMATE0000037561.75281.b3

MacClune, K., Allan, C., Venkateswaran, K., & Sabbag, L. (2014). *Floods in Boulder: A study of resilience*. ISET-International. http://i-s-e-t.org/resources/case-studies/floods-in-boulder.html

Nordgren, J., Stults, M., & Meerow, S. (2016). Supporting local climate change adaptation: Where we are and where we need to go. *Environmental Science & Policy, 66*, 344–352. https://doi.org/10.1016/j.envsci.2016.05.006

O'Brien, K., Eriksen, S., Nygaard, L., & Schjolden, A. (2007). Why different interpretations of vulnerability matter in climate change discourses. *Climate Policy, 7*, 73–88.

Patt, A., & Dessai, S. (2005). Communicating uncertainty: Lessons learned and suggestions for climate change assessment. *Comptes Rendus Geoscience, 337*, 425–441. https://doi.org/10.1016/j.crte.2004.10.004

Patt, A., Klein, R., & de la Vega-Leinert, A. (2005). Taking the uncertainty in climate-change vulnerability assessment seriously. *Comptes Rendus Geoscience, 337*, 411–424. https://doi.org/10.1016/j.crte.2004.11.006

Patwardhan, A., Downing, T., Leary, N., & Wilbanks, T. (2009). Towards an integrated agenda for adaptation research: Theory, practice and policy. *Current Opinion in Environment Sustainability, 1*, 219–225. https://doi.org/10.1016/j.cosust.2009.10.010

Pedde, S., Kok, K., Onigkeit, J., Brown, C., Holman, I., & Harrison, P. (2019). Bridging uncertainty concepts across narratives and simulations in environmental scenarios. *Regional Environmental Change, 19*, 655–666. https://doi.org/10.1007/s10113-018-1338-2

Peduzzi, P., Dao, H., Herold, C., & Mouton, F. (2009). Assessing global exposure and vulnerability towards natural hazards: The disaster risk index. *Natural Hazards and Earth System Sciences, 9*, 1149–1159. https://doi.org/10.5194/nhess-9-1149-2009

Regan, H., & Colyvan, M. (2000). Fuzzy sets and threatened species classification. *Conservation Biology, 14*, 1197–1199. https://doi.org/10.1046/j.1523-1739.2000.99130.x

Renn, O. (2008). Review of psychological, social and cultural factors of risk perception. In O. Renn (Ed.), *Risk governance: Coping with uncertainty in a complex world* (pp. 98–148). Earthscan.

Renn, O., Klinke, A., & van Asselt, M. (2011). Coping with complexity, uncertainty and ambiguity in risk governance: A synthesis. *Ambio, 40*, 231–246. https://doi.org/10.1007/s13280-010-0134-0

Ribot, J. (2010). Vulnerability does not just fall from the sky: Toward multi-scale pro-poor climate policy. In R. Mearns & A. Norton (Eds.), *Social dimensions of climate change: Equity and vulnerability in a warming world* (pp. 164–199). The World Bank.

Rohat, G., Wilhelmi, O., Flacke, J., Monaghan, A., Gao, J., Dao, H., & van Maarseveen, M. (2019). Characterizing the role of socioeconomic pathways in shaping future urban heat-related challenges. *Science of the Total Environment, 695*, 133941. https://doi.org/10.1016/j.scitotenv.2019.133941

Romero-Lankao, P., Gnatz, D., & Sperling, J. (2016). Examining urban inequality and vulnerability to enhance resilience: Insights from Mumbai, India. *Climate Change, 139*, 351–365. https://doi.org/10.1007/s10584-016-1813-z

Romero-Lankao, P., & Norton, R. (2018). Interdependencies and risk to people and critical food, energy, and water systems: 2013 flood, Boulder, Colorado, USA. *Earth's Future, 6*, 1616–1629. https://doi.org/10.1029/2018EF000984

Romero-Lankao, P., Qin, H., & Dickinson, K. (2012). Urban vulnerability to temperature-related hazards: A meta-analysis and meta-knowledge approach. *Global Environmental Change, 22*, 670–683. https://doi.org/10.1016/j.gloenvcha.2012.04.002

Romieu, E., Welle, T., Schneiderbauer, S., Pelling, M., & Vinchon, C. (2010). Vulnerability assessment within climate change and natural hazard contexts: Revealing gaps and synergies through coastal applications. *Sustainability Science, 5*, 159–170. https://doi.org/10.1007/s11625-010-0112-2

Rufat, S. E., Tate, C. T., Emrich, & Antolini, F. (2019). How valid are social vulnerability models? *Annals of the American Association of Geographers, 109*, 1131–1153. https://doi.org/10.1080/24694452.2018.1535887

Scheuer, S., Haase, D., & Meyer, V. (2011). Exploring multicriteria flood vulnerability by integrating economic, social and ecological dimensions of flood risk and coping capacity: From a starting point view towards an end point view of vulnerability. *Natural Hazards, 58*, 731–751. https://doi.org/10.1007/s11069-010-9666-7

Schipper, E. L. F., Thomalla, F., Vulturius, G., Davis, M., & Johnson, K. (2016). Linking disaster risk reduction, climate change and development. *International Journal of Disaster Resilience in the Built Environment, 7*, 216–228.

Schmidtlein, M., Shafer, J., Berry, M., & Cutter, S. (2011). Modeled earthquake losses and social vulnerability in Charleston, South Carolina. *Applied Geography, 31*, 269–281. https://doi.org/10.1016/j.apgeog.2010.06.001

Schuck, P. H. (1983). *Suing government: Citizen remedies for official wrongs* (p. 264). Yale University Press.

Scott, C. A., Kurian, M., & Wescoat, J. L. (2015). The water-energy-food nexus: Enhancing adaptive capacity to complex global challenges. In M. Kurian & R. Ardakanian (Eds.), *Governing the nexus: Water, soil and waste resources considering global change* (pp. 15–38). Springer International Publishing.

Smith, K. A., Wilby, R. L., Broderick, C., Prudhomme, C., Matthews, T., Harrigan, S., & Murphy, C. (2018). Navigating cascades of uncertainty—As easy as ABC? Not quite.... *Journal of Extreme Events, 5*, 70. https://doi.org/10.1142/S2345737618500070

Sword-Daniels, V., Eriksen, C., Hudson-Doyle, E., Alaniz, R., Adler, C., Schenk, T., & Vallance, S. (2018). Embodied uncertainty: Living with complexity and natural hazards. *Journal of Risk Research, 21*, 290–307. https://doi.org/10.1080/13669877.2016.1200659

Tate, E. (2012). Social vulnerability indices: A comparative assessment using uncertainty and sensitivity analysis. *Natural Hazards, 63*, 325–347. https://doi.org/10.1007/s11069-012-0152-2

Tate, E. (2013). Uncertainty analysis for a social vulnerability index. *Annals of the Association of American Geographers, 103*, 526–543. https://doi.org/10.1080/00045608.2012.700616

Tate, E., Burton, C., Berry, M., Emrich, C., & Cutter, S. (2011). Integrated hazards mapping tool. *Transactions in GIS, 15*, 689–706. https://doi.org/10.1111/j.1467-9671.2011.01284.x

Thomas, D., Jang, S., & Scandlyn, J. (2020). The CHASMS conceptual model of cascading disasters and social vulnerability: The COVID-19 case example. *International Journal of Disaster Risk Reduction, 51*, 101828. https://doi.org/10.1016/j.ijdrr.2020.101828

Turner, B., et al. (2003). A framework for vulnerability analysis in sustainability science. *Proceedings of the National Academy of Sciences of the United States of America, 100*, 8074–8079. https://doi.org/10.1073/pnas.1231335100

Vincent, K. (2007). Uncertainty in adaptive capacity and the importance of scale. *Global Environmental Change, 17*, 12–24. https://doi.org/10.1016/j.gloenvcha.2006.11.009

Wilby, R. L. (2017). What shapes climate vulnerability? In R. L. Wilby (Ed.), *Climate change in practice* (pp. 103–123). Cambridge University Press.

Wilhelmi, O., & Morss, R. (2013). Integrated analysis of societal vulnerability in an extreme precipitation event: A Fort Collins case study. *Environmental Science & Policy, 26*, 49–62. https://doi.org/10.1016/j.envsci.2012.07.005

Yu, D., et al. (2020). Disruption of emergency response to vulnerable populations during floods. *Nature Sustainability, 3*, 728–736. https://doi.org/10.1038/s41893-020-0516-7

Open Access This chapter is licensed under the terms of the Creative Commons Attribution 4.0 International License (http://creativecommons.org/licenses/by/4.0/), which permits use, sharing, adaptation, distribution and reproduction in any medium or format, as long as you give appropriate credit to the original author(s) and the source, provide a link to the Creative Commons license and indicate if changes were made.

The images or other third party material in this chapter are included in the chapter's Creative Commons license, unless indicated otherwise in a credit line to the material. If material is not included in the chapter's Creative Commons license and your intended use is not permitted by statutory regulation or exceeds the permitted use, you will need to obtain permission directly from the copyright holder.

Climate/Earth System Projections and Their Uncertainties: An Overview

15

Chris E. Forest and William D. Collins

15.1 Prologue

This book section is focused on the uncertainty of climate models, which were typically called General Circulation Models (GCMs) of the atmosphere in the early years and have been renamed Global Climate Models or Earth System Models in the past few decades. The models were developed from the governing equations for the atmosphere, land system, oceans, sea ice, and ice sheets. Additionally, the models include the chemical cycles for carbon cycle, atmospheric aerosols, and other chemicals appearing in the different components of the model. These models represent the dynamics of the individual systems at appropriate spatial and temporal scales. These models are the workhorses for understanding climate and earth system behavior over the past 60 years.

15.2 Part 1: A Brief History of Climate Modeling and Assessments

Earth System Models are the fundamental tools that we use both to understand the history of past global climate change and to predict future global climate changes. To provide an introduction to the next few chapters in this book, we will set the scene on how we determine and quantify uncertainties in projections of future climate. Without going into a long and detailed history of climate change (e.g., Weart, 2003), we want to identify a starting point to acknowledge the early projects in the United States and in Europe where the first computers were working to forecast weather. In turn, this started the climate modeling activities, which had roots in the weather forecasting models after the World War II era. For this history, the developments of climate models were an extension from the first computers used to simulate atmospheric dynamics using the computers developed at the Institute of Advanced Studies.

The first numerical models investigating the climate system simulated the vertical structure of the atmosphere and focused on mechanisms relating to the energy budgets and radiative transfer within the atmosphere. These numerical computing models were developed at the Institute of Advanced Studies (IAS) in Princeton, NJ (a precursor to the Geophysical Fluid Dynamics Laboratory (GFDL) climate modeling teams). Using the first computers developed at IAS (Weart, 2003), some of the first quantitative simulations of atmospheric circulations were being developed by Jule Charney (MIT & IAS), Norman Philips (IAS), and Arnt Eliassen (IAS) to apply both the atmospheric equations for weather applications, while the numerical methods were being developed for the computer, known as the "IAS machine." Advances in applied mathematics and computational modeling were closely linked to the development of Geophysical Fluid Dynamics during this era in the late 1940s and early 1950s.

Three examples are Charney and Eliassen (1949), Eliassen (1952), and Charney and Phillips (1953) which were published from the work developed at the IAS. Given the developments of the numerical weather and climate models based at the Institute of Advanced Studies, the development of both atmospheric and climate models have continued to be developed to the present day given the computational platforms for the numerical models that are now available. (NOTE: For additional details, please see the latest set of models in the 6th Intergovernmental Panel on Climate Change Assessment Report (AR6) (ipcc.ch).)

Next, we fast-forward two decades to one of the first assessments to document how climate models were used to

C. E. Forest (✉)
Department of Meteorology and Atmospheric Science, The Pennsylvania State University, University Park, PA, USA

W. D. Collins
Lawrence Berkeley National Laboratory, Berkeley, CA, USA

© The Author(s) 2025
L. O. Mearns et al. (eds.), *Uncertainty in Climate Change Research*,
https://doi.org/10.1007/978-3-031-85542-9_15

assess the impacts of climate change provided in "Carbon Dioxide and Climate: A Scientific Assessment" Report from the National Academy of Science (National Research Council and Climate (1979); hereafter, the Charney Report). This is the first quantitative assessment on the estimate of equilibrium climate sensitivity (the climate response to a doubling of carbon dioxide concentrations). Based on the limited available models, the Charney Report concluded the best estimate of the equilibrium sensitivity to changes in carbon dioxide concentrations is: 3.0°C, with a range of 1.5–4.5°C. At the time, there were multiple climate models that contributed to the report with at least three (3) climate system models that included an atmosphere, ocean, land, and ice components. In the Charney Report, multiple models were considered to assess uncertainties in the feedback processes that could be addressed.

The Charney Report focused primarily on the atmospheric feedback processes that would respond to changes in temperature. Ramanathan and Coakley (1978) reported a climate sensitivity using a one-dimensional radiative-convective model with fixed specific humidity profile, fixed lapse rate, and fixed cloud cover and heights. This model was unable to address uncertainties related to changes in: cloud cover, cloud height, relative humidity, and lapse rates. Other groups (Lian and Cess (1977) and others) were assessing feedback from changing albedo due to sea ice and snow cover changes and additional changes from cloud albedos. Among atmospheric feedbacks, cloud feedbacks were the most difficult, given that cloud model parameterizations of that era were simple compared with present-day models. Still to this day, the cloud feedbacks continue to be a key uncertain feedback as high-resolution cloud models become more integrated within the global scale climate models. To represent clouds in climate models, modelers must address feedback of the short-wave and long-wave effects, which were discussed in Section 3.2 in the Charney Report.

In Section 3.2 of the Charney Report, the authors focused on ocean processes that slowed the rate of warming. At the time, the numerical models were not representing the ocean dynamics that would alter the poleward heat flux. Instead, they were imposing poleward fluxes rather than explicitly modeling the western boundary currents. Additionally, the models were using approximations to represent the heat exchange between the atmosphere and ocean without calculating the energy transfer across the ocean surface and deeper into the thermocline region. Uncertainties in the rate of heat uptake would determine whether the energy absorbed by the ocean would be lower than observed and result with more energy stored in the land surface or upper oceans. Alternatively, it could be higher than observed and lead to more energy stored in deeper regions of the ocean basins and keeping the surface relatively cooler. As a result, the rate of ocean warming can strongly influence the rates of climate change. Similarly, rates of ocean carbon uptake strongly depend on the ocean temperature. Another aspect of ocean dynamics is how the scaling of the equations of ocean dynamics compare with those for atmospheric dynamics. Given the smaller length scales for baroclinic processes, convection, and mixing processes, ocean climate models require finer grids and thus, higher computational costs to resolve specific processes such as wind mixing, thermal mixing, and evaporative processes.[1]

The last section of the Charney Report addressed the capabilities of the three-dimensional general circulation models (which should be considered as the state of the science models of the era). The models available in the late 1970s for the report were: three models from the NOAA Geophysical Fluid Dynamics Laboratory (M1, M2, M3, led by S. Manabe and colleagues (Manabe and Wetherald, 1975, 1980; Manabe and Stouffer, 1980)) and two models from the NASA Goddard Institute for Space Studies (H1 and H2, led by J. Hansen and colleagues Hansen et al. (1983, n.b., final model description)). Additionally, the British Meteorological Office model (Mitchell, 1979) was available, but the numerical experiments were based on fixed surface conditions that were different enough to not be included with the five models from the two US groups. The five (5) models available were considered to be independently developed, used different parameterizations for key aspects of the structures, and included components of the models that simulated sea surface temperatures. In today's parlance, each of these models could be considered an independent model with different dynamics, physics, and representations of the key climate components (Atmosphere, Ocean, Land Surface, Ice) although each would be very simplified compared to the current IPCC-class models as used for the IPCC Assessment Reports. Key differences listed are "geographies, seasonal changes, cloud feedback, snow and ice properties, and horizontal and vertical resolutions." Overall, the five distinct models produced results that led to the results discussed in Section 4 of the Charney Report.

Both the GFDL models and NASA models provided results that accounted for "swamp oceans" v. "mixed layer oceans," geographic representation of land-ocean characteristics but all were at low resolution when compared with current climate models. A significant difference from current models is that neither model included interactive heat transports in the ocean. In addition, the NASA models accounted for a larger number of processes including: "ground heat storage, sea-ice leads, and dependence of snow-ice albedo on snow age." In the late 1970s, the simulation experiments

---

[1] At the time, we note that current modern climate topics were not considered in the Charney Report such as biotic and abiotic components of the ocean flora and fauna or more comprehensive carbon cycle estimates.

included scenarios for both "doubled" and "quadrupled" CO2 concentrations, which have continued to be one of the standardized estimates for IPCC Assessment Reports which started in 1990 with the First Assessment Report (AR1) (Houghton et al., 1990). At this time, the IPCC has produced and published six (6) major IPCC Assessment Reports and other targeted scientific reports on climate change for the United Nations.

15.3 Part 2: What Is the Path We Have Traveled?

Since the initial climate model assessment with the Charney Report (National Research Council and Climate, 1979) and the first IPCC Assessment Report in 1990, the climate modeling community has expanded considerably. After the 1990 report, there were 11 modeling groups and 26 models that provided inputs to the IPCC AR1 Report. In the latest IPCC Report, there were 28 modeling groups and 59 models included in the AR6 Report (with additional models for both higher resolution atmosphere models and cryosphere models), and the development of a whole set of associated downscaling techniques to allow the assessment of regional impacts (Chap. 18). Based on the approach set forth by the Charney Report National Research Council and Climate (1979) from the US National Academy of Science, the IPCC assessments of climate change and the evaluation of the climate model projections have been critical scientific components for informing the international negotiations related to the UNFCCC process.

Developing better climate science has expanded from a fundamental question: What will global temperature change be in a future with a doubled (or quadrupled) concentration of carbon dioxide? This has evolved toward the assessment of a large number of emissions and concentration scenarios today (Chap. 16). Based on the UNFCCC goals, the scientific community has continuously adapted the science to learn how climate change information can be helpful for informing relevant and important decisions. Specifically, how will we use the climate modeling expertise to analyze proposals to move toward net-zero forcing? This is a critical question for the climate impact assessment community. We need to learn how we can address these challenges and provide scenarios to guide problem solving for future climate change managers. Understanding this uncertainty will be helpful to guide decision-makers on how to prepare for alternative scenarios or for specific risks.

The IPCC process provided a platform for developing Climate/Earth System Models to inform the scientific community to develop and document the current science related to drivers of climate change, the projections of climate change, and the impacts of climate change. The First IPCC Report (1990) was a single volume with eleven (11) chapters with 365 pages. Compared with the latest IPCC Report (Masson-Delmotte et al., 2021), respectively, the three working group reports together have 7345 pp [WG1 (2049p), WG2 (3068p), WG3 (2042p), Synthesis Report (186p)], and they staggered the release for each Working Group and the Synthesis Report over the period from August 2021 to March 2023. Based on these six IPCC Reports to date, we can focus on only a few of the issues related to uncertainty in the climate modeling research area.

First, using the six IPCC Assessment Reports over the past 34 years, we can provide a table for the Equilibrium Climate Sensitivity (ECS) across the set of reports (Table 15.1).

At face value, this table suggests that we have learned very little about the equilibrium climate sensitivity, but the Global Climate Models have continued to improve by identifying what regions (or processes) require specific attention to improve the climate projections. During the 1980s and 1990s, fixing the climate models appeared like a "whack-a-mole" game, where each modeling group fixed one model component that affected other processes, which then needed attention. With limits on computing capacity, within each modeling group, developers could choose to focus on clouds, atmosphere-ocean fluxes, or aerosol processes, and this community approach worked well.

15.4 Part 3: What Critical Climate Observations Are Required?

To improve climate models, one critical issue is maintaining and archiving climate data sets across all model components. *From the bottom of the ocean to the top of the atmosphere*, the climate data observations have been archived and provide multiple ways to test and improve the model processes, if not providing additional predictive skill. Based on the synoptic measurements for weather forecasting, the climate community has used the long records of surface temperatures and upper-air radiosonde archives to provide diagnostics on the atmospheric and land surface processes. The global ocean data has shorter data sets for deep temperature and salinity data with technology developed in the post-World War II era (e.g., Swallow 1955; Swallow and Worthington 1961). The transects of the ocean data have been developed since the 1970s (e.g., https://cchdo.ucsd.edu/ or https://joa.ucsd.edu/Data_homepage). The ARGO float fleet (Wong and Coauthors 2020) has provided an alternative data acquisition which augments the ship-based transects. The coverage of the oceans depths to 2000m has been available since 2016 (Riser et al. 2016). The combined WOCE and ARGO data sets have provided the deep sea data for the past 50 years, while some ocean data in the nineteenth century (e.g., Gleckler et al. 2016) has provided some insights from single ships.

Table 15.1 Estimated Equilibrium Temperature for a doubling of $[CO_2]$ concentrations from the preindustrial era. These estimates are based on the Charney Report (1979) and the six major IPCC Assessment Reports https://ipcc.ch/reports/

Assessment report	Best value	Likely range	Very likely range	Number of models
Charney Report (1979)	2.4°C	1.5–4.5°C	1.5–4.5°C	$N = 5$
IPCC AR1 (1990)	3.0 °C	1.5–4.5°C	1.5–4.5°C	$N = 26$
IPCC AR2 (1995)	3.0°C	1.5–4.5°C	1.5–4.5°C	$N = 16$
IPCC AR3 (2001)	3.0°C	1.5–4.5°C	1.5–4.5°C	$N = 31$
IPCC AR4 (2007)	3.0°C	1.5–4.5°C	1.5–4.5°C	$N = 23$
IPCC AR5 (2013)	3.0°C	1.5–4.5°C	1.5–4.5°C	$N = 50$
IPCC AR6 (2021)	3.0°C	2.5–4.0°C	2.0–5.0°C	$N = 59$

With the advances in data collection in the oceans since ~2000, the climate data archives have provided the tools to test climate model response to the historical anthropogenic forcing data by combining the longer historical surface and upper-air data. This provides the starting point for testing climate model response to the human impacts on the climate system. Starting in IPCC AR4, multiple groups were able to use Monte Carlo methods (Andronova and Schlesinger, 2001; Forest et al., 2002; Knutti et al., 2002) to derive limits on the Equilibrium Climate Sensitivity (ECS) and the Transient Climate Response (TCR). The changes in the surface temperature and the ocean heat content over multiple decades provide an observations-based approach to generate climate models that are consistent with uncertainty in the model realizations and the data sets. These early attempts to quantify the range of ECS and TCR using the observational data were possible due to the climate data, climate models, and statistical tools. In the AR6, these basic approaches are still being used but with more computational power to run full Earth System Models. One critical set of uncertainties related to the projections will be to test short-lived climate forcings driven by transient emissions, both natural and anthropogenic sources.

15.5 Part 4: Model Uncertainty Quantification Limits Based on the Computational Constraints

To begin understanding the uncertainty in climate projections, we start with the development of climate models. This has always translated the core equations (fluid, mass, radiation, etc.) using numerical methods that discretize the climate model equations onto spatial grids (or onto spherical harmonics) on the sphere. This requires choices on how to limit the computations based on the memory limits of the "computer du jour." Similarly, we also have time constraints when considering how many simulations are possible based on the computational speed of calculations (i.e., floating point operations per second (flops)). Both limits, on the speed and the memory, determine how many climate simulations can be generated based on the computational limits. In a general sense, these computational limits still set the fundamental limits to the quality of our climate simulations. Over the past seven decades since the 1950s, we can track the advances in memory that allow scientists to create smaller grids, which, in turn, require both smaller time steps and require higher speeds (i.e., requires more flops) when we increase the grid resolutions.

An additional source of uncertainty for climate simulations is the sensitivity to initial conditions across the set of climate state-vector variables (see Chap. 17). Often, this is discussed as the limits of discretization in representing the equations of motion using a specific grid and time step.

The transition from a continuum derivative for a nonlinear equation $\frac{\partial f(v_x \cdot E, t)}{\partial t}$ to a discretization requires expanding the terms in the following equation:

$$\begin{aligned}
\frac{\partial f(v_x \cdot E, t_o)}{\partial t} &= \frac{\partial v_x}{\partial t} \cdot E + \frac{\partial E}{\partial t} \cdot v_x \\
&= \frac{v_x(t_o + \Delta t) - v_x(t_o)}{\Delta t} \cdot E(x, t_o) \\
&\quad + \frac{E(t_o + \Delta t) - E(t_o)}{\Delta t} \cdot v_x(x, t_o) \\
&= \frac{v_x(t_o + \Delta t)}{\Delta t} \cdot E(x, t_o) - \frac{v_x(t_o)}{\Delta t} \cdot E(x, t_o) \\
&\quad + \frac{E(t_o + \Delta t)}{\Delta t} \cdot v_x(x, t_o) - \frac{E(t_o)}{\Delta t} \cdot v_x(x, t_o) \\
&= \frac{v_x(t_o + \Delta t)}{\Delta t} \cdot E(x, t_o)) \\
&\quad - 2 \frac{v_x(t_o) \cdot E(t_o)}{\Delta t} \\
&\quad + \frac{E(t_o + \Delta t)}{\Delta t} \cdot v_x(x, t_o),
\end{aligned}$$

which adds small errors in the calculation of $\frac{\partial f(v_x \cdot E, t)}{\partial t}$. As the climate modeling community has learned to simulate climate, we recognize how projections of climate change will be sensitive to the accuracy of estimating the derivatives, $\frac{\partial f(v_x \cdot E, t)}{\partial t}$. To improve the numerical accuracy, we require higher-order numerical methods, but the higher accuracy of the derivatives will increase the computational costs.

So, better representation of the nonlinear discretization will require good choices when developing codes to discretize the governing equations.

15.6 Part 5: "The Climate Modeling Paradigm": How Many Ensemble Members Can We Afford?

Given a specific computer resource, we can identify (and track) the computational costs that limit our assessments of uncertainties in climate change projections. As climate models have developed and become more complex over the past 60 years (Fig. 15.1), we have improved our ability to compare climate model simulations with observational data sets. Given the uncertainties in the set of climate equations, typically, we compare an ensemble of model simulations to a single set of observational data. If we see that the observational data is within the uncertainty range of our climate model simulations, this is a first step to quantify whether the climate model is behaving well. As the complexity of a model increase, we can then proceed to identify the interactions across the different model components, and this requires examining how the communication between the model components is also aligned with the observational data. After this last part, we learn how the feedback processes within each of the different subsystems can influence the internal variability as well as the response of the climate system components. If simpler numerical discretizations lead to poorer accuracy in the models, this can lead to poorer estimates of the simulation results, which in turn will limit our interpretations of the climate change over the next century and beyond.

As models improve, the estimated outputs provide opportunities to test how well the models compare against internal variability of the climate system for the individual components (atmosphere, ocean, land, or ice), as well as the interactions among the climate model components. An alternative approach would be to analyze the interacting budgets across the component models (energy, momentum, angular momentum, trace gases, etc.).

As an example, based on the terms of the energy budget equations within an atmospheric global model, we can estimate the individual terms of the energy flux budgets from the models and estimate the interactions across model components. Uncertainties in the model components are contributing to the internal variability and the prediction uncertainty of the Earth System Models. We are only able to assess the uncertainty if we are able to compare model outputs to the observational data sets that are available. The long-term monitoring of atmospheric state variables (e.g., radiation, temperature, winds, or precipitation fields) is an essential task to provide a useful but not sufficient to test the

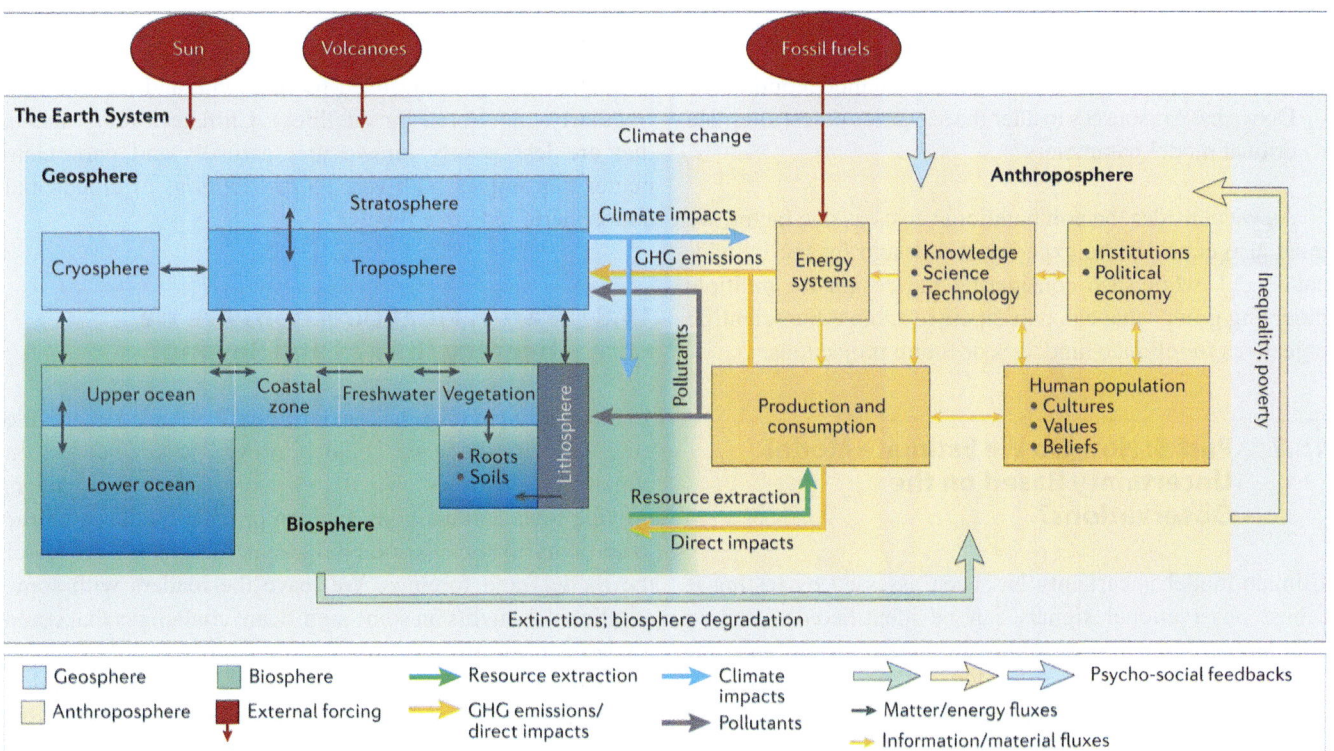

Fig. 15.1 This presents the components and the interactions of the climate systems required to incorporate the physical and human components of the Earth System Models. (Based on the original figure in: NRC, 1986, doi:10.17226/19210; [(Steffen et al., 2020), Springer Nature])

transports and fluxes. The spatio-temporal analyses require to go further into the analyzed fields that are operating across daily, seasonally, and long-term climatological time-space scales. The synoptic weather data are just the beginning of the data that is required as we focus on climate hindcasts or projections into the next centuries.

In general, the uncertainty characteristics of a model are being set by the individual components of the specific models that are currently being evaluated. A holistic approach to quantifying uncertainties would require to incorporate diagnostics that would have cross-over covariance diagnostics that inform how a coupled model could depend on other components within the full model interactions. The use of causal inference (Pearl, 2009; Barnes et al., 2019, 2020) could be a key element to include in additional machine learning tools.

We have standard practices to define the next set of simulations based on what model components are involved and how the models fail to represent the statistics of the observations. When we start the design phase, we need to consider both the computational limits that set the computational design and the experimental design for a set of climate simulations to answer the scientific objectives. Here are some questions that need to be considered:

- Do we have sufficient processors to run high-resolution simulations for both smaller time steps and grid sizes?
- Do we have a computational platform that has sufficient memory (RAM) for the computations?
- Do we have disk storage for the outputs?
- Do we have resources to run alternative model options?
- Do we have resources to alter the scenarios or test alternate critical model parameters?

As we consider the computational costs as one factor, we must also consider the experimental design for the individual model, while also considering the design for multiple modeling groups that can participate to address the scientific objectives for climate hindcasts or future projections.

15.7 Part 6: How Do We Estimate Model Uncertainty Based on the Observations?

Climate model uncertainty has been assessed by testing if robust observational signals can be identified/detected in simulations of the historical climate change era. Because climate models have multiple simulations of the historical past, the climate model outputs can provide uncertainty estimates using standard statistical methods (e.g., regressions, correlations, or other statistics which are derived from observational data). Additionally, the output of a climate model can be compared against output from other climate models to assess the similarities or differences among available climate simulations. In early years of the IPCC process, the development of climate assessment tools were developed largely from the community and were consolidated at the Program for Climate Model Diagnosis and Intercomparisons (PCMDI, est. 1989) https://pcmdi.llnl.gov/ and https://pcmdi.llnl.gov/metrics/ at the DOE Lawrence Livermore National Laboratory (LLNL). PCMDI has contributed to every IPCC report from AR1 (in 1990) to AR6 (in 2021). In the last decade, many international climate research laboratories have been developing suites of statistical tools, managing data archives, and climate model outputs, which have developed and extended the capacity to work with climate model outputs. (Nota bene: Climate modeling teams and their data sets have always pushed the envelope of computational science capacity and standards throughout the climate science era.)

In the latest AR6 assessment, climate diagnostics have been in the public domain, and programs are available via GitHub: https://github.com/IPCC-WG1. Critical variables that have direct connections to observations include:

- Surface and upper-air temperatures
- Land surface temperatures
- Precipitation
- Aerosols, and
- Ocean heat content

In addition, the climate driving agents have long-term records for: (1) Heat trapping gasses are long-lived and well-monitored and (2) long-lived forcings via land-use changes (regularly, monitored by satellites). Climate forcing agents that are less poorly known are: aerosols and particulate matter that impact both vertical distributions of clouds and atmospheric radiative fluxes.

15.8 Part 7: What Directions Are We Heading in 2025 and Beyond?

To move forward on reducing uncertainty, we need to provide guidance on where we could improve our predictions on climate change timescales. In this book section, the chapters provide tools and methods that can provide guidance on the uncertainty of the climate predictions for specific setups of the models and forcings. We leave the readers with some additional thoughts on some significant challenges that could guide the next directions for climate prediction.

As a scientific community, we lack a first-principles theory of climate—so we do not know the necessary and sufficient criteria for a Climate/Earth system model to be useful as predictors. This has consequences as we move forward on developing ESMs on how we are adjusting to Edward

Lorenz's butterfly wings driving unexpected consequences across the model projections. We are limited by too few ways to test our nonlinear interactions in our models that are being used to guide climate policy in the coming decades.

As a grand challenge for testing climate models in dramatically different climates, we are highly limited based on the historical records of the present-day and proxy data from paleoclimate eras. As an example, the interglacial periods during the ice ages are not included, and the proxy records for the interglacials are limited. To test climate models using the paleoclimate proxy observations, we need to provide larger sets of spatio-temporal data that can be used to rule out possible model configurations. For the recent past, we have higher resolution records of the recent few centuries, but we would like to have better data resolution of the ice-age cycles. The proxies, so far, are not providing strong evidence to rule out possible responses or feedback processes.

Since the first insights into climate models (e.g., Foote 1856; Croll 1875; Arrhenius 1896), we have seen Manabe et al. (1965); Manabe and Wetherald (1967) move our field into the present era, where we have started from a top-down approach where the incoming solar radiation is balanced by the outgoing long-wave radiation to space. While the climate modeling development has been driven from the top-down rather than the bottom up, we are limited by the open system approach. To develop a "first-principles" approach, we need to develop tools and diagnostics from the fundamental components of the physical processes driving the budget equations for Momentum, Energy, Continuity, Angular Momentum, Mass, etc. This could provide a Hierarchy of Physics where we can develop theory for the transitions to physical/dynamical scaling.

As one example, for fundamental fluid dynamics, we consider microscopic fluid dynamics, which sets a path to boundary-layer turbulence to 3D convection, and eventually, to equations for the geophysical fluid dynamics. As a second example, we could consider the interactions related to microphysics of water droplets and aerosols that drive cloud formation and, in turn, have set the structure of the stratocumulus clouds that lead to cloud reflectivity that can modulate the incoming solar radiation. These fundamental physics and chemistry can also connect to the biosphere where we could discuss the role of the Amazon Rainforest (evapotranspiration and cloud condensation nuclei) or the Tropical Ocean Biosphere (biomolecular dynamics producing aerosol precursors for cloud condensation nuclei).

Systems with very long memory are another major challenge for predictions. Two examples are how we use the observations of the ice sheets or the land surface. These are critical parts of the present-day climate system yet these are central to understanding the climate services we obtain from the terrestrial carbon cycle or understanding the risks of sea-level rise (Chap. 19).

As we look forward to the chapters in this section, the authors have addressed the uncertainty for each climate component and addressed the key items that can be helpful for acknowledging where significant uncertainties still exist.

References

Andronova, N. G., & Schlesinger, M. E. (2001). Objective estimation of the probability density function for climate sensitivity. *Journal of Geophysical Research, 106*(D19), 22605–22611.

Arrhenius, S. (1896). Über den einfluss des atmosphärischen kohlensäuregehalts auf die temperatur der erdoberfläche. *Bihang Till K. Svenska Vet. -Akad. Handlingar, 1*(22), 1–102.

Barnes, E. A., Hurrell, J. W., Ebert-Uphoff, I., Anderson, C., & Anderson, D. (2019). Viewing forced climate patterns through an AI lens. *Geophysical Research Letters, 46*(22), 13389–13398.

Barnes, E. A., Toms, B., Hurrell, J. W., Ebert-Uphoff, I., Anderson, C., & Anderson, D. (2020). Indicator patterns of forced change learned by an artificial neural network. *Journal of Advances in Modeling Earth Systems, 12*(9), e2020MS002195.

Charney, J. G. & Eliassen, A. (1949). A numerical method for predicting the perturbations of the middle latitude westerlies. *Tellus A, 1*, 38–54.

Croll, J. (1875). *Climate and time in their geological relations: A theory of secular changes of the earth's climate* (Vol. XII(295)).

Eliassen, A. (1952). Symposium on numerical forecasting: Simplified dynamic models of the atmosphere, designed for the purpose of numerical weather prediction. *Tellus, 4*(3), 145–156.

Foote, E. (1856). Circumstances affecting the heat of the sun's rays. *The American Journal of Science and Arts, XXXI*(66), 382–383.

Forest, C. E., Stone, P. H., Sokolov, A. P., Allen, M. R., & Webster, M. D. (2002). Quantifying uncertainties in climate system properties with the use of recent climate observations. *Science, 295*(5552), 113–117.

Gleckler, P. J., Durack, P. J., Stouffer, R. J., Johnson, G. C., & Forest, C. E. (2016). Industrial-era global ocean heat uptake doubles in recent decades. *Nature Climate Change, 6*(4), 394–398.

Hansen, J., Russell, G., Rind, D., Stone, P., Lacis, A., Lebedeff, S., Ruedy, R., & Travis, L. (1983). Efficient three-dimensional global models for climate studies: Models I and II. *Monthly Weather Review, 111*, 609–662.

Houghton, J. T., Jenkins, G. J., & Ephraums, J. J. (Eds.). (1990). *Climate change: The IPCC scientific assessment*. Cambridge University Press.

Charney, J. G., & Phillips, N. A. (1953). Numerical integration of the quasigeostrophic equations for barotropic and simple baroclinic flows. *Journal of Atmospheric Sciences, 10*(2), 71–99.

Knutti, R., Stocker, T. F., Joos, F., & Plattner, G.-K. (2002). Constraints on radiative forcing and future climate change from observations and climate model ensembles. *Nature, 416*(6882), 719–723.

Lian, M. S. & Cess, R. D. (1977). Energy balance climate models: A reappraisal of ice-albedo feedback. *Journal of the Atmospheric Sciences, 34*, 1058–1062.

Manabe, S., & Stouffer, R. J. (1980). Sensitivity of a global climate model to an increase of CO_2 concentration in the atmosphere. *Journal of Geophysical Research, 85*, 5554.

Manabe, S. & Wetherald, R. T. (1967). Thermal equilibrium of the atmosphere with a given distribution of relative humidity. *Journal of the Atmospheric Sciences, 24*(3), 241–259.

Manabe, S. & Wetherald, R. T. (1975). The effects of doubling the CO_2 concentration on the climate of a general circulation model. *Journal of the Atmospheric Sciences, 32*, 1–15.

Manabe, S., & Wetherald, R. T. (1980). On the distribution of climate change resulting from an increase of CO_2 content in the atmosphere. *Journal of the Atmospheric Sciences, 37*, 118.

Manabe, S., Smagorinsky, J., & Strickler, R. F. (1965). Simulated climatology of a general circulation model with a hydrologic cycle. *Monthly Weather Review, 93*(12), 769–798.

Masson-Delmotte, V., Zhai, P., Pirani, A., Connors, S. L., Péan, C., Berger, S., Caud, N., Chen, Y., Goldfarb, L., Gomis, M. I., Huang, M., Leitzell, K., Lonnoy, E., Matthews, J. B. R., Maycock, T. K., Waterfield, T., Yelekçi, O., Yu, R., & Zhou, B. (Eds.) (2021). *IPCC, 2021: Climate Change 2021: The Physical Science Basis. Contribution of Working Group I to the Sixth Assessment Report of the Intergovernmental Panel on Climate Change.* Cambridge University Press.

Mitchell, J. F. B. (1979). Preliminary report on the numerical study of the effect on climate of increasing atmospheric carbon dioxide. Technical Report II/137, Meteorological Office, Bracknell, Berkshire, United Kingdom.

Ad Hoc Study Group on Carbon Dioxide National Research Council and Climate. (1979). *Carbon dioxide and climate: A scientific assessment.* The National Academies Press. NRC Panel: J. G. Charney, A. Arakawa, D. J. Baker, B. Bolin, R. E. Dickinson, R. M. Goody, C. E. Leith, H. M. Stommel, C. Wunsch.

Pearl, J. (2009). *Causality: Models, reasoning and inference* (2nd ed.). Cambridge University Press.

Ramanathan, V. & Coakley, J. A. Jr. (1978). Climate modeling through radiative-convective models. *Reviews of Geophysics, 16*(4), 465–489.

Riser, S. C., Freeland, H. J., Roemmich, D., Wijffels, S., Troisi, A., Belbéoch, M., Gilbert, D., Xu, J., Pouliquen, S., Thresher, A., & Others. (2016). Fifteen years of ocean observations with the global Argo array. *Nature Climate Change, 6*(2), 145–153.

Steffen, W., Richardson, K., Rockström, J., Schellnhuber, H. J., Dube, O. P., Dutreuil, S., Lenton, T. M., & Lubchenco, J. (2020). The emergence and evolution of earth system science. *Nature Reviews Earth and Environment, 1*(1), 54–63.

Swallow, J. C. (1955). A neutral-buoyancy float for measuring deep currents. *Deep Sea Research, 3*(1), 74–81.

Swallow, J. C., & Worthington, L. V. (1961). An observation of a deep countercurrent in the Western North Atlantic. *Deep Sea Research (1953), 8*(1), 1–IN3.

Weart, S. R. (2003). *The discovery of global warming.* Harvard University Press.

Wong, A. P. S. & Co-authors. (2020). Argo data 1999–2019: Two million temperature-salinity profiles and subsurface velocity observations from a global array of profiling floats. *Frontiers in Marine Science, 7*, 700.

Open Access This chapter is licensed under the terms of the Creative Commons Attribution 4.0 International License (http://creativecommons.org/licenses/by/4.0/), which permits use, sharing, adaptation, distribution and reproduction in any medium or format, as long as you give appropriate credit to the original author(s) and the source, provide a link to the Creative Commons license and indicate if changes were made.

The images or other third party material in this chapter are included in the chapter's Creative Commons license, unless indicated otherwise in a credit line to the material. If material is not included in the chapter's Creative Commons license and your intended use is not permitted by statutory regulation or exceeds the permitted use, you will need to obtain permission directly from the copyright holder.

Emissions and Concentration Scenarios

Jennifer Morris and John M. Reilly

16.1 Introduction

Climate impacts depend on the magnitude of climate change, as well as planning practices and the level and effectiveness of adaptive responses. In turn, the magnitude of future climate change is uncertain, depending not only on uncertainty in the Earth system and natural variability but also on uncertainty in future emissions and their underlying human system drivers. As such, consideration of different potential emissions futures is essential for understanding and preparing for future climate change.

This chapter briefly reviews the role of uncertainty in emissions projections. It then addresses the key drivers of emissions of greenhouse gases and other pollutants and how uncertainty in those drivers affects projections of future emissions. The Intergovernmental Panel on Climate Change (IPCC) has played a dominant role in developing emissions and concentration scenarios that are used by the climate community, and ultimately, these provide a range of climate scenarios of potential use to those assessing climate impacts. The IPCC approach has focused on scenario development, eschewing formal uncertainty analysis. We review the IPCC approach to scenario development and how it has changed through successive assessment reports (AR). We then describe approaches that use formal uncertainty analysis techniques and how this differs from scenario analysis, as well as methods that can connect probability distributions back to individual scenarios of interest. Finally, we summarize lessons learned from more than 40 years of socioeconomic effort to create scenarios to support climate change research.

J. Morris (✉) · J. M. Reilly
Center for Sustainability Science and Strategy, Massachusetts Institute of Technology, Cambridge, MA, USA
e-mail: holak@mit.edu

16.2 The Relative Role of Socioeconomic and Earth System Uncertainties in Climate Projections

There is substantial uncertainty about future emissions, driven by underlying uncertainty in socioeconomic drivers of emissions, which contributes to the uncertainty in projections of climate change. There is also uncertainty about how the Earth system responds to emissions (see Chap. 15). In addition, there is widely recognized natural variability in the climate system, captured as internal variability in general circulation models of the Earth system (see Chap. 17). Efforts to partition climate projection uncertainty (e.g. Hawkins & Sutton, 2009; Lehner et al., 2020) have found that internal variability and uncertainty in the climate system response and modeling of it ("model" uncertainty) are the largest drivers in the near-term. However, beyond the next decade or two, uncertainty in human systems and resulting emissions ("scenario" uncertainty) becomes an increasingly important contributor to uncertainty in climate change projections. Model uncertainty remains important, but the influence of socioeconomic-driven scenario uncertainty grows over time and perhaps becomes the largest source of uncertainty after 50–75 years (Fig. 16.1).

Figure 16.1 is but one attempt to partition uncertainty. Different approaches to estimate underlying scientific and socioeconomic uncertainty can give widely varying estimates of uncertainty in scenarios, leading to different assessments of how different sources of uncertainty contribute to the overall uncertainty in climate outcomes. For example, Gillingham et al. (2018), in a multi-model comparison, found an interquartile range of CO_2 concentrations from integrated assessment models of 200 ppm to 400 ppm just from within model differences (all models used the same input uncertainty distributions). A key uncertain input was labor productivity growth obtained by averaging results of

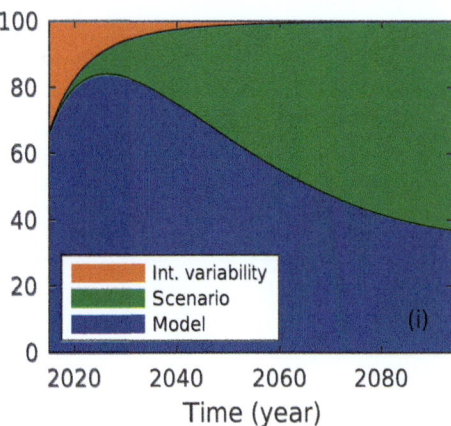

Fig. 16.1 Fractional contribution of individual sources of uncertainty to total uncertainty in global mean temperature, based on model simulations from the sixth Coupled Model Intercomparison Project (CMIP6). Adapted from Lehner et al. (2020)

expert elicitation. The individual distributions from experts varied dramatically, and so if each were used separately, that would produce a much wider uncertainty range for resulting emissions and concentrations. Estimates of uncertainty in Earth system response also vary widely (Knutti et al., 2017). Also, the internal variability in models may not capture all of natural variability, such as the full effect of ENSO (El Niño-Southern Oscillation) events or the effects of volcanic eruptions.

Any effort to assign some fraction of uncertainty to each of these sources will depend on estimates of the individual uncertainties. However, the basic conclusion of Lehner et al. (2020)—that natural variability is a critical source of uncertainty in climate projections 10–20 years into the future, Earth system uncertainty dominates projections 20–60 years out, and scenario uncertainty grows over time becoming a key driver of projections 30 or more years into the future—is relatively robust. The characteristic of this partitioning follows directly from the cumulative nature of the climate problem, stemming from the fact that most greenhouse gases are long-lived in the atmosphere. This means that even pretty large differences in emissions for a few years have a small effect on near-term concentrations and climate. It is only with scenarios differing emissions for decades that we begin to see significant differences in concentrations and the resulting climate. Similarly, the effect of Earth system uncertainties on projections relative to natural variability depends on the change in the concentration level—the uncertainty in resulting climate is bigger if the concentration change is larger.

The general partitioning pattern has important implications for both climate policy and climate impact assessment. For climate mitigation policy, the implication is that mitigation efforts need to be pursued for decades before significant benefits for climate are observed. Short-term climate variability may give false diagnoses of the relationship between emissions, concentrations, and climate. For example, the so-called warming hiatus—a decade or so from 1998 to 2008—showed little atmospheric warming. Careful diagnoses eventually showed that the nature of ENSO events and other short-term variations had offset what otherwise would have been continued atmospheric warming (Kopp & Lean, 2011). For climate impact analysis, and any conclusions regarding the effectiveness of adaptation measures, one needs to be cognizant of the fact that, again, in the short-term, natural variability may dominate and needs to be considered. Mistaking short-term variability for a long-term climate trend can lead to maladaptation (e.g., Schipper, 2020). Many early (circa IPCC AR1 to AR4) studies of climate impacts focused on end-of-century climate projections because the signal of climate change was more evident. However, adaptation policy- and decision-making generally are focused on what should be done now or in the next few years, recognizing that the climate is changing but also that both short-term and long-term projections are quite uncertain. Unless the decision in question is with regard to very long-lived investments, scenario uncertainty may have little effect on decisions of what to do now with regard to adaptation. For example, whether the world achieves net-zero emissions by 2050 or not will not have much effect on a decision about what crop to plant this season or even whether to install irrigation equipment in the next few years. Those decisions are largely affected by past emissions and the resulting concentrations, and associated climate uncertainty is driven by scientific uncertainty in Earth system response to those concentrations, with the decision complicated by the noise of natural variation.

It is also essential to understand that different sources of uncertainty (and related risks) can interact with one another. Terms such as "cascading," "propagating," and "compounding" have been used in literature to describe the interacting uncertainties and risks related to climate change. Simpson et al. (2021) present a framework for complex climate change risk assessment that accounts for different kinds of interactions between multiple risks, including aggregate risk (the accumulation of multiple risks), compound risk (the interaction of multiple risks), and cascading risk (causal relationships between multiple risks). This type of framing is further emphasized in Reed et al. (2022). For climate impact assessment, we must consider uncertainty in underlying socioeconomic drivers that leads to uncertainty in emissions, uncertainty in concentrations because of uncertainty in the carbon cycle and in the fate of other gases, uncertainty in the radiative effects of these gases, uncertainties in climate feedbacks, natural variability, and finally uncertainty in how a sector or system responds.

If these uncertainties were simply additive, it would seem that uncertainty would grow unbounded. However, in formal

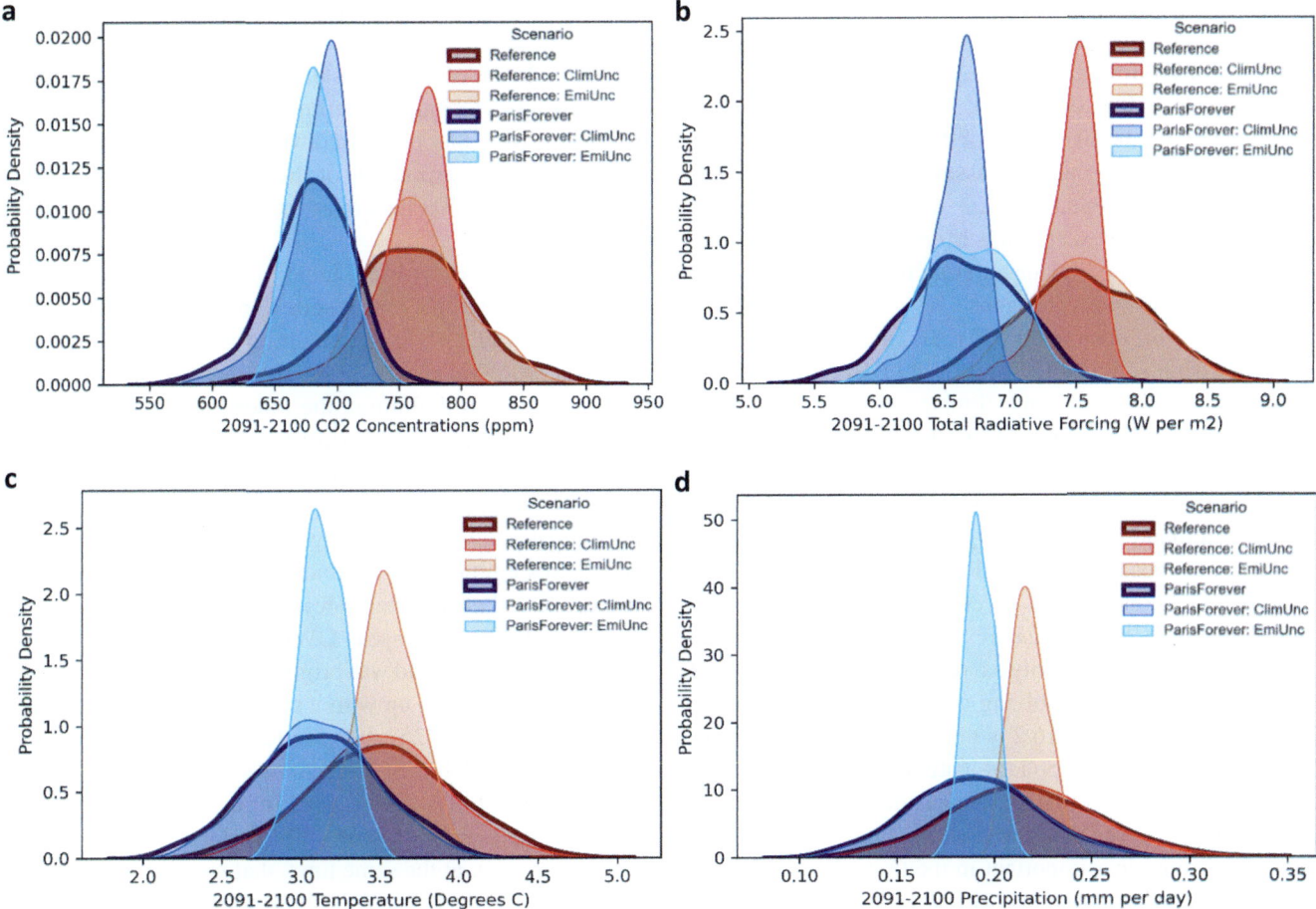

Fig. 16.2 Frequency distributions for climate outcomes in 2091–2100 relative to 1861–1880 for *Reference* and *ParisForever* ensembles with emissions uncertainty plus median climate (EmiUnc, orange for *Reference* and light blue for *ParisForever*) vs. ensembles with median emissions plus climate uncertainty (ClimUnc, red for *Reference* and blue for *ParisForever*) vs. ensembles with both emissions and climate uncertainty (dark blue for *Reference* and dark red for *ParisForever*, bold lines). (**a**) CO_2 concentrations, (**b**) total radiative forcing, (**c**) temperature, and (**d**) precipitation. From Morris et al. (2025)

uncertainty analyses, where uncertain values of different parameters are chosen randomly, uncertainty ranges do not grow in an unbounded fashion, as many of the sampled uncertainties end up offsetting one another. This is illustrated in results from Morris et al. (2025) in the series of panels in Fig. 16.2. This analysis produced Monte Carlo simulations where both climate response and socioeconomic drivers of emissions were uncertain, comparing them to results when only one or the other type of uncertainty was considered, and doing so under different climate policy assumptions. Focusing on panel (c) showing uncertainty in temperature projections, the combined effect of uncertainty in climate response and emissions is almost identical to climate response alone. Adding emissions uncertainty barely changes the overall uncertainty. However, there is still considerable temperature uncertainty resulting from emissions scenario uncertainty alone; hence, the uncertainties are less than "additive."

These results appear to contradict the results of Fig. 16.1 because even the end-of-century results appear to show climate response uncertainty as dominating the temperature uncertainty in either of the policy scenarios. However, one reason for this is the often "conditional" nature of uncertainty projections, here conditional on a policy assumption—whether the near-term Paris targets are met or not. While results in Fig. 16.1 were based on sets of emissions scenarios that included scenarios that at least implicitly had different policy outcomes and so were more unconditional in nature, the distributions in Fig. 16.2 are explicitly conditional on policy goals. When making investments in adaptation measures, investments today are not conditional[1] and so it is necessary to make a judgment about how likely it is that a particular pol-

---

[1] Although one may be able to space out in time the full investment to get better information on the total investment needed as the problem evolves, e.g., a dike could be built today, with additions added in the future if needed, though it is likely necessary to build it initially in such way that it can support additions.

icy goal will be met or not. The climate distribution of interest for impact analysis would be a combination of the two policy scenario distributions in Fig. 16.2 (and others representing other possible policy outcomes), weighting them by some judgement about the likelihood of countries adhering to any particular mitigation goal. Such assessments of the likelihood of different policy outcomes are uncommon in scientific literature. A notable exception is Morris et al. (2022), which formally represents policy uncertainty in an analysis of near-term energy investment decisions.

Another point illustrated by Fig. 16.2, and relevant to our later discussion of the IPCC approach to scenario development, is how emissions and climate response uncertainty jointly affect concentration uncertainty as shown in panel (a). Even for a certain emissions scenario, there is considerable uncertainty in concentrations because of uncertainty in ocean and land uptake of CO_2 as well as natural sources of other gases because of their dependence on climate. And perhaps an obvious point—if emissions are more tightly controlled (as in the "ParisForever" policy scenario), then climate response uncertainty is relatively more important. Finally, while the time evolving contribution of uncertainty in the study behind Fig. 16.2 is not shown here, the general pattern of Fig. 16.1 still applies—natural variability dominating in the near term, climate response uncertainty significant and growing over the first few decades, and emissions scenario uncertainty less important in the first few decades and becoming more important over time.

Given the importance of emissions, it is necessary to understand which emissions matter, where they come from, what factors influence their trajectory, how we model their uncertainty, and how we get from emissions to concentrations to radiative forcing to climate change and impacts.

16.3 Emissions, Sources, and Drivers

The major greenhouse gases (GHGs) emitted by human activities include carbon dioxide (CO_2), methane (CH_4), nitrous oxide (N_2O), and three fluorinated gases (or F-gases): hydrofluorocarbons (HFCs), perfluorocarbons (PFCs), and sulfur hexafluoride (SF6). There are also a variety of other anthropogenic emissions/changes that affect the climate directly and indirectly. The major GHGs have relatively long atmospheric lifetimes and thus are well-mixed in the atmosphere. That said, the radiative forcing (direct heat-trapping) effect and atmospheric "lifetimes" vary greatly among these gases. The non-CO_2 gases are gradually oxidized through various processes, and so the term lifetime or half-life is appropriate. Emissions of CO_2 are partitioned among the atmosphere, ocean, and terrestrial vegetation (and a small amount into rocks through mineralization), and so it is more appropriate to speak of the fate of emitted CO_2—how much ends up in each of these compartments. Since the lifetimes/fate and radiative forcing of these gases will vary over time and depend on levels of other substances in the atmosphere, there is not a single unchanging lifetime or radiative forcing effect. Nevertheless, it has proved convenient to approximate the climate effects of emissions of the major, well-mixed gases through an index that integrates their radiative forcing effect over time.

The most widely used index is the 100-year global warming potential (GWP) reported in the first IPCC report (Houghton et al., 1990), originally proposed by Lashof and Ahuja (1990). Values are regularly updated in subsequent IPCC reports. By definition, the index value of CO_2 is 1.0, and all other gases are valued relative to CO_2. In general, the other gases are much more effective at trapping heat and/or last longer in the atmosphere, and so their values are much greater than 1.0, meaning that even though the tons of these gases emitted to the atmosphere are small relative to tons of CO_2, their integrated warming effect is much larger. While the IPCC's 100-year GWP is widely used, it is not without controversy and was originally offered to indicate the difficulties in coming up with a simple index (Shine, 2009). The IPCC calculates 25-, 100-, and 500-year GWPs (number of years over which they integrate warming). Those focused on short-term effects have argued for the 25-year GWP as it increases the weight of short-lived methane (e.g., Howarth et al., 2012), but, at the same time, that would further reduce the incentive to decrease emissions of some of the gases that are nearly permanent, lasting 1000's of years. Others have argued the focus should be on comparing the temperature effect over time, not stopping at radiative forcing (Shine et al., 2005). Still others have argued that ultimately the index should be based on damages (Manne & Richels, 2001; Reilly & Richards, 1993). Ideally, atmospheric models take as input the individual gases and substances and include radiative forcing codes for each substance so the GWP calculation is irrelevant for that purpose. However, GWPs come into play in emissions control efforts, where such policies allow flexibility of which gas to abate, the trade-off among them depending on which index is used. Despite the purely physical science nature of the GWP calculation, Sarofim and Giordano (2018) suggest accounting of temperature effects through to damages and including economic discounting roughly supports use of a 100-year GWP given discount rates typically used in climate damage assessment.

CO_2 is the most common of the long-lived, well-mixed GHGs, comprising about 75% of global emissions (IPCC, 2014). Its primary source is fossil fuel use (i.e., the combustion of coal, oil and natural gas for electricity, industry, transportation, buildings, etc.). CO_2 is also emitted through some industrial processes, such as making cement. Another important source, or sink, of CO_2 is land use change. Trees, plants, and soil store carbon. When land with a lot of carbon storage (such as forest) is converted to another land use that stores less carbon (such as crop land or pasture), the reduced store of carbon in organic matter is released into the atmosphere through decomposition as CO_2. Conversely, converting land to uses with greater carbon storage (such as reforestation) removes CO_2 from the atmosphere as it is accumulated in biomass and organic matter, resulting in a carbon sink (i.e., negative CO_2 emissions).

The main anthropogenic source of CH_4 (16% of global emissions) is agriculture, particularly animal waste and rice cultivation. Waste management, energy use (particularly natural gas), and biomass burning also contribute to CH4 emissions. There are also natural sources of CH_4, including CH_4 released from melting permafrost. The main anthropogenic source of N_2O (6% of global emissions) is also agriculture, particularly fertilizer use. Fossil fuel combustion also generates N_2O. F-gases (2% of global emissions) are from refrigeration, industrial processes, a variety of consumer products, and the transmission and distribution of electricity.

In terms of economic sectors, electricity and heat production is the largest source of global GHGs (25%), followed by agriculture, forestry and other land use (24%), industry (21%), transportation (14%), other energy (e.g., the extraction, refining, processing and transportation of fuels) (10%), and buildings (6%) (IPCC, 2014). In terms of countries/regions, the top contributors to global emissions are China (24%), the United States (13%), the European Union (EU-28) (9%), India (6%), Russia (5%), Japan (3%), and Brazil (2.5%) (International Energy Agency (IEA), 2021). Together, they are responsible for almost two-thirds of global GHGs.

Beyond the long-lived and well-mixed GHGs, short-lived local air pollutants and other anthropogenic activities are also important climate forcers. These include carbon monoxide (CO), volatile organic compounds (VOCs), nitrous oxide (NOx), sulfur dioxide (SO_2), ammonia (NH_3), black carbon (BC), and organic carbon (OC). Sulfate aerosols (SO_2, BC), most of which are released from the burning of carbon-based fuels, actually have a negative forcing or cooling effect as they reflect incoming solar radiation. Black carbon and organic carbon absorb heat in the lower atmosphere and thus can affect the distribution of heat in the troposphere. Tropospheric ozone is also an important greenhouse gas and human activity contributes to precursor emissions. Jet contrails have also been identified as a source of warming. In addition, changes in the land surface that alter albedo or hydrology/evaporation can contribute to climate change.

For the most part, GWPs are not calculated for these substances as there are various problems that arise. A GWP for sulfate aerosols would suggest increasing these would be a good thing, offsetting positive radiative forcing, but that would ignore the serious health consequences of these aerosols. In addition, these substances often form in the atmosphere, through complex and nonlinear processes, so it is not straightforward to calculate how an emission of, for example, an ozone precursor becomes ozone and then has a radiative effect. As regularly reviewed, various studies have attempted to examine how long-term changes in these substances have affected climate (e.g., IPCC, 2023).

Future emissions depend on several factors. Key drivers include economic growth and development, population growth, the cost and availability of low-carbon technologies, resource availability, land use changes, and policy. These are all uncertain and vary by region and sector. Emissions and energy futures in Africa and other developing/emerging economies are particularly critical to global outcomes.

Policy is a critical driver of emissions, yet future policy is deeply uncertain. Carbon pricing (via carbon tax or cap-and-trade system), regulations (e.g., renewable portfolio standards, fuel economy standards, bans on new coal generation, air quality standards, energy efficiency standards), government expenditures (e.g., research and development, tax credits, loan guarantees), trade policy (e.g., tariffs on embedded carbon in traded goods), and land use policy can all lead to emissions reductions. However, the extent to which these will be employed by governments in the future is highly uncertain.

Ultimately, it is the concentrations of emissions in the atmosphere and their radiative forcing that matter for climate change. Concentrations depend not only on emissions but also on the rate of carbon uptake by the ocean and terrestrial ecosystems, which is also uncertain. Total radiative forcing, which is the sum of the effects of all long-lived greenhouse gases plus tropospheric ozone and aerosols, is driven by concentrations as well as the strength of sulfate aerosol forcing, which is uncertain. How total radiative forcing translates to global temperature change depends on climate sensitivity (how responsive the Earth system is to forcings), which is also uncertain. How temperature and other climatic changes (e.g., precipitation) translate into impacts (e.g. sea-level rise, land/crop productivity changes, labor productivity impacts, etc.), particularly at regional and local levels is further uncertain, and the implications of those impacts depend in turn on adaptive responses.

In the context of uncertainty about future emissions and their drivers, scenarios are often employed to explore plausible pathways of how economies, population, energy, technology, resource availability, land use changes, emissions, concentrations, radiative forcing, and/or global temperature may change over time. Some scenarios are designed to explore how emissions might evolve in the absence of climate policy, while others are expressed in terms specific targets, such as limiting the increase in global surface air temperature relative to preindustrial levels to below a certain level (e.g., 2 °C or 1.5 °C) or stabilizing concentrations at certain levels or achieve net-zero emissions by a certain year.

16.4 Emissions and Concentration Scenarios

Community efforts to support the Intergovernmental Panel on Climate Change (IPCC) have employed standard sets of scenarios to represent a range of plausible future emissions or concentration pathways to be used as inputs into climate models to explore a range of future climate projections. Climate impact researchers can then investigate impacts across the range of climate scenarios. The range of climate projections is further broadened by simulating climate pathways with varying climate models or by altering parameters of a climate model to represent Earth system response.

The approach to scenario development, as first developed under the IPCC umbrella, utilized storylines around various drivers of emissions growth (e.g., economic and population growth, technology and resource availability, amount of cooperation among countries, concern for the environment, etc.) with the goal of creating a large spread in emissions. This approach of using narratives or storylines is often referred to as the Shell Scenario approach, which dates to the 1960s (Postma & Liebl, 2005), and is motivated by the goal of getting decision-makers to imagine quite different ways in which the world could develop and not get locked into simple extrapolations of current trends or focusing on just a single "best estimate" of the future. In that sense, the approach recognizes that the future is uncertain. However, IPCC reports explicitly cautioned against attaching likelihoods to the scenarios, identifying them as all "equally plausible" (e.g., Nakicenovic, 2000).

Over time, there has been a shift in the focus of scenarios developed in support of the IPCC. Initial sets of scenarios focused on projections of greenhouse gas and air pollution emissions. Following sets of scenarios identified representative concentration pathways (RCPs), with the latest scenarios focusing again on emissions targets (e.g., net-zero targets).

Widely used IPCC scenarios are summarized in Table 16.1, and many are shown in Fig. 16.3.

The sets of scenarios used in the first three IPCC assessment reports and the Special Report on Emissions Scenarios (SA90, IS92, and SRES) are emission-based scenarios (Bretherton et al., 1990; Leggett et al., 1992; Nakicenovic, 2000). They utilize increasingly complex storylines—while the SA90 scenarios are based on world population projections, SRES scenarios lay out assumptions about demographics, economic characteristics, international trade, technology, and other factors. The assumptions driving the storylines were run through human system and integrated assessment models (IAMs) to project emissions. Many IAMs were used for SRES, resulting in multiple emissions scenarios for each storyline. For each of six storylines, one emissions scenario was chosen to be the "marker" scenario to be used as inputs to climate models.

The IPCC Fifth Assessment Report (AR5; IPCC, 2014) focused on Representative Concentration Pathways (RCPs, Moss, 2010; van Vuuren, 2011). Unlike the earlier sets of standard scenarios, modeling teams generated multi-gas concentration levels in 2100 consistent with four radiative forcing targets (RCP2.6 RCP4.5, RCP6.0, and RCP8.5, with the numbers indicating the watts per meter squared forcing increase relative to preindustrial levels). Rather than starting with detailed socioeconomic storylines to generate emissions and climate scenarios, the RCPs instead start with radiative forcing to pass to climate models and were not initially associated with any particular socioeconomic or emissions scenarios. They were designed to produce different end-of-century climate results, spanning the range of outcomes from emissions and concentrations projections in the literature at the time. As such, IAMs needed to work backwards to derive a range of emissions trajectories (and corresponding assumptions about e.g. policies and technology strategies) for

Table 16.1 Summary of widely used IPCC scenarios

Name of scenario set	Number of main scenarios	Individual scenario names	Year Introduced	IPCC report using scenarios	Scenarios defined by
SA90	4	Scenario A–D	1990	First assessment report (FAR)	Emissions
IS92	6	IS92a-f	1992	Second and third assessment reports (SAR and TAR)	Emissions
SRES	6	A1B, A1T, A1FI, A2, B1, B2	2000	Special report on emissions scenarios; third and fourth assessment reports (TAR and AR4)	Emissions
RCP (representative concentration pathways)	4	RCP2.6, RCP4.5, RCP6.0, RCP8.5	2010	Fifth assessment report (AR5)	Concentrations
SSP-RCP (shared socioeconomic pathways—representative concentration pathways)	7	SSP1–1.9, SSP1–2.6, SSP4–3.4, SSP2–4.5, SSP4–6.0, SSP3–7.0, SSP5–8.5	2016	Sixth assessment report (AR6)	Concentrations
1.5 °C (net-zero emissions)	4	P1, P2, P3, P4	2018	Special report on global warming of 1.5 °C	Emissions

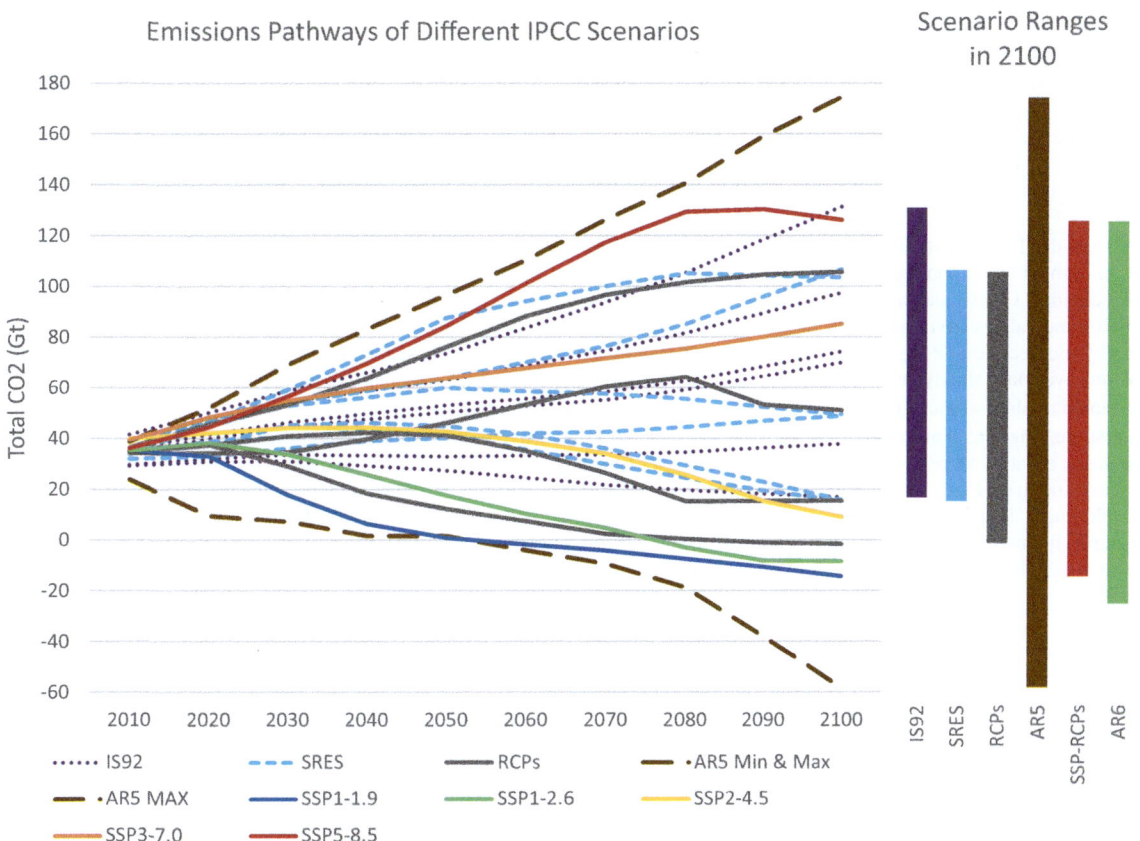

Fig. 16.3 Emissions pathways and ranges in 2100 of different IPCC scenarios. Based on data from publicly available databases: IS92 (https://sedac.ciesin.columbia.edu/ddc/is92/index.html); SRES (http://sres.ciesin.org/final_data.html); RCPs (https://tntcat.iiasa.ac.at/RcpDb); AR5 (https://tntcat.iiasa.ac.at/AR5DB); SSP-RCPs (https://tntcat.iiasa.ac.at/SspDb); AR6 (https://data.ene.iiasa.ac.at/ar6)

each RCP that would achieve concentrations that generated the specified level of radiative forcing. For each RCP, one of the emissions pathways (in terms of GHGs, aerosols, air pollutants, and other short-lived species) was selected as the main scenario to be used in climate model simulations.

Another difference between RCPs and previous scenarios is that while none of the SRES scenarios included explicit policies and measures to limit climate forcing, the three lower RCP scenarios (2.6, 4.5, and 6.0) are all climate policy scenarios. RCP8.5 is the only non-mitigation scenario considered. It was intended to be a "very high baseline emission scenario" representing the 90th percentile of no-policy baseline scenarios available at the time (van Vuuren, 2011). Many authors have used the RCP8.5 scenario as a reference "no-policy" baseline against which climate action is measured. However, given economic growth trends, falling costs of low-carbon energy options, and government interventions worldwide (e.g. to expand renewables and reduce emissions), many now believe the RCP8.5 "no-policy" scenario to be highly unlikely (Hausfather & Peters, 2020).

Independently of the RCPs, a broad range of socioeconomic scenarios were developed—the Shared Socioeconomic Pathways (SSPs, O'Neill, 2014; Riahi, 2017). The SSPs are based on five narratives describing how socioeconomic trends (e.g., population, economic development, technology deployment, education, globalization, etc.) might evolve in the absence of climate policies (though they do consider non-climate policies, such as those related to development, technology, environmental protection, etc.). They focus on challenges to mitigation and adaptation, and are intended to span the range of plausible socioeconomic futures. The narratives for the SSPs are described in Table 16.2. As no-(climate) policy baselines, the SSPs lead to different emissions and temperature outcomes, with 2100 warming ranging from 3.1 °C to 5.1 °C above preindustrial levels.

SSPs can also be combined with mitigation targets consistent with the RCPs to explore how different levels of climate change mitigation could be achieved given different socioeconomic assumptions (i.e., baselines). The mitigation challenge varies significantly depending on the SSP baseline assumed, with some SSPs being incompatible with RCPs that limit warming to 2 °C or 1.5 °C above preindustrial levels. For example, within the SSP database, of the models that ran SSP3, none could successfully achieve RCP2.6 and models had difficulty achieving RCP1.9 for all but SSP1 (Rogelj, 2018).

Table 16.2 Summary of SSP narratives (from Riahi, 2017)

SSP1	Sustainability—taking the green road (low challenges to mitigation and adaptation)
	The world shifts gradually, but pervasively, toward a more sustainable path, emphasizing more inclusive development that respects perceived environmental boundaries. Management of the global commons slowly improves, educational and health investments accelerate the demographic transition, and the emphasis on economic growth shifts toward a broader emphasis on human well-being. Driven by an increasing commitment to achieving development goals, inequality is reduced both across and within countries. Consumption is oriented toward low material growth and lower resource and energy intensity
SSP2	Middle of the road (medium challenges to mitigation and adaptation)
	The world follows a path in which social, economic, and technological trends do not shift markedly from historical patterns. Development and income growth proceed unevenly, with some countries making relatively good progress while others fall short of expectations. Global and national institutions work toward but make slow progress in achieving sustainable development goals. Environmental systems experience degradation, although there are some improvements, and overall, the intensity of resource and energy use declines. Global population growth is moderate and levels off in the second half of the century. Income inequality persists or improves only slowly and challenges to reducing vulnerability to societal and environmental changes remain
SSP3	Regional rivalry—a rocky road (high challenges to mitigation and adaptation)
	A resurgent nationalism, concerns about competitiveness and security, and regional conflicts push countries to increasingly focus on domestic or, at most, regional issues. Policies shift over time to become increasingly oriented toward national and regional security issues. Countries focus on achieving energy and food security goals within their own regions at the expense of broader-based development. Investments in education and technological development decline. Economic development is slow, consumption is material-intensive, and inequalities persist or worsen over time. Population growth is low in industrialized and high in developing countries. A low international priority for addressing environmental concerns leads to strong environmental degradation in some regions
SSP4	Inequality—a road divided (low challenges to mitigation, high challenges to adaptation)
	Highly unequal investments in human capital, combined with increasing disparities in economic opportunity and political power, lead to increasing inequalities and stratification both across and within countries. Over time, a gap widens between an internationally connected society that contributes to knowledge- and capital-intensive sectors of the global economy and a fragmented collection of lower-income, poorly educated societies that work in a labor intensive, low-tech economy. Social cohesion degrades and conflict and unrest become increasingly common. Technology development is high in the high-tech economy and sectors. The globally connected energy sector diversifies, with investments in both carbon-intensive fuels like coal and unconventional oil but also low-carbon energy sources. Environmental policies focus on local issues around middle and high-income areas
SSP5	Fossil-fueled development—taking the highway (high challenges to mitigation, low challenges to adaptation)
	This world places increasing faith in competitive markets, innovation, and participatory societies to produce rapid technological progress and development of human capital as the path to sustainable development. Global markets are increasingly integrated. There are also strong investments in health, education, and institutions to enhance human and social capital. At the same time, the push for economic and social development is coupled with the exploitation of abundant fossil fuel resources and the adoption of resource and energy intensive lifestyles around the world. All these factors lead to rapid growth of the global economy, while the global population peaks and declines in the twenty-first century. Local environmental problems like air pollution are successfully managed. There is faith in the ability to effectively manage social and ecological systems, including by geo-engineering if necessary

The IPCC Sixth Assessment Report (AR6; IPCC, 2023) featured SSP-RCP combinations, which incorporate two additional RCPs—RCP1.9 and RCP7. The report highlights seven combinations: SSP1–1.9, SSP1–2.6, SSP4–3.4, SSP2–4.5, SSP4–6.0, SSP3–7.0, and SSP5–8.5. For these scenarios, the total radiative forcing level from the RCPs is the target defining the scenarios, and emissions pathways are developed by applying mitigation policy as needed given the assumed socioeconomic baseline assumptions dictated by the SSPs such that the resultant emissions, when run through a climate model, achieve the target radiative forcing.

AR6 Working Group II also introduced a new categorization of scenarios, defined by their likelihood of exceeding global warming levels (at peak and in 2100) (IPCC, 2023):

- Category C1 comprises modelled scenarios that limit warming to 1.5 °C in 2100 with a likelihood of greater than 50% and reach or exceed warming of 1.5 °C during the twenty-first century with a likelihood of 67% or less. It is referred to as 1.5 °C (>50%) with no or limited overshoot. Limited overshoot refers to exceeding 1.5 °C global warming by up to about 0.1 °C and for up to several decades.
- Category C2 comprises modelled scenarios that limit warming to 1.5 °C in 2100 with a likelihood of greater than 50%, and exceed warming of 1.5 °C during the twenty-first century with a likelihood of greater than 67%. It is referred to as 1.5 °C (>50%) after a high overshoot. High overshoot refers to temporarily exceeding 1.5 °C global warming by 0.1–0.3 °C for up to several decades.
- Category C3 comprises modelled scenarios that limit peak warming to 2 °C throughout the twenty-first century with a likelihood of greater than 67%. It is referred to as 2 °C (>67%).
- Categories C4, C5, C6, and C7 comprise modelled scenarios that limit warming to 2 °C, 2.5 °C, 3 °C, and 4 °C, respectively, throughout the twenty-first century with a likelihood of greater than 50%. In some scenarios in C4 and many scenarios in C5–C7, warming continues beyond the twenty-first century.

AR6 pulls from a database of almost two thousand global scenarios produced in different ways by a wide range of models and, for those that project climate outcomes, assigns each to the C1-C8 categories. It also introduced a set of illustrative mitigation pathways (IMPs) and two pathways illustrative of high emissions. These were selected from the larger set of scenarios to illustrate a range of different mitigation strategies that would be consistent with different warming levels. The five IMPs illustrate pathways that achieve deep and rapid emissions reductions through different types of mitigation strategies—one focusing on renewables, one on carbon dioxide removal and net negative emissions, one on low demand for resources, one on less rapid mitigation initially followed by gradual strengthening, and one on sustainable development and reducing inequality.

The IPCC Special Report on Warming of 1.5 °C (SR1.5) focused on scenarios that could achieve climate stabilization at 1.5 °C, identifying a range of emissions pathways that could do so (IPCC, 2018). SR1.5 also emphasized the importance of achieving net-zero emissions at some point, ideally by mid-century, in order to achieve the 1.5 °C temperature stabilization goal. This report strongly influenced research agendas, beginning a heavy focus on net-zero emission (NZE) scenarios in the scientific community (e.g., Morris et al., 2023; International Energy Agency (IEA), 2021; Azevedo et al., 2021). This marks a shift back toward emissions-focused scenarios.

Net-zero emissions means that released emissions need to equal negative emissions (emissions removed from the atmosphere), and for atmospheric stabilization that would need to include all greenhouse gases and all sources. However, with regard to policy pledges and scenarios explored, there remain important questions about what exactly "net zero" means. First, to which emissions does the net-zero target apply? Sometimes the target is applied to CO_2 only, typically meaning fossil energy and industrial process CO_2 emissions. Sometimes it is applied to all greenhouse gases and all sources. CO_2 emissions from land use and land cover changes (LULCC) may or may not be included, or a net-zero goal may allow credits from verified reduction in land use emissions or increases in sinks. Second, what about the pathway to net zero? A net-zero goal by a prespecified date, such as 2050, does not identify the path by which it will be achieved—different paths can lead to the same end goal, but with quite different cumulative emissions, which ultimately are what is important in determining the climate outcome. Third, for any given pathway to net zero, there is also a question of how much positive and negative emissions should be involved. The net-zero target does not determine the gross emissions that remain (e.g., from fossil fuel combustion). A net-zero target can, in principle, be achieved with deep reductions in emissions and small amounts of negative emissions, or quite large, continued emissions and reliance on fossil fuel combustion (without carbon capture and storage) with large offsetting amounts of negative emissions. These strategies, and those in between, can result in the same net emissions path and thus the same or similar climate implications but quite different implications for technologies, cost of achieving the target, and other areas such as health (e.g., air pollutant emissions associated with fossil fuel combustion) and land use change (e.g., for afforestation or biomass for negative emissions).

Although NZE by 2050 has become an ubiquitous target and used in many scenarios, the necessary timing for net-zero emissions in order to achieve a given temperature target is highly uncertain. SR1.5C included 90 individual scenarios with a 50% chance of 1.5 °C in 2100 (IPCC, 2018). Only 18 of those (20%) have net-zero CO_2 emissions (energy sector and industrial process CO_2) in 2050. In AR6, while all of the scenarios that limit warming to 1.5 °C with at least a 50% chance and no or limited overshoot reach net-zero CO_2 emissions, the timing ranges from 2035 to 2070 and only half of the pathways reach net-zero GHG emissions at any point during the second half of the twenty-first century (IPCC, 2023).

Related to net-zero emissions, SR1.5 also emphasized a carbon budget approach—that achieving a specific temperature stabilization target, such as 1.5 °C, at a given probability requires limiting the total cumulative global anthropogenic emissions of CO_2 since the preindustrial period to a certain amount. In turn, efforts have been made to estimate the remaining carbon budget—how much CO_2 can be released from today to the time CO_2 emissions reach net zero (e.g., Mengis et al., 2018; Millar, 2017; Rogelj et al., 2016, 2019; Tokarska et al., 2018). However, these estimates are highly uncertain and wide ranges appear in scientific literature. For example, values for limiting warming at 1.5 °C with 50% probability, expressed as cumulative carbon emissions from the beginning of 2018, vary from slightly negative to 900 $GtCO_2$ (Rogelj et al., 2019). Reasons for the differences include the use of different definitions of carbon budget, such as threshold exceedance budget (TEB) versus threshold avoidance budget (TAB) (Rogelj et al., 2016) and different assumptions about non-CO_2 emissions/forcing (e.g., Mengis et al., 2018; Millar, 2017; Tokarska et al., 2018), particularly aerosol forcing (e.g., Samset et al., 2018) and the role of methane (Rogelj et al., 2015) and other non-CO_2 GHGs. Carbon budget estimates also depend on how the probability of temperature staying below a chosen level is calculated and the model used, including assumptions about climate sensitivity, ocean heat uptake, and the strength of aerosol forcing, which determine the transient climate response to cumulative carbon emissions (TCRE). Accounting for all the uncertainties, the range of carbon budgets that could be considered consistent with a given temperature target is large.

Throughout much of the history of the IPCC scenario process, the approach has been to target an endpoint—e.g., emissions or total radiative forcing in 2100, or net-zero emissions by a given date. Storylines were developed around different pathways to motivate them. Multiple modeling teams with a variety of models independently altered parameters of their models in an attempt to hit or approximately hit the target level of emissions or concentrations. In principle, this approach illustrated that there were different underlying assumptions about basic drivers of emissions that could lead to similar paths. One might contrast this approach with a more conventional scenario approach, such as the Energy Information Administration's long-term energy projections (US EIA, 2023), where analysts form judgments about basic parameters of the model (e.g., population growth, labor productivity/GDP growth, technology cost/availability, etc.) and then run the model to determine what level of emissions result.

One needs to see the IPCC approach as responding to the broader intergovernmental policy process. Early on, there was no governmental consensus on policy goals or targets, and the scenario approach avoided suggesting specific climate policies or targets. As such, early IPCC scenarios did not include explicit climate policies. The storylines developed for some of the emissions paths included attention or concern for the environment and sustainability but without a specific set of policy instruments through which such concerns would be addressed. However, implicitly, to address such concerns would likely mean some type of concerted policy direction, and observers generally interpret the lower emissions scenarios as likely only if there were policies, for example, those targeted at reducing fossil fuel use.

More generally, early IPCC efforts eschewed any effort to create consensus projections, which likely would have been nearly impossible or very controversial given the state of international politics on climate change at the time. As such, the process was carefully designed to avoid creating a central scenario that could be interpreted as "most likely." Instead, recognizing that climate models and impact assessors would want to contrast scenarios with more or less climate change, the process was designed to ensure a wide range of emissions/concentrations pathways, which were explicitly not assigned likelihoods. By targeting a limited set of emission or radiative forcing endpoints, there were then a manageable number of scenarios that could be simulated with computationally intensive climate models.

As the intergovernmental policy process identified agreed-upon policy goals such as staying below 2 or 1.5 °C from preindustrial levels or getting to net-zero emissions, the IPCC scenario approach added scenarios with specific policies and targets that achieved these goals. However, avoidance of assigning likelihoods to scenarios has continued throughout IPCC assessments. Despite changes in the focus of scenarios over time, the basic scenario range did not change much until the most recent rounds when new, much lower emissions/radiative forcing scenarios were added that included specific climate policy goals (see Fig. 16.3).

There have been a variety of critiques of the IPCC process. Many question whether the few storylines are enough or if they overly limit possible outcomes. Others have suggested that the focus on "consistent" storylines of the future narrow thinking too much and, as a result, fail to consider really surprising developments such as radical new technologies (Postma & Liebl, 2005). Until recently, the lack of explicit policy in scenarios and even the refusal to suggest that to achieve relatively low emissions scenarios would require explicit policy to reduce emissions had been frustrating to many. Many have also been frustrated by the refusal to identify likelihoods to scenarios. As an example consequence, some analysts have argued the RCP 8.5 scenario is now highly unlikely but is still used as a "reference" case for much climate impacts work. Relatedly, studies or policy decision-making processes may want to consider "low probability, high consequence" outcomes, but the IPCC does not provide insight into which scenarios would produce those outcomes. For this reason, we turn to methods that attempt to explicitly quantify the likelihood of specific emissions/climate outcomes.

16.5 Uncertainty Quantification

While scenarios have long been used to explore plausible future emissions, a scenario approach can only explore a limited number of combinations of assumptions and provides no quantitative interpretation of likelihood. Further, the space explored through scenarios is ultimately a judgment, and the limited set of scenarios considered may have severe biases or miss important areas of the uncertainty space. In contrast, uncertainty quantification via a probabilistic Monte Carlo approach can more fully and systematically explore the uncertainty space while also putting error bars on projections. Monte Carlo analysis involves creating distributions for underlying input parameters, sampling from them, and then simulating the model hundreds or thousands of times to generate probability distributions of outcomes.

Morris et al. (2025) and Morris et al. (2022) take a probabilistic ensemble approach to representing a comprehensive set of both socioeconomic and climate uncertainties. That work advances an approach used by Webster (2012) and Sokolov (2009), employing an updated and improved version of the MIT Integrated Global System Model (IGSM) and a significant reassessment of uncertainty in input parameters. The result is a consistent framework for uncertainty quantification in coupled human-Earth system models, which supports a broad exploration of global-change uncertainty. The

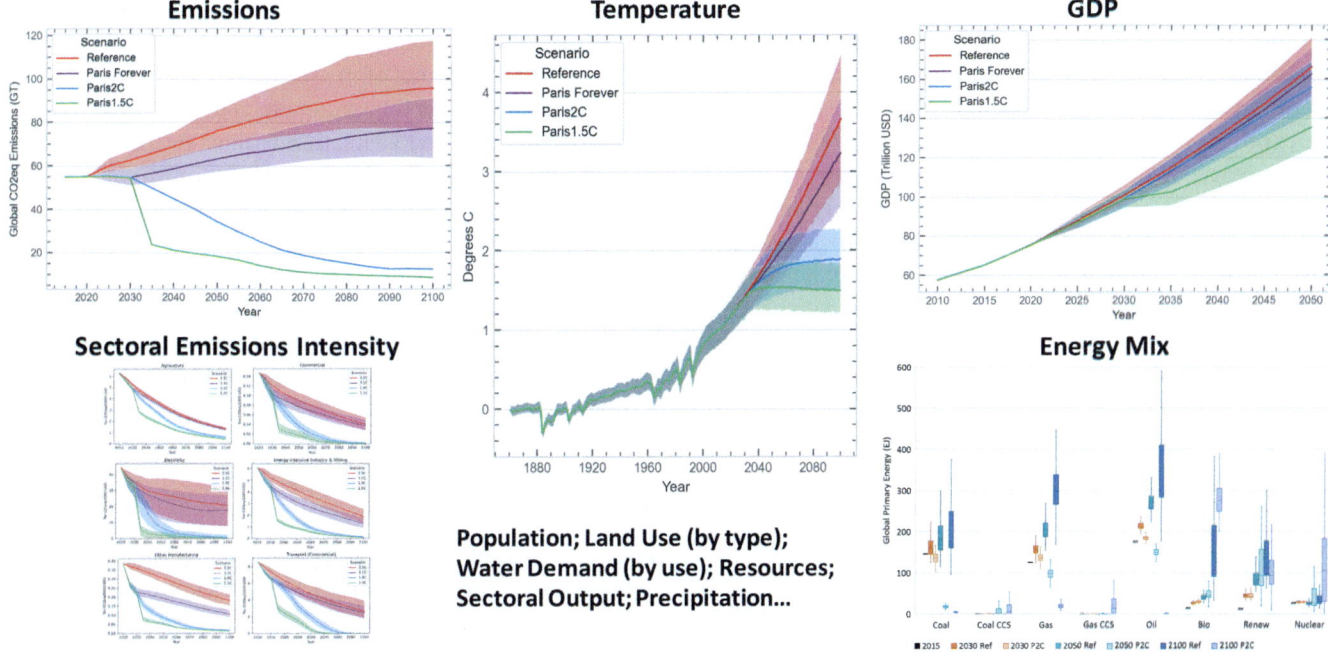

Fig. 16.4 Example probabilistic outcomes from a Monte Carlo uncertainty quantification approach applied to a coupled human-Earth system model. Adapted from Morris et al. (2022)

resulting integrated, probabilistic socioeconomic and climate projections provide insight into the probability of outcomes of interest, including emissions, CO_2 concentrations, temperature, precipitation, gross domestic product (GDP), and energy use, among many others (see Fig. 16.4). These probability distributions of human and Earth system outcomes can serve as a basis for risk-based decision-making.

While there are growing calls for more formal probabilistic, risk-based approaches to inform discussions about mitigation and adaptation research (Gillingham et al., 2018; Hausfather & Peters, 2020; Rose & Scott, 2018), there have also been various critiques of probabilistic approaches to uncertainty quantification (e.g. Katzav et al., 2021; Morgan, 2018; Weyant, 2017). An alternative approach is robust decision-making (e.g., Lempert, 2019), which explores multiple representations of the future (without assigning likelihoods) and uses robustness, rather than optimality, as a decision criterion (i.e., to identify decisions that are resilient to uncertain changes in future circumstances). This often involves avoiding worst case outcomes, using formulations such as minimizing the maximum loss or, alternatively, maximizing minimum utility to guide decision-making.

Scenario discovery is a model-based approach for scenario development aimed at finding areas of interest within large, multidimensional databases of simulation model results (see, e.g., Bryant & Lempert, 2010). It involves screening databases of model simulations (through statistical, machine learning or data-mining algorithms) to identify outcomes of interest and their conditions for occurring. It can then inform the development of specific individual scenarios to explore in depth. This is in contrast to IPCC-style scenario development (e.g., the SSPs) which creates storyline narratives. A danger with the storyline approach is that one may be overconfident that only one narrow set of outcomes can be part of a consistent story. Scenario discovery approaches can avoid this error and can identify variables associated with given outcomes of interest without defining a priori which variables are most important. In this way, it can "discover" different scenario pathways to a given outcome and identify individual scenarios that are particularly relevant or interesting.

While scenario discovery is commonly paired with robust decision making and therefore does not assign probability distributions for sampling input parameters, the approaches can also be applied to probabilistic ensembles (see, e.g., Morris et al., 2022; Rozenberg et al., 2014; Guivarch et al., 2016). Similarly, while the focus of scenario discovery has mostly been to identify values of input assumptions that tend to drive an outcome of interest, it can also be used to investigate combinations of endogenous outcomes in an ensemble of projections. By combining scenario discovery techniques with a probabilistic Monte Carlo ensemble, one can explore how different endogenous outcomes are related and potential tradeoffs, if there are prevailing storylines behind outcomes of interest, and identify individual scenarios that are defined by specific combinations of outcomes.

There are also methods that explicitly factor uncertainty about the future into near-term decision-making. Many such approaches (e.g., stochastic dynamic programming, stochastic optimization) build on Bellman (1957) and attach specific likelihoods to different future outcomes and then identify a near-term decision that, for example, minimizes the sum of the expected policy costs plus any remaining expected climate damage. While formally constructing a fully quantified policy optimization problem of this type is rarely pursued in practice, one way or another, any decision process tends to attach weights to different outcomes, generally placing less weight on outcomes perceived to have almost no chance of occurring. Of course, there are possible exceptions to this, such as the fat tail hypothesis (Weitzman, 2014) that basically argues that deep uncertainty about climate change creates a finite probability of catastrophic outcomes, and the expected value of damage is essentially unbounded. As a result, any estimate of expected damage is dominated by unquantifiable small, but finite risk. Various critiques of this hypothesis exist, and the policy implication—that we must avoid climate change at all costs—is at best difficult to implement. Does it mean we must turn off all lights immediately and get rid of all methane belching ruminants and rice paddies tomorrow or proceed with all due haste phasing these out over a decade (or two or three)?

Interactive approaches where scenario developers interact with decision-makers, or where the decision-makers can actually use simplified modeling tools to help them learn about how different actions might lead to different results tend to be more successful in terms of having scenario information factor more prominently into decision-making. Examples of models that can be used by decision-makers are those developed under the Climate Interactive (2023) effort.

16.6 Summary

With 40+ years of scenario development for climate analysis, there are some key takeaways:

- Scenario uncertainty can be as important as uncertainty in the science of climate change, but its contribution to overall climate uncertainty doesn't emerge as a main driver until the second half of the century.
- Various scenario approaches have been developed to capture a wide range of future emissions and concentration outcomes.
- The IPCC scenario process has had a strong influence on the scenario development community.
- The IPCC process has eschewed formal uncertainty techniques but has looked to the broader literature to develop target or marker scenarios. That literature also includes formal uncertainty analysis.
- Scenarios are often conditioned on specific assumptions, such as specific policy goals or broader storylines that define some elements of the scenario.
- Expectations about future climate and what it means for climate impacts and adaptation require further judgment about the likelihoods of different scenarios/their various conditional assumptions, such as whether policy goals will be achieved.
- Applying newer approaches like scenario discovery with probabilistic approaches can combine the benefits of distinct scenarios with likelihoods of outcomes of interest, providing a richer foundation for mitigation and adaptation studies.

References

Azevedo, I., Bataille, C., Bistline, J., Clarke, L., & Davis, S. (2021). Net-zero emissions energy systems: What we know and do not know. *Energy and Climate Change, 2*, 100049. https://doi.org/10.1016/J.EGYCC.2021.100049

Bellman, R. E. (1957). *Dynamic programming*. Princeton University Press.

Bretherton, F. P., Bryan, K., & Woods, J. D. (1990). Time-dependent greenhouse-gas-induced climate change. In *Climate change. The IPCC scientific assessment* (pp. 173–194). IPCC.

Bryant, B. P., & Lempert, R. J. (2010). Thinking inside the box: A participatory, computer-assisted approach to scenario discovery. *Technological Forecasting and Social Change, 77*, 34–49. https://doi.org/10.1016/j.techfore.2009.08.002

Climate Interactive. (2023). *En-roads*. Retrieved March 20, 2023, from https://www.climateinteractive.org/

Gillingham, K., Nordhaus, W., Anthoff, D., Blanford, G., Bosetti, V., Christensen, P., McJeon, H., & Reilly, J. (2018). Modeling uncertainty in integrated assessment of climate change: A multimodel comparison. *Journal of the Association of Environmental and Resource Economists, 4*, 698910. https://doi.org/10.1086/698910

Guivarch, C., Rozenberg, J., & Schweizer, V. (2016). The diversity of socio-economic pathways and CO_2 emissions scenarios: Insights from the investigation of a scenarios database. *Environmental Modelling and Software, 80*, 336–353. https://doi.org/10.1016/j.envsoft.2016.03.006

Hausfather, Z., & Peters, P. (2020). Emissions—The 'business as usual' story is misleading. *Nature, 577*, 618–620. Retrieved https://www.nature.com/articles/d41586-020-00177-3

Hawkins, E., & Sutton, R. (2009). The potential to narrow uncertainty in regional climate predictions. *Bulletin of the American Meteorological Society, 90*, 1095–1107. https://doi.org/10.1175/2009BAMS2607.1

Houghton, J. T., Jenkins, G. T., & Ephraums, J. J. (Eds.). (1990). *Climate change: The IPCC scientific assessment* (p. 364). Cambridge University Press.

Howarth, R. W., Santoro, R., & Ingraffea, A. (2012). Venting and leaking of methane from shale gas development: Response to Cathles et al. *Climatic Change, 113*, 537–549. https://doi.org/10.1007/s10584-012-0401-0

International Energy Agency (IEA). (2021). *Net zero by 2050*. Retrieved September 20, 2024, from https://www.iea.org/reports/net-zero-by-2050

IPCC. (2014). Climate change 2014: Synthesis report. In R. K. Pachauri & L. A. Meyer (Eds.), *Contribution of working groups I, II and III to the fifth assessment report of the Intergovernmental Panel on Climate Change* (p. 151). IPCC.

IPCC. (2018). Global warming of 1.5°C. In V. Masson-Delmotte (Ed.), *An IPCC special report on the impacts of global warming of 1.5°C above pre-industrial levels and related global greenhouse gas emission pathways.* IPCC.

IPCC. (2023). Climate change 2023: Synthesis report. In H. Lee & J. Romero (Eds.), *Contribution of working groups I, II and III to the sixth assessment report of the Intergovernmental Panel on Climate Change* (pp. 35–115). IPCC. https://doi.org/10.59327/IPCC/AR6-9789291691647

Katzav, J., Thompson, E. L., Risbey, J., Stainforth, D. A., Bradley, S., & Mathias, F. (2021). On the appropriate and inappropriate uses of probability distributions in climate projections and some alternatives. *Climatic Change, 169*, 15. https://doi.org/10.1007/s10584-021-03267-x

Knutti, R., Rugenstein, M., & Hegerl, G. (2017). Beyond equilibrium climate sensitivity. *Nature Geoscience, 10*, 727–736. https://doi.org/10.1038/ngeo3017

Kopp, G., & Lean, J. L. (2011). A new, lower value of total solar irradiance: Evidence and climate significance. *Geophysical Research Letters, 38*, L01706. https://doi.org/10.1029/2010GL045777

Lashof, D., & Ahuja, D. (1990). Relative contributions of greenhouse gas emissions to global warming. *Nature, 344*, 529–531. https://doi.org/10.1038/344529a0

Leggett, J., Pepper, W. J., Swart, R. J., Edmonds, J., Meira Filho, L. G., Mintzer, I., Wang, M. X., & Wasson, J. (1992). Emissions scenarios for the IPCC: An update. In *Climate change* (pp. 75–95). IPCC. Retrieved September 20, 2024, from https://citeseerx.ist.psu.edu/document?repid=rep1&type=pdf&doi=40b8a5ba6902ac1be8575606ba7faef0ba209575

Lehner, F., Deser, C., Maher, N., Marotzke, J., Fischer, E. M., Brunner, L., Knutti, R., & Hawkins, E. (2020). Partitioning climate projection uncertainty with multiple large ensembles and CMIP5/6. *Earth System Dynamics, 11*, 491–508. https://doi.org/10.5194/esd-11-491-2020

Lempert, R. J. (2019). Robust decision making (RDM). In V. A. W. J. Marchau, W. E. Walker, P. J. T. M. Bloemen, & S. W. Popper (Eds.), *Decision making under deep uncertainty: From theory to practice.* Springer.

Manne, A. S., & Richels, R. G. (2001). An alternative approach to establishing trade-offs among greenhouse gases. *Nature, 410*, 675–677. https://doi.org/10.1038/35070541

Mengis, N., Partanen, A.-I., Jalbert, J., & Matthews, H. D. (2018). 1.5 °C carbon budget dependent on carbon cycle uncertainty and future non-CO_2 forcing. *Scientific Reports, 8*, 5831. https://doi.org/10.1038/s41598-018-24241-1

Millar, R. (2017). Emission budgets and pathways consistent with limiting warming to 1.5 °C. *Nature Geoscience, 10*, 741–747. https://doi.org/10.1038/ngeo3031

Morgan, M. G. (2018). Uncertainty in long-run forecasts of quantities such as per capita gross domestic product. *Proceedings of the National Academy of Sciences, 115*, 5314–5316. https://doi.org/10.1073/pnas.1805767115

Morris, J., Chen, Y.-H. H., Gurgel, A., Reilly, J., & Sokolov, A. (2023). Net zero emissions of greenhouse gases by 2050: Achievable and at what cost? *Climate Change Economics, 14*, 23400002. https://doi.org/10.1142/S201000782340002X

Morris, J., Reilly, J., Paltsev, S., Sokolov, A., & Cox, K. (2022). Representing socio-economic uncertainty in human system models. *Earth's Future, 10*, e2021EF002239. https://doi.org/10.1029/2021EF002239

Morris, J., Sokolov, A., Reilly, J., Libardoni, A., Forest, C. S., Paltsev, S., Schlosser, C. A., Prinn, R., & Jacoby, H. (2025). Quantifying both socioeconomic and climate uncertainty in coupled human-Earth systems analysis. *Nature Communications, 16*(2703). https://doi.org/10.1038/s41467-025-57897-1

Moss, R. H. (2010). The next generation of scenarios for climate change research and assessment. *Nature, 463*, 747–756. https://doi.org/10.1038/nature08823

Nakicenovic, N. (2000). Special report on emissions scenarios. In *A special report of working group III of the Intergovernmental Panel on Climate Change* (p. 599). Cambridge University Press. Retrieved https://archive.ipcc.ch/pdf/special-reports/emissions_scenarios.pdf

O'Neill, B. C. (2014). A new scenario framework for climate change research: The concept of shared socioeconomic pathways. *Climatic Change, 122*, 387–400. https://doi.org/10.1007/s10584-013-0905-2

Postma, T. J. B. M., & Liebl, F. (2005). How to improve scenario analysis as a strategic management tool? *Technological Forecasting and Social Change, 72*, 161–173. https://doi.org/10.1016/j.techfore.2003.11.005

Reed, P. M., Hadjimichael, A., Moss, R. H., Brelsford, C., Burleyson, C. D., Cohen, S., et al. (2022). Multisector dynamics: Advancing the science of complex adaptive human-Earth systems. *Earth's Future, 10*, e2021EF002621. https://doi.org/10.1029/2021EF002621

Reilly, J. M., & Richards, K. R. (1993). Climate change damage and the trace gas index issue. *Environmental and Resource Economics, 3*, 41–61. https://doi.org/10.1007/BF00338319

Riahi, K. (2017). The shared socioeconomic pathways and their energy, land use, and greenhouse gas emissions implications: An overview. *Global Environmental Change, 42*, 153–168. https://doi.org/10.1016/j.gloenvcha.2016.05.009

Rogelj, J. (2018). Scenarios towards limiting global mean temperature increase below 1.5 °C. *Nature Climate Change, 8*, 325–332. https://doi.org/10.1038/s41558-018-0091-3

Rogelj, J., Forster, P. M., Kriegler, E., Smith, C. J., & Séférian, R. (2019). Estimating and tracking the remaining carbon budget for stringent climate targets. *Nature, 571*, 335–342. https://doi.org/10.1038/s41586-019-1368-z

Rogelj, J., Meinshausen, M., Schaeffer, M., Knutti, R., & Riahi, K. (2015). Impact of short-lived non-CO_2 mitigation on carbon budgets for stabilizing global warming. *Environmental Research Letters, 10*, 075001. https://doi.org/10.1088/1748-9326/10/7/075001

Rogelj, J., Schaeffer, M., Friedlingstein, P., Gillett, N. P., van Vuuren, D. P., Riahi, K., Allen, M., & Knutti, R. (2016). Differences between carbon budget estimates unravelled. *Nature Climate Change, 6*, 245–252. https://doi.org/10.1038/nclimate2868

Rose, S., & Scott, M. (2018). *Grounding decisions: A scientific foundation for companies considering global climate scenarios and greenhouse gas goals.* EPRI. Retrieved https://www.epri.com/research/products/000000003002014510

Rozenberg, J., Guivarch, C., Lempert, R., & Hallegatte, S. (2014). Building SSPs for climate policy analysis: A scenario elicitation methodology to map the space of possible future challenges to mitigation and adaptation. *Climatic Change, 122*, 509–522. https://doi.org/10.1007/s10584-013-0904-3

Samset, B., Sand, M., Smith, C. J., Bauer, S. E., Forster, P. M., Fuglestvedt, J. S., Osprey, S., & Schleussner, C.-F. (2018). Climate impacts from a removal of anthropogenic aerosol emissions. *Geophysical Research Letters, 45*, 1020–1029. https://doi.org/10.1002/2017GL076079

Sarofim, M. C., & Giordano, M. R. (2018). A quantitative approach to evaluating the GWP timescale through implicit discount rates. *Earth System Dynamics, 9*, 1013–1024. https://doi.org/10.5194/esd-9-1013-2018

Schipper, E. L. F. (2020). Maladaptation: When adaptation to climate change goes very wrong. *One Earth, 3*, 409–414. https://doi.org/10.1016/j.oneear.2020.09.014

Shine, K. P. (2009). The global warming potential—The need for an interdisciplinary retrial. *Climatic Change, 96*, 467–472. https://doi.org/10.1007/s10584-009-9647-6

Shine, K. P., Fuglestvedt, J. S., Hailemariam, K., & Stuber, N. (2005). Alternatives to the global warming potential for comparing climate impacts of emissions of greenhouse gases. *Climatic Change, 68*, 281–302. https://doi.org/10.1007/s10584-005-1146-9

Simpson, N. P., et al. (2021). A framework for complex climate change risk assessment. *One Earth 4*(4), 489–501. https://doi.org/10.1016/j.oneear.2021.03.005

Sokolov, A. P. (2009). Probabilistic forecast for 21st century climate based on uncertainties in emissions (without policy) and climate parameters. *Journal of Climate, 22*, 5175–5204. https://doi.org/10.1175/2009JCLI2863.1

Tokarska, K., Gillett, N., Arora, V., Lee, W., & Zickfeld, K. (2018). The influence of non-CO_2 forcings on cumulative carbon emissions budgets. *Environmental Research Letters, 13*, 034039. https://doi.org/10.1088/1748-9326/aaafdd

US EIA. (2023). *Annual energy outlook 2023*. Retrieved September 20, 2024, from https://www.eia.gov/outlooks/aeo/

van Vuuren, D. P. (2011). The representative concentration pathways: An overview. *Climatic Change, 109*, 5–31. https://doi.org/10.1007/s10584-011-0148-z

Webster, M. (2012). Analysis of climate policy targets under uncertainty. *Climatic Change, 112*, 569–583. https://doi.org/10.1007/s10584-011-0260-0

Weitzman, M. L. (2014). Fat tails and the social cost of carbon. *American Economic Review, 104*, 544–546. https://doi.org/10.1257/aer.104.5.544

Weyant, J. (2017). Some contributions of integrated assessment models of global climate change. *Review of Environmental Economics and Policy, 11*, 115–137. https://doi.org/10.1093/reep/rew018

Open Access This chapter is licensed under the terms of the Creative Commons Attribution 4.0 International License (http://creativecommons.org/licenses/by/4.0/), which permits use, sharing, adaptation, distribution and reproduction in any medium or format, as long as you give appropriate credit to the original author(s) and the source, provide a link to the Creative Commons license and indicate if changes were made.

The images or other third party material in this chapter are included in the chapter's Creative Commons license, unless indicated otherwise in a credit line to the material. If material is not included in the chapter's Creative Commons license and your intended use is not permitted by statutory regulation or exceeds the permitted use, you will need to obtain permission directly from the copyright holder.

The Importance of Internal Variability for the Uncertainty in Climate Change Projections and Decision-Making

17

Flavio Lehner and Clara Deser

17.1 Introduction

17.1.1 What Is Internal Climate Variability and Where Does It Come From?

Internal variability refers to fluctuations that arise intrinsically in a nonlinear dynamical system, even when that system is closed (energy, mass, and momentum are conserved) and not subject to any changes in external forcing. In the case of Earth's climate, internal variability arises primarily from the uneven distribution of energy across the planet at any given time. Physical processes, such as oceanic or atmospheric heat transport or radiation to space, act to balance this unevenness, but they do so at different temporal and spatial scales. Together with Earth's rotation and uneven solar radiation, this means that an even energy distribution is never reached, leaving Earth's climate to shift energy around perpetually.

Internal variability affects virtually every aspect of the climate system and is manifested, for example, as variations in temperature from one day to the next but also by strengthening and weakening of ocean currents from one decade to the next. Importantly, internal variability occurs around a mean state, which is dictated by the long-term balance of the system at hand. Internal variability thus almost always occurs within a certain range (e.g., a range of temperature values) and does not drive the system unidirectionally. In Lorenz' seminal papers on chaos, a set of differential equations describes a nonlinear system with one or more attractors around which the system varies (Lorenz, 1963). In a simplified analogy, Earth's long-term average climate is such an attractor, and variations around it represent internal variability.

The Lorenz model offers insight into a key feature of internal variability: its inherent unpredictability. It was found that even in a deterministic system such as the Lorenz model, miniscule differences in initial conditions used to solve the equations can lead to very different trajectories after a while. This phenomenon exists in reality as well, where fluid mediums, such as the atmosphere and ocean, are intrinsically unstable to tiny perturbations (e.g., the baroclinically unstable atmospheric jet stream or tropical convection patterns). Because the initial conditions of a system under prediction (e.g., weather) are never known perfectly, this inevitably leads to a gradual divergence between the prediction (e.g., a weather forecast) and the actual outcome.

17.1.2 Why Care About Internal Variability?

The example of weather forecasting and initial condition predictability suggests that internal variability arises from small spatial and temporal scales and grows to encompass global and decadal scales. This leads to important issues for climate science. The historical climate record reflects not only a response to various external forcings affecting its energy balance (such as changes in orbital parameters, solar irradiance, greenhouse gas concentrations, or aerosols), but it has also embedded fluctuations that are the expression of a unique realization of internal variability. "Unique" here means that it is just one of many possible and equally plausible realizations that could have occurred. Taking this insight to the extreme has led to the analogy that the flap of a butterfly could cause a hurricane. While proving such causality is beyond our reach,

F. Lehner (✉)
Department of Earth and Atmospheric Sciences, Cornell University, Ithaca, NY, USA

Climate and Global Dynamics Laboratory, National Center for Atmospheric Research, Boulder, CO, USA

Polar Bears International – US, Bozeman, MT, USA
e-mail: flavio.lehner@cornell.edu

C. Deser
Climate and Global Dynamics Laboratory, National Center for Atmospheric Research, Boulder, CO, USA
e-mail: cdeser@ucar.edu

this concept of "contingency" (Gould, 1989) is useful when thinking about the possible consequences of even the smallest perturbations to the atmosphere and, once they grow, to the larger climate system—or, conversely, how the system would have evolved had that butterfly not flapped its wing.

This superposition of external and internal influences complicates the interpretation of the historical climate record, as our understanding of climate variability and change relies on a robust separation of these influences. Like the past, future climate will also reflect the combined influence of external forcing (e.g., an increase in greenhouse gas concentrations) and internal variability. Precise prediction of future climate for lead times longer than a few years is thus not possible. This poses a communication challenge and calls for probabilistic predictions—again very similar to weather forecasts, where a *chance* of rain rather than a binary forecast of rain or no rain is given.

This chapter will discuss the role of internal variability in climate projections using the example of air temperature, give an overview of methods used to separate and quantify external and internal drivers of climate variability, and discuss outstanding scientific challenges in this field. This chapter draws from a recent perspective paper on the origin, importance, and predictive limits of internal climate variability (Lehner & Deser, 2023).

17.2 The Role of Internal Variability in Climate Projections

17.2.1 An Example: Temperature Projections

To illustrate the influence of internal variability on climate projections, we show projections of wintertime air temperature over North America from many different simulations with the same climate model under the same radiative forcing scenario (Fig. 17.1). As with the Lorenz model, any individual simulation is started from slightly different initial conditions (here, roundoff errors in atmospheric temperatures at year 1920 of each simulation—the butterfly wing flap), while the rest of the model setup is kept identical between the different simulations (Kay et al., 2015). Specifically, each simulation is subject to the same historical external forcing, such as volcanic eruptions or changes in greenhouse gas concentrations, as well as an emissions scenario going out to year 2100. Such a setup is called a single-model initial-condition large ensemble (SMILE). The initial perturbations to each simulation are so small that the weather on the model planet over the first few days looks almost identical between simulations. Then, gradually, the different simulations diverge and eventually become decorrelated with regard to their interannual variability. At that point, the climate system has "forgotten" about the initial conditions and the resulting range of temperatures represents the many possible trajectories of climate around its climatological attractor.

Here, the different trajectories are illustrated by linear trends over the next 30 years (2021–2050) from each of the 30 SMILE ensemble members (Fig. 17.1). All ensemble members show widespread increases in temperature, consistent with the expectation of warming with future greenhouse gas emissions. The spatial pattern of this warming, however, shows substantial variations—there are even ensemble members with regional cooling trends over 2021–2050. Due to the identical experimental setup in all ensemble members, any member-to-member differences are attributable solely to internal variability.

The average of the 30 trend maps is shown in the bottom left panel of Fig. 17.1. This "ensemble mean" trend pattern represents the response of the climate model to the external forcing imposed to each simulation, as the differences between the individual ensemble members are largely averaged out. In other words, the forced response or "signal" common to all simulations is distilled from the "noise" of internal variability through the averaging process. This "signal" shows a poleward-amplified pattern of warming throughout North America. The "noise" pattern, quantified by computing the standard deviation across the 30 individual maps, also shows a poleward-amplified pattern (bottom middle panel of Fig. 17.1). As a result, the signal-to-noise map shows a more amorphous pattern than either the signal or the noise (bottom right panel of Fig. 17.1). Importantly, the signal-to-noise values are larger than 1 over almost all of North America—this indicates the emergence of the forced response from the background climate noise.

Assuming that the internal variability of this particular climate model is realistic (a point discussed later), any projection of temperature for the real world is subject to irreducible uncertainty of the magnitude shown in Fig. 17.1. In other words, the real world could end up looking like any one of the maps in Fig. 17.1, and we would not be able to predict ahead of time which one it might resemble.

17.2.2 Contribution of Internal Variability to Total Projection Uncertainty

Besides the largely irreducible uncertainty from internal variability, other sources of uncertainty are important for climate change projections: model response uncertainty and scenario or socioeconomic uncertainty (see Chaps. 15 and 16). How does internal variability compare to these other sources of uncertainty? A common framework to quantify this is to calculate a "total projection uncertainty" from a collection

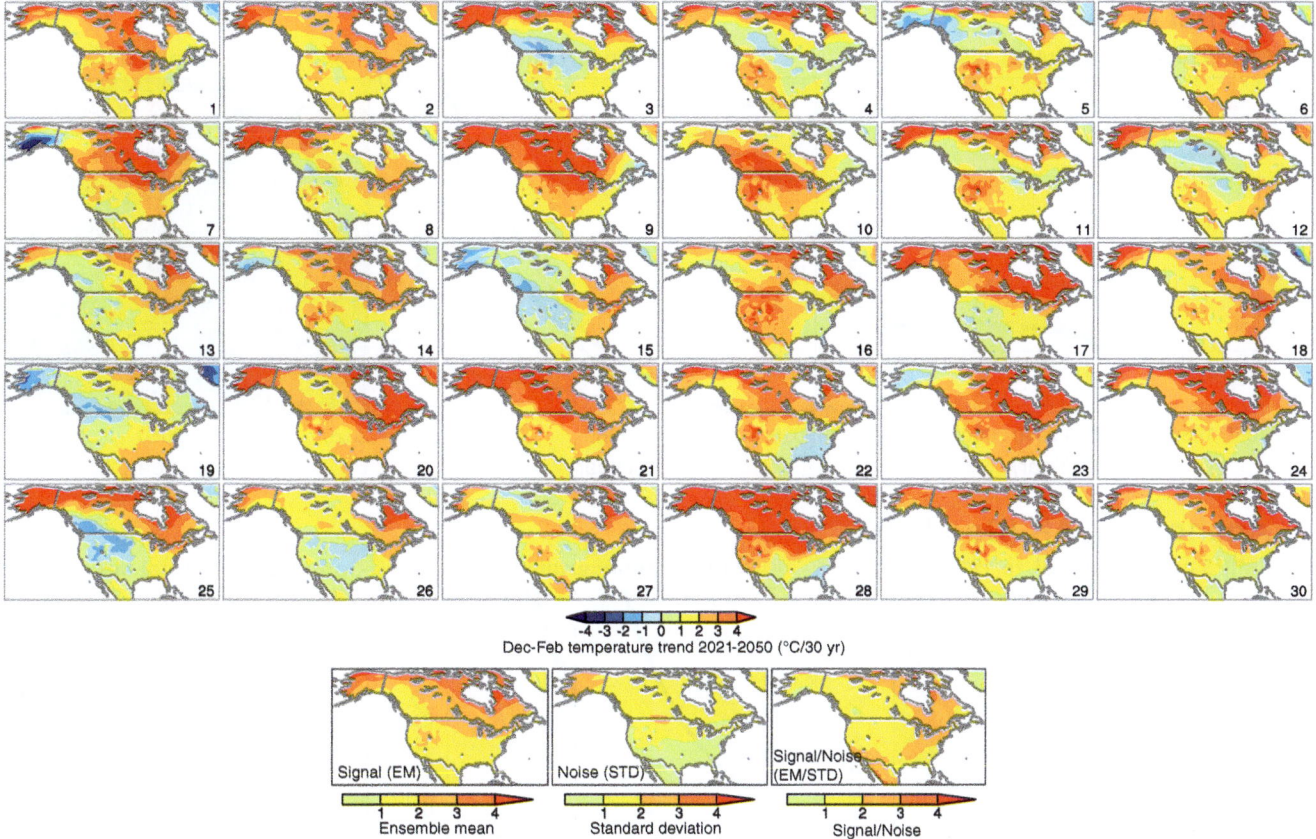

Fig. 17.1 Linear trend in winter (December–February) mean temperature over the period 2021–2050 from 30 individual CESM1 Large Ensemble simulations (panels labeled 1–30), all under the identical forcing protocol of RCP8.5 (Kay et al., 2015). The bottom panels (from left to right) show the ensemble mean (EM) trend, the standard deviation (STD) across the 30 trend maps, and the signal/noise (EM/STD). Figure adapted from Lehner and Deser (2023)

of simulations from different modeling groups run under a common set of emissions scenarios (Hawkins & Sutton, 2009).

Taking all available models from the fifth Coupled Model Intercomparison Project (CMIP5) that provide simulations for multiple emissions scenarios results in a subset of 28 models and three different emissions scenarios (Lehner et al., 2020). For each model and scenario, the forced response can be estimated either by averaging all of the available simulations, or—as is more common due to the lack of multiple simulations from some models—it can be estimated as a statistical fit to each model's simulation (e.g., a fourth-order polynomial can be fit to a temperature time series as done in Hawkins & Sutton, 2009). The residual from this fit provides an estimate of internal variability (quantified by computing the variance of the residual time series). The variance across the estimated forced responses in each model for a given emissions scenario constitutes an estimate of model uncertainty. Finally, averaging the forced responses across all the models for each scenario separately and then calculating the variance across the scenarios constitutes an estimate of the scenario uncertainty. These uncertainties are approximately additive (Yip et al., 2011), such that the total uncertainty can be estimated as the sum of the individual sources of uncertainty.

Each source of uncertainty can now be expressed as a time-varying fraction of total uncertainty. For example, for projections of decadal global mean temperature (Fig. 17.2a), internal variability and model uncertainty initially contribute about 50% each to the total uncertainty, while scenario uncertainty is zero because the different scenarios have not diverged yet (Fig. 17.2b, c). Moving further into the future, internal variability contributes increasingly less to total uncertainty, while model uncertainty and eventually scenario uncertainty become the dominant sources of uncertainty (Fig. 17.2b, c). At local scales, such as for a grid cell near Anchorage, AK, internal variability initially dominates total uncertainty and remains important for a longer time, as climate tends to be more variable at smaller spatial scales (Fig. 17.2d–f).

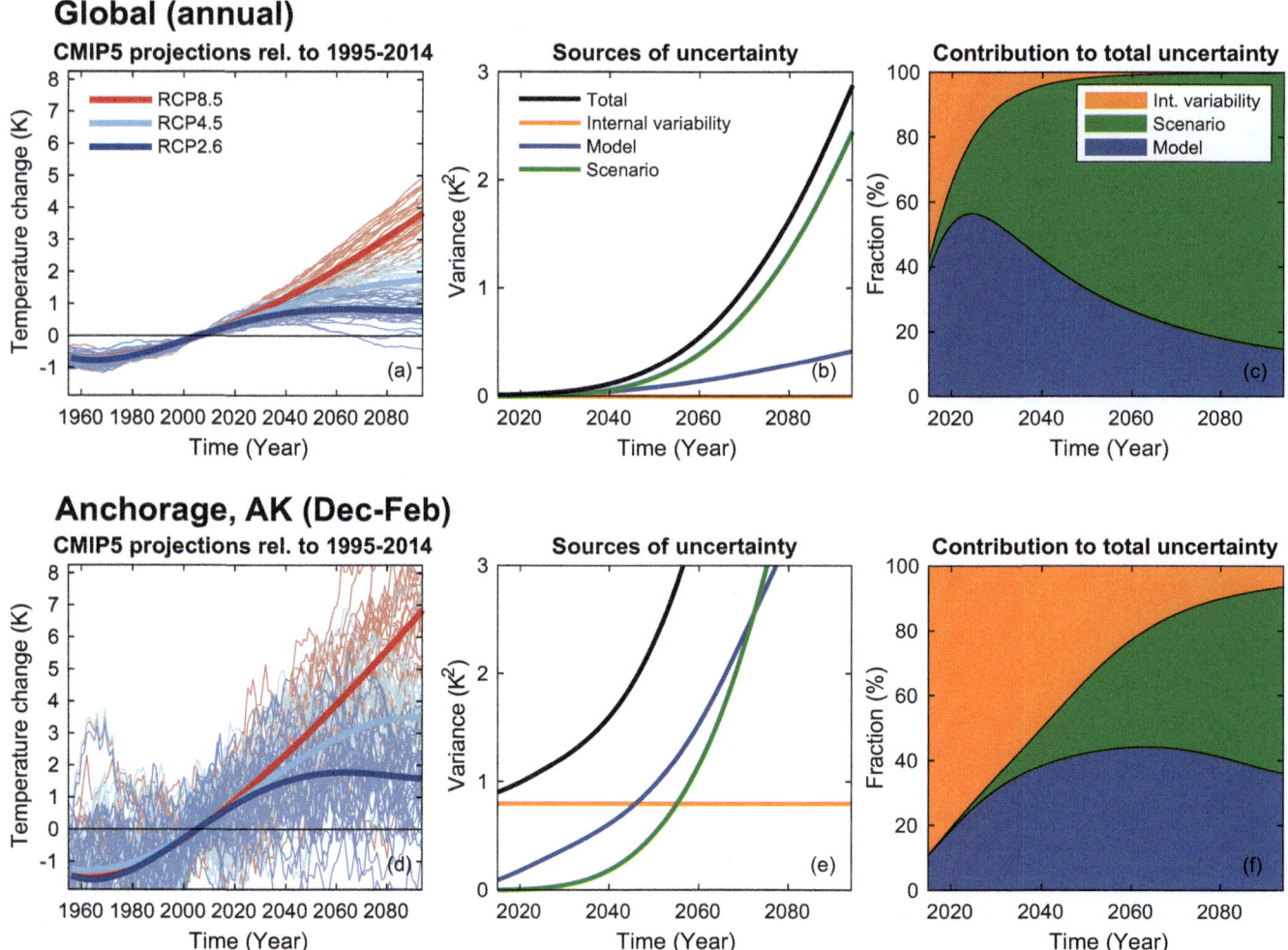

Fig. 17.2 Projections, sources of uncertainty and relative contribution of sources of uncertainty for decadal mean temperature from 28 CMIP5 models for (**a–c**) annual global mean and (**d–f**) winter (Dec–Feb) mean over a grid cell near Anchorage, AK. See text for details. Figure modified from Lehner and Deser (2023)

17.3 Drivers of Internal Variability

17.3.1 Drivers of Temperature Variability

Generally, temperature changes as shown in Fig. 17.1 can be thermodynamically- and/or dynamically induced (Wallace et al., 2015). "Thermodynamically induced" refers to changes caused by time-varying radiative fluxes or changes in sensible and latent heat fluxes at the Earth's surface. Formally, this excludes any concomitant changes in the atmospheric circulation. "Dynamically induced" refers to temperature changes attributable to changes in the atmospheric circulation, irrespective of their cause.

Together with the presence of external forcing, the cause for a given temperature trend can be partitioned into four categories: forced-thermodynamic, forced-dynamic, unforced-thermodynamic, and unforced-dynamic. "Forced" refers to the externally forced radiative imbalance of Earth (e.g., due to greenhouse gases), and "forced-thermodynamic" then indicates that the external forcing impacts temperature via a thermodynamic process (e.g., changes to radiative fluxes caused by increasing greenhouse gases). Conversely, "unforced-thermodynamic" also refers to a thermodynamic process influencing temperature, except this process is not caused by external forcing but just arises from unforced internal variability (e.g., changes in surface fluxes due to intermittent states of snow cover or soil moisture). Finally, "forced-dynamic" and "unforced-dynamic" refer to changes in temperature via forced or unforced changes in atmospheric circulation variability (e.g., changes in sea-level pressure). It is important to note that such a separation is empirical and inevitably imperfect, as there are feedbacks between terms and across time scales, such as circulation influencing snow cover, which can then influence circulation again.

17.3.2 Decomposition of Drivers

The categories laid out above enable a decomposition of the relative contributions of internal variability and forced response to a given temperature trend—for example, the 2021–2050 winter trend pattern over North America in one of the simulations shown in Fig. 17.1—and also a means of diagnosing the processes through which internal variability and forcing exert influence (Fig. 17.3). In practice, several steps are needed to conduct the partitioning quantitatively. As illustrated in Figs. 17.1 and 17.2, the forced contribution can be estimated either by using the ensemble mean of a SMILE or by making statistical assumptions about the forced response, with the residual representing the internal (or "unforced") contribution.

To further partition the forced and unforced components into thermodynamic and dynamic contributions, the method of dynamical adjustment is used (Deser et al., 2016; Sippel et al., 2019; Smoliak et al., 2015). This method aims to estimate the contribution of the atmospheric circulation to a given temperature pattern. Briefly, using the example of monthly mean data, the atmospheric circulation (e.g., sea-level pressure) from a given target month is reconstructed from other months' circulation data, for example, using regression or analogs. Each circulation reconstruction is associated with a temperature reconstruction, which thus gives an estimate of the typical temperature pattern that occurs with this type of circulation pattern. Once repeated for all the months in the dataset, one obtains an estimate of the role of atmospheric circulation in bringing about the temperature changes seen in the original data—the dynamic contribution. The residual between the original data and the dynamic contribution is an estimate of the thermodynamic contribution. Further decomposition into forced-dynamic and forced-thermodynamic is achieved by conducting the above analysis for each member of a SMILE and then averaging the respective estimates.

The pattern seen in Fig. 17.3a can now be decomposed completely within the limits of the ensemble size. This particular ensemble member was chosen for its interesting pattern of cooling over Western North America (which occurs despite the projected increase in greenhouse gases over the next 30 years), along with pronounced warming over Eastern North America. The decomposition reveals that internal variability contributes substantially to this cooling (Fig. 17.3b), in particular when compared to the total forced response which shows ubiquitous warming (Fig. 17.3c). Most of the cooling occurs due to atmospheric circulation, i.e., the dynamic contribution, specifically a strong trend toward atmospheric ridging off the west coast (Fig. 17.3d). This dynamically induced cooling is entirely internal (Fig. 17.3e) as there is essentially no forced trend in atmospheric circulation (Fig. 17.3f). On the other hand, the warming over

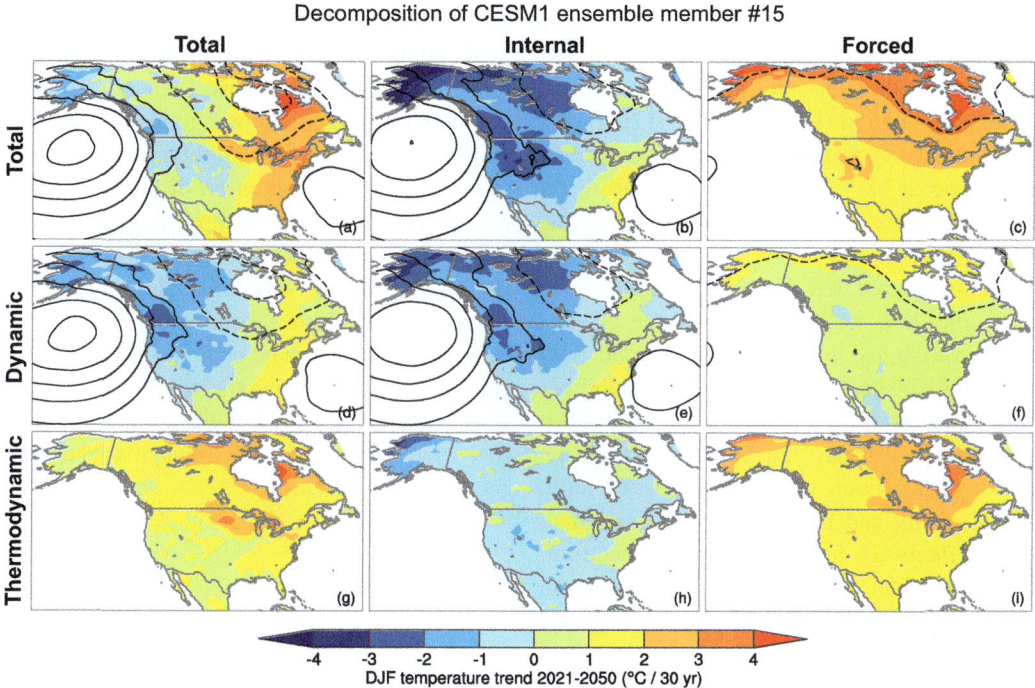

Fig. 17.3 Decomposition of the winter (Dec–Feb) mean temperature (color shading) and sea-level pressure (contours: increment of 2 hPa/30 year, starting at ±1 hPa/30 year) trend over the period 2021–2050 in one particular CESM1 ensemble member (#15). (**a**) The total trend and its contributions from (**b**) internal variability and (**c**) the forced response. (**d**) The trend attributable to dynamics and its contributions from (**e**) internal variability and (**f**) the forced response. (**g–i**) Same as (**d–f**) but for the trend attributable to thermodynamic processes. See text and Deser et al. (2016) for details. Figure adapted from Lehner and Deser (2023)

Eastern North America is almost entirely due to thermodynamic processes (Fig. 17.3g), of which the forced component dominates (Fig. 17.3i); indeed, the internal-thermodynamic component drives a weak cooling trend (Fig. 17.3h).

More generally, Fig. 17.3 shows that, for western North America, changes in atmospheric circulation are of almost negligible importance compared to thermodynamic processes for understanding regional temperature trends due to anthropogenic forcing. However, large internal variability of the atmospheric circulation can at times overwhelm the anthropogenic response in a given realization.

17.4 Summary and Discussion

This chapter provides an overview of the role of internal variability in climate projections using the example of wintertime temperature trends for the next 30 years over North America. The significant influence of internal variability, especially at regional scales, is illustrated by the range of possible future temperature trends simulated by the members of a single-model initial-condition large ensemble (Fig. 17.1). The influence of internal variability is also shown to be important when compared to other sources of uncertainty such as structural differences across models or the choice of future emissions scenario, although its relative importance decreases with lead time (Fig. 17.2). The chaotic nature of the atmospheric circulation is a key driver of internal variability of regional temperature trends over western North America (Fig. 17.3). Despite the focus on temperature, the lessons-learned apply equally to other variables and research questions, for example, precipitation (Deser et al., 2012; Guo et al., 2019; Lehner et al., 2018), response to volcanic eruptions (Lehner et al., 2016), ocean biogeochemical tracers (Rodgers et al., 2015), and even climate-related health impacts (Garcia-Menendez et al., 2017).

There are topics not covered here that nonetheless deserve attention. For example, how well do we know the characteristics of internal variability in the real world, and how faithful are models in their simulation of internal variability? The instrumental record is often too short to meaningfully constrain low-frequency (decadal and beyond) variability, and paleoclimate reconstructions, while of sufficient length, are subject to data quality and interpretation issues which limit their ability to effectively quantify the power spectrum. This data challenge has fueled continued disagreement on whether climate models are able to correctly simulate internal variability (Ault et al., 2013; Laepple & Huybers, 2014; Zhu et al., 2019). The search for the true magnitude of internal variability will continue for a long time, but it is indeed on the topic of model validation where progress can be made today. The advent of multiple SMILEs (Deser et al., 2020) and the development of observational large ensembles (McKinnon & Deser, 2018) enable a probabilistic comparison of models and observations with the goal not of identifying which model is right or wrong, but which ones are more or less plausible. Such nuances are surprisingly important as we rethink model evaluation in light of internal variability. Model behavior that was previously viewed as inconsistent with observations is suddenly compatible, as the latter is revealed to under-sample variability (Deser et al., 2018). In turn, if needed, certain models can now be rejected or downweighted more confidently thanks to the larger sample sizes (Van Oldenborgh et al., 2020).

Further, internal variability itself can change with external forcing, due to mean state changes and feedbacks between climate system components. For example, precipitation variability is expected to increase with warming due to air's growing moisture-holding capacity (Pendergrass et al., 2017); mid-latitude winter temperature variability is expected to decrease due to a reduction of albedo from a shrinking cryosphere (Screen, 2014); and the future of the El Niño-Southern Oscillation remains uncertain (Maher et al., 2023). With the increasing number of SMILEs and their public availability (Deser et al., 2020), quantitative assessments of changes in variability become possible, also for typically under-sampled phenomena such as extreme events. It does, however, require confidence in the models' ability to represent internal variability itself.

Uncertainty from internal variability has here been portrayed as being irreducible—and for good reason (Hawkins et al., 2016). Still, efforts of decadal prediction via initialized climate model simulations exist but so far show limited skill for lead times beyond about two years (Yeager et al., 2018). Gradual progress might be possible as climate models improve their representation of atmosphere-ocean coupling among other aspects (Simpson et al., 2019), but decadal prediction skill is likely to remain limited for many areas of the globe. Internal variability and the uncertainty it injects into projections for the coming decades will thus continue to accompany any climate change impact assessment. This, in turn, suggests that a perspective focused on risk and robust decision-making (Reed et al., 2022; Sutton, 2019) might be more pertinent to users of climate model information than the focus on a "best estimate."

References

Ault, T. R., Deser, C., Newman, M., & Emile-Geay, J. (2013). Characterizing decadal to centennial variability in the equatorial Pacific during the last millennium. *Geophysical Research Letters, 40*, 3450–3456. https://doi.org/10.1002/grl.50647

Deser, C., Phillips, A., Bourdette, V., & Teng, H. (2012). Uncertainty in climate change projections: The role of internal variability. *Climate Dynamics, 38*, 527–546. https://doi.org/10.1007/s00382-010-0977-x

Deser, C., Simpson, I. R., Adam, S., Phillips, & McKinnon, K. A. (2018). How well do we know ENSO's climate impacts over North America, and how do we evaluate models accordingly? *Journal of Climate, 31*, 4991–5014. https://doi.org/10.1175/JCLI-D-17-0783.1

Deser, C., Terray, L., & Phillips, A. S. (2016). Forced and internal components of winter air temperature trends over North America during the past 50 years: Mechanisms and implications. *Journal of Climate, 29*, 2237–2258. https://doi.org/10.1175/JCLI-D-15-0304.1

Deser, C., et al. (2020). Insights from earth system model initial-condition large ensembles and future prospects. *Nature Climate Change*. https://doi.org/10.1038/s41558-020-0731-2

Garcia-Menendez, F., Monier, E., & Selin, N. E. (2017). The role of natural variability in projections of climate change impacts on U.S. ozone pollution. *Geophysical Research Letters*. https://doi.org/10.1002/2016GL071565

Gould, S. J. (1989). *Wonderful life: The burgess shale and the nature of history* (p. 347). W. W. Norton.

Guo, R., Deser, C., Terray, L., & Lehner, F. (2019). Human influence on winter precipitation trends (1921–2015) over North America and Eurasia revealed by dynamical adjustment. *Geophysical Research Letters, 46*, 3426–3434. https://doi.org/10.1029/2018GL081316

Hawkins, E., Smith, R. S., Gregory, J. M., & Stainforth, D. A. (2016). Irreducible uncertainty in near-term climate projections. *Climate Dynamics, 46*, 3807–3819. https://doi.org/10.1007/s00382-015-2806-8

Hawkins, E., & Sutton, R. (2009). The potential to narrow uncertainty in regional climate predictions. *Bulletin of the American Meteorological Society, 90*, 1095–1107. https://doi.org/10.1175/2009BAMS2607.1

Kay, J. E., et al. (2015). The community earth system model (CESM) large ensemble project: A community resource for studying climate change in the presence of internal climate variability. *Bulletin of the American Meteorological Society, 96*, 1333–1349. https://doi.org/10.1175/BAMS-D-13-00255.1

Laepple, T., & Huybers, P. (2014). Ocean surface temperature variability: Large model-data differences at decadal and longer periods. *Proceedings of the National Academy of Sciences of the United States of America, 111*, 16682–16687. https://doi.org/10.1073/pnas.1412077111

Lehner, F., & Deser, C. (2023). Origin, importance, and predictive limits of internal climate variability. *Environmental Research: Climate, 2*, 023001. https://doi.org/10.1088/2752-5295/accf30

Lehner, F., Deser, C., Maher, N., Marotzke, J., Fischer, E., Brunner, L., Knutti, R., & Hawkins, E. (2020). Partitioning climate projection uncertainty with multiple large ensembles and CMIP5/6. *Earth System Dynamics Discussions, 11*, 491–508. https://doi.org/10.5194/esd-11-491-2020

Lehner, F., Deser, C., Simpson, I. R., & Terray, L. (2018). Attributing the U.S. Southwest's recent shift into drier conditions. *Geophysical Research Letters, 45*, 6251–6261. https://doi.org/10.1029/2018GL078312

Lehner, F., Schurer, A. P., Hegerl, G. C., Deser, C., & Frölicher, T. L. (2016). The importance of ENSO phase during volcanic eruptions for detection and attribution. *Geophysical Research Letters, 43*, 2851–2858. http://dx.doi.org/10.1002/2016GL067935

Lorenz, E. N. (1963). Deterministic nonperiodic flow. *Journal of the Atmospheric Sciences, 20*, 130–141. https://doi.org/10.1175/1520-0469(1963)020<0130:dnf>2.0.co;2

Maher, N., et al. (2023). The future of the El Niño–Southern oscillation: Using large ensembles to illuminate time-varying responses and inter-model differences. *Earth System Dynamics, 14*, 413–431. https://doi.org/10.5194/esd-14-413-2023

McKinnon, K., & Deser, C. (2018). Internal variability and regional climate trends in an observational large ensemble. *Journal of Climate, 31*, 6783–6802. https://doi.org/10.1175/JCLI-D-17-0901.1

Pendergrass, A. G., Knutti, R., Lehner, F., Deser, C., & Sanderson, B. M. (2017). Precipitation variability increases in a warmer climate. *Scientific Reports, 7*, 17966. https://doi.org/10.1038/s41598-017-17966-y

Reed, P. M., et al. (2022). *MultiSector dynamics: Scientific challenges and a research vision for 2030*. United States Department of Energy's Office of Science. https://doi.org/10.5281/zenodo.5825890

Rodgers, K. B., Lin, J., & Frölicher, T. L. (2015). Emergence of multiple ocean ecosystem drivers in a large ensemble suite with an earth system model. *Biogeosciences, 12*, 3301–3320. https://doi.org/10.5194/bg-12-3301-2015

Screen, J. A. (2014). Arctic amplification decreases temperature variance in northern mid- to high-latitudes. *Nature Climate Change, 4*, 577–582. https://doi.org/10.1038/nclimate2268

Simpson, I. R., Yeager, S. G., McKinnon, K. A., & Deser, C. (2019). Decadal predictability of late winter precipitation in western Europe through an ocean-jet stream connection. *Nature Geoscience, 12*, 613–619. https://doi.org/10.1038/s41561-019-0391-x

Sippel, S., Meinshausen, N., Merrifield, A., Lehner, F., Pendergrass, A. G., Fischer, E., & Knutti, R. (2019). Uncovering the forced climate response from a single ensemble member using statistical learning. *Journal of Climate, 32*, 5677–5699. https://doi.org/10.1175/JCLI-D-18-0882.1

Smoliak, B. V., Wallace, J. M., Lin, P., & Fu, Q. (2015). Dynamical adjustment of the northern hemisphere surface air temperature field: Methodology and application to observations. *Journal of Climate, 28*, 1613–1629. https://doi.org/10.1175/JCLI-D-14-00111.1

Sutton, R. T. (2019). Climate science needs to take risk assessment much more seriously. *Bulletin of the American Meteorological Society, 100*, 1637–1642. https://doi.org/10.1175/BAMS-D-18-0280.1

Van Oldenborgh, G. J., et al. (2020). Attribution of the Australian bushfire risk to anthropogenic climate change. *Natural Hazards and Earth System Sciences, 21*, 941–960. https://doi.org/10.5194/nhess-21-941-2021

Wallace, J. M., Deser, C., Smoliak, B. V., & Phillips, A. S. (2015). Attribution of climate change in the presence of internal variability. In C.-P. Chang, M. Ghil, M. Latif, & J. M. Wallace (Eds.), *Climate change: Multidecadal and beyond* (pp. 1–29). World Scientific.

Yeager, S. G., et al. (2018). Predicting near-term changes in the earth system: A large ensemble of initialized decadal prediction simulations using the community earth system model. *Bulletin of the American Meteorological Society, 99*, 1867–1886. https://doi.org/10.1175/BAMS-D-17-0098.1

Yip, S., Ferro, C. A. T., Stephenson, D. B., & Hawkins, E. (2011). A simple, coherent framework for partitioning uncertainty in climate predictions. *Journal of Climate, 24*, 4634–4643. https://doi.org/10.1175/2011JCLI4085.1

Zhu, F., et al. (2019). Climate models can correctly simulate the continuum of global-average temperature variability. *Proceedings of the National Academy of Sciences of the United States of America, 116*, 8728–8733. https://doi.org/10.1073/pnas.1809959116

Open Access This chapter is licensed under the terms of the Creative Commons Attribution 4.0 International License (http://creativecommons.org/licenses/by/4.0/), which permits use, sharing, adaptation, distribution and reproduction in any medium or format, as long as you give appropriate credit to the original author(s) and the source, provide a link to the Creative Commons license and indicate if changes were made.

The images or other third party material in this chapter are included in the chapter's Creative Commons license, unless indicated otherwise in a credit line to the material. If material is not included in the chapter's Creative Commons license and your intended use is not permitted by statutory regulation or exceeds the permitted use, you will need to obtain permission directly from the copyright holder.

Downscaling Future Climate Projections: Compounding Uncertainty But Adding Value?

Hayley J. Fowler, Linda O. Mearns, and Robert L. Wilby

18.1 What Is Downscaling?

'Downscaling' describes methods that take climate information at large scales from Global Climate Models (GCMs) to derive climate change effects at local scales. GCMs are physically based numerical models that represent important earth system processes (see Chap. 15, this volume). However, even state-of-the-art GCMs are too coarse to resolve important processes or provide relevant information at scales typically needed for climate impact and adaptation studies. Bridging this gap has focused community effort on the development of various downscaling approaches, reviewed elsewhere (see: Fowler et al., 2007; Maraun et al., 2010; Wilby & Wigley, 1997).

Downscaling techniques are conventionally divided into dynamical and statistical methods (Fig. 18.1). Dynamical downscaling uses large-scale boundary conditions from a GCM to drive a higher-resolution Regional Climate Model (RCM) for a limited area which enables better representation of surface topography and atmospheric processes. Statistical downscaling comprises a wide range of techniques that build empirical relationships between large-scale climate patterns better resolved by GCMs and observed local climate variables. Both RCM and statistical downscaling methods can then be applied to GCM projections of the future climate to generate local climate change scenarios. Additionally, some studies now apply hybrid dynamical-statistical downscaling, whereby the GCM is first downscaled by an RCM to 10–50 km resolution (or even 2–4 km via convection-permitting models [CPMs]), then statistically adjusted to correct for remaining biases at the local scale (e.g. Casanueva et al., 2020).

This chapter examines the key uncertainties inherent to both dynamical and statistical downscaling. These include the choice of GCM (or RCM) and emissions scenarios driving the boundary conditions; the size and location of downscaling domains; downscaling method and parameterization of sub-grid processes; predictor variable selection; and representation of climate variability. We show how intercomparison and benchmarking studies help to evaluate the merits and added value of various downscaling techniques. Finally, we call for greater application of downscaling in decision-making contexts and explain how this can be achieved despite myriad uncertainties.

18.2 Dynamical Downscaling Uncertainties

As mentioned before, RCMs are dynamical models like GCMs but applied over a limited area at finer horizontal resolution than standard GCMs (Ekström et al., 2015; McGregor, 2015). They are widely used to produce sub-continental climate information. The traditional nested regional modelling technique consists of running the RCM over a domain driven by initial conditions and lateral boundary conditions from observational re-analyses or GCMs. Information about lateral conditions—such as wind components, temperature, water vapour, and pressure—is typically updated every 6 h (Fig. 18.2). Key sources of uncertainty in dynamical downscaling are discussed below.

Linda O. Mearns has died before the publication of this book.

H. J. Fowler (✉)
School of Engineering, Newcastle University, Newcastle upon Tyne, UK
e-mail: hayley.fowler@newcastle.ac.uk

L. O. Mearns (deceased)

R. L. Wilby
Department of Geography and Environment, Loughborough University, Loughborough, UK

© The Author(s) 2025
L. O. Mearns et al. (eds.), *Uncertainty in Climate Change Research*,
https://doi.org/10.1007/978-3-031-85542-9_18

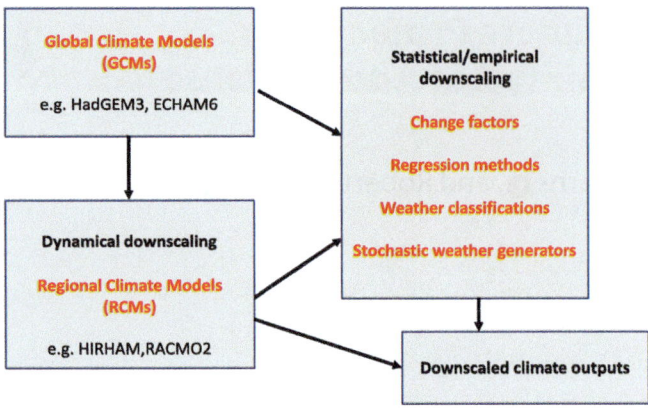

Fig. 18.1 A downscaling framework showing dynamical and statistical/empirical methods and possibility of hybrid methods which use a combination of both

18.2.1 Boundary Conditions

When RCMs are nested within GCMs, uncertainties from GCMs are propagated via the boundary conditions from the GCMs. This includes uncertainties due to model structure (Chap. 15, this volume), future scenarios of greenhouse gas emissions (Chap. 16, this volume), initial conditions and internal climate variability (Chap. 17, this volume). Importantly, the present generation of RCMs has one-way nesting, such that the RCM does not feed back into the GCM. Therefore, local-scale atmospheric responses to land-surface changes or higher-resolution topography are not able to influence large-scale circulation dynamics and atmospheric moisture.

18.2.2 Model Structure

RCMs typically represent atmospheric and land components of the climate system, but do not always include other processes such as those in a full ocean model and the effects of aerosols. This inherently leaves some uncertainties unexplored, although more complete Earth System regional models have been developed (Giorgi & Gao, 2018). Like GCMs, there are numerous RCMs that use different ways of modelling the climate system. For the most part, these different models, when used with driving conditions of different GCMs, expand the uncertainty present in the global models (Doblas-Reyes et al., 2021).

18.2.3 Spatial and Temporal Resolution

RCMs can now operate at convection-permitting scales (e.g. Kendon et al., 2014, 2017; Prein et al., 2015). Although RCMs with resolutions of 10–50 km can be run for up to 150 years, 2–4 km CPMs are generally only run for 10–20 years and over limited domains due to the high computational demand. Finer resolution RCMs have smaller biases for some variables, particularly precipitation (when compared with observations), and can produce somewhat different climate change projections, especially for extreme events at sub-daily timescales (e.g. Kendon et al., 2014). Over the past decade, numerous experiments have been performed at very high resolutions, where the parameterization of deep convection is switched off. Even so, convective processes operating at 1 km resolution or less must still be parameterized through shallow convection schemes. Although running RCMs at convection-permitting scales remains challenging for century-long simulations, some examples are now available. For example, UKCP Local is based on 2.2 km CPM transient simulations run for the UK over 1980–2080 from a 12-member perturbed physics ensemble (Kendon et al., 2023). However, lower resolution models are still applied for long simulations, such as those used in CORDEX (Giorgi et al., 2009).

18.2.4 Means of Ingesting Boundary Conditions

How boundary conditions from the driving GCM are communicated to the RCM is an additional source of uncertainty. This can be done in two main ways:

Pseudo Global Warming Experiments If one assumes that changes in regional climate are dominated by thermodynamic rather than by circulation changes, then pseudo global warming (PGW) experiments may be conducted to mitigate the effect of circulation biases in GCMs (Rasmussen et al., 2011; Sato et al., 2007; Schär et al., 1996). Typically, boundary conditions for the RCM are taken from re-analysis data and then modified by the thermodynamic (temperature, humidity, and atmospheric stability) signals of climate change for the future GCM simulation. In this case only uncertainty from thermodynamic changes is accounted for in the future simulation.

Spectral Nudging This technique allows forcing of part of the spectrum of a model solution with the equivalent part of the driving GCM. When large-scale dynamics are accurately captured by GCMs, then spectral nudging can help constrain synoptic-scale processes in large-domain RCM simulations (Miguez-Macho et al., 2004; von Storch et al., 2000). However, if there are large errors in the GCM dynamics, this can degrade the quality of the RCM simulation. Spectral nudging has been explored in systematic ways with different RCMs (e.g. Spero et al., 2018).

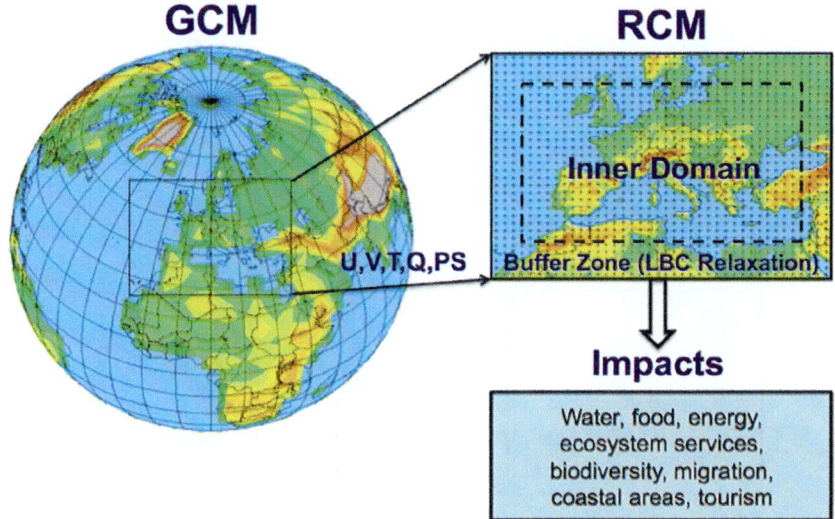

Fig. 18.2 Representation of a basic regional modelling nesting configuration and typical impact domains using results from RCM simulations. Source: Giorgi (2019)

18.2.5 GCM-RCM Pairs

Uncertainties in RCM ensembles can be explored by examining matrices of paired GCM-RCM simulations. These ensembles represent the combined uncertainty of driving GCMs and downscaling RCMs. One of the first such experiments for Europe determined that uncertainty from the boundary forcing (by GCMs) was, in general, greater than that from the RCMs (for temperature and precipitation) (Déqué et al., 2007). However, results from the North American Regional Climate Change Assessment Program established that GCMs contributed more to the uncertainty in projections of winter temperature, but RCMs contributed more to uncertainty in the projections of summer precipitation changes (Mearns et al., 2013).

The CORDEX program is the most extensive exploration of uncertainty represented by ensembles of RCMs simulations to date (Giorgi et al., 2009; Gutowski et al., 2016). Under CORDEX, multiple RCM experiments were performed at 10–50 km for 1950–2100 under various emissions scenarios, for most world regions. So far, Euro-CORDEX has produced the largest ensemble of 55 members, from permutations of 8 GCMs and 11 RCMs (Coppola et al., 2020; Vautard et al., 2020). This allows better characterization of uncertainty across simulations as well as evaluation of the robustness of climate change signals (Coppola et al., 2020). Other regions have also been assessed such as CORDEX-SEA (Southeast Asia), where 11 GCMs were downscaled by 7 RCMs, mostly at 25 km resolution (Tangang et al., 2020). This region is particularly apposite for RCMs, given the topographic complexity and archipelagic features. The higher resolution RCMs add value to future climate simulations (Fig. 18.3), but the partitioning of uncertainty is challenging given the complexity of the overall region.

Experiments have now been performed using multiple GCMs to run multiple RCMs and then CPMs. For example, the CORDEX-Flagship Pilot Study produced the first multi-model ensemble of 10-year current, mid-twenty-first century, and end-of-century simulations with 12 CPMs (Pichelli et al., 2021). This involved first downscaling CMIP5 GCMs to an intermediate resolution (12 km), then using results from those simulations to downscale to km-scale over an Alpine domain. The km-scale simulations improved the fine-scale details of precipitation, including the diurnal cycle of convection and representation of extreme events, particularly sub-daily. Also, the uncertainty across CPM simulations was reduced for some seasons and parameters related to precipitation.

18.2.6 Initial-Condition and Climate Sensitivity

Large initial-condition ensembles can be created by using different GCM ensemble members to drive a particular RCM (e.g. Fyfe et al., 2017; Leduc et al., 2019; Mizuta et al., 2017). The value of this type of experiment is to: (1) more robustly assess the effect of climate change on the intensity and occurrence of extremes and (2) better sample climate variability. These GCM ensemble experiments show how natural variability can lead to very different near-term climate projections (e.g. Deser et al., 2012). Hence, a large number of ensemble members are necessary to obtain statistically significant signals for short-term projections, especially for precipitation (e.g. Fyfe et al., 2017). More realistic simulations of extreme rainfall intensities are also found in

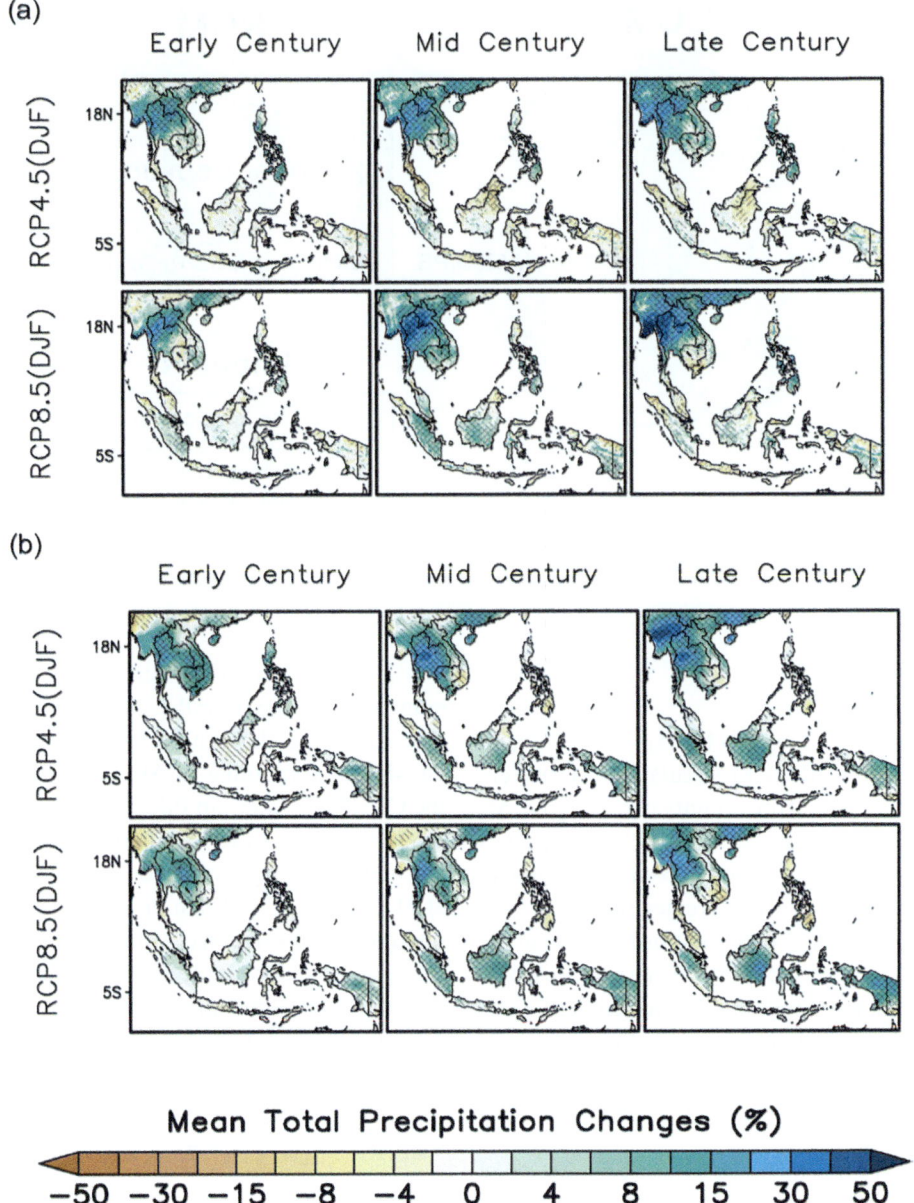

Fig. 18.3 Change in precipitation for three time periods in winter (DJF) for (**a**) RCMs and (**b**) GCMs from the CORDEX-SEA experiments for RCP4.5 and RCP8.5. Source: Tangang et al. (2020)

these studies. Furthermore, GCMs can be selected with low, medium and high climate sensitivity, thus representing an important component of uncertainty when driving RCMs with GCM output (e.g. Teichmann et al., 2021).

18.3 Statistical Downscaling Uncertainties

Using different statistical downscaling methods expands uncertainty in downscaled outputs. Many methods have been developed, ranging from simple techniques such as spatial and temporal analogues, through delta change methods, to highly sophisticated weather generators or machine learning approaches. Much statistical downscaling focuses on single-site (i.e. point scale) variables but methods have also been developed for downscaling variables over large domains. Likewise, although many methods downscale single variables, weather generators are capable of outputting multiple consistent variables (e.g. Kilsby et al., 2007). However, most statistical downscaling methods relate a combination of large-scale climate fields (e.g. mean sea level pressure, moisture at different atmospheric levels, etc.) from GCMs/RCMs (the predictors) to local-scale variables, such as temperature and/or precipitation (the predictands) (Fig. 18.4). The following sections consider the main sources of uncertainty affecting statistical downscaling methods.

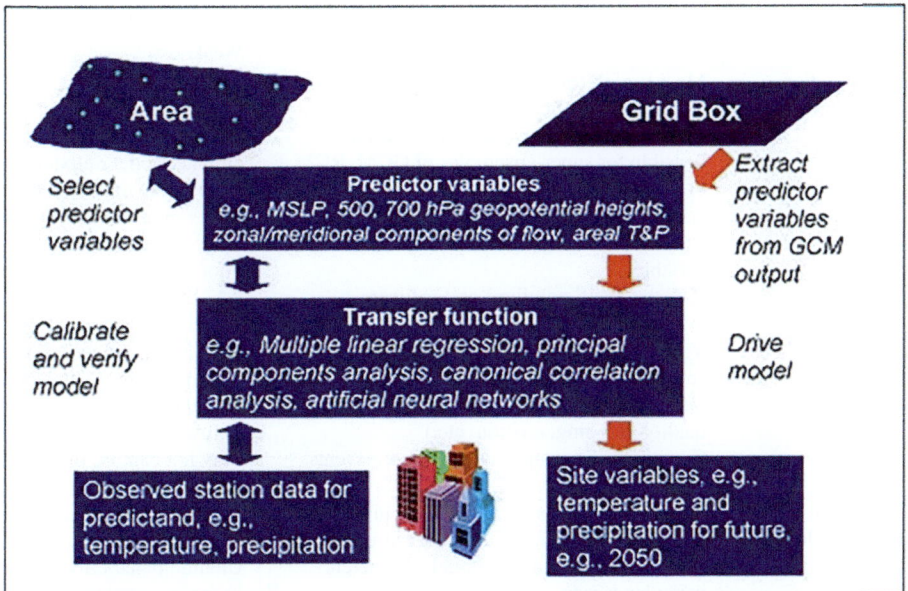

Fig. 18.4 Statistical downscaling framework for relating large-scale predictors through a transfer function to local scale predictands (normally temperature and precipitation observations). Source: Suryanarayana and Mistry (2016)

18.3.1 Downscaling Technique

Different downscaling methods have various strengths and weaknesses, and associated uncertainties (Table 18.1). Some of the most common techniques and attendant uncertainties are outlined below:

Delta Change and Bias-Correction The simplest statistical downscaling method is to apply GCM- or RCM-scale projections in the form of change factors (CFs)—also called the 'delta-change' approach. The method first establishes differences between the control and future climate model simulations. These are applied to baseline observations by simply adding (for temperature) or scaling (for precipitation) the mean CF to each day. The method can be rapidly applied to produce a range of climate scenarios from different GCMs/RCMs. However, there are several caveats. First, it asserts that GCMs more accurately simulate relative change than absolute values by assuming constant biases through time. Second, simple CFs only scale the mean, maxima, and minima of climatic variables, ignoring changes in variability whilst assuming that the spatial and temporal patterns of weather remain constant. Nonetheless, due to its simplicity, CF is still a method widely used by practitioners (Diaz-Nieto & Wilby, 2005).

To reduce biases, climate model output is commonly post-processed by bias correction (often called bias adjustment) methods. These range from simple adjustments to the mean through to flexible, multivariate, and quantile mapping approaches (e.g. Maraun et al., 2010; Piani et al., 2010). Simple mean bias correction estimates the bias as the difference (or ratio for precipitation) between simulated and observed means over a reference historical period and then adjusts the model-simulated time series over a scenario period by the estimated bias (by subtracting, or rescaling). Quantile mapping adjusts each quantile individually. The transfer functions are then applied to climate change simulations under the assumption that they are time-invariant. However, many problems have been identified with bias-correction methods, including the generation of implausible climate change signals when applied without considering underlying processes (Maraun et al., 2017). Hence, bias correction can be a major source of uncertainty, and naive application could result in ill-informed adaptation decisions.

Transfer Functions In its simplest form, a 'transfer function' between the predictand and predictor variable(s) can be described by a multiple regression model, or via more complex methods such as Principal Components Analysis (PCA) or Artificial Neural Networks (ANNs) (Giorgi et al., 2001). A recognized limitation of all regression-based methods is the underprediction of observed variance (von Storch, 1999). The problem is particularly acute for daily precipitation downscaling because of the relatively low predictability of local amounts from large-scale forcing alone (Bürger, 2002).

Weather Typing Weather typing, or classification scheme, relates the occurrence of particular 'weather patterns'—atmospheric data grouped into discrete weather types, circulation patterns, or 'states' according to synoptic similarity—to a local climate variable. Weather patterns are typically defined by applying objective classifications,

Table 18.1 Strengths and weaknesses of the main statistical downscaling methods. Updated from Wilby and Fowler (2010)

Method	Strengths	Weaknesses
Change factors (e.g. bias correction, delta change, quantile mapping)	Most straightforward to apply Versatile as applicable to any variable output by a GCM with equivalent observations Feasible for super-ensembles of GCM, RCM, or CPM output	Assumes climate model biases cancel over different periods Assumes accuracy of climate change signal in climate model output Adjusts the mean but seldom higher moments
Transfer functions (e.g. linear regression, neural networks, canonical correlation analysis, kriging)	Relatively straightforward to apply Employs full range of available predictor variables 'Off-the-shelf' solutions and software available	Poor representation of observed variance May assume linearity and/or normality of data Poor representation of extreme events
Weather typing (e.g. analogue method, hybrid approaches, fuzzy classification, self-organizing maps, Monte Carlo methods)	Yields physically interpretable linkages to surface climate Versatile (e.g. applicable to surface climate, air quality, flooding, erosion, etc.) Compositing for analysis of extreme events	Requires additional task of weather classification Circulation-based schemes can be insensitive to future climate forcing May not capture intra-type variations in surface climate
Weather generators (e.g. Markov chains, stochastic models, spell length methods, storm arrival times, mixture modelling)	Production of large ensembles for uncertainty analysis or long simulations for extremes Spatial interpolation of model parameters using landscape Can generate sub-daily information	Arbitrary adjustment of parameters for future climate Unanticipated effects to secondary variables of changing precipitation parameters

cluster or variance reduction techniques to atmospheric pressure fields (e.g. Jones et al., 1993). Local variable(s) of interest are then assigned to the weather pattern and replicated under changed climate conditions by re-sampling or regression functions (e.g. Corte-Real et al., 1999). Uncertainties in this method arise from the grouping of patterns and the representativeness of such groupings for future climate because climate change is evaluated via changes in the frequency of the weather patterns simulated by GCMs.

Weather Generators Weather generators (WGs) are designed to replicate the statistical attributes of a local variable (such as the mean, variance, or autocorrelation) but are not expected to reproduce exact sequences of observed events (Wilks & Wilby, 1999). The simplest WGs simulate precipitation occurrence via a first-order Markov process for wet-day/dry-day transitions. However, much more sophisticated WGs exist. For example, Kilsby et al. (2007) took change factors from RCMs and GCMs to condition observed data and then fit the WG. In this case, uncertainties result from the WG used as well as the RCM or GCM and the quality of the observed series that are adjusted. A key advantage of WGs is that they can be used to generate large ensembles of stochastic weather sequences as well as long simulations for stress testing systems.

18.3.2 Predictor Variables

Statistical downscaling predictor sets are typically derived from sea level pressure, geopotential height, wind fields, absolute or relative humidity, and temperature variables, amongst others. These data are archived at the grid resolution of operational and re-analysis climate models (such as ERA and NCEP), which is typically 50–100 km. Statistical downscaling essentially assumes that the regional climate is conditioned by the large-scale climate state in the form $R = f(X)$, where R is the local climate variable being downscaled, X is a set of large-scale climate variables, and f is a transfer function. This is typically established by training and validating models using point meteorological data or gridded re-analysis. Performance of these methods in reproducing observed or re-analysis statistics is normally measured using correlation coefficients, distance measures such as root mean squared error, or explained variance. However, the optimal predictor set X is highly site, season, and period dependent (e.g. González-Rojí et al., 2019).

18.3.3 Predictor-Predictand Relationships

Several key assumptions are inherent to the relationships embedded in statistical downscaling models. First, the methods assume that predictor variables X are physically meaningful, well-simulated by the climate model, and able to reflect the processes responsible for climatic variability over a range of timescales. Second, the predictor–predictand relationship f is presumed to be fixed in time, even under changed climate conditions (the stationarity assumption). Ideally, both assumptions would be tested before using statistical downscaling methods, but this is seldom done in practice (Fowler et al., 2007). The choice of predictor variable can be crucial: a predictor may not appear significant when developing a downscaling model under the present climate, but future changes in that predictor may be critical in determining climate change. The choice of predictor domain is also important and provides additional uncertainty, as the best predictor field may not coincide with the region of the local-scale predictand (Wigley & Wigley, 2000).

18.3.4 Dynamical Versus Statistical Downscaling

An issue that remains partly unresolved is the relative credibility of different downscaling methods, particularly statistical versus dynamical downscaling (Table 18.2). The EU STARDEX project[1] was among the first to systematically compare statistical, dynamical, and statistical-dynamical downscaling methods, focusing on extremes. Across all methodologies (dynamical and statistical), downscaling of precipitation extremes was found to be more skilful for winter than for summer, and more credible for indices of rainfall occurrence than amounts (Haylock et al., 2006). Statistical and dynamical methods often produce different results (such as changes to precipitation), and the VALUE project provided a framework for validation and comparison (Maraun et al., 2015). There is considerable uncertainty across the methods, but explaining exactly why the results differ is not straightforward (Fowler et al., 2007; Fowler & Wilby, 2007). Thus, it is difficult to determine which results are more credible, so this uncertainty persists.

Statistical methods cannot adequately capture mesoscale signals of change. For example, Salathé et al. (2007) noted that their statistical downscaling approach applied to the Pacific northwest of the United States could not pick up important mesoscale signals of change, whereas simulations at 15 km with an RCM could capture the land surface and topographic structures controlling mesoscale climate. Similarly, Schmidli et al. (2007) compared the responses of six statistical downscaling methods and results from three RCMs for the European Alps. They found that uncertainty in the regional climate scenarios due to downscaling method was substantial in summer because of the differing simulations of important mesoscale processes.

Although there have been few comparisons of statistical and dynamical downscaling models on a deep, process-based level, some basic, statistical comparisons have been produced (e.g. Gutmann et al., 2012; Hay & Clark, 2003; Mearns et al., 1999; Salathé et al., 2007; Schmidli et al., 2007; Spak et al., 2007; Tang et al., 2016; Wilby et al., 2000; Wood et al., 2004). Most show that results vary between methods (dynamical and statistical) but do not provide deep analysis of the reasons. This is not easy to do.

Nevertheless, more recent studies have made progress in this area. For example, Dixon et al. (2016) developed a 'perfect model' experimental design to test the stationarity assumption of a given statistical downscaling technique. They found that the assumption was invalidated (for maximum temperature) in summer along coasts and under conditions of greater warming. Likewise, Walton et al. (2020) analyzed future climate projections produced using the LOCA statistical downscaling method and the WRF regional climate model over the Sierra Nevada, California. LOCA was used to downscale the GCM but was also applied to the WRF results. The PGW approach was used to dynamically downscale the GCM using WRF. They found that only via dynamical downscaling could physically consistent regional springtime warming patterns be obtained. However, they also noted the benefits of hybrid approaches that blended the two downscaling approaches. Kotamarthi et al. (2021) provide an overview of the appropriate use of different downscaling methods based on the variables of interest, their statistical moments, and spatial scale. Their analysis, while useful, provides only general guidance.

18.4 Discussion

The significance of uncertainty sources in downscaling has long been recognized (Benestad et al., 2008; Pielke & Wilby, 2012; Wilby et al., 2004). The six most significant uncertainties arise from:

1. *GCM boundary forcing* which affects all downscaling methods due to the influence of initial conditions and emissions scenarios used in GCM experiments, grid resolution, and parameterization of missing/sub-grid scale processes.
2. *Downscaling method* whether RCMs (due to model structure, parameterizations) or statistical (due to the different types of transfer function, spatial and temporal resolution, and whether outputs are univariate, multivariate, single site, multi-site, or area averages).
3. *Boundary information* used for RCMs and statistical downscaling model calibration due to non-homogeneity or gaps in observational data, measurement errors and biases, and source of re-analysis predictor variables.
4. *Predictor variable set* used to calibrate statistical downscaling models, which typically depends on the objective function(s) and time step(s) used for goodness of fit and data availability.
5. *Stationarity assumption* as applied to predictor-predictand relationships linking the calibration period and (future) climate—a premise that applies to both statistical and dynamical downscaling due to model parameterization in both cases.

---

[1] http://www.cru.uea.ac.uk/projects/stardex/.

Table 18.2 Main strengths and weakness of statistical and dynamical downscaling. Updated from Wilby and Fowler (2010)

	Statistical downscaling	Dynamical downscaling
Strengths	• Provides point-scale climate information from GCM-scale output • Cheap, computationally undemanding and readily transferable • Ensembles of climate scenarios permit risk/uncertainty analyses • Applicable to 'exotic' predictands such as air quality and wave heights	• Provides 10–50 km (or even 2–4 km from CPMs) climate information from GCM-scale output • Respond in physically consistent ways to different external forcings • Resolve atmospheric processes such as orographic precipitation • Consistency with GCM
Weaknesses	• Dependent on the realism of GCM (or RCM) boundary forcing • Choice of domain size and location affects results • Requires high-quality data for model calibration • Predictor-predict and relationships may be non-stationary • Choice of predictor variables affects results • Low-frequency climate variability problematic • Always applied off-line, therefore, results do not feedback into the host GCM	• Dependent on the realism of GCM (or RCM) boundary forcing • Choice of domain size and location affects results • Requires significant computing resources • Initial boundary conditions affect results • Choice of cloud/convection scheme affects (precipitation) results • Not readily transferred to new regions or domains • Typically applied off-line, therefore, results do not always feedback into the host GCM

6. *Land-surface feedbacks* such as from local soil moisture, sensible heating, or changes in land cover and/or snow and ice cover, and irrigation which are not 'seen' by the GCM downscaling predictor suite but nonetheless are drivers of local meteorological behaviour—again applicable to both RCMs and statistical methods.

Numerous downscaling inter-comparison studies have helped to characterize the above uncertainties and to show how they ultimately affect climate impact assessment (Fowler et al., 2007). Progress is being made in reducing uncertainties in the downscaling boundary forcing from GCMs by constraining model ensembles to those that best simulate observed processes (Chap. 15, this volume). Uncertainties linked to downscaling model calibration data, predictor selection, and non-stationarity can be reduced in various ways, including by homogeneity testing and detrending input data; optimizing the physical basis, number and mix of predictors; and changing the predictand or domain of boundary forcing (Benestad et al., 2008). However, the surge in downscaling studies over the last 20 years brings two major challenges.

First, by the early 2020s, the research community was publishing more than 900 papers per year on climate downscaling, of which approximately 300/year were devoted to method development and testing (Fig. 18.5). Some are beginning to question the added value by all this effort, increased complexity and volume of output (e.g. Di Luca et al., 2015; Manzanas et al., 2018). Early statistical-dynamical intercomparison studies established that inter-model differences between downscaled changes in future heavy precipitation can be at least as large as the emission scenario uncertainty when downscaling from a single method (Haylock et al., 2006). Hence, expanding the pool of available downscaling techniques without considering—often incremental—benefits is, in effect, expanding ensemble uncertainty.

Second, relatively few statistical downscaling studies are informing decisions—the widely stated rationale for producing high-resolution, climate change scenarios. A survey by Wilby and Dawson (2023) showed that 30% of statistical downscaling papers were devoted to methodological development and intercomparison; 50% to climate change scenario generation and impact assessment; 15% to uncertainty estimation; and just 5% to decision-making or climate adaptation (Fig. 18.6). Nonetheless, this latter group of decision-centric studies shows how uncertainties inherent to downscaling can be usefully deployed to stress-test infrastructure performance or adaptation options under plausible ranges of conditions (e.g. Kristvik et al., 2018; Whitehead et al., 2006; Yates et al., 2015). In these bottom-up studies, uncertainty is embraced.

This still leaves a bewildering number of downscaling methods and a lack of protocols to guide their consistent and robust application. Elsewhere, steps have been taken to develop climate service standards (e.g. Climate Sense, 2022). These include quality benchmarks to reduce the prevalence or use of services that do not achieve a standard. This builds confidence that products and services offered will match the needs of the user. Previous efforts to benchmark downscaling methods have focused on applying consistent sets of model diagnostics and reference data (e.g. González-Rojí et al., 2019; Haylock et al., 2006). However, the whole statistical downscaling workflow should be covered, spanning ideally: justification of reference downscaling method, input data checks, predictor selection, calibration procedures, and performance metrics. Moreover, the workflow should be entirely transparent.

The choice of a reference downscaling method is likely value-laden, often a matter of convenience, and hence con-

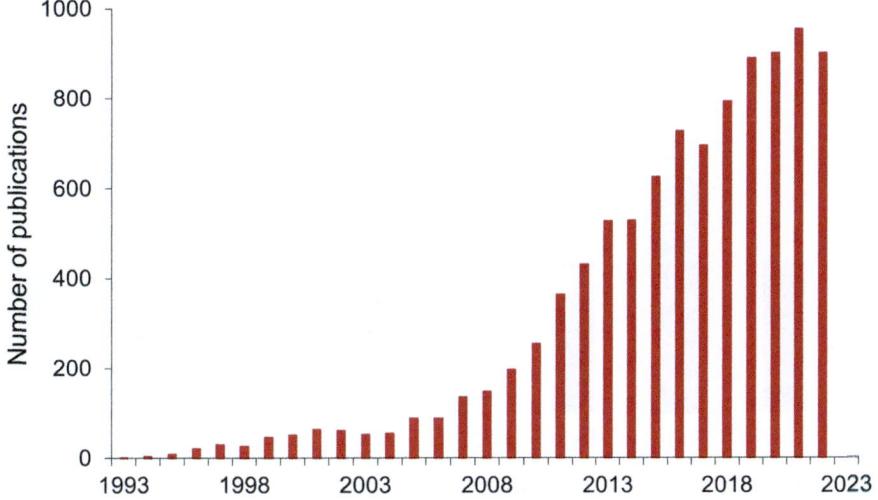

Fig. 18.5 Annual count of peer-reviewed publications on 'climate downscaling method'. Data source: Web of Science

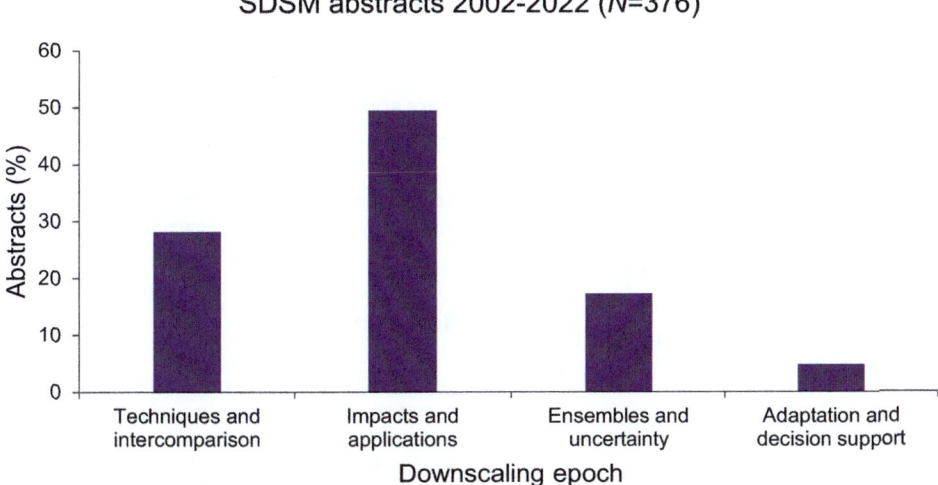

Fig. 18.6 Meta-analysis of SDSM abstracts in the Web of Science. Each abstract was assigned to one of four primary downscaling study types based; these are likely partially overlapping categories but treated as mutually exclusive here. Adaptation studies are under-represented because grey literature was not searched exhaustively. Hence, these results are only indicative of research topic preferences. Source: Wilby and Dawson (2023)

tentious. By benchmarking against a low-complexity, limited data demanding, and widely adopted model, the challenge for new entrants is to demonstrate added value relative to such a denominator. One such benchmarking study evaluated the skill of eight machine learning (ML) downscaling methods using eight diagnostics of daily precipitation at eight sites in northern Serbia (Wilby et al., 2023). The benchmark was a manually (and recursively) calibrated Statistical Downscaling Model (SDSM). The contender ML models were Least Absolute Shrinkage and Selection Operator (LASSO) with least angle regression optimization (LassoLarsCV), or ridge regression (RidgeCV); a Bayesian regression model, Automatic Relevance Determination (ARD); ensemble models with homogeneous decision tree base learners, random forest (RF), bagging regressor, Xtreme Gradient Boosting (XGBoost), and Adaptive Boosting (AdaBoost).

Despite its relative simplicity, the automatically calibrated SDSM_A was the most skilful model in 21/64 tests (i.e. over 8 diagnostics × 8 sites) against independent data (Fig. 18.7). Next were XGBoost (18/64 tests), the manually calibrated SDSM_M (16/64 tests), and AdaBoost (7/64 tests). Overall, the ML models demonstrated added value in 42% of the tests relative to the SDSM_A and SDSM_M benchmarks. More importantly, the testing exposed fundamental limitations in the ML models' ability to simulate wet-day occurrence, giving clear directions for future refinement of these algorithms. With expected advances in AI approaches, hybrid methods (Slater et al., 2023), and end-to-end downscaling, the case for benchmarking will become even stronger (Fouotsa Manfouo

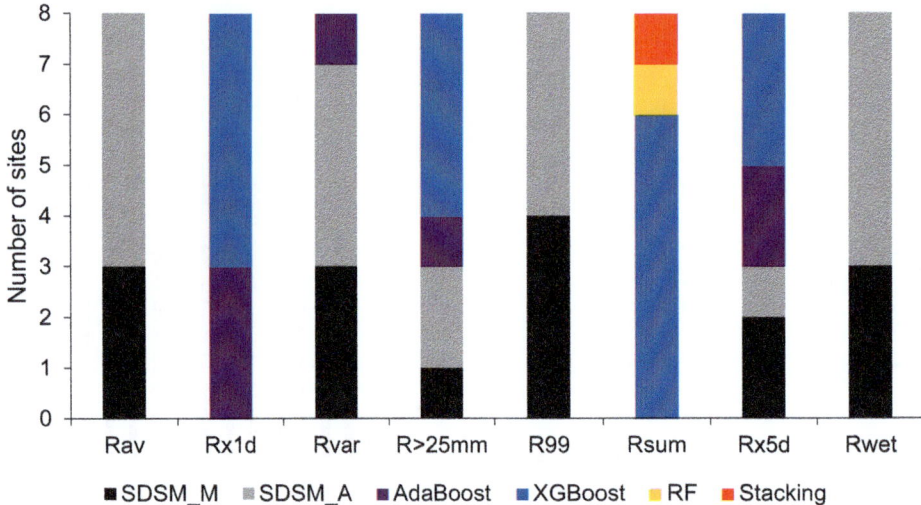

Fig. 18.7 Frequency of downscaling method selection based on lowest Mean Absolute Error (%) for eight diagnostics at eight sites in the test periods. The diagnostics are mean daily wet-day amount (Rav); maximum 1-day rainfall amount (Rx1d); variance of daily wet-day amount (Rvar); frequency of daily rainfall greater than 25 mm ($R > 25$ mm); 99th percentile wet-day amount (R99); mean annual rainfall total (Rsum); maximum 5-day rainfall amount (Rx5d); and wet-day percentage (Rwet). Adapted from: Wilby et al. (2023)

et al., 2023). Other benchmarks (beyond SDSM) could be applied, such as spatial interpolation between GCM/RCM grid points, Model Output Statistics, or simple bias correction methods.

Whatever the workflow, benchmark model(s), or reference data, transparency is essential for reporting the amount of uncertainty and added value of future downscaling method developments. Ideally, this would be a requirement on authors and perhaps one way of curtailing the proliferation of studies.

18.5 Conclusions

This chapter has summarized key uncertainties driving and resulting from downscaling methods. Downscaling research has proliferated over the last 20 years, but still with little applied focus. Uncertainties resulting from downscaling methods are large and the use of different methods, climate models, regional domains, model resolutions etc., has made it very difficult to compare methods and determine the most 'robust' downscaling protocol, or indeed how much uncertainties from different sources affect outcomes. Depending on the region analysed, the greatest source of uncertainty arises from the global or regional climate model, or even the statistical downscaling method. Therefore, the authors urge caution and thought in using these downscaling methods and, above all, the use of different methods and models to explore as much of the uncertainty space as possible, especially when evaluating real-life adaptation options. Benchmarking should become standard practice to establish the added value by more sophisticated downscaling methods compared with less resource-intensive counterparts.

Acknowledgements The authors thank Stephen Blenkinsop for his constructive and timely review of this chapter.

References

Benestad, R. E., Chen, D., & Hanssen-Bauer, I. (2008). *Empirical-statistical downscaling* (p. 228). World Scientific Publishing Company.

Bürger, G. (2002). Selected precipitation scenarios across Europe. *Journal of Hydrology, 262*, 99–110. https://doi.org/10.1016/S0022-1694(02)00014-8

Casanueva, A., Herrera, S., Iturbide, M., Lange, S., Jury, M., Dosio, A., Maraun, D., & Gutiérrez, J. M. (2020). Testing bias adjustment methods for regional climate change applications under observational uncertainty and resolution mismatch. *Atmospheric Science Letters, 21*, e978. https://doi.org/10.1002/asl.978

Climate Sense. (2022). *Climate services: Principles, requirements, and guidelines* (p. 36). UK Climate Resilience Programme.

Coppola, E., et al. (2020). Assessment of the European climate projections as simulated by the large EURO-CORDEX regional and global climate model ensemble. *JGR Atmospheres, 126*, e2019JD032356. https://doi.org/10.1029/2019JD032356

Corte-Real, J., Qian, B., & Xu, H. (1999). Circulation patterns, daily precipitation in Portugal and implications for climate change simulated by the second Hadley Centre GCM. *Climate Dynamics, 15*, 921–935. https://doi.org/10.1007/s003820050322

Déqué, M., et al. (2007). An intercomparison of regional climate simulations for Europe: Assessing uncertainties in model projections. *Climatic Change, 81*, 53–70. https://doi.org/10.1007/s10584-006-9228-x

Deser, C., Phillips, A., Bourdette, V., & Teng, H. (2012). Uncertainty in climate change projections: The role of internal variability. *Climate Dynamics, 38*, 527–546. https://doi.org/10.1007/s00382-010-0977-x

Di Luca, A., de Elía, R., & Laprise, R. (2015). Challenges in the quest for added value of regional climate dynamical downscaling. *Current Climate Change Reports, 1*, 10–21. https://doi.org/10.1007/s40641-015-0003-9

Diaz-Nieto, J., & Wilby, R. L. (2005). A comparison of statistical downscaling and climate change factor methods: Impacts on low flows in the river Thames, United Kingdom. *Climatic Change, 69*, 245–268. https://doi.org/10.1007/s10584-005-1157-6

Dixon, K. W., Lanzante, J. R., Nath, M. J., Hayhoe, K., Stoner, A., Radhakrishnan, A., Balaji, V., & Gaitán, C. F. (2016). Evaluating the stationarity assumption in statistically downscaled climate projections: Is past performance an indicator of future results? *Climatic Change, 135*, 395–408. https://doi.org/10.1007/s10584-016-1598-0

Doblas-Reyes, A., et al. (2021). Linking global to regional climate change. In V. Masson-Delmotte (Ed.), *Climate change 2021: The physical science basis*. Cambridge University Press.

Ekström, M., Grose, M. R., & Whetton, P. H. (2015). An appraisal of downscaling methods used in climate change research. *WIREs Climate Change, 6*, 301–319. https://doi.org/10.1002/wcc.339

Fouotsa Manfouo, N. C., Potgieter, L., Watson, A., & Nel, J. H. (2023). A comparison of the statistical downscaling and long-short-term-memory artificial neural network models for long-term temperature and precipitations forecasting. *Atmosphere, 14*, 708. https://doi.org/10.3390/atmos14040708

Fowler, H. J., Blenkinsop, S., & Tebaldi, C. (2007). Linking climate change modelling to impact studies: Recent advances in downscaling techniques for hydrological modelling. *International Journal of Climatology, 27*, 1547–1578. https://doi.org/10.1002/joc.1556

Fowler, H. J., & Wilby, R. L. (2007). Beyond the downscaling comparison study. *International Journal of Climatology, 27*, 1543–1545. https://doi.org/10.1002/joc.1616

Fyfe, J., et al. (2017). Large near-term projected snowpack loss over the western United States. *Nature Communications, 8*, 14996. https://doi.org/10.1038/ncomms14996

Giorgi, F. (2019). Thirty years of regional climate modeling: Where are we and where are we going next? *JGR Atmospheres, 124*, 5696–5723. https://doi.org/10.1029/2018JD030094

Giorgi, F., & Gao, X.-J. (2018). Regional earth system modeling: Review and future directions. *Atmospheric and Oceanic Science Letters, 11*, 189–197. https://doi.org/10.1080/16742834.2018.1452520

Giorgi, F., Jones, C., & Asrar, G. R. (2009). Addressing climate information needs at the regional level: The CORDEX framework. *WMO Bulletin, 58*, 175–183.

Giorgi, F., et al. (2001). Regional climate information—Evaluation and projections. In J. T. Houghton, Y. Ding, & D. J. Griggs (Eds.), *Climate change 2001: The scientific basis*. Cambridge University Press.

González-Rojí, S. J., Wilby, R. L., Sáenz, J., & Ibarra-Berastegi, G. (2019). Configuring regional downscaling models for more robust comparison of SDSM and WRF over the Iberian Peninsula. *Climate Dynamics, 53*, 1413–1433. https://doi.org/10.1007/s00382-019-04673-9

Gutmann, E. D., Rasmussen, R. M., Liu, C., Ikeda, K., Gochis, D. J., Clark, M. P., & Dudhia, J. (2012). A comparison of statistical and dynamical downscaling of winter precipitation over complex terrain. *Journal of Climate, 25*, 262–281. https://doi.org/10.1175/2011JCLI4109.1

Gutowski, W. J., et al. (2016). WCRP coordinated regional downscaling EXperiment (CORDEX): A diagnostic MIP to CMIP6. *Geoscientific Model Development, 11*, 4087–4095. https://doi.org/10.5194/gmd-9-4087-2016

Hay, L. E., & Clark, M. P. (2003). Use of statistically and dynamically downscaled atmospheric model output for hydrologic simulations in three mountainous basins in the western United States. *Journal of Hydrology, 282*, 56–75. https://doi.org/10.1016/S0022-1694(03)00252-X

Haylock, M. R., Cawley, G. C., Harpham, C., Wilby, R. L., & Goodess, C. M. (2006). Downscaling heavy precipitation over the UK: A comparison of dynamical and statistical methods and their future scenarios. *International Journal of Climatology, 26*, 1397–1415. https://doi.org/10.1002/joc.1318

Jones, P. D., Hulme, M., & Briffa, K. R. (1993). A comparison of lamb circulation types with an objective classification scheme. *International Journal of Climatology, 13*, 655–663. https://doi.org/10.1002/joc.3370130606

Kendon, E., Roberts, N., Fowler, H. J., Roberts, M., Chan, S. C., & Senior, C. A. (2014). Heavier summer downpours with climate change revealed by weather forecast resolution model. *Nature Climate Change, 4*, 570–576. https://doi.org/10.1038/nclimate2258

Kendon, E. J., Fischer, E. M., & Short, C. J. (2023). Variability conceals 100-year trend in 100 year projections of UK local hourly rainfall extremes. *Nature Communications, 14*, 1133. https://doi.org/10.1038/s41467-023-36499-9

Kendon, E. J., et al. (2017). Do convection-permitting regional climate models improve projections of future precipitation change? *Bulletin of the American Meteorological Society, 98*, 79–93. https://doi.org/10.1175/BAMS-D-15-0004.1

Kilsby, C. G., et al. (2007). A daily weather generator for use in climate change studies. *Environmental Modelling and Software, 22*, 1705–1719. https://doi.org/10.1016/j.envsoft.2007.02.005

Kotamarthi, R., Hayhoe, K., Mearns, L. O., Wuebbles, D., Jacobs, J., & Jurardo, J. (2021). *Downscaling techniques for high-resolution climate projections*. Cambridge University Press.

Kristvik, E., Kleiven, G. H., Lohne, J., & Muthanna, T. M. (2018). Assessing the robustness of raingardens under climate change using SDSM and temporal downscaling. *Water Science and Technology, 77*, 640–1650. https://doi.org/10.2166/wst.2018.043

Leduc, M., et al. (2019). The ClimEx project: A 50-member ensemble of climate change projections at 12-km resolution over Europe and northeastern North America with the Canadian regional climate model (CRCM5). *Journal of Applied Meteorology and Climatology, 58*, 663–693. https://doi.org/10.1175/JAMC-D-18-0021.1

Manzanas, R., Gutiérrez, J. M., Fernández, J., Van Meijgaard, E., Calmanti, S., Magariño, M. E., Cofiño, A. S., & Herrera, S. (2018). Dynamical and statistical downscaling of seasonal temperature forecasts in Europe: Added value for user applications. *Climate Services, 9*, 44–56. https://doi.org/10.1016/j.cliser.2017.06.004

Maraun, D., et al. (2010). Precipitation downscaling under climate change: Recent developments to bridge the gap between dynamical models and the end user. *Reviews of Geophysics, 48*, 314. https://doi.org/10.1029/2009RG000314

Maraun, D., et al. (2015). Value: A framework to validate downscaling approaches for climate change studies. *Earth's Future, 3*, 1–14. https://doi.org/10.1002/2014EF000259

Maraun, D., et al. (2017). Towards process-informed bias correction of climate change simulations. *Nature Climate Change, 7*, 764–773. https://doi.org/10.1038/nclimate3418

McGregor, J. L. (2015). Recent developments in variable-resolution global climate modelling. *Climatic Change, 129*, 369–380. https://doi.org/10.1007/s10584-013-0866-5

Mearns, L. O., Bogardi, I., Giorgi, F., Matyasovszky, I., & Palecki, M. (1999). Comparison of climate change scenarios generated from regional climate model experiments and statistical downscaling. *Journal of Geophysical Research: Atmospheres, 104*, 6603–6621. https://doi.org/10.1029/1998JD200042

Mearns, L. O., et al. (2013). Climate change projections of the north American regional climate change assessment program (NARCCAP). *Climatic Change, 120*, 965–975. https://doi.org/10.1007/s10584-013-0831-3

Miguez-Macho, G., Stenchikov, G. L., & Robock, A. (2004). Spectral nudging to eliminate the effects of domain position and geometry in regional climate model simulations. *Journal of Geophysical Research Atmospheres, 109*, D13104. https://doi.org/10.1029/2003JD004495

Mizuta, R., et al. (2017). Over 5,000 years of ensemble future climate simulations by 60-km global and 20-km regional atmospheric models. *Bulletin of the American Meteorological Society, 98*, 1383–1398. https://doi.org/10.1175/BAMS-D-16-0099.1

Piani, C., Haerter, J. O., & Coppola, E. (2010). Statistical bias correction for daily precipitation in regional climate models over Europe. *Theoretical and Applied Climatology, 99*, 187–192. https://doi.org/10.1007/s00704-009-0134-9

Pichelli, E., et al. (2021). The first multi-model ensemble of regional climate simulations at kilometer-scale resolution. Part 2: Historical and future simulations of precipitation. *Climate Dynamics, 56*, 3581–3602. https://doi.org/10.1007/s00382-021-05657-4

Pielke, R. A., & Wilby, R. L. (2012). Regional climate downscaling—What's the point? *Eos, 93*, 52–53. https://doi.org/10.1029/2012EO050008

Prein, A., et al. (2015). A review on regional convection-permitting climate modeling: Demonstrations, prospects, and challenges. *Reviews of Geophysics, 53*, 323–361. https://doi.org/10.1002/2014RG000475

Rasmussen, R. M., et al. (2011). High resolution coupled climate runoff simulations of seasonal snowfall over Colorado. *Journal of Climate, 24*, 3015–3048. https://doi.org/10.1175/2010JCLI3985.1

Salathé, E. P., Mote, P. W., & Wiley, M. W. (2007). Review of scenario selection and downscaling methods for the assessment of climate change impacts on hydrology in the United States pacific northwest. *International Journal of Climatology, 27*, 1611–1621. https://doi.org/10.1002/joc.1540

Sato, T., Kimura, F., & Kitoh, A. (2007). Projection of global warming onto regional precipitation over Mongolia using a regional climate model. *Journal of Hydrology, 333*, 144–154. https://doi.org/10.1016/j.jhydrol.2006.07.023

Schär, C., Frie, C., Luthi, D., & Davies, H. C. (1996). Surrogate climate-change scenarios for regional climate models. *Geophysical Research Letters, 23*, 669–672. https://doi.org/10.1029/96GL00265

Schmidli, J., Goodess, C. M., Frei, C., Haylock, M. R., Hundecha, Y., Ribalaygua, J., & Schmith, T. (2007). Statistical and dynamical downscaling of precipitation: An evaluation and comparison of scenarios for the European Alps. *Journal of Geophysical Research, 112*, 105. https://doi.org/10.1029/2005JD007026

Slater, L. J., et al. (2023). Hybrid forecasting: Blending climate predictions with AI models. *Hydrology and Earth System Sciences, 27*, 1865–1889. https://doi.org/10.5194/hess-27-1865-2023

Spak, S., Holloway, T., Lynn, B., & Goldberg, R. (2007). A comparison of statistical and dynamical downscaling for surface temperature in North America. *Journal of Geophysical Research: Atmospheres, 112*, 2005JD006712. https://doi.org/10.1029/2005JD006712

Spero, T. L., Nolte, C. G., Mallard, M. S., & Bowden, J. H. (2018). A maieutic exploration of nudging strategies for regional climate applications using the WRF model. *Journal of Applied Meteorology and Climatology, 57*, 1883–1906. https://doi.org/10.1175/JAMC-D-17-0360.1

Suryanarayana, T. M. V., & Mistry, P. B. (2016). Principal component regression for crop yield estimation. *SpringerBriefs in Applied Sciences and Technology.* https://doi.org/10.1007/978-981-10-0663-0

Tang, J., Niu, X., Wang, S., Gao, H., Wang, X., & Wu, J. (2016). Statistical downscaling and dynamical downscaling of regional climate in China: Present climate evaluations and future climate projections. *Journal of Geophysical Research: Atmospheres, 121*, 2110–2129. https://doi.org/10.1002/2015JD023977

Tangang, F., et al. (2020). Projected future changes in rainfall in Southeast Asia based on CORDEX–SEA multi-model simulations. *Climate Dynamics, 55*, 1247–1267. https://doi.org/10.1007/s00382-020-05322-2

Teichmann, C., et al. (2021). Assessing mean climate change signals in the global CORDEX-CORE ensemble. *Climate Dynamics, 57*, 1269–1292. https://doi.org/10.1007/s00382-020-05494-x

Vautard, R., et al. (2020). Evaluation of the large EURO-CORDEX regional model ensemble. *Journal of Geophysical Research: Atmospheres, 126*, e2019JD032344. https://doi.org/10.1029/2019JD032344

von Storch, H. (1999). On the use of 'inflation' in statistical downscaling. *Journal of Climate, 12*, 3505–3506. https://doi.org/10.1175/1520-0442(1999)012<3505:OTUOII>2.0.CO;2

von Storch, H., Langenberg, H., & Feser, F. (2000). A spectral nudging technique for dynamical downscaling purposes. *Monthly Weather Review, 128*, 3664–3673. https://doi.org/10.1175/1520-0493(2000)128<3664:asntfd>2.0.co;2

Walton, D., Berg, N., Pierce, D., Maurer, E., Hall, A., Lin, Y.-H., Rahimi, S., & Cayan, D. (2020). Understanding differences in California climate projections produced by dynamical and statistical downscaling. *Journal of Geophysical Research: Atmospheres, 125*, e2020JD032812. https://doi.org/10.1029/2020JD032812

Whitehead, P. G., Wilby, R. L., Butterfield, D., & Wade, A. J. (2006). Impacts of climate change on nitrogen in a lowland chalk stream: An appraisal of adaptation strategies. *Science of the Total Environment, 365*, 260–273. https://doi.org/10.1016/j.scitotenv.2006.02.040

Wigley, R. L., & Wigley, T. M. L. (2000). Precipitation predictors for downscaling: Observed and general circulation model relationships. *International Journal of Climatology, 20*, 641–661. https://doi.org/10.1002/(SICI)1097-0088(200005)20:6<641::AID-JOC501>3.0.CO;2-1

Wilby, R. L., Basarin, B., Boateng, D., & Josić, M. (2023). A workflow for benchmarking added value by new statistical downscaling methods. *Journal of Extreme Events.*

Wilby, R. L., Charles, S., Mearns, L. O., Whetton, P., Zorito, E., & Timbal, B.. (2004). *Guidelines for use of climate scenarios developed from statistical downscaling methods.* IPCC task group on scenarios for climate impact assessment (TGCIA). Retrieved March 11, 2023, from http://www.ipcc-data.org/guidelines/dgm_no2_v1_09_2004.pdf

Wilby, R. L., & Dawson, C. W. (2023). Conceptual development and use of downscaled climate model information. *EGU General Assembly.* https://doi.org/10.5194/egusphere-egu23-3343

Wilby, R. L., & Fowler, H. J. (2010). Regional climate downscaling. In F. Fai, A. Lopez, & M. New (Eds.), *Modelling the impact of climate change on water resources* (p. 34). Blackwell Publishing Ltd.

Wilby, R. L., Hay, L. E., Gutowski, W. J., Arritt, R. W., Takle, E. S., Pan, Z., Leavesley, G. H., & Clark, M. P. (2000). Hydrological responses to dynamically and statistically downscaled climate model output. *Geophysical Research Letters, 27*, 1199–1202. https://doi.org/10.1029/1999GL006078

Wilby, R. L., & Wigley, T. M. L. (1997). Downscaling general circulation model output: A review of methods and limitations. *Progress in Physical Geography, 21*, 530–548. https://doi.org/10.1177/030913339702100403

Wilks, D. S., & Wilby, R. L. (1999). The weather generation game: A review of stochastic weather models. *Progress in Physical Geography, 23*, 329–357. https://doi.org/10.1177/030913339902300302

Wood, A. W., Leung, L. R., Sridhar, V., & Lettenmaier, D. P. (2004). Hydrologic implications of dynamical and statistical approaches to downscaling climate model outputs. *Climatic Change, 62*, 189–216. https://doi.org/10.1023/B:CLIM.0000013685.99609.9e

Yates, D., Miller, K. A., Wilby, R. L., & Kaatz, L. (2015). Decision-centric adaptation appraisal for water management across Colorado's continental divide. *Climate Risk Management, 10*, 35–50. https://doi.org/10.1016/j.crm.2015.06.001

Open Access This chapter is licensed under the terms of the Creative Commons Attribution 4.0 International License (http://creativecommons.org/licenses/by/4.0/), which permits use, sharing, adaptation, distribution and reproduction in any medium or format, as long as you give appropriate credit to the original author(s) and the source, provide a link to the Creative Commons license and indicate if changes were made.

The images or other third party material in this chapter are included in the chapter's Creative Commons license, unless indicated otherwise in a credit line to the material. If material is not included in the chapter's Creative Commons license and your intended use is not permitted by statutory regulation or exceeds the permitted use, you will need to obtain permission directly from the copyright holder.

19. Characterizing the Uncertainty Surrounding Sea-Level Projections to Inform Decisions

Tony E. Wong, Ryan L. Sriver, and Andra J. Garner

19.1 Overview

Sea-level changes are a hallmark of Earth's climate. Global mean sea level (GMSL) can vary by more than 100 m on glacial-interglacial timescales (approx. 100,000 years). Around 50 million years ago, as more water became "locked up" in polar land ice, GMSL decreased overall by nearly 200 m. According to the Intergovernmental Panel on Climate Change (IPCC) Sixth Assessment Report (AR6; Fox-Kemper et al., 2021), GMSL rose >80 mm from 1993 to 2018 at a rate of about 3.25 mm/year, and there is strong evidence that anthropogenic global warming is accelerating the rates at which GMSL rise occurs, to levels not experienced in thousands of years. GMSL change depends on multiple factors, including thermal expansion due to ocean warming, melting of land ice (glaciers and ice sheets), and changes in land water storage. Simulating these processes and coupled interactions poses major challenges for the current generation of climate models used to analyze sea-level changes and variability.

GMSL rise poses considerable societal, economic, and ecological risks to Earth's coasts. Many areas around the world are already experiencing the effects of higher sea levels, including increases in the number and severity of coastal floods and extreme sea levels. These impacts are expected to worsen with continued warming (Hermans et al., 2023; Tebaldi et al., 2021; Vousdoukas et al., 2018). Reliable projections of GMSL and careful characterization of the uncertainties are critical for regional coastal flood risk assessments and adaptation planning. Major challenges now facing the scientific community are how to characterize and communicate the uncertainties effectively, and how to work with stakeholders and planners to develop robust strategies to protect coastal investments and residents.

Future projections of GMSL face many intrinsic uncertainties. Major sources of uncertainty in global estimates arise due to:

1. Differences in numerical and physical representations of sea-level processes in climate models;
2. Uncertainty in future greenhouse gas emissions; and
3. Incomplete representation of
   (a) Ice sheet dynamics (e.g., Greenland and West Antarctic regions) and
   (b) Internal climate feedbacks (e.g., ocean heat).

Additional factors contribute to spatial variability within local and regional sea-level patterns, including:

1. Ocean circulation changes that affect dynamic sea-level shifts;
2. Vertical land motion, such as tectonic uplift or land subsidence due to glacial isostatic adjustment or groundwater withdrawal; and
3. Changes in frequency, severity, and duration of coastal storms, which vary across regions and contribute to flooding associated with extreme sea levels.

Best estimates of GMSL change in the year 2100 suggest a GMSL rise of around 2/3 of a meter (Fox-Kemper et al., 2021). The IPCC AR6 reported a "likely" range for GMSL rise between 0.55 and 0.90 m by 2100 (relative to 1995–2014)

T. E. Wong (✉)
School of Mathematics and Statistics, Rochester Institute of Technology, Rochester, NY, USA
e-mail: tony.wong@rit.edu

R. L. Sriver
Department of Atmospheric Sciences, University of Illinois, Urbana, IL, USA
e-mail: rsriver@illinois.edu

A. J. Garner
Department of Environmental Science, Rowan University, Glassboro, NJ, USA
e-mail: garnera@rowan.edu

for a high-end emissions projection scenario (e.g., continuation of current trends in fossil fuel consumption). Here, the term "likely" is meant to represent at least a 2/3 probability that GMSL rise will be within this range. The "likely" range does not fully capture the tails or the potentially extreme GMSL rise scenarios up to and exceeding 2 m by 2100 (Bamber et al., 2019), although these more extreme scenarios also pose the greatest risk and the potential for more severe floods. For that reason, the IPCC AR6 also reported future GMSL projections for a set of high-end, "low-confidence" scenarios, which indicate a possible GMSL rise of 0.63–1.61 m by 2100 (2/3 probability). These "low-confidence" projections incorporate mechanisms of GMSL rise for which the report has assessed "low-confidence," such as the possibility for more rapid ice sheet melt than is currently captured in most global climate models. While confidence in such processes is lower than that of processes included in primary projections from the IPCC AR6, the possibility of these more extreme and highly damaging GMSL rise scenarios cannot be ruled out. Note that IPCC terminology (such as "likely" and "unlikely") has been carefully crafted and defined so as not to be misinterpreted as probabilistic estimates—a driving factor in why more comprehensive methods are frequently needed for use by decision-makers.

Threshold responses in GMSL changes, such as abrupt GMSL change due to ice sheet disintegration, have occurred in the past during glacial-interglacial cycles, causing global sea levels to vary by tens to hundreds of meters over hundreds of thousands of years (Dutton & Lambeck, 2012). On these time scales, ice sheets tend to melt abruptly when triggered by warming and internal climate feedbacks (such as the ice-albedo feedback), but can take thousands of years to reform. There is strong evidence that ice sheet mass loss is accelerating, and that current warming could trigger a catastrophic ice sheet response (DeConto & Pollard, 2016). Recent analyses of GMSL change estimate that around 35% of GMSL rise since 2006 is due to ice sheet mass loss, whereas thermal expansion and glaciers dominated GMSL budgets during the previous 100 years (Fox-Kemper et al., 2021).

GMSL change represents an average over the entire planet. This change is positive under most contemporary future scenarios, hence the term "GMSL *rise*." However, there is large spatial variability in sea-level patterns due to underlying physical processes such as ocean dynamics and solid-Earth processes (e.g., Horton et al., 2018). As compared to the globally aggregated changes in GMSL, the term *sea-level change* is more general and can refer to changes in actual sea level at any given location. The variability in sea-level patterns can cause actual sea-level change to differ significantly from the global average. Furthermore, local flood risk assessments consider additional factors beyond mean sea level, such as frequency and severity of storms (e.g., hurricanes), ocean circulation, coastal geography, land use and flood protection, surface hydrology, and river dynamics (e.g., Hall et al., 2019; Kopp et al., 2019). Many of these regional effects co-vary with GMSL. For example, global warming will likely contribute to an increase in the severity of both coastal storms and ocean heat content, as well as causing GMSL to rise. Estimating the combined effects of such environmental hazards requires fine-scale and highly localized coupled analysis specific to each region.

Several different approaches are commonly used to characterize future sea-level change uncertainties in regional flood hazard assessments. One common method is to define deterministic sea-level change scenarios based on community assessments of best-estimates, worst-case scenarios, and physically plausible upper-bounds. While simple in concept, this approach is limited by the subjectivity of the decision-makers. For instance, how does one define a worst-case scenario, and what are the tradeoffs for different stakeholder groups?

Probabilistic assessments are another useful set of tools to characterize sea-level uncertainties for decision-making. These assessments combine projections from multiple sea-level change data and model products to produce probability density functions of sea-level change for different forcing scenarios. The probability estimates can, in turn, be used with a risk-based decision framework such as cost-benefit analysis (Sriver et al., 2018; Tiggeloven et al., 2020), analyzing tradeoffs between upfront investments and long-term flood costs (Lincke & Hinkel, 2018; Wong et al., 2022). Armed with probabilistic information and/or a menu of scenarios, decision-making strategies, such as robust decision-making, can yield additional insight about pros and cons of potential investment decisions under deep uncertainty (e.g., Hinkel et al., 2019; Sriver et al., 2018). This technique analyzes different proposed decisions across a wide range of potential scenarios, helping to expose vulnerabilities in the system and minimize overall risk across a set of possible choices and different stakeholders/objectives.

The operational utility of these tools depends on having reliable estimates of the different contributions to sea-level change over time and space. What constitutes "reliable," in turn, depends on both the time domain and the spatial scales over which these projections are needed. Beginning with semi-empirical sea-level projections from the 1970s and 1980s, both technology and scientific understandings of the processes that contribute to sea-level change have improved over time. More recent sea-level projections combine multiple approaches, including process-based models, expert judgments (where projections are based on responses to broad surveys of active sea-level experts), and probabilistic

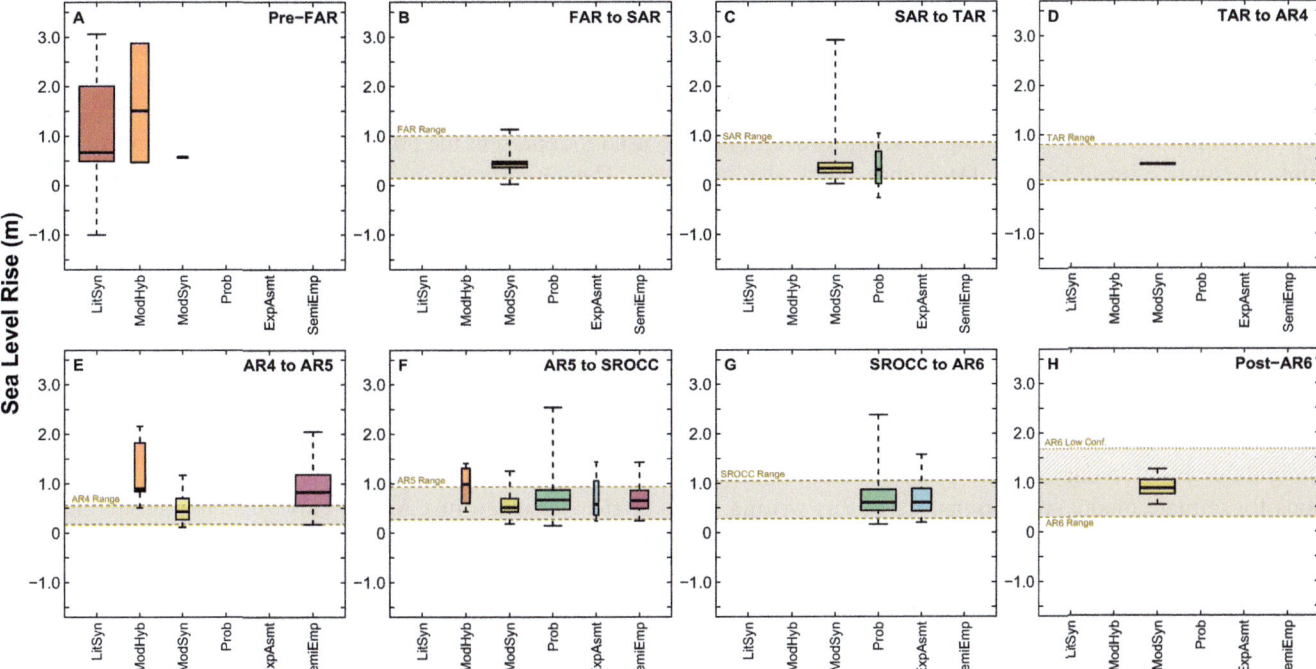

Fig. 19.1 Evolution of the ranges of GMSL projections from 1983 to 2021. Box and whisker plots depicting ranges of projected sea-level rise from (**a**) prior to the publication of the IPCC First Assessment Report (FAR) in 1990; (**b**) between the publication of FAR and the IPCC Second Assessment Report (SAR) in 1996; (**c**) between the publication of SAR and the IPCC Third Assessment Report (TAR) in 2001; (**d**) between the publication of TAR and the IPCC Fourth Assessment Report (AR4) in 2007; (**e**) between the publication of AR4 and the IPCC Fifth Assessment Report (AR5) in 2013; (**f**) between the publication of AR5 and the IPCC Special Report on the Oceans and Cryosphere in a Changing Climate (SROCC) in 2019; (**g**) between SROCC and the IPCC Sixth Assessment Report (AR6) in 2021; (**h**) after the publication of AR6. Separate box and whisker plots are provided for projections made using each methodological approach represented during each time period, including literature synthesis (LitSyn; red), model hybrid (ModHyb; orange), model synthesis (ModSyn; yellow), probabilistic (Prob; green), expert assessment (ExpAsmt; blue), and semi-empirical (SemiEmp; pink). Tan-shaded regions and dashed lines represent the ranges of SLR from the IPCC reports, as follows: the extreme range of projections for (**b**) IPCC FAR and (**c**) IPCC SAR, the range of all AOGCMs and SRES scenarios for (**d**) IPCC TAR, the 5–95% range across SRES scenarios for (**e**) IPCC AR4 (does not include dynamic ice sheet response), the likely (17th to 83rd percentiles) range in 2100 from process-based models for (**f**) IPCC AR5, the likely range of SLR in 2100 from (**g**) SROCC, and the likely range across all SSP Scenarios (not including low-confidence processes) for (**h**) IPCC AR6. Additionally, in (**h**), dotted line and hatching represent the upper bound of the likely range of the low-confidence SSP5–8.5 projections for AR6. Note that the width of each box and whisker plot corresponds to the relative number of projections that are represented within the plot. Box edges extend from the 25th to 75th percentiles; the solid line in each box shows the 50th percentile. Whiskers extend to data extremes, essentially ranging from 0 to 100th percentiles, to show the full range of SLR projections in each case

assessments. Despite nearly 40 years of GMSL projections and significant advances in methodological approaches to estimating future sea levels, the upper bound of future GMSL rise remains deeply uncertain (e.g., Garner et al., 2018). Both methodologies and ranges of future sea levels have evolved in recent decades (Fig. 19.1), and this highlights how sea-level projections from individual studies compare to the ranges of GMSL rise provided in IPCC reports.

In this chapter, we outline key sources of GMSL uncertainties, current approaches to sea-level projections, and ways forward to inform regional to local flood risk assessments. In particular, we highlight key challenges in quantifying shallow (quantifiable) and deep (generally unquantifiable) uncertainties in contributions to GMSL change and how they contribute to overall uncertainty in sea-level projections and local risk analysis.

19.2 Major Sources of Uncertainty

The main sources of uncertainty in sea-level change can be divided into two main categories: *shallow* and *deep* uncertainties (Cox, 2012; Stein & Stein, 2013). Shallow uncertainties involve situations and processes where the probability distributions of the outcomes of interest are reasonably well-known or well-agreed upon. In instances of shallow uncertainty, past observations are typically a good indicator of future behavior. Some examples of shallow uncertainties include the thermosteric contribution to sea-level change and natural variability (Sriver et al., 2018).

By contrast, deep uncertainty describes a situation in which experts disagree on the probability distributions or the set of outcomes itself. With deep uncertainties, the past is not necessarily a good (sole) predictor of future behavior.

For example, the accelerated melting of the major ice sheets, Greenland and Antarctica, has no analog previously in the instrumental period (DeConto & Pollard, 2016; Dutton et al., 2015; Hofer et al., 2020). Future greenhouse gas emissions depend on human decision-making, leading to deep uncertainty in future climate forcing. Deep uncertainty also lurks around uncertain climate thresholds and tipping points such as the timing of rapid ice loss from the Antarctic ice sheet and the potential shut/slow-down of the Atlantic Meridional Overturning Circulation (Lenton et al., 2008; Steffen et al., 2018).

Climate scientists can estimate these uncertainties with a variety of tools. On the modeling side, deep uncertainty in future climate forcing is frequently characterized by using a standard set of scenarios that provide boundary conditions for climate model simulations. As of this writing, the latest generation of these scenarios includes combinations of Representative Concentration Pathways (RCP, Moss et al., 2010), which describe radiative forcing, and Shared Socioeconomic Pathways (SSP, Riahi et al., 2017), which describe global technological and socioeconomic development, including narratives surrounding greenhouse gas emissions reduction and mitigation.

Climate models are based on the sets of coupled differential equations that any given model uses to represent the physics of the Earth system. While the underlying physical principles (conservation equations) are relatively well understood, the implementation of these equations using numerical algorithms requires careful treatment of the relevant time and space scales, which may vary between climate system components. There can be deep uncertainty surrounding what is the "right" model structure for these equations and solution methods deployed to solve them, including internal feedbacks within the coupled climate system. Modelers can characterize deep model structural uncertainty and/or standardize model experiments to account for inter-model differences by using multiple model experiments, subject to the same, controlled boundary condition scenarios (e.g., the RCP-SSP scenarios). The internal spread in a multiple model ensemble is also sometimes used to characterize natural variability in the climate system, as the differences between models arise from structural differences in their model physics. The Coupled Model Intercomparison Project Phases 5 and 6 (CMIP5/CMIP6) are some of the most widely used multiple model ensembles (Eyring et al., 2016), and they have paved the way for a variety of similar model intercomparison projects focusing on sea-level rise contributions (e.g., Edwards et al., 2021; Payne et al., 2021).

While multiple model ensembles can characterize variability across models, even when conditioning experiments on the use of a single model structure, there are additional uncertainties within a single model. This arises due to uncertainty in the initial conditions used to start the model ("what was the land surface temperature on Day 1 of our model simulation?"), uncertainty in the exogenous forcing used to run the model ("how much CO_2 goes into the atmosphere each year?"), and uncertainty in the parameters and equations used to represent the physics of the Earth system.

The representation of physics and hydrology within a climate model can affect model output for sea-level changes. This includes direct effects and indirect effects. Direct effects include causes like ice melt from glaciers on land making its way into the ocean or melting from the Greenland and Antarctic ice sheets. Regarding the major ice sheets, it is important to note that due to Archimedes' Principle, ice that is already in the ocean is already contributing to current sea levels. However, ice that is being buttressed by calving icebergs, and then subsequently enters into the ocean, does lead to changes in GMSL. This "disintegration" can lead to sizable GMSL contributions from the major ice sheets. Indirect effects on GMSL changes include the transfer of heat into the ocean at the ocean-air interface, which causes thermal expansion of the ocean. Changes to local sea levels are also driven by changes in atmospheric circulation—that is, wind and sea-level pressure. Additionally, as the major ice sheets melt, their reduced mass means their gravitational pull is lower too. This leads to a decrease in local mean sea level close to the ice sheets and increases further away.

Over the last 30 years, thermal expansion has contributed roughly 40% of the total GMSL rise while melting ice sheets have contributed around 25% (Fox-Kemper et al., 2021). Ice sheets are projected to contribute more substantially to GMSL rise over the next 100 years, contributing more than 70% for high-GMSL scenarios (Vega-Westhoff et al., 2019). Whether due to distant changes in ice sheets and glaciers or more local variations, rising sea levels can drive substantial risks for coastal communities. Even small changes in regional sea levels can have major impacts on flooding in areas already below local mean sea level like New Orleans, Louisiana, or the Netherlands. Changes in wind patterns and sea-level pressure lead to changes in sea-level extremes, such as storm surges, which are a key factor in local flood risk. Changes in precipitation patterns and streamflow also contribute to changes in the total overall flood hazard from these multiple correlated drivers of risk, or the compound hazard. These changes are difficult to resolve in models, as individual storms are often at the limits of what global climate models can represent spatially. Further, changes in streamflow are often driven by erosion and sediment transport, which may not be well-represented in models larger than basin scale. Environmental extremes, including storm surges and heavy precipitation, are also difficult to estimate because extremes, by their very definition, are rare. This inherent limitation of data poses a key challenge for modeling the risks posed by environmental extremes.

19.3 Current Approaches

The challenge of paucity of data is just one of many computational- and data-related scientific challenges of sea-level science. Characterizing uncertainty in future GMSL is critical for managing risk in the world's heavily populated coastal zones. This generally involves running models for future GMSL, and computing statistics based on either ensembles of model simulations (e.g., multiple model, initial condition, and perturbed parameter ensembles) or a few detailed simulations from regional or global climate models (GCMs). GCMs, however, are computationally expensive. On the other hand, semi-empirical models (SEMs) or emulators to represent specific processes provide a coarse representation of the physical system but are relatively less expensive. Given a fixed computational budget, there is a trade-off between the level of detail in the physical model and the level of detail in the statistical model that can be employed (Fig. 19.2).

Navigating this trade-off means scientists must select appropriate models and statistical tools for their particular application. For managing on-the-ground flood risks and informing local coastal adaptation decisions, high-resolution coastal models (e.g., Fischbach et al., 2012) are an appropriate choice. These detailed models resolve features for a specific geographic region and specific drivers of risk well. However, given a fixed overall computational budget, it is often infeasible to represent these details in a wider spatiotemporal-scale model, or with a computationally expensive statistical model (Fig. 19.2). On the other hand, for examining sensitivities and relationships among geophysical drivers of risk (e.g., greenhouse gas emissions) and local impacts (e.g., changes in local mean sea level), coarse semi-empirical models and detailed statistical modeling may be an appropriate choice because large numbers of samples are made possible by the low computational expense of these approaches. New methods seek to combine multiple sources of observational and modeling information to analyze local flood risk with more robust characterizations of relevant uncertainties.

The characterization of uncertainty that is made possible through statistical modeling is a useful avenue for studying drivers of local risks related to sea-level rise. These methods can include both probabilistic projections of sea-level rise (e.g., Kopp et al., 2014; Mengel et al., 2016; Nauels et al., 2017; Wong et al., 2017) and possibilistic projections (e.g., Le Cozannet et al., 2017). Probabilistic projections can be useful in that they assign probabilities to sea-level rise outcomes, and provide a direct representation of uncertainties across different models and data products. However, these models typically do not yet sample the full range of uncertainties, which could lead to overconfidence in the projections (e.g., under-sampling of potentially extreme sea-level scenarios). Additionally, these methods are generally conditioned on the data sets and models employed (both physical and statistical models). This model and data dependence leads to deep structural uncertainties that make it difficult to combine multiple probabilistic projections in a way that well represents those deep uncertainties. Possibilistic projections, on the other hand, can combine probabilistic information to construct upper and lower bounding cumulative distribution functions. These distributions form a probability-box, or p-box, around an unknown cumulative distribution function that reconciles the differences among the constituent distributions. Possibilistic methods do not provide as precise of a description of sea-level hazards as probabilistic methods, but in cases of deep uncertainty (e.g., multiple models, multiple calibration data sets), this can sometimes be preferable.

Probabilistic assessments are frequently used to produce probability distribution functions of local mean sea-level change and local extreme sea levels. Typically, resolving the tails of these probability distributions requires large sample sizes and computationally efficient models. To achieve this, emulators and semi-empirical models are commonly applied with probabilistic methods. While these modeling approaches are suitable for uncertainty characterization and sensitivity analysis, those benefits trade off against resolving more detailed Earth system processes.

Coupled Earth system models provide an avenue to represent more Earth system dynamics, and an opportunity to relate how uncertainties and sensitivities propagate across components of the coupled system. These models range from relatively simple and/or stylized (e.g., Urban & Keller, 2010; Vega-Westhoff et al., 2019; Wong et al., 2022) to

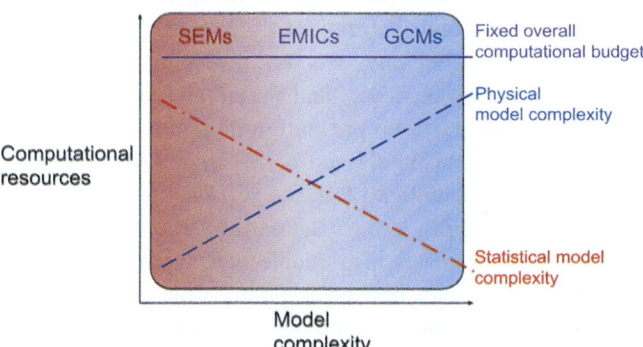

Fig. 19.2 Tradeoff between physical and statistical model complexity (conceptual framework). Given a fixed overall computational budget, as the computational resources allocated to running a more detailed physical model increases, the resources available for characterizing/quantifying uncertainty decreases. This diagram conceptualizes these tradeoffs for three modeling approaches ranging from: relatively coarse representations of the physical system (SEMs); to intermediate representation (EMICs–earth system models of intermediate complexity); to detailed representation (GCMs). Varying spatiotemporal scales of interest constitute another trade-off in computational resources, which could be envisioned as a third axis in this conceptual diagram

computationally expensive and highly detailed in terms of the specific Earth system processes that are resolved in the models (e.g., Danabasoglu et al., 2020; E3SM Project, DOE, 2021). These more detailed models, especially GCMs, include explicit representations of more processes than the stylized SEMs and emulators. This makes coupled Earth system models important tools for more closely examining specific processes, such as ice sheet dynamics, and exploring how uncertainties propagate across components of the Earth system. However, this also raises important considerations about how uncertainties are characterized, especially in light of computational constraints (Fig. 19.2). Owing to these constraints, Earth system models of intermediate complexity (EMICs) use simplified representations of processes and/or relationships between Earth system components (Claussen et al., 2002; Eby et al., 2013). These features enable EMICs to complement GCMs and SEMs in representing specific components of the Earth system.

Research questions and decision-maker objectives can inform how to balance the computational expense of running more detailed physical models against the benefits from using sophisticated statistical models to characterize and quantify uncertainties. For example, decision-making paradigms, such as robust decision-making, can yield insight into the pros and cons of potential adaptation investment decisions under deeply uncertain future sea-level scenarios. This technique analyzes different proposed decisions across a range of scenarios, exposing potential vulnerabilities in the system. Overall decisions can be made to maximize utility, minimize regret, or other characterizations of uncertainty that represent decision-maker tolerances for risk. Decision-making frameworks can also be useful to explore what scenarios (e.g., policy decisions or geophysical uncertainties) drive high-end GMSL rise outcomes. In those cases, larger sample sizes may well be needed to adequately resolve the high-risk upper tails of the probability distributions of outcomes. Sensitivity analyses and supervised machine learning methods lend themselves well to identify scenario features and processes that are associated with negative outcomes, and on what time scales. However, these analyses generally require large sample sizes to explore the space of plausible outcomes.

19.4 Ways Forward

Future sea levels are deeply uncertain, and there is a significant possibility of catastrophic outcomes for coastal areas around the world in the next several centuries (Strauss et al., 2021). While new methods described above are being developed to characterize sea-level uncertainties and assess changing coastal flood risks with global warming, there is a strong need for new innovations to trace uncertainties and sensitivities across coupled models of the human-Earth system (Srikrishnan et al., 2022; Wong et al., 2022). Representing these couplings is a challenge that the modeling community has largely met. However, resolving feedbacks between coupled model components still poses a challenge. For example, greenhouse gas emissions may serve as a boundary condition for a climate model. If greenhouse gas emissions increase, then there is generally an increase in coastal hazards, stemming from rising sea levels. Current models do not well represent the human component, wherein, as realized risks surpass our decision-maker tolerances for risk, action to alter human emissions of greenhouse gases may then decrease in response.

New methods are also emerging to analyze coastal flood risk within multi-sectoral complex adaptive systems. These methods consider not only the environmental hazards but also aspects of human systems and decision-making. One example is hierarchical modeling frameworks that combine multiple different sources of sea-level data with statistical estimation and calibration for highly localized flood risk assessments (Vousdoukas et al., 2018). Sea-level information can come from different sources featuring different spatial and temporal scales, such as Earth system models, tide gauge records, high-resolution regional flood models, storm statistics, and semi-empirical time series models. The result is fine-scale flood risk assessment for urban planning, insurance pricing, and coastal preservation.

Hybrid data-model methods are also useful for examining compound flood events. Compound events are extreme floods resulting from multiple contributing factors occurring at or around the same time. The relevant environmental factors can include storm surge, extreme precipitation (pluvial flood hazard), and extreme streamflow (fluvial flood hazard). One example is a land falling hurricane, in which case local flooding depends on the amount of storm surge on top of uncertain sea-level change, amount of precipitation, surface hydrology and river flows, and potential failures in built infrastructure (e.g., flood prevention measures). Combining different sources of model and observational data can help estimate the probability of these tail events and identify potential vulnerabilities (e.g., Bates et al., 2021). Hybrid stat-model-data approaches also enable new types of information and perspectives beyond standard projections, such as estimating the timing of future sea-level exceedances (e.g., "when will sea-level rise reach or exceed 1 m?"), as opposed to estimating the precise magnitude of sea-level change at a specific point in time (Vega-Westhoff et al., 2020). Such estimates for the timing of GMSL rise are presented in the most recent IPCC report (Fox-Kemper et al., 2021). This type of information can be more useful to decision-makers, for example when considering coastal infrastructure investments, but it is also more difficult to estimate this timing using standard numerical climate models.

Future changes in sea levels are inherently uncertain. This poses a particular challenge for policy development, which typically aims to provide certainty for coastal communities and the citizens who live in them. Ultimately, sea-level change affects multiple sectors of society, which means that adaptation planning will require multidisciplinary efforts between scientists, engineers, economists, city planners, and the community. Such multidisciplinary collaborations are needed to create robust adaptation strategies and flexible policies that can be adjusted ahead of damages occurring, as the state of our knowledge of future sea-level change continues to evolve. Finally, in high-end emissions scenarios, even the best adaptation policies may result in unacceptably high losses, meaning that successful management of future SLR also requires aggressive mitigation of greenhouse gas emissions, in order to avoid worst-case future scenarios.

References

Bamber, J. L., Oppenheimer, M., Kopp, R. E., Aspinall, W. P., & Cooke, R. M. (2019). Ice sheet contributions to future sea-level rise from structured expert judgment. *Proceedings of the National Academy of Sciences, 116*, 11195–11200. https://doi.org/10.1073/pnas.1817205116

Bates, P. D., et al. (2021). Combined modeling of US fluvial, pluvial, and coastal flood hazard under current and future climates. *Water Resources Research, 57*, e2020WR028673. https://doi.org/10.1029/2020WR028673

Claussen, M., et al. (2002). Earth system models of intermediate complexity: Closing the gap in the spectrum of climate system models. *Climate Dynamics, 18*, 579–586. https://doi.org/10.1007/s00382-001-0200-1

Cox, L. A. (2012). Confronting deep uncertainties in risk analysis. *Risk Analysis, 32*, 1607–1629. https://doi.org/10.1111/j.1539-6924.2012.01792.x

Danabasoglu, G., et al. (2020). The community earth system model version 2 (CESM2). *Journal of Advances in Modeling Earth Systems, 12*, e2019MS001916. https://doi.org/10.1029/2019MS001916

DeConto, R. M., & Pollard, D. (2016). Contribution of Antarctica to past and future sea-level rise. *Nature, 531*, 591–597. https://doi.org/10.1038/nature17145

Dutton, A., & Lambeck, K. (2012). Ice volume and sea level during the last interglacial. *Science, 337*, 216–219. https://doi.org/10.1126/science.1205749

Dutton, A., et al. (2015). Sea-level rise due to polar ice-sheet mass loss during past warm periods. *Science, 349*, aaa4019. https://doi.org/10.1126/science.aaa4019

E3SM Project, DOE. (2021). *Energy exascale earth system model v2.0.* https://doi.org/10.11578/E3SM/DC.20210927.1

Eby, M., et al. (2013). Historical and idealized climate model experiments: An intercomparison of earth system models of intermediate complexity. *Climate of the Past, 9*, 1111–1140. https://doi.org/10.5194/cp-9-1111-2013

Edwards, T. L., et al. (2021). Projected land ice contributions to twenty-first-century sea level rise. *Nature, 593*, 74–82. https://doi.org/10.1038/s41586-021-03302-y

Eyring, V., Bony, S., Meehl, G. A., Senior, C. A., Stevens, B., Stouffer, R. J., & Taylor, K. E. (2016). Overview of the coupled model intercomparison project phase 6 (CMIP6) experimental design and organization. *Geoscientific Model Development, 9*, 1937–1958. https://doi.org/10.5194/gmd-9-1937-2016

Fischbach, J. R., Johnson, D. R., Ortiz, D. S., Bryant, B. P., Hoover, M., & Ostwald, J. (2012). *Coastal Louisiana risk assessment model: Technical description and 2012 coastal master plan analysis results.* RAND Corporation. Retrieved February 27, 2022, from https://www.rand.org/pubs/technical_reports/TR1259.html

Fox-Kemper, B., et al. (2021). Ocean, cryosphere and sea level change. In *The working group I contribution to the sixth assessment report addresses the most up-to-date physical understanding of the climate system and climate change* (pp. 1211–1362). IPCC. https://doi.org/10.1017/9781009157896.011

Garner, A. J., Weiss, J. L., Parris, A., Kopp, R. E., Horton, R. M., Overpeck, J. T., & Horton, B. P. (2018). Evolution of 21st century sea level rise projections. *Earth's Future, 6*, 1603–1615. https://doi.org/10.1029/2018EF000991

Hall, J. A., et al. (2019). Rising sea levels: Helping decision-makers confront the inevitable. *Coastal Management, 47*, 127–150. https://doi.org/10.1080/08920753.2019.1551012

Hermans, T. H. J., et al. (2023). The timing of decreasing coastal flood protection due to sea-level rise. *Nature Climate Change, 13*, 359–366. https://doi.org/10.1038/s41558-023-01616-5

Hinkel, J., et al. (2019). Meeting user needs for sea level rise information: A decision analysis perspective. *Earth's Future, 7*, 320–337. https://doi.org/10.1029/2018EF001071

Hofer, S., Lang, C., Amory, C., Kittel, C., Delhasse, A., Tedstone, A., & Fettweis, X. (2020). Greater Greenland ice sheet contribution to global sea level rise in CMIP6. *Nature Communications, 11*, 6289. https://doi.org/10.1038/s41467-020-20011-8

Horton, B. P., Kopp, R. E., Garner, A. J., Hay, C. C., Khan, N. S., Roy, K., & Shaw, T. A. (2018). Mapping Sea-level change in time, space, and probability. *Annual Review of Environment and Resources, 43*, 481–521. https://doi.org/10.1146/annurev-environ-102017-025826

Kopp, R. E., Gilmore, E. A., Little, C. M., Lorenzo-Trueba, J., Ramenzoni, V. C., & Sweet, W. V. (2019). Usable science for managing the risks of sea-level rise. *Earth's Future, 7*, 1235–1269. https://doi.org/10.1029/2018EF001145

Kopp, R. E., Horton, R. M., Little, C. M., Mitrovica, J. X., Oppenheimer, M., Rasmussen, D. J., Strauss, B. H., & Tebaldi, C. (2014). Probabilistic 21st and 22nd century sea-level projections at a global network of tide-gauge sites. *Earth's Future, 2*, 239. https://doi.org/10.1002/2014EF000239

Le Cozannet, G., Manceau, J. C., & Rohmer, J. (2017). Bounding probabilistic sea-level projections within the framework of the possibility theory. *Environmental Research Letters, 12*, 014012. https://doi.org/10.1088/1748-9326/aa5528

Lenton, T. M., Held, H., Kriegler, E., Hall, J. W., Lucht, W., Rahmstorf, S., & Schellnhuber, H. J. (2008). Tipping elements in the Earth's climate system. *Proceedings of the National Academy of Sciences, 105*, 1786–1793. https://doi.org/10.1073/pnas.0705414105

Lincke, D., & Hinkel, J. (2018). Economically robust protection against 21st century sea-level rise. *Global Environmental Change, 51*, 67–73. https://doi.org/10.1016/j.gloenvcha.2018.05.003

Mengel, M., Levermann, A., Frieler, K., Robinson, A., Marzeion, B., & Winkelmann, R. (2016). Future Sea level rise constrained by observations and long-term commitment. *Proceedings of the National Academy of Sciences of the United States of America, 113*, 2597–2602. https://doi.org/10.1073/pnas.1500515113

Moss, R. H., et al. (2010). The next generation of scenarios for climate change research and assessment. *Nature, 463*, 747–756. https://doi.org/10.1038/nature08823

Nauels, A., Meinshausen, M., Mengel, M., Lorbacher, K., & Wigley, T. M. L. (2017). Synthesizing long-term sea level rise projections - The MAGICC Sea level model v2.0. *Geoscientific Model Development, 10*, 2495–2524. https://doi.org/10.5194/gmd-10-2495-2017

Payne, A. J., et al. (2021). Future Sea level change under coupled model intercomparison project phase 5 and phase 6 scenarios from the

Greenland and Antarctic ice sheets. *Geophysical Research Letters, 48*, e2020GL091741. https://doi.org/10.1029/2020GL091741

Riahi, K., et al. (2017). The shared socioeconomic pathways and their energy, land use, and greenhouse gas emissions implications: An overview. *Global Environmental Change, 42*, 153–168. https://doi.org/10.1016/j.gloenvcha.2016.05.009

Srikrishnan, V., et al. (2022). Uncertainty analysis in multi-sector systems: Considerations for risk analysis, projection, and planning for complex systems. *Earth's Future, 10*, e2021EF002644. https://doi.org/10.1029/2021EF002644

Sriver, R. L., Lempert, R. J., Wikman-Svahn, P., & Keller, K. (2018). Characterizing uncertain sea-level rise projections to support investment decisions. *PLoS One, 13*, e0190641. https://doi.org/10.1371/journal.pone.0190641

Steffen, W., et al. (2018). Trajectories of the earth system in the anthropocene. *Proceedings of the National Academy of Sciences, 115*, 8252–8259. https://doi.org/10.1073/pnas.1810141115

Stein, S., & Stein, J. L. (2013). Shallow versus deep uncertainties in natural hazard assessments. *Eos, Transactions of the American Geophysical Union, 94*, 133–134. https://doi.org/10.1002/2013EO140001

Strauss, B. H., Kulp, S. A., Rasmussen, D. J., & Levermann, A. (2021). Unprecedented threats to cities from multi-century sea level rise. *Environmental Research Letters, 16*, 114015. https://doi.org/10.1088/1748-9326/ac2e6b

Tebaldi, C., et al. (2021). Extreme sea levels at different global warming levels. *Nature Climate Change, 11*, 746–751. https://doi.org/10.1038/s41558-021-01127-1

Tiggeloven, T., et al. (2020). Global-scale benefit–cost analysis of coastal flood adaptation to different flood risk drivers using structural measures. *Natural Hazards and Earth System Sciences, 20*, 1025–1044. https://doi.org/10.5194/nhess-20-1025-2020

Urban, N. M., & Keller, K. (2010). Probabilistic hindcasts and projections of the coupled climate, carbon cycle and Atlantic meridional overturning circulation system: A Bayesian fusion of century-scale observations with a simple model. *Tellus A, 62*, 737–750. https://doi.org/10.1111/j.1600-0870.2010.00471.x

Vega-Westhoff, B., Sriver, R. L., Hartin, C., Wong, T. E., & Keller, K. (2020). The role of climate sensitivity in upper-tail sea level rise projections. *Geophysical Research Letters, 47*, e2019GL085792. https://doi.org/10.1029/2019GL085792

Vega-Westhoff, B., Sriver, R. L., Hartin, C. A., Wong, T. E., & Keller, K. (2019). Impacts of observational constraints related to sea level on estimates of climate sensitivity. *Earth's Future, 7*, 677–690. https://doi.org/10.1029/2018EF001082

Vousdoukas, M. I., Mentaschi, L., Voukouvalas, E., Verlaan, M., Jevrejeva, S., Jackson, L. P., & Feyen, L. (2018). Global probabilistic projections of extreme sea levels show intensification of coastal flood hazard. *Nature Communications, 9*, 2360. https://doi.org/10.1038/s41467-018-04692-w

Wong, T. E., Bakker, A. M. R., & Keller, K. (2017). Impacts of Antarctic fast dynamics on sea-level projections and coastal flood defense. *Climatic Change, 144*, 347–364. https://doi.org/10.1007/s10584-017-2039-4

Wong, T. E., Ledna, C., Rennels, L., Sheets, H., Errickson, F. C., Diaz, D., & Anthoff, D. (2022). Sea level and socioeconomic uncertainty drives high-end coastal adaptation costs. *Earth's Future, 10*, e2022EF003061. https://doi.org/10.1029/2022EF003061

Open Access This chapter is licensed under the terms of the Creative Commons Attribution 4.0 International License (http://creativecommons.org/licenses/by/4.0/), which permits use, sharing, adaptation, distribution and reproduction in any medium or format, as long as you give appropriate credit to the original author(s) and the source, provide a link to the Creative Commons license and indicate if changes were made.

The images or other third party material in this chapter are included in the chapter's Creative Commons license, unless indicated otherwise in a credit line to the material. If material is not included in the chapter's Creative Commons license and your intended use is not permitted by statutory regulation or exceeds the permitted use, you will need to obtain permission directly from the copyright holder.

Uncertainty Quantification: A Statistical Perspective

Stephan R. Sain and William Kleiber

20.1 Introduction

Astronomy might provide one of the earliest examples of the development of statistical tools for uncertainty quantification (UQ) as early researchers struggled with how to handle measurement error. By the early 1600s, researchers were observing and making different kinds of measurements on the Moon, the planets of the Solar System, and other objects. These researchers wrestled with different approaches to experimental design (a single, well-constructed measurement versus multiple measurements), how to combine multiple measurements, and what a distribution for measurement error would look like. Statistical innovations followed such as the use of the mean versus the median, least squares as an objective way to provide an optimal estimate of central tendency, the Gaussian distribution as a model for measurement error, and even the Central Limit Theorem (Stahl, 2006). As we fast forward another few centuries, the UQ toolbox has expanded dramatically to include statistical inference, Bayesian statistics, modern experimental design, linear and generalized linear models, machine learning, and much more.

UQ has emerged as the field devoted to characterizing uncertainty. UQ sits at the intersection of computational science, applied mathematics, statistical science, and a wide variety of scientific domains where numerical models, fueled by increasing computational capability and increasing amounts of data, have become virtual laboratories for scientific discovery. While measurement error that was the focus of early astronomers continues to be a crucial component with the breadth of today's observation systems, the scope of possible sources of uncertainties has greatly expanded; such is the focus of the field of UQ.

As we narrow the focus to climate modeling, climate change, and ensuing impacts, there are many different sources of uncertainty. There is the inherent or natural variability in the climate system, complex physical processes that exist on vastly different spatial and temporal scales, issues that arise with observational systems, the size and scope of available datasets, climate model bias, construction of climate model ensembles, etc. The list is quite large and cannot possibly be covered in one review paper. The focus here is on statistical frameworks for working with general computer models (Sect. 20.2), emulators (Sect. 20.3), and ensembles of climate models (Sect. 20.4). Some final comments and additional directions are outlined in (Sect. 20.5).

20.2 Design and Analysis of Computer Experiments

Many of the statistical approaches to UQ stem from the early work in the design and analysis of computer experiments (DACE; Sacks et al., 1989, Santner et al., 2003). The National Research Council report on verification, validation, and UQ for complex computer models (National Research Council, 2012) also contains many excellent references.

Computer models begin with a mathematical representation of a physical system that is then implemented in a computational framework in order to simulate the physical system under different conditions. Experiments can then be run with these models at a selected set of inputs to test the physical system's response. These experiments are particularly critical in Earth system models where direct experimentation is limited or even impossible. Moreover, for realistic models of Earth's climate, significant computing resources are needed, necessitating high-performance computing systems. It is important to remember that these models are not reality, and various approximations are often used, inducing discrepancies from reality which are sometimes large. Nonetheless, these models have become integral tools in the study of many natural phenomena.

The goals of computer experiments can be loosely classified into forward and inverse problems. The forward problem typically is thought of as exploring the output of the computer model as a function of the input(s). Examining how the output of different climate models varies in response to different greenhouse gas forcings can be considered a forward problem. The inverse problem looks in the other direction, seeking to estimate parameters or other model inputs that yield model output consistent with observations. Both forward and inverse problems require well-defined sets of inputs and their corresponding outputs, beckoning the use of modern statistical design of experiments.

20.2.1 Statistical Design of Experiments

Many point to research on agricultural experiments in the early 1900s as the beginning of a formalized statistical approach to the design of experiments. This expanded to industrial experiments in the 1950s, computer experiments in the 1980s, and even online experiments in e-commerce in the 2010s, along with many other application areas. Box et al. (1978) is a classic text and a fantastic introduction to experimental design, and Sacks et al. (1989) and Santner et al. (2003) are excellent resources for applications to computer experiments.

One goal of conducting experiments, even computer experiments, is determining what inputs have the largest impact or effect on the results or output of an experiment. In the design of experiments literature, inputs in this context are referred to as factors. If multiple factors are involved, a particular factor can impact the output individually or in a way that depends on the settings of other factors; in the statistical literature, this is known as an interaction. Perhaps the key feature of the statistical approach to the design of experiments is the focus on manipulating multiple factors at the same time, often through factorial designs, which allows for the assessment of interactions, as opposed to the less-efficient one-factor-at-a-time approach.

Although historically the design of experiments has progressed in agriculture and industrial settings, the area has found new importance and challenges in helping to design climate model experiments. For example, the North American Regional Climate Change Assessment Program (NARCCAP; Mearns et al. 2012, Mearns et al. 2013) is an experiment in which atmosphere-ocean general circulation models are used to drive a collection of regional climate models to explore uncertainties in dynamic downscaling over North America. This is a noteworthy example in that the design is based on a formal fractional factorial experiment. Fractional factorial experiments are useful when resource constraints may make it difficult to conduct the full factorial experiment, in which all possible combinations of the (discrete) levels of the factors are included in the setup of the design. In such cases, only a portion of the full factorial experiment is conducted, and the design is constructed to preserve main effects and, in some cases, the typically larger lower-order interactions.

Alternatively, it is often a goal to fit a surface to the model output as a function of the inputs, so designs that span the space of inputs are useful. Perhaps the most simple such design is randomly selecting points from a uniform distribution across the input space. However, such designs are fairly inefficient as they lead to portions of the input space that are oversampled and portions that are undersampled. See, for example, the left frame of Fig. 20.1.

As opposed to fully randomized designs, space-filling designs seek to spread out design points over the input space while maintaining approximately uniform sampling of the design points on the individual inputs (Santner et al., 2003). For example, Williamson et al. (2013) describe the creation of a large, perturbed physics ensemble with the third Hadley Centre Climate Model (HadCM3) by using a Latin hypercube sampling. A simple Latin hypercube sampling is shown for two factors in the middle frame of Fig. 20.1. In this case, the range of each of the two inputs is divided into an equal number of segments, effectively laying a grid over the input space. The design points are laid out randomly into each segment marginally so that there is only one point per row and column. Further, points are randomly chosen within an individual segment. Kleiber et al. (2013b) utilize an alternative type of space-filling design in a calibration experiment for a model of the magnetosphere. An example of such a design is shown in right frame of Fig. 20.1, where a distance criterion is used to ensure that points are not too close together nor too far apart in the design.

20.2.2 Gaussian Processes and Emulators

It has become commonplace to use stochastic models in the analysis of deterministic computer experiments. Typically, the output of the computer model is viewed as a partial realization of a stochastic process. While it may seem odd to statistically represent the output of a deterministic computer model—for a fixed input, the model output is always the same—the rationalization comes from the fact that we only record a subset of possible outputs, and thus there is uncertainty regarding the model output at unsampled inputs.

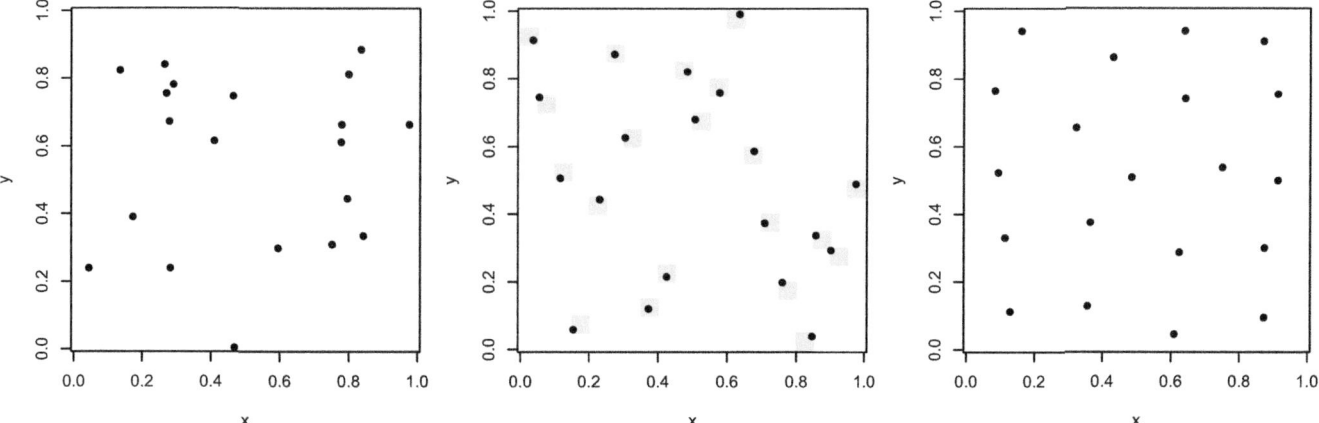

Fig. 20.1 Examples of space-filling designs. The left frame shows a random design, the middle frame a Latin hypercube, and the right frame a criterion-based design

To illustrate some common themes, suppose $Z(\mathbf{s})$ is a stochastic process with index $\mathbf{s} \in \Re^d$. Examples might include time series or spatial processes. Simplifying assumptions about the process are often made; for example, weak stationarity assumes the mean is constant over $\mathbf{s}$, and the covariance of the random variables at two indices is simply a function of their displacement, that is, $\text{Cov}[Z(\mathbf{s}), Z(\mathbf{s} + \mathbf{h})] = C(\mathbf{h})$. The covariance function C is a positive definite function and may also be isotropic (when the value of C depends only on the distance between two points and not their direction).

An important special case of stochastic processes is the Gaussian process or the Gaussian random field. A Gaussian process is simply a stochastic process where, for any finite collection of indices $\mathbf{s}_1, \ldots, \mathbf{s}_k$, the vector $[Z(\mathbf{s}_1), \ldots, Z(\mathbf{s}_k)]'$ has a multivariate Gaussian distribution, where $\prime$ denotes transpose. A Gaussian process is conveniently parameterized by its mean and covariance function, leading to its widespread use in UQ applications.

A Gaussian process can be used as an approximation for how the output of a computer model varies with respect to the inputs, and some basic probability theory results can be utilized to define a simple emulator. For example, assume some values of inputs $\mathbf{s}_1, \ldots, \mathbf{s}_n$ and the corresponding (univariate) outputs $\mathbf{Z} = [Z_1, \ldots, Z_n]'$ where $Z_i = Z(\mathbf{s}_i)$ are available. Further assume that the mean and covariance functions are known (or have been estimated), so that the joint distribution of the output of a computer model at a new input (i.e., $Z_0 = Z(\mathbf{s}_0)$) and $\mathbf{Z}$ can be given by

$$\begin{bmatrix} Z_0 \\ \mathbf{Z} \end{bmatrix} \sim \mathcal{N}\left(\begin{pmatrix} \mu_0 \\ \boldsymbol{\mu} \end{pmatrix}, \begin{pmatrix} \Sigma_{00} & \Sigma_{01} \\ \Sigma_{10} & \Sigma_{11} \end{pmatrix} \right),$$

where $\mu_i = \mu(\mathbf{s}_i)$, $\boldsymbol{\mu} = [\mu_1, \ldots, \mu_n]$, $\Sigma_{00} = \text{Cov}[Z_0]$, $\Sigma_{11} = \text{Cov}[\mathbf{Z}]$, $\Sigma_{01} = \text{Cov}[Z_0, \mathbf{Z}]$, and $\Sigma_{10} = \Sigma_{01}'$. Then a prediction of Z_0 can be found via the conditional expectation given by

$$E[Z_0|\mathbf{Z}] = \mu_0 + \Sigma_{01} \Sigma_{11}^{-1} (\mathbf{Z} - \boldsymbol{\mu}) \quad (1)$$

with the variance of that prediction given by

$$\text{Var}[Z_0|\mathbf{Z}] = \Sigma_{00} - \Sigma_{01} \Sigma_{11}^{-1} \Sigma_{10}. \quad (2)$$

Moreover, $[Z_0|\mathbf{Z}]$ is also normally distributed with the above conditional mean and variance.

Figure 20.2 shows a simple example of a Gaussian process emulator. The black dots represent the results of an experiment $\mathbf{Z} = [Z_1, \ldots, Z_5]'$ where the computer model was run for a set of input values $\mathbf{s}_1, \ldots, \mathbf{s}_5$ that happen to be real valued and in $[0, 1]$. The solid black line represents the predictions at potential input values where the computer model was not evaluated and is based on the conditional expectation given in (1). Uncertainty about the conditional expectation can be expressed by constructing pointwise prediction intervals based on the conditional variance in (2), shown by the shaded regions in the figure. Note that these intervals are widest between the design points, and go to zero at the design points where the computer model was actually evaluated. Alternatively, uncertainty could be evaluated through an ensemble created by conditional simulation, denoted by the dashed lines in Fig. 20.2. The ensemble members are simulated from the Gaussian process conditional on the values represented by the black dots and represent plausible traces of the computer model across the input space based on the limited information from the results of the experiment.

Kennedy and O'Hagan (2001) explore the use of Gaussian processes in a Bayesian framework for calibration, i.e., estimating unknown model parameters. See also Higdon et al. (2008) and Kleiber et al. (2013b) for extensions to multivariate settings. This framework for calibration has been

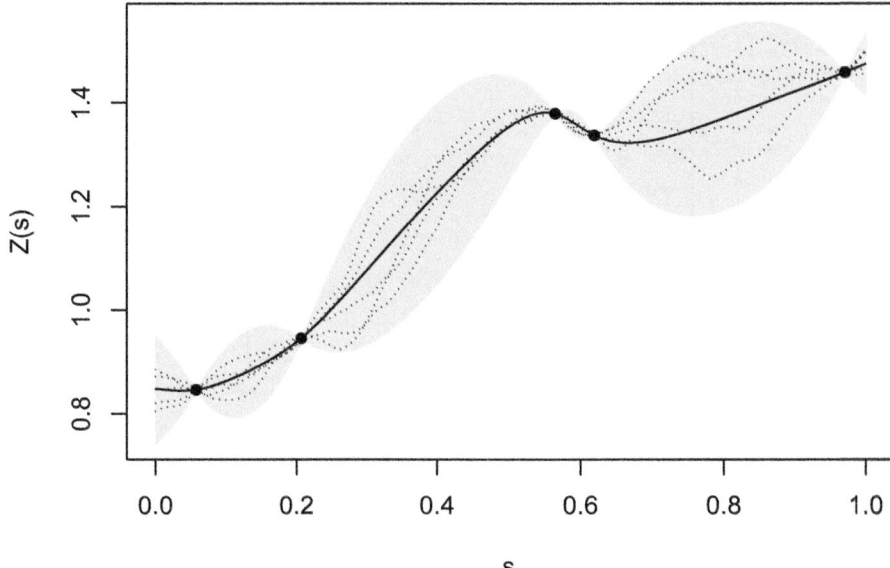

Fig. 20.2 An example of an application of a Gaussian process as an emulator. The solid black dots represent the outputs of the computer model $Z(\mathbf{s})$, while the solid black line represents the conditional mean $E[Z|\mathbf{Z}]$. Prediction intervals are given by the shaded gray regions, and plausible realizations of the computer model are indicated by the dashed lines

used with climate models. See, for example, Forest et al. (2008), Sanso and Forest (2009), or Chang et al. (2014). In a different application, Castruccio and Stein (2013) use Gaussian processes for climate model emulation.

20.3 Other Emulators

While many applications of emulators focus on the inverse problem or calibration/parameter estimation, one example of the use of emulators in the forward problem is producing projections of future climate. Climate models are often run for a collection of greenhouse gas emissions scenarios, but climate models are resource intensive and expensive to run, typically requiring significant time on high-performance computing systems. It is not always possible to run a particular model for all scenarios of interest, and cheaper and faster alternatives are needed.

Pattern scaling, introduced by Santer et al. (1990), implemented in the MAGICC/SCENGEN software package (magicc.org), and further assessed and developed by Mitchell et al. (1999) and Mitchell (2003), has been used in a wide variety of applications. Essentially, pattern scaling relies on the idea that many measures of climate change produced by a climate model can be related linearly to that models time-varying global temperature change. Following Tebaldi and Arblaster (2014), this relation can be written for a specific climate model as

$$Y(t, x, y, s) = T(t)y(x, y, s),$$

where Y represents some climate change parameters such as temperature or precipitation measured at some time t and locations x and y. The index s can specify a particular time of year or perhaps a season. T is the global average temperature change and y is a time-invariant spatial pattern of change for the climate parameter of interest. Tebaldi and Arblaster (2014) and others have shown that pattern scaling can perform reasonably well for annual and seasonal summaries of temperature and precipitation (although to a lesser degree). Tebaldi et al. (2020) also explore the performance of pattern scaling on climate extreme indices. Alexeeff et al. (2018) also include uncertainty estimates via resampling, an aspect important for many applications, including impact studies.

There are also other statistical approaches to producing emulators to examine the climate response to different greenhouse gas forcings. Castruccio et al. (2014) is an excellent example and discusses some of the issues to alternative approaches such as those using Gaussian processes (Williamson et al., 2012) and empirical orthogonal functions (Holden and Edwards, 2010).

20.4 Statistical Analysis of Ensembles

Climate model ensembles are excellent tools for exploring different kinds of uncertainty. From initial condition ensembles (e.g., Kay et al., 2015) to perturbed physics ensembles (e.g., Stainforth et al., 2005) to multi-model ensembles (e.g., Eyring et al., 2016), each of these constructions can be used to gain insights into the uncertainties that surround climate projections. Often, these ensembles, particularly multi-model ensembles, are used to construct a probabilistic projection of climate change. Many of the approaches used to produce these projections are Bayesian in nature.

Bayesian statistics differs from frequentist approaches primarily by allowing a subjective view of probability and allowing probability statements about statistical model parameters rather than considering them as fixed, unknown quantities. Bayesian inference starts with specifying a prior distribution on statistical model parameters that represents a belief about a parameter before seeing any data. Stated more formally, we begin by assuming the vector of statistical parameters θ follows a so-called prior distribution, typically denoted $\pi(\theta)$. Next, a data model that describes how random data depend on θ is specified as $f(y|\theta)$, a conditional probability distribution for the random variable Y conditional on the value of the parameter θ. Finally, after observing observations $Y_1, \ldots, Y_n$, the posterior distribution is calculated as $f(\theta|Y_1, \ldots, Y_n)$, which is the conditional distribution of θ given the data $Y_1, \ldots, Y_n$, and reflects the updated uncertainty about θ in the face of data. This is summarized succinctly in Bayes' Theorem which states

$$f(\theta|Y_1, \ldots, Y_n) \propto f(Y_1, \ldots, Y_n|\theta)\pi(\theta), \quad (3)$$

where $f(Y_1, \ldots, Y_n|\theta)$ is referred to as the likelihood. We see that the posterior distribution is proportional to the data likelihood times the prior.

20.4.1 Combining Models Through Bayesian Hierarchy

Bayesian methods have become fairly common in climate research and more broadly across the geosciences, often through Bayesian hierarchical modeling. Expanding on the conditioning inherent in Bayes' Theorem, hierarchical modeling involves breaking complex problems into smaller, more manageable pieces which can allow for improved handling of different sources of uncertainty and the incorporation of physical knowledge (see, for example, Berliner, 1996, 2003, and the excellent introduction in Cressie and Wikle, 2011).

A basic representation of a hierarchical model, breaking down the joint distribution into the common three-level hierarchy, is often written generically as

Data model: $[\mathbf{Y}|\mathbf{Z}, \theta]$

Process model: $[\mathbf{Z}|\theta]$

Parameter model: $[\theta]$,

where $\mathbf{Y}$ represents the data, $\mathbf{Z}$ represents the possibly hidden process which is often the target for such modeling, θ represent model parameters, and the notation $[\cdot]$ and $[\cdot|\cdot]$ denote a probability distribution and conditional probability distribution, respectively. For example, $\mathbf{Y}$ might represent in situ temperature measurements from a set of thermometers which may be subject to measurement error, while $\mathbf{Z}$ represents the true temperature values. Bayes' Theorem yields the posterior distribution $[\mathbf{Z}, \theta|\mathbf{Y}] \propto [\mathbf{Y}|\mathbf{Z}, \theta][\mathbf{Z}|\theta][\theta]$ from which inferences can then be drawn on the hidden process $\mathbf{Z}$ and θ. Two early examples applications of hierarchical modeling in the geosciences are Berliner et al. (2000) and Wikle et al. (2001).

Tebaldi et al. (2004, 2005) and Smith et al. (2009) introduce a hierarchical modeling approach to produce a probabilistic assessment of climate change on the basis of a multi-model ensemble. Focusing on seasonal and regional 30-year averages of current and future temperature and precipitation from the different models in the ensemble, the hierarchical model assumes these averages follow a Gaussian distribution centered on the true, but unknown climate values. The authors also include observations of the current climate to help constrain the estimates. A posterior distribution on the difference of the future and the current values is easily obtained via a computational approach to drawing samples from the posterior known as Markov chain Monte Carlo (MCMC), details of which are laid out in Tebaldi et al. (2005) and Smith et al. (2009).

Furrer et al. (2007b) and Furrer et al. (2007a) extended the use of hierarchical modeling for use with multi-model ensembles to include not just regional averages, but the entire spatial grid of the models. Incorporating ideas from spatial statistics (Cressie, 1993, Schabenberger and Gotway, 2004, Cressie and Wikle, 2011), this work expands the two-dimensional field of spatial differences as $\mathbf{D}_i = \boldsymbol{\mu} + \boldsymbol{\epsilon}_i$, where $\mathbf{D}_i = \mathbf{Y}_i - \mathbf{X}_i$, $\mathbf{X}_i$, and $\mathbf{Y}_i$ represent spatial fields summarizing a control run and a transient run from the ith model in the ensemble, respectively. The mean difference field $\boldsymbol{\mu}$ is expanded as a linear combination of basis functions capturing large-scale spatial effects, and the error term $\boldsymbol{\epsilon}_i$ captures the smaller scale spatial structure via an isotropic and stationary process on the sphere. (It should be noted that these spatial models share connections with the Gaussian processes discussed previously; for example, Schabenberger and Gotway, 2004.) MCMC is again used to sample the posterior distribution of the mean difference fields yielding insights to the spatial distribution of climate change for seasonal temperature and precipitation.

These early papers spawned much work on using Bayesian methods to combine the output from multi-model ensembles. Many of the methods discussed have relied on certain assumptions. Among these are fundamental assumptions about the nature of multi-model ensembles, including the basic fact that multi-model ensembles do not represent the statistical idea of a simple random sample and are not representative of some super-population of all climate models. There is also the issue of the climate model output being centered on some true climate or whether the true climate should simply be considered another model in the ensemble. And, of course, there is the issue of model bias. Few approaches have tried to deal with all of these issues. However, Chandler (2013) provides an excellent discussion and a potential solution, again through a Bayesian approach.

20.4.2 Model Weighting

An alternative to using Bayesian approaches to produce a posterior distribution for climate change based on an ensemble of models is using a direct approach for weighting the different models in a multi-model ensemble. There are a number of strategies and algorithms for explicitly determining these weights models, and there is overlap with some of the methods outlined above (see, for example, Tebaldi et al., 2005). These algorithms seek to determine weights, one for each ensemble member, that are then used to compute different relevant statistics. For example, a weighted ensemble mean would be computed as

$$\bar{X}_{\text{weighted}} = \sum_{i=1}^{n} w_i X_i,$$

where $X_i, i = 1, \ldots, n$, denotes a quantity computed from the ith member of the n member ensemble (e.g., global or regional temperature or temperature change, precipitation or precipitation change, etc.) and w_i is the weight assigned to each ensemble member. Typically, the weights are constrained so that $\sum_i w_i = 1$. Uncertainty can be captured through the weighted version of the standard deviation given by

$$s_{\text{weighted}} = \sqrt{\sum_{i=1}^{n} w_i (X_i - \bar{X}_{\text{weighted}})^2}.$$

Similar ideas can be constructed for quantiles, which can be used to capture central tendency (e.g., the median) and uncertainty (e.g., interquartile range).

Brunner et al. (2020b) provide an excellent overview of a number of model weighting schemes in an attempt at an objective comparison framework. Two of those will be briefly discussed here. Both of these methods produce weights that are a product of two competing scores. Letting $w_i = A_i B_i$, the first score A_i represents the skill of the model as measured against observations and the second B_i is a measure of the similarity between models. However, these scores are parameterized and trained quite differently in the two methods.

The first of these methods is the reliability ensemble averaging (REA) method outlined in Giorgi and Mearns (2002, 2003). The weights in REA are defined through a product of reliability factors R_i, where $w_i = R_i / \sum_i R_i$ and

$$R_i = \left(R_{B,i}^m \times R_{D,i}^n \right)^{1/(m \times n)}.$$

These reliability factors are scaled functions of skill or model performance ($R_{B,i}$) and model convergence ($R_{D,i}$) where model performance is assessed by comparing model output to observations and model convergence is assessed by examining how close a given model is to the other models in the ensemble. The parameters m and n are adjusted to emphasize one or the other of the two factors and are typically chosen as $m = n = 1$, and the factors are scaled such that a model with bias and convergence consistent with internal variability would have a maximum reliability factor of $R_i = 1$.

Giorgi and Mearns (2002) study a multi-model ensemble with $n = 9$ models and focus on changes in seasonal temperature and precipitation over 22 land regions over the globe, computed as differences between a future period from 2071 to 2100 and a current period from 1961 to 1990. The bias factor compares the current period of the models against observations over the same period. An iterative approach is used for the convergence factor, comparing each individual model difference against the weighted ensemble mean. The final results show some differences between the weighted ensemble mean and the unweighted mean, and generally lower uncertainty ranges attributed to minimizing the influence of poorly performing models and models that had more extreme or outlying differences. This explanation is reinforced by Nychka and Tebaldi (2003), who offer a more formal statistical interpretation of the convergence criterion and show that the convergence criterion is consistent with a robust regression methodology (e.g., an L_1 regression for $m = n = 1$) that minimizes the influence of heavy tails and the likely presence of models that might be considered outliers.

The second weighting algorithm is referred to as ClimWIP (github.com/lukasbrunner/ClimWIP) and attempts to balance model performance and dependence between models. The approach is used in Lorenz et al. (2018), Merrifield et al. (2020), Brunner et al. (2019), and Brunner et al. (2020a) and builds on the earlier work of Knutti et al. (2017) and Sanderson et al. (2015a,b). Weights in ClimWIP are defined as

$$w_i = \left(e^{-\left(\frac{D_i}{\sigma_D}\right)^2} \right) \times \left(1 + \sum_{j \neq i}^{n} e^{-\left(\frac{S_{ij}}{\sigma_S}\right)^2} \right)^{-1},$$

where D_i is a measure of the distance to observations for the ith model and S_{ij} is a measure of the distance between the ith and jth models. The tuning parameters σ_D and σ_S influence the relative contribution of model performance versus dependence. Larger values of the tuning parameters will approximate equal weighting, while smaller values will lead to fewer models having the most influence. Weights are normalized to sum to one.

Lorenz et al. (2018) examine summer (June, July, and August) temperatures over North and Central North America and examine the use of multiple diagnostic variables in the construction of the weights. Lorenz et al. (2018) provide a recipe on how to choose diagnostics as well as their number. The authors note that a single diagnostic might lead to weights that are overconfident, yet too many will diminish the effectiveness of the weights. They also note that the spread of the multi-model ensemble is less about the uncertainty in the models and more about the ensemble design as often multi-model ensembles are ensembles of convenience rather than on the basis of some well-defined statistical design. However, they suggested that the weighted spread can be interpreted in almost a Bayesian-like way, reflecting uncertainty given everything that is known. Additional research in this thread focuses on different ensembles and different regions as well as methodological advances. For example, Brunner et al. (2019) incorporate multiple observational datasets to account for uncertainty in the historical reconstructions, while Merrifield et al. (2020) incorporate multiple ensembles, including the growing number of initial condition ensembles to address internal variability. Min et al. (2007) discuss an alternative approach to weighting ensembles using Bayesian model averaging which relies on combining probability density functions associated with each ensemble member.

20.5 Final Remarks

Uncertainty quantification continues to grow as an important part of research in climate and the geosciences. Both the Society for Industrial and Applied Mathematics (SIAM) and the American Statistical Association (ASA) have groups focused on uncertainty quantification as well as a jointly sponsored journal (SIAM/ASA Journal on Uncertainty Quantification). SIAM also sponsors a popular conference devoted to uncertainty quantification, and both the American Geophysical Union Fall Meeting and the American Meteorological Society Annual Meeting have many sessions and presentations with a focus on uncertainty quantification.

The statistical frameworks underlying uncertainty quantification also continue to grow, and there are many researchers across the world that are engaged in this research. However, many of these ideas and their application in the geosciences were pioneered through the now-defunct Geophysical Statistics Project (GSP) which was housed at the National Center for Atmospheric Research in Boulder, CO, USA (Hering & Cooley, 2019). The research that was conducted through GSP led to a number of contributions in both statistical methodology and different application areas. Many of these focused on improvements in spatial modeling, including methods for nonstationarity (Nychka et al., 2002, Kleiber and Nychka, 2012), methods for large spatial datasets (Furrer et al., 2006, Nychka et al., 2015), and even alternative models for multivariate spatial data (Sain et al., 2011a). Additional methods for the analysis of climate model ensembles, in particular ensembles of regional climate models, include functional analysis of variance (ANOVA) quantifying different sources of uncertainty (Kaufman and Sain, 2010, Sain et al., 2011b), alternatives to ANOVA including latent variable modeling (Christensen and Sain, 2012), merging information from regional models, global models, and multiple datasets (Heaton et al., 2013), and the value-added associated from downscaling (Parker et al., 2015). In addition to some of the application areas already mentioned here include, climate reconstructions (Oh et al., 2003), paleoclimate reconstructions (Li et al., 2007), spatial extremes (Cooley and Sain, 2010), and weather generators (Kleiber et al., 2012, Kleiber et al., 2013a).

References

Alexeeff, S. E., Nychka, D., Sain, S. R., & Tebaldi, C. (2018). Emulating mean patterns and variability of temperature across and within scenarios in anthropogenic climate change experiments. *Climatic Change, 146*, 319–333. https://doi.org/10.1007/s10584-016-1809-8

Berliner, L. M. (1996). Hierarchical bayesian time series models. In K. M. Hanson & R. N. Silver (Eds.), *Maximum Entropy and Bayesian Methods*. Springer Netherlands (pp. 15–22).

Berliner, L. M. (2003). Physical-statistical modeling in geophysics. *Journal of Geophysical Research-Atmospheres, 108*, 8776. https://doi.org/https://doi.org/10.1029/2002JD002865

Berliner, L. M., Wikle, C. K., & Cressie, N. (2000). Long-lead prediction of Pacific SSTs via bayesian dynamic modeling. *Journal of Climate, 13*, 3953–3968. https://doi.org/10.1175/1520-0442(2001)013<3953:LLPOPS>2.0.CO;2

Box, G. E. P., Hunter, W. G., & Hunter, J. S. (1978). *Statistics for experimenters*. Wiley (653 pp).

Brunner, L., Lorenz, R., Zumwald, M., & Knutti, R. (2019). Quantifying uncertainty in European climate projections using combined performance-independence weighting. *Environmental Research Letters, 14*, 124010. https://doi.org/10.1088/1748-9326/ab492f

Brunner, L., Pendergrass, A. G., Lehner, F., Merrifield, A. L., Lorenz, R., & Knutti, R. (2020a). Reduced global warming from CMIP6 projections when weighting models by performance and independence. *Earth System Dynamics, 11*, 995–1012. https://doi.org/10.5194/esd-11-995-2020

Brunner, L., & Coauthors (2020b). Comparing methods to constrain future European climate projections using a consistent framework. *Journal of Climate, 33*, 8671–8692. https://doi.org/10.1175/JCLI-D-19-0953.1

Castruccio, S., McInerney, D. J., Stein, M. L., Crouch, F. L., Jacob, R. L., & Moyer, E. J. (2014). Statistical emulation of climate model projections based on precomputed GCM runs. *Journal of Climate, 27*, 1829–1844. https://doi.org/10.1175/JCLI-D-13-00099.1

Castruccio, S., & Stein, M. L. (2013). Global space-time models for climate ensembles. *The Annals of Applied Statistics, 7*, 1593–1611. https://doi.org/10.1214/13-AOAS656

Chandler, R. E. (2013). Exploiting strength, discounting weakness: Combining information from multiple climate simulators. *Philosophical Transactions of the Royal Society A, 371*, 20120388. https://doi.org/10.1098/rsta.2012.0388

Chang, W., Haran, M., Olson, R., & Keller, K. (2014). Fast dimension-reduced climate model calibration and the effect of data aggregation. *The Annals of Applied Statistics, 8*, 649–673. https://doi.org/10.1214/14-AOAS733

Christensen, W. F., & Sain, S. R. (2012). Latent variable modeling for integrating output from¬†multiple climate models. *Mathematical Geoscience, 44*, 395–410. https://doi.org/10.1007/s11004-011-9321-1

Cooley, D., & Sain, S. R. (2010). Spatial hierarchical modeling of precipitation extremes from a regional climate model. *Journal of Agricultural, Biological, and Environmental Statistics, 15*, 381–402. https://doi.org/10.1007/s13253-010-0023-9

Cressie, N. (1993). *Statistics for spatial data* (928 pp). Wiley.

Cressie, N., & Wikle, C. (2011). *Statistics for spatio-temporal data* (624 pp). Wiley.

Eyring, V., Bony, S., Meehl, G. A., Senior, C. A., Stevens, B., Stouffer, R. J., & Taylor, K. E. (2016). Overview of the Coupled Model Intercomparison Project Phase 6 (CMIP6) experimental design and organization. *Geoscientific Model Development, 9*, 1937–1958. https://doi.org/10.5194/gmd-9-1937-2016

Forest, C. E., Sanso, B., & Zantedeschi, D. (2008). Inferring climate system properties using a computer model. *Bayesian Analysis, 3*, 1–37. https://doi.org/10.1214/08-BA301

Furrer, R., Genton, M. G., & Nychka, D. (2006). Covariance tapering for interpolation of large spatial datasets. *Journal of Computational and Graphical Statistics, 15*, 502–523. https://doi.org/10.1198/106186006X132178

Furrer, R., Knutti, R., Sain, S. R., Nychka, D. W., & Meehl, G. A. (2007a). Spatial patterns of probabilistic temperature change projections from a multivariate bayesian analysis. *Geophysical Research Letters, 34*, L06711. https://doi.org/10.1029/2006GL027754

Furrer, R., Sain, S. R., Nychka, D., & Meehl, G. A. (2007b). Multivariate bayesian analysis of atmosphere-ocean general circulation models. *Environmental and Ecological Statistics, 14*, 249–266. https://doi.org/10.1007/s10651-007-0018-z

Giorgi, F., & Mearns, L. O. (2002). Calculation of average, uncertainty range, and reliability of regional climate changes from AOGCM simulations via the reliability ensemble averaging (REA) method. *Journal of Climate, 15*, 1141–1158. https://doi.org/10.1175/1520-0442(2002)015<1141:COAURA>2.0.CO;2

Giorgi, F., & Mearns, L. O. (2003). Probability of regional climate change based on the reliability ensemble averaging (REA) method. *Geophysical Research Letters, 30*, 1629. https://doi.org/10.1029/2003GL017130

Heaton, M. J., Greasby, T. A., & Sain, S. R. (2013). Modeling uncertainty in climate using ensembles of regional and global climate models and multiple observation-based data sets. *SIAM/ASA Journal on Uncertainty Quantification, 1*, 535–559. https://doi.org/10.1137/12088505X

Hering, A. S., & Cooley, D. (2019). 20 years of statistics at the National Center for Atmospheric Research. *CHANCE, 32*, 40–43. https://doi.org/10.1080/09332480.2019.1695440

Higdon, D., Gattiker, J., Williams, B., & Rightley, M. (2008). Computer model calibration using high-dimensional output. *Journal of the American Statistical Association, 103*, 570–583. https://doi.org/10.1198/016214507000000888

Holden, P. B., & Edwards, N. R. (2010). Dimensionally reduced emulation of an AOGCM for application to integrated assessment modelling. *Geophysical Research Letters, 37*. https://doi.org/10.1029/2010GL045137

Kaufman, C. G., & Sain, S. R. (2010). Bayesian functional ANOVA modeling using Gaussian process prior distributions. *Bayesian Analysis, 5*, 123–149. https://doi.org/10.1214/10-BA505

Kay, J. E., & Coauthors (2015). The community earth system model (CESM) large ensemble project: A community resource for studying climate change in the presence of internal climate variability. *Bulletin of the American Meteorological Society, 96*, 1333–1349. https://doi.org/10.1175/BAMS-D-13-00255.1

Kennedy, M. C., & O'Hagan, A. (2001). Bayesian calibration of computer models. *Journal of the Royal Statistical Society Series B (Statistical Methodology), 63*, 425–464. https://doi.org/10.1111/1467-9868.00294

Kleiber, W., Katz, R. W., & Rajagopalan, B. (2012). Daily spatiotemporal precipitation simulation using latent and transformed Gaussian processes. *Water Resources Research, 48*. https://doi.org/10.1029/2011WR011105

Kleiber, W., Katz, R. W., & Rajagopalan, B. (2013a). Daily minimum and maximum temperature simulation over complex terrain. *The Annals of Applied Statistics, 7*, 588–612. https://doi.org/10.1214/12-AOAS602

Kleiber, W., & Nychka, D. (2012). Nonstationary modeling for multivariate spatial processes. *Journal of Multivariate Analysis, 112*, 76–91. https://doi.org/10.1016/j.jmva.2012.05.011

Kleiber, W., Sain, S. R., Heaton, M. J., Wiltberger, M., Reese, C. S., & Bingham, D. (2013b). Parameter tuning for a multi-fidelity dynamical model of the magnetosphere. *The Annals of Applied Statistics, 7*, 1286–1310. https://doi.org/10.1214/13-AOAS651

Knutti, R., Sedláček, J., Sanderson, B. M., Lorenz, R., Fischer, E. M., & Eyring, V. (2017). A climate model projection weighting scheme accounting for performance and interdependence. *Geophysical Research Letters, 44*, 1909–1918. https://doi.org/10.1002/2016GL072012

Li, B., Nychka, D. W., & Ammann, C. M. (2007). The 'hockey stick' and the 1990s: A statistical perspective on reconstructing hemispheric temperatures. *Tellus A: Dynamic Meteorology, 59*, 591–598. https://doi.org/10.1111/j.1600-0870.2007.00270.x

Lorenz, R., Herger, N., Sedláček, J., Eyring, V., Fischer, E. M., & Knutti, R. (2018). Prospects and caveats of weighting climate models for summer maximum temperature projections over North America. *Journal of Geophysical Research: Atmospheres, 123*. 4509–4526. https://doi.org/10.1029/2017JD027992

Mearns, L. O., & Coauthors (2012). The North American Regional Climate Change Assessment Program: Overview of Phase I results. *Bulletin of the American Meteorological Society, 93*, 1337–1362. https://doi.org/10.1175/BAMS-D-11-00223.1

Mearns, L. O., & Coauthors (2013). Climate change projections of the North American regional climate change assessment program (NARCCAP). *Climatic Change, 120*, 965–975. https://doi.org/10.1007/s10584-013-0831-3

Merrifield, A. L., Brunner, L., Lorenz, R., Medhaug, I., & Knutti, R. (2020). An investigation of weighting schemes suitable for incorporating large ensembles into multi-model ensembles. *Earth System Dynamics, 11*, 807–834. https://doi.org/10.5194/esd-11-807-2020

Min, S.-K., Simonis, D., & Hense, A. (2007). Probabilistic climate change predictions applying bayesian model averaging. *Philosophical Transactions of the Royal Society of London, Series A, 365*, 2103–2116. https://doi.org/10.1098/rsta.2007.2070

Mitchell, J. F. B., Johns, T. C., Eagles, M., Ingram, W. J., & Davis, R. A. (1999). Towards the construction of climate change scenarios. *Climate Change, 41*, 547–581. https://doi.org/10.1023/A:1005466909820

Mitchell, T. D. (2003). Pattern scaling: An examination of the accuracy of the technique for describing future climates. *Climatic Change, 60*, 217–242. https://doi.org/10.1023/A:1026035305597

National Research Council (2012). *Assessing the reliability of complex models: Mathematical and statistical foundations of verification, validation, and uncertainty quantification*. The National Academies Press. https://doi.org/10.17226/13395

Nychka, D., Bandyopadhyay, S., Hammerling, D., Lindgren, F., & Sain, S. (2015). A multiresolution gaussian process model for the analysis of large spatial datasets. *Journal of Computational and Graphical Statistics, 24*, 579–599. https://doi.org/10.1080/10618600.2014.914946

Nychka, D., & Tebaldi, C. (2003). Comments on Calculation of Average, Uncertainty Range, and Reliability of Regional Climate Changes from AOGCM Simulations via the Reliability Ensemble Averaging (REA) Method. *Journal of Climate, 16*, 883–884. https://doi.org/10.1175/1520-0442(2003)016<0883:COCOAU>2.0.CO;2

Nychka, D., Wikle, C., & Royle, J. A. (2002). Multiresolution models for nonstationary spatial covariance functions. *Statistical Modelling, 2*, 315–331. https://doi.org/10.1191/1471082x02st037

Oh, H.-S., Ammann, C. M., Naveau, P., Nychka, D., & Otto, B., Bliesner, L. (2003). Multi-resolution time series analysis applied to solar irradiance and climate reconstructions. *Journal of Atmospheric and Solar - Terrestrial Physics, 65*, 191–201. https://doi.org/10.1016/S1364-6826(02)00291-2

Parker, R. J., Reich, B. J., & Sain, S. R. (2015). A multiresolution approach to estimating the value added by regional climate models. *Journal of Climate, 28*, 8873–8887. https://doi.org/10.1175/JCLI-D-14-00557.1

Sacks, J., Welch, W., Mitchell, T., & Wynn, H. (1989). Design and analysis of computer experiments. *Statistical Science, 4*, 409–435. https://doi.org/10.1214/ss/1177012413

Sain, S. R., Furrer, R., & Cressie, N. (2011a). A spatial analysis of multivariate output from regional climate models. *The Annals of Applied Statistics, 5*, 150–175. https://doi.org/10.1214/10-AOAS369

Sain, S. R., Nychka, D., & Mearns, L. (2011b). Functional ANOVA and regional climate experiments: A statistical analysis of dynamic downscaling. *Environmetrics, 22*, 700–711. https://doi.org/10.1002/env.1068

Sanderson, B. M., Knutti, R., & Caldwell, P. (2015a). Addressing interdependency in a multimodel ensemble by interpolation of model properties. *Journal of Climate, 28*, 5150–5170. https://doi.org/10.1175/JCLI-D-14-00361.1

Sanderson, B. M., Knutti, R., & Caldwell, P. (2015b). A representative democracy to reduce interdependency in a multimodel ensemble. *Journal of Climate, 28*, 5171–5194. https://doi.org/10.1175/JCLI-D-14-00362.1

Sanso, B., & Forest, C. (2009). Statistical calibration of climate system properties. *Journal of the Royal Statistical Society, Series C, 58*, 485–503. https://doi.org/https://doi.org/10.1111/j.1467-9876.2009.00669.x

Santer, B., Wigley, T., Schlesinger, M., & Mitchell, J. (1990). *Developing climate scenarios from equilibrium GCM results, MPI Report Number 47*. Hamburg, Germany.

Santner, T. J., Williams, B. J., & Notz, W. I. (2003). *The design and analysis of computer experiments* (284 pp). Springer.

Schabenberger, O., & Gotway, C. (2004). *Statistical methods for spatial data analysis* (504 pp). Taylor & Francis.

Smith, R. L., Tebaldi, C., Nychka, D., & Mearns, L. O. (2009). Bayesian modeling of uncertainty in ensembles of climate models. *Journal of the American Statistical Association, 104*, 97–116. https://doi.org/10.1198/jasa.2009.0007

Stahl, S. (2006). The evolution of the normal distribution. *Mathematics Magazine, 79*, 96–113. https://doi.org/10.1080/0025570X.2006.11953386

Stainforth, D. A., & Coauthors (2005). Uncertainty in predictions of the climate response to rising levels of greenhouse gases. *Nature, 433*, 403–406. https://doi.org/10.1038/nature03301

Tebaldi, C., & Arblaster, J. M. (2014). Pattern scaling: Its strengths and limitations, and an update on the latest model simulations. *Climate Change, 122*, 459–471. https://doi.org/10.1007/s10584-013-1032-9

Tebaldi, C., Armbruster, A., Engler, H. P., & Link, R. (2020). Emulating climate extreme indices. *Environmental Research Letters, 15*, 074006. https://doi.org/10.1088/1748-9326/ab8332

Tebaldi, C., Mearns, L. O., Nychka, D., & Smith, R. L. (2004). Regional probabilities of precipitation change: A bayesian analysis of multimodel simulations. *Geophysical Research Letters, 31*, L24213. https://doi.org/10.1029/2004GL021276

Tebaldi, C., Smith, R. L., Nychka, D., & Mearns, L. O. (2005). Quantifying uncertainty in projections of regional climate change: A bayesian approach to the analysis of multimodel ensembles. *Journal of Climate, 18*, 1524–1540. https://doi.org/10.1175/JCLI3363.1

Wikle, C. K., Milliff, R. F., Nychka, D., & Berliner, L. M. (2001). Spatiotemporal hierarchical bayesian modeling tropical ocean surface winds. *Journal of the American Statistical Association, 96*, 382–397. https://doi.org/10.1198/016214501753168109

Williamson, D., Goldstein, M., Allison, L., Blaker, A., Challenor, P., Jackson, L., & Yamazaki, K. (2013). History matching for exploring and reducing climate model parameter space using observations and a large perturbed physics ensemble. *Climate Dynamics, 41*, 1703–1729. https://doi.org/10.1007/s00382-013-1896-4

Williamson, D., Goldstein, M., & Blaker, A. (2012). Fast linked analyses for scenario-based hierarchies. *Journal. Royal Statistical Society, 61*, 665–691. https://doi.org/10.1111/j.1467-9876.2012.01042.x

Open Access This chapter is licensed under the terms of the Creative Commons Attribution 4.0 International License (http://creativecommons.org/licenses/by/4.0/), which permits use, sharing, adaptation, distribution and reproduction in any medium or format, as long as you give appropriate credit to the original author(s) and the source, provide a link to the Creative Commons license and indicate if changes were made.

The images or other third party material in this chapter are included in the chapter's Creative Commons license, unless indicated otherwise in a credit line to the material. If material is not included in the chapter's Creative Commons license and your intended use is not permitted by statutory regulation or exceeds the permitted use, you will need to obtain permission directly from the copyright holder.

Uncertainty and Extremes

Mark D. Risser and Claudia Tebaldi

21.1 Introduction

An extreme event is an episode of unusual weather whose timescale can be as short as several hours or as long as multiple years, often related to temperature (e.g., heat or cold extremes), precipitation (e.g., heavy downpours or droughts), or storms (e.g., tropical cyclones, atmospheric rivers, mesoscale convective systems). The important aspects of a particular event include its duration (or length) and intensity (magnitude or severity), as well as its expected frequency of occurrence. Although extreme events are infrequent by definition, they can have large impacts on human and natural systems causing damages to infrastructure, loss of property, habitat, and life. Multivariate extreme events (e.g., a period that is unusually hot and dry) or concurrent extreme events, either in space or time (e.g., heat waves affecting at the same time multiple regions, or multiple heavy rainfall events happening in close temporal succession) can have even larger impacts.

The European heat wave of July and August, 2003 was the hottest summer on record in Europe since at least 1540, and had a death toll of over 70,000 (Stott et al., 2004). A severe storm system brought 17 inches of rain in the course of 4 days, over September 9–12, 2013, to Boulder County in northern Colorado, where the average annual rainfall is only 20.7 inches. The resulting flooding led to over $2 billion in damages, including homes, roads, and other infrastructure, and agricultural losses (Pall et al., 2017). A heat wave in Pakistan in 2015 was characterized by unusually high temperatures (in excess of 120°F) and high relative humidity, leading to over 2000 deaths (Wehner et al., 2016). In August 2017, a total of at least 735 mm of precipitation fell on the Houston, Texas, region over 7 days during Hurricane Harvey, which more than doubled the previous record seven-day storm total over the same region (dating back to 1950; Risser & Wehner, 2017). The rainfall contributed to almost 60 deaths (Jonkman et al., 2018) and up to $90 billion USD in losses (Frame et al., 2020); we return to consider Hurricane Harvey precipitation further in Sect. 21.2. Each of these events was characterized by an episode of unusual weather and had significant detrimental effects on social and natural systems. For an up-to-date summary of other extreme events, also placed in a historical context, the National Oceanic and Atmospheric Administration's National Center for Environmental Information now publishes a State of the Climate report that summarizes extreme events globally, including heat waves, major storm systems, droughts, heavy precipitation, sea ice extent, and wildfires (see https://www.ncdc.noaa.gov/sotc/).

Given the importance of extreme events in driving climate and weather-related impacts, any quantitative analysis of the risk of those impacts (currently and in the future) hinges on a characterization of the statistics of extreme events, including, importantly, their uncertainty. And while uncertainty quantification is critical and often challenging for any statistical analysis of data, it is particularly important when dealing with extremes, since extremes are by definition rare, resulting in a reduced number of measurements and therefore less robust estimates. Extreme Value Analysis (EVA) aims specifically at addressing the characterization of the behavior of rare events from the tail of a distribution.

The chapter proceeds as follows: First, in Sect. 21.2, we present a motivating example for the value of EVA; in Sect. 21.3 we describe a variety of metrics for characterizing extremes. Then, in Sect. 21.4 we discuss several widely used methods for univariate EVA and introduce corresponding

M. D. Risser (✉)
Earth & Environmental Sciences, Lawrence Berkeley National Laboratory, Berkeley, CA, USA
e-mail: mdrisser@lbl.gov

C. Tebaldi
Joint Global Change Research Institute, Pacific Northwest National Laboratory, College Park, MD, USA
e-mail: claudia.tebaldi@pnnl.gov

© The Author(s) 2025
L. O. Mearns et al. (eds.), *Uncertainty in Climate Change Research*,
https://doi.org/10.1007/978-3-031-85542-9_21

methods for characterizing uncertainty in Sect. 21.5. Finally, we conclude with a discussion of extreme event attribution in Sect. 21.6, an area of climate research which seeks to quantify the role of anthropogenic forcings (e.g., greenhouse gas emissions) in increasing (or decreasing) the chances of a particular extreme weather event occurring.

21.2 Motivating Example

Before describing specific tools and methods, we first illustrate the importance and value of using extreme value analysis for the statistical characterization of a rare event by considering a specific example. Hurricane Harvey made landfall on the Texas coast in late August 2017, as a category 4 storm. Instead of moving inland, the storm stalled with a portion of the system remaining over the warm Gulf of Mexico waters for four more days, leading to an unprecedented amount of rainfall across the greater Houston area (Kunkel & Champion, 2019). Beaumont, Texas, was one of the many urban areas in the path of the storm; a Global Historical Climate Network (GHCN) weather station positioned in the city (station USC00410611, located at 30.0969°N and 94.0997°W) recorded a daily total of 368.3 mm on August 30, 2017, and this amount fell on top of precipitation from the three previous days totalling 125.7 mm, 258.6 mm, and 116.8 mm, respectively. To put the storm in historical context, the GHCN also provides daily data from the Beaumont, Texas, station over the past century, from 1902 to 2016 (Menne et al., 2012). Daily totals from hurricane season (July to November only) over this time period are shown in Fig. 21.1; note that we have specifically excluded 2017, the year of Hurricane Harvey. Over the past 115 years, approximately 71.8% of days experience no rain; on days with nonzero rainfall, the average (median) daily total is 14.5 mm (6.9 mm), and the largest daily total on record prior to 2017 is 297.2 mm.

In the aftermath of the storm, there was great public interest in quantifying exactly how unusual it was to receive 368.3 mm of precipitation on a single day in Beaumont, Texas. Using the historical data from the GHCN record, we can estimate the probability of experiencing more than 368.3 mm on a single day; formally, we need to answer the question "what is the probability that the maximum daily precipitation total during hurricane season is larger than 368.3 mm?" Given that the previous record for this weather station was 297.2 mm, answering the question means having to extrapolate into the tail of the distribution of hurricane season daily precipitation. We have two choices in using the time series from Beaumont to answer this question: Either (1) use all of the (nonzero) data to characterize the entire distribution of daily precipitation or (2) utilize only the "extreme" data, focusing our analysis on the tail of that distribution.

Modeling All Precipitation Data First, we can use all of the daily precipitation data from July to November. However, we first need to account for the fact that many days have zero precipitation; as such, we focus on the approximately 28.2% of days with nonzero precipitation. For these data, a common statistical model is to assume that the nonzero daily precipitation measurements follow a Gamma distribution, which is a probability distribution for nonnegative quantities (i.e., greater than zero) and right-skewed with a long right tail. The Gamma distribution is characterized by two parameters that define its mean and standard deviation. Using maximum likelihood, i.e., computing the probability of the data as a function of the Gamma parameters, and finding the values of those parameters that maximize that probability, we can estimate these parameters for the nonzero precipitation measurements, resulting in a mean of 14.50 mm and a standard deviation of 17.65 mm. A histogram of the nonzero values with the fitted Gamma distribution is shown in Fig. 21.2a.

Returning to our question of interest, recall that we want to estimate the probability that the maximum daily precipitation total during hurricane season is larger than 368.3 mm. Using the notation $P(A)$ to represent the probability of a particular event A occurring, using the fitted Gamma distribution, and defining X_t to be the daily precipitation during hurricane season on an arbitrary day t, we can first calculate the probability of a particular day exceeding 368.3 mm:

$$P(X_t > 368.3 \text{ mm}) = P(X_t > 368.3 \text{mm}|$$

non-zero precipitation) $\times P($ non-zero precipitation$)$

(here, "|" means "conditional on")

$$= (1 - F(368.3)) \times 0.282 = (1.048 \times 10^{-8}) \times 0.282$$

$$= 2.961 \times 10^{-9}$$

(here, F is the cumulative distribution function of the fitted Gamma distribution, and recall that 28.2% of days have nonzero precipitation). Clearly, it is very unlikely to experience this daily rainfall total on any particular day. However, to answer our question about the largest daily rainfall total over an entire 153-day hurricane season (there are 153 days from July to November), we are not quite finished:

$$P(\text{seasonal maximum} > 368.3\text{mm})$$
$$= 1 - P(\text{all 153 days are } < 368.3)$$
$$= 1 - (1 - P(X_t > 368.3\text{mm}))^{153}$$
$$= 1 - (1 - 2.961 \times 10^{-9})^{153} = 4.531 \times 10^{-7}$$

(note that this calculation assumes independence of daily observations). To put things into perspective, this means

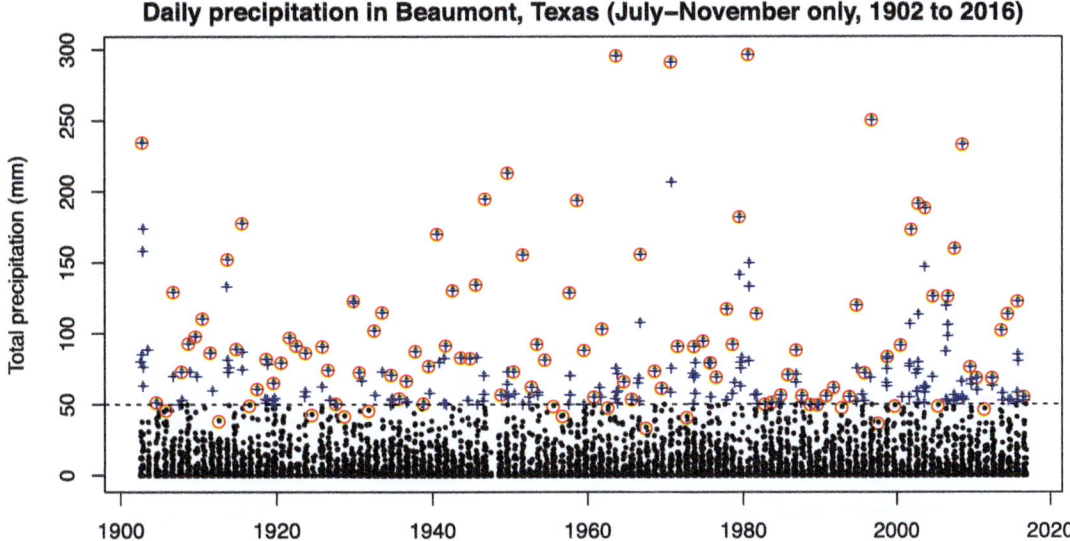

Fig. 21.1 Daily precipitation totals (mm) in hurricane season (July–November) for 1902 to 2016 at a Global Historical Climate Network (GHCN) station located in Beaumont, Texas (station USC00410611, located at 30.0969°N and 94.0997°W). Circled points represent the largest value in each season-year; points with a "+" represent all values larger than a cutoff of 50 mm

that we would expect a hurricane season daily maximum of ≥ 368.3 mm to occur once in every $1/(4.531 \times 10^{-7}) = 2{,}207{,}260$ years. Even after accounting for the entire season of measurements, this event is still extremely unusual!

A quantile-quantile (or Q-Q) plot is a useful tool for assessing the goodness of the Gamma distribution fit to these data: See Fig. 21.2c. If the statistical model is appropriate, the points will fall along the 45° line; the model appears to do a pretty good job for daily measurements up to about 60 mm (note that almost 97% of the data are less than 60 mm). However, for measurements larger than 60 mm, the points quickly diverge from the 45° line, meaning that the Gamma distribution provides a very poor fit for the upper tail of the distribution, and hence our estimate of the probability of exceeding 368.3 mm is likely wrong.

Modeling Hurricane Season Maxima Alternatively, we can fit a statistical model to the largest daily precipitation total from each hurricane season. These seasonal maxima are highlighted with red circles in Fig. 21.1, and a histogram of these values is shown in Fig. 21.2b. We can then fit the Generalized Extreme Value (GEV) distribution to these data (we will discuss in Sect. 21.4 why this is the correct distribution), which is also shown in Fig. 21.2b. Using the fitted GEV distribution, we now estimate that the probability of the seasonal maximum exceeding 368.3 mm is 0.0147, meaning that we would expect this to occur once in every $1/0.0147 \approx 68$ years. The Q-Q plot for the seasonal maxima versus the fitted GEV distribution is shown in Fig. 21.2d; clearly, the GEV distribution provides a good fit for the seasonal maxima, which gives us confidence that this much larger estimate of the probability is more likely to be correct.

In conclusion, using only the "extreme" measurements of daily precipitation leads to a much more reasonable estimate of experiencing the rainfall that actually occurred in Beaumont, TX, during Hurricane Harvey in 2017. Why is this the right strategy? Intuitively, one explanation is that the phenomena that generate extreme rainfall totals (e.g., major hurricanes) are fundamentally different than those that generate typical daily measurements. As such, we should focus on measurements of the phenomena of interest to quantify probabilities of extreme events occurring.

21.3 Metrics for Characterizing Extremes

In Sect. 21.2 we encountered a general principle of analyzing the extreme values of a physical process: "let the tails speak for themselves." In other words, it is important to use only the extreme (very large or very small) measurements when calculating probabilities related to extreme events, because the processes that drive extreme events are often quite different from the processes that drive more typical weather and climate events. We now describe several ways of selecting and summarizing the extreme values from a time series of interest.

21.3.1 Derived Numerical Summaries

The availability of daily or even subdaily data (observations or model output) allows for a complete freedom of choice, in terms of extreme characterization. Extreme temperature events can be defined in terms of high (or low) quantiles of

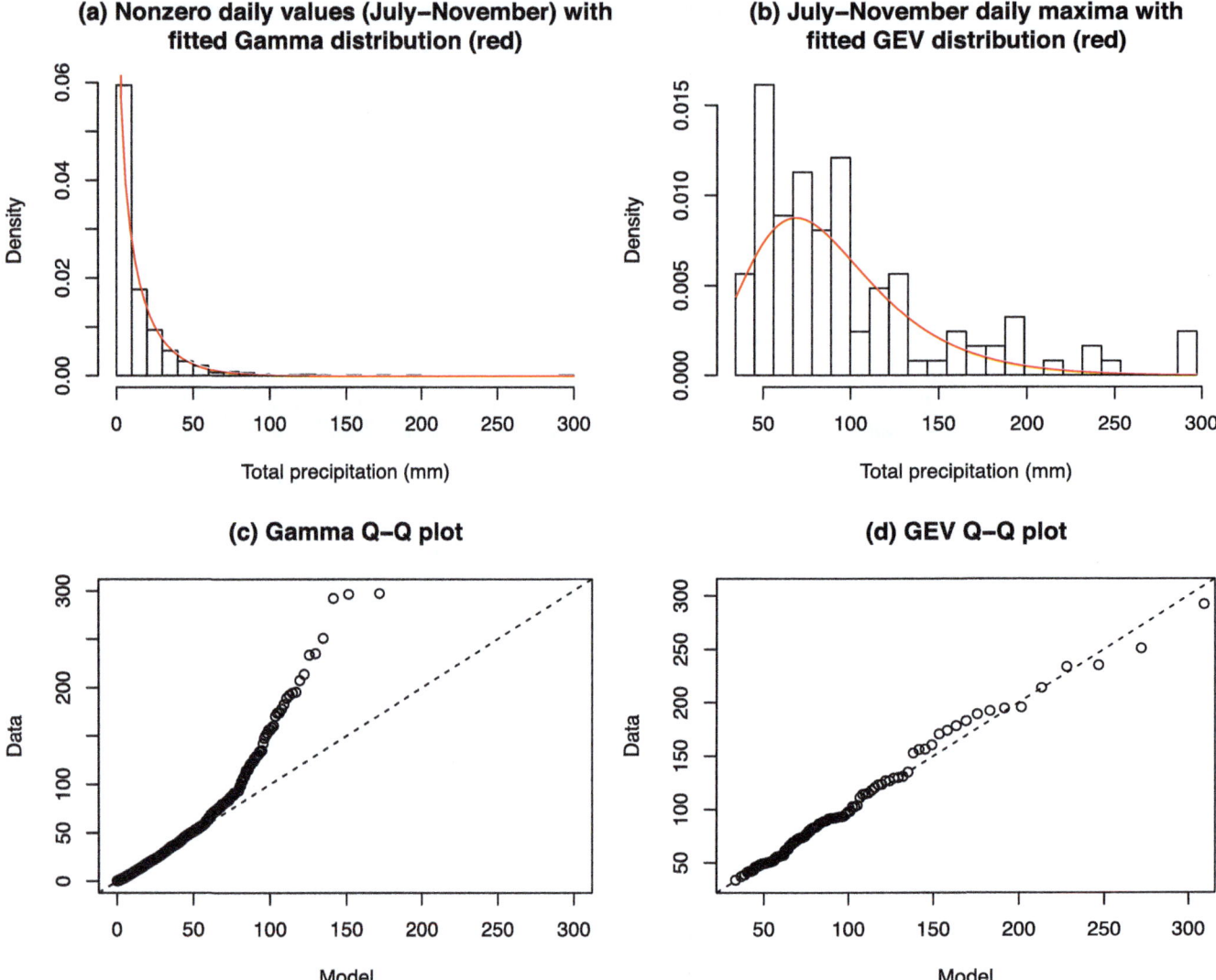

Fig. 21.2 Histograms, fitted statistical models, and quantile-quantile plots for daily precipitation totals during hurricane season (July to November) in Beaumont, TX, fitting a Gamma distribution to all measurements greater than zero (panels (**a**) and (**c**)) versus fitting a generalized extreme value (GEV) distribution to the seasonal maxima (panels (**b**) and (**d**)). (**a**) Nonzero daily values (July–November) with fitted Gamma distribution (red). (**b**) July-November daily maxima with fitted GEV distribution (red). (**c**) Gamma Q-Q plot. (**d**) GEV Q-Q plot

the distribution of daily temperatures, or of that of multiday temperature spells. EVA analysis can be applied, and any statistics of the tail of the distribution can be computed. But at times data availability is an issue, especially when the granularity required is as fine as daily. For some national weather services, daily data is considered proprietary. In those circumstances, it has been found useful to define metrics of extreme that are summary of daily behavior and can efficiently be recorded, stored, and shared. This was the motivation of the Expert Team on Climate Change Detection and Indices (ETCCDI) effort (Alexander, 2016) through which a set of metrics of extremes was defined, and incentives to record, archive, and share them across the world fostered. At the same time, ETCCDI indices have become popular object of analysis for climate model projections (Tebaldi et al., 2006; Sillmann et al., 2013; Kharin et al., 2007; Aerenson et al., 2018), in not small part due to the possibility of confronting model output with observational analogues with good coverage over the world. ETCCDI indices were defined with attention to impacts (on human health, agriculture, water resources). A complete list of the metrics can be found at http://etccdi.pacificclimate.org/list_27_indices.shtml, and indices computed for a number of experiments are available at https://www.climdex.org/. Incidentally, some of these indices are defined as block maxima, or minima, over a season or a year, and are therefore amenable to a treatment similar to our example in Sect. 21.2, where we fitted a GEV to the seasonal maxima of precipitation (see later Sects. 21.3.3 and 21.4.1).

21.3.2 Counting Methods

The most straightforward way to estimate probabilities related to extreme events is to use a counting or Binomial approach. This method requires essentially no assumptions on the underlying data, other than the fact that there must be a set of replicated data that are independent and arise from a single distribution. For example, this requirement is satisfied when considering a large ensemble of climate model runs, where each ensemble member has perturbed initial conditions such that (after a suitable amount of spin-up time) the members represent different plausible (and independent) realizations of the climate system. This condition might also be satisfied when considering temporally averaged quantities, e.g., monthly or annual precipitation or temperature, although in this case there might be other considerations such as seasonality or long-term trends to worry about. For now, let us suppose that we have n realizations of a particular variable of interest that can plausibly be considered independent and arising from a single distribution. We can then estimate the probability of an extreme event by simply counting the number of realizations for which a particular event occurs and dividing by the sample size. For example, suppose that we have an ensemble of $n = 100$ climate model runs from a particular year, and we are interested in the probability of the global mean temperature exceeding 22°C. Then, we can simply count the number of ensembles that have a global average temperature in excess of 22°C, say 9, such that our estimate of the global mean temperature exceeding 22°C is $9/100 = 0.09$. This approach is facilitated, of late, by the availability of a number of the so-called large initial condition ensembles that several modeling centers have produced under standardized, and therefore comparable, scenarios (Deser et al., 2020).

21.3.3 Block Maxima

A more traditional way of summarizing extremes that is related to the generalized extreme value distribution (see Sect. 21.4.1) is the block maximum, i.e., the largest measurement over a prespecified "block" or fixed window of time. Usually, the measurements correspond to some relatively high-frequency time period (e.g., a day), and the block is defined as a much lower-frequency time period (e.g., a season or year). Indeed, the annual daily maxima is the same as Rx1Day defined in Sect. 21.3.1; however, this could also be the largest hourly measurement from a week, or alternatively the largest monthly average from a decade. In general, a block maximum is often used when the "block" consists of a very large number of the higher-frequency measurements (e.g., 365 days in a year).

21.3.4 Exceedances of a Threshold

Another method for extracting the extreme values of a time series is to select the exceedances of a high threshold. This approach is also related to traditional extreme value analysis methods (see Sect. 21.4.2). Often, the threshold is set to be some large percentile of the measurements of interest, for example, the 90th, 95th, or 98th percentile. The threshold could also be chosen to represent a value that is relevant for design standards or climate impacts, for example, a daily precipitation total that is known to cause flooding when exceeded, or a value of temperature that has been linked to crop failure or human health impacts.

21.4 Extreme Value Analysis

Extreme value analysis, or the study of rare events, is a branch of mathematical statistics that seeks to provide a formal framework for characterizing extremes and their uncertainty. The goal of an extreme value analysis (EVA) is to quantify the magnitude or severity of a worst-case scenario, which often requires extrapolation to events that have not actually occurred. EVA is applied in a wide range of disciplines, including hydrology (stream and river flows, flooding), weather and climate (as described in Sect. 21.1), finance, insurance, and engineering (structural design, failure). While "ordinary" statistics most often seek to characterize the mean or center of a distribution, EVA instead sets out to characterize the "tails" of a distribution, i.e., the very small and very large values. EVA does this by deriving theoretical results and formulas for how to properly carry out the extrapolation needed for a particular study. However, it is important to note that each of the following techniques arise as mathematical results in an "asymptotic" or artificial case where an unlimited amount of data are available; on the other hand, for any specific analysis, we of course have only a finite, limited amount of data. As such, it is extremely important to ensure that the underlying mathematical assumptions are reasonably satisfied when applying these methods to real data.

We now describe two traditional methods for analyzing the extreme values of a general physical process. For a full treatment, we refer the interested reader to Coles (2001). Note that both of the following methods apply to extremely large values of a given process; however, both can also be applied to extremely small values. Furthermore, we emphasize that the methods described in this section are for univariate extremes (i.e., the extremes of a single variable of interest) for which the underlying data exhibit minimal autocorrelation. Analyzing multivariable extremes (e.g., simultaneously high temperature and relative humidity), co-occurring extremes (e.g., when two nearby weather stations both experience large daily precipitation totals), or the extremes of a variable with strong temporal autocorrelation is a much more difficult task and is beyond the scope of this chapter.

21.4.1 Generalized Extreme Value Distribution

The generalized extreme value (GEV) distribution is the cornerstone of extreme value theory and provides a statistical model for maximum of a sequence of independent values that arise from a common underlying distribution. Referring back to Sect. 21.3.3, we use the GEV distribution for characterizing block maxima (as was done in the example in Sect. 21.2).

Before describing why the GEV distribution is the correct distribution for maxima, first recall that much of statistical theory is based on the Central Limit Theorem, which states that the Normal distribution is the right distribution to use as a statistical model for averages. The critical point is that regardless of the underlying distribution that generates a set of measurements, the sampling distribution for the average of the measurements is well approximated by a Normal distribution (assuming certain conditions are met). The Normal distribution is then characterized by two statistical parameters: the mean (which corresponds to an average value) and the standard deviation (which describes the variability about the mean).

The corresponding result for the maximum of a set of measurements is the Extremal Types Theorem (see, e.g., Section 3.1.2 of Coles, 2001), which states that the GEV distribution is the correct theoretical distribution for maxima. While the Normal distribution is characterized by two statistical parameters (the mean and standard deviation), the GEV family of distributions is characterized by three parameters. The first parameter is called the *location parameter* (often denoted with μ), which describes the center of the distribution; the value of the location parameter is a "typical" maxima. The second parameter is called the *scale parameter* (often denoted with σ), which describes the spread or width of the distribution. Finally, the third parameter is called the *shape parameter* (often denoted with ξ), which is a unitless quantity that describes the upper tail behavior of the GEV distribution. If $\xi < 0$, the distribution has a finite upper bound; if $\xi > 0$, the distribution has no upper limit and a very heavy upper tail; if $\xi = 0$, the distribution is again unbounded but has a "lighter" tail. Referring back to the example in Sect. 21.2, for the Beaumont, TX, time series, we estimated that the location parameter is 68.53 mm (meaning that, on average, a typical annual maximum precipitation total is approximately 70 mm), the scale parameter is 28.71 mm (meaning that the year-to-year variability is approximately 30 mm), and the shape parameter is 0.38 (meaning that the distribution of hurricane season maxima is unbounded with a heavy upper tail).

While the statistical parameters of the GEV distribution are sometimes of direct interest, we are often more interested in summaries of the distribution. The form of the GEV distribution allows us to write down formulas for each of the following based on the statistical parameters $\{\mu, \sigma, \xi\}$ (see Chapter 3 of Coles, 2001):

1. **Return value** (sometimes referred to as a **return level**): a particular quantile or percentile of the extreme value distribution; in other words, a daily precipitation total x such that the probability of exceeding x is 0.1, 0.05, or 0.01.
2. **Return probability:** the chance of a exceeding a particular value; for example, what is the probability of exceeding 100 mm day^{-1}?
3. **Return period:** one divided by the return probability, i.e., how often (on average) a particular value will be exceeded; for example, if the probability of exceeding 100 mm day^{-1} is $0.05 = 1/20$, then 100 mm day^{-1} is the 20-year return period.

As with all statistical methods (but particularly for extreme value analysis), it is extremely important to ensure that the underlying assumptions are sufficiently met before proceeding with a data analysis. When applying the GEV distribution, there are three conditions that must be satisfied: First, much like the Central Limit Theorem, the GEV distribution applies when the sample size is large—here, the sample size refers to the number of measurements in each "block." What is considered "large enough" is example-specific and depends on the similarity of the distribution of the data (e.g., the daily measurements) with the GEV distribution. Second, the GEV can be applied when the measurements that comprise each block are independent, meaning there is no (or minimal) autocorrelation. Third, the measurements in the block must be "identically distributed," meaning that they all arise from a common distribution. This final requirement is likely to be violated for measurements of daily precipitation in locations where there is strong seasonality or large variation in the phenomena that cause extreme events (e.g., hurricanes versus mesoscale convective systems). Of course, for any real data application, it is unlikely that these assumptions will be perfectly met, but it is nonetheless important to be aware of the required assumptions, ensure that they are sufficiently met, and clearly communicate the assumptions when reporting any results.

21.4.2 Generalized Pareto Distribution

While the GEV distribution is an extremely useful tool for extreme value analysis, the fact that it is based on block maxima means that a significant amount of data regarding the extreme behavior of interest is thrown away. For example, when extracting the largest daily precipitation total in a year (Rx1Day), the second, third, fourth, and fifth largest daily precipitation totals are ignored. An alternative approach is the one described in Sect. 21.3.4, where we regard all events that exceed a particular threshold as "extreme." In this case, the appropriate distribution for statistically modeling threshold exceedances is the generalized Pareto distribution (GPD), which is defined by two statistical parameters in addition to the specification of the threshold (see Chapter 4 of Coles, 2001, for further information). These two parameters are the scale and shape which, similarly to the GEV distribution, control the spread of the distribution and the upper tail behavior of the GPD, which can be bounded or unbounded with either a light or heavy tail. Also like the GEV distribution, the GPD distribution can be used to estimate return values, return probabilities, and return periods.

An important distinction is that the GPD distribution uses all "extreme" values to estimate the extreme statistics of the process of interest, while the GEV distribution uses a single maxima from each block. As such, the GPD approach is preferred since it uses more of the data and hence should provide a reduction in the uncertainty. However, for this approach to be appropriately applied, it is very important that the threshold is chosen carefully: If the threshold is too small, values that are not actually extreme will be included in the analysis which will bias the resulting extreme statistics. On the other hand, if the threshold is too large, there will be very few exceedances resulting in large uncertainty. The literature around the GPD provides several ways to choose the threshold, all of which specify as low a threshold as possible subject to the quality of the resulting statistical fit. Two traditional approaches are the mean residual life plot and the threshold stability plot, described in detail in Coles (2001), both of which involve choosing a threshold and assessing qualitative aspects of the fitted distribution. Other approaches seek to estimate this threshold directly from the data in conjunction with fitting the GPD distribution parameters (see, e.g., Scarrott & MacDonald, 2012). However, many of these approaches involve a subjective choice, and ultimately the best strategy is to conduct a sensitivity analysis wherein one fits a GPD for a range of thresholds to make sure the final conclusions do not depend on a specific threshold choice.

One final consideration when using the GPD to analyze threshold exceedances has to do with dealing with clusters of extreme events, e.g., when a time series of daily precipitation has large values on successive days, most often from a single storm that persists for more than one day. The theory underlying the GPD approach requires that the exceedances used to estimate the extreme statistics are independent (meaning that they do not contain autocorrelation). The case in which a single weather station experiences an extreme storm for more than one day is a case where the successive large daily measurements are *not* independent since they arose from the same event. As such, a preprocessing step that should be done in advance of any GPD analysis is to "de-cluster" the threshold exceedances in order to identify independent exceedances. Among the variety of approaches for declustering, the simplest solution involves selecting the largest daily value from any consecutive run of exceedances: For example, if the daily time series of precipitation over 10 days is {0, 0, 4.3, 5.1, 2.1, 0, 0, 6.4, 0, 1.2} and the threshold is set to 2, the "independent" exceedances would be {5.1, 6.4}.

21.4.3 Extensions

An exhaustive survey of the many extensions to these basic treatment for applications in environmental studies is beyond the scope of our chapter. Here we note that frequently these parametric fits have been carried out by incorporating a time-varying location parameter or threshold (e.g., Behrens et al. 2004; Huerta and Sansó 2007), or by modeling the parameters of distributions fitted at nearby locations through spatial statistical models (e.g., Tye and Cooley 2015; Hewitt et al. 2019). These are natural modifications of the application of EVA—which was developed under the assumption of stationarity and for individual records—to problems where the temporal and spatial dimensions are often central to the analysis. For example, one would want to characterize changing extremes in a warming climate by modeling the location of a GEV, or the threshold of a GPD as function of time. If the interest is in characterizing the behavior of extremes over a complex geographic domain, incorporating spatial covariates and fitting a Gaussian process to the parameters of extreme value distribution can be carried out, but in these cases a robust estimation of the often challenging to characterize (Hüsler & Reiss, 1989) tail dependence in co-occurring extremes becomes critical and methods to do that are an active subject of research Huang et al. (2019).

21.5 Quantifying Uncertainty

While quantification of uncertainty is an important component of any statistical analysis, it takes on a particular urgency when considering extremes since (by definition) the events of interest are rare and hence the available data is diminished. Fortunately, there are a diverse set of tools and formal methods for uncertainty quantification for each of the extreme metrics described in Sect. 21.3 as well as the extreme value analyses described in Sect. 21.4, including bootstrap-based approaches, asymptotic approaches based on mathematical theory, as well as Bayesian methods.

The simplest and often most intuitive way to quantify uncertainty is via the bootstrap, which is based on resampling the data with replacement. As a concrete example, suppose that we have a set of data $\mathbf{x} = \{x_1, \ldots, x_n\}$ which could represent a set of block maxima (Sect. 21.3.3) or one of the ETCCDI indices (Sect. 21.3.1), and that we wish to quantify the uncertainty in some function, say $f(\mathbf{x})$, of these values. For example, we might simply wish to estimate the uncertainty of the mean of these data, $f(\mathbf{x}) = \frac{1}{n}\sum_{i=1}^n x_i$. The bootstrap constructs a sampling distribution for $f(\mathbf{x})$ by resampling the vector $\{x_1, \ldots, x_n\}$ a large number of times, say B (where $B = 250, 500,$ or 1000) *with replacement* and calculating the function of interest for each bootstrap sample. In other words, for $b = 1, \ldots, B$:

1. Sample n values from $\{x_1, \ldots, x_n\}$ with replacement, meaning that certain values might be included more than once and other values might be excluded. Call this set of values $\mathbf{x}_b$.
2. Calculate the function of interest via $f(\mathbf{x}_b)$.

Then, one can estimate the uncertainty of $f(\mathbf{x})$ by taking the standard deviation of the $\{f(\mathbf{x}_b)\}$ or alternatively use the bootstrap values to calculate a confidence interval. Note that this procedure works just as well if $f(\cdot)$ is a known function (like the mean) or if $f(\cdot)$ involves estimating a quantity like a return value using the GEV distribution. While the bootstrap values can be used to calculate a confidence interval, it should be emphasized that the bootstrap values should *not* be treated as Bayesian posterior samples: In other words, the bootstrap values summarize the sampling distribution of the estimate as opposed to a probability distribution of the quantity summarized by $f(\cdot)$. So, for example, bootstrap values should always be summarized with error bars and not boxplots or histograms, since the latter imply a probability distribution.

Alternatively, when one is using one of the EVA approaches described in Sect. 21.4, the software packages used to fit, e.g., a GEV distribution, usually provide an accompanying standard error for each of the statistical parameters that define a GEV distribution (the location, scale, and shape). This standard error is based on widely used statistical theory that derives an approximation to the uncertainty of each statistical parameter. When the GEV parameters are themselves of interest, these standard errors can be used to further calculate confidence intervals using the usual Normal assumptions (e.g., a 95% confidence interval is the estimate $\pm 1.96 \times$ standard error). However, if the uncertainty is needed for a function of the GEV parameters, e.g., the return value, one can instead use the Delta method (see, e.g., Section 3.3.3 of Coles, 2001), which is a mathematical formula for calculating the standard error of a function of statistical parameters. The standard error for return values or return periods is also commonly provided as an output of statistical software. This standard error can similarly be used to calculate a confidence interval using assumptions of Normality (e.g., the estimate $\pm 1.96 \times$ standard error).

Finally, an entirely different paradigm for quantifying uncertainty is to use a Bayesian framework. There are two primary differences between the Bayesian approach and more traditional approaches, commonly referred to as "Frequentist" (which encompasses the previous two uncertainty quantification methods). First, while a Frequentist approach assumes that an unknown statistical parameter (e.g., the GEV location parameter) has a fixed but unknown value, the Bayesian approach assumes that parameters are themselves random and have a probability distribution. This results in the Bayesian interpretation being somewhat more intuitive since one can safely make statements like "there is a 0.95 probability that the location parameter lies between 67 mm and 84 mm," which is technically the incorrect interpretation of the usual Frequentist confidence interval. Furthermore, in Bayesian hypothesis testing, one can make probability statements about the null hypothesis being true/false, which again cannot technically be done in a Frequentist setting. The second major difference in a Bayesian analysis is that all unknown quantities or parameters must be assigned a "prior" distribution, which summarizes all preexisting knowledge and expectations regarding the unknown quantities. These prior distributions can include either very specific information (e.g., specifying an extremely narrow range for what the parameter might be) or can instead be "noninformative" and impose little to no restrictions on the unknown quantity. A Bayesian analysis consists of a prior distribution and a statistical model for the data (often referred to as the "likelihood"), which are combined via Bayes' Theorem to arrive at the so-called posterior distribution, which summarizes all knowledge about the unknown quantities or parameters after updating the prior distribution based on observed data. In some simple examples, it is possible to derive the posterior distribution in closed form; otherwise, most Bayesian analyses require simulation- or Monte Carlo-based approaches to draw samples from the posterior distribution. Measures of uncertainty (including the posterior standard deviation or a "credible interval," the Bayesian equivalent of a confidence interval) can be derived either from the closed-form posterior distribution or from the Monte Carlo posterior samples. For a more thorough introduction to Bayesian analysis, the interested reader is referred to Gelman et al. (2013) and the references therein.

In the context of extreme value analysis, a Bayesian analysis can be helpful to provide meaningful restrictions on certain statistical parameters for which the data may have limited information. An important example is the shape parameter in either the GEV or GPD distribution: The estimate of the shape parameter has very important implications on extrapolation to very low-probability events. For the GEV distribution, recall that if the shape parameter is negative, the resulting fitted distribution has a finite upper bound; on the other hand, when the shape parameter is estimated to be positive, the fitted distribution is unbounded. Unfortunately, in most cases, the limited measurements or observations of extremes mean the (Frequentist) uncertainty is usually quite large for the shape parameter. For a specific application, a prior distribution can be used to appropriately constrain the range of values the shape parameter might take on. For example, when analyzing extreme precipitation, the shape parameter should almost certainly be positive since precipitation extremes tend to have a long upper tail. It is very difficult for a Frequentist analysis to include bounds; on the other hand, it would be very straightforward in a Bayesian analysis to require the shape parameter to be positive via the prior distribution.

21.6 Extreme Event Attribution

One important area of application for the tools of extreme value analysis is extreme event attribution (EA). The goal of EA is to understand and describe the influence of the various drivers behind a specific observed extreme weather event, with a focus on disentangling the role of greenhouse gas emissions and other human activities from that of internal variability (National Academies of Sciences, Engineering, and Medicine, 2016). Usually, EA is exercised on events that cause significant impacts and are therefore objects of the public and policy makers' interest. Until about 10 years ago the question about the role of anthropogenic climate change on individual events was met with the statement that no attribution was possible. Studies of EA, often conducted in a timely fashion, have started to address the question more satisfactorily, by quantifying the role of anthropogenic forcings in making the type of event at the center of attention more (or less) likely. This often puts EA in the media spotlight but also opens up application of its finding in legal actions against polluters, and in disaster adaptation efforts and policy discussions (Arent et al., 2014; Smith et al., 2014).

A traditional EA study takes a risk-based approach, wherein one quantifies the effect of anthropogenic factors on weather by comparing two scenarios: the "factual" or real-world climate scenario (sometimes referred to as "the world as it is") with a "counterfactual," non-anthropogenically forced climate scenario (sometimes referred to as "the world as it might have been") where all natural external forcings are at play, but no increase in greenhouse gases has happened. These risk-based studies use a probabilistic framework (Allen, 2003; Stone and Allen, 2005; Hansen et al., 2014) to compare the probabilities of a predefined unusual weather event (chosen to be descriptive of the actual event of interest, but not so closely matched to it to make the event unrepeated) in the two climate scenarios and estimate how much more or less likely that extreme event is in the anthropogenically influenced world than it would have been otherwise. For notation, let $p_F = $ be the probability for the factual world and $p_C = $ be the probability for the counterfactual world: The anthropogenic influence is then quantified using either the ratio of these probabilities, i.e., the "risk ratio"

$$RR = p_F/p_C$$

or the fraction of attributable risk

$$FAR = 1 - p_C/p_F.$$

Typically, the probabilities for each of these scenarios (and hence the RR or FAR) are estimated from simulations of climate models. The interpretation of the risk ratio is as follows: If $RR = 1$, then there is no difference in the probability of an extreme event; if $RR < 1$, then extreme events are more common in the counterfactual world; finally, if $RR > 1$, then extreme events are more common in the factual world. The fraction of attributable risk is interpreted as the fraction of the chance of experiencing an event that can be attributed to the anthropogenically induced changes in the factual scenario.

Risk-based EA studies can be either targeted or systematic. Targeted studies explore in detail the meteorological mechanisms involved in a specific, single event and how the anthropogenic influence is transmitted through them, and they are generally reactive in the sense that they are conducted for an event that has actually occurred (e.g. Stott et al., 2004; Pall et al., 2011). Systematic studies, on the other hand, cover a much larger number of events using an identical method for all events. This approach may be less suitable for any given event (Angélil et al., 2017), but, since it does not necessarily depend on the specific characteristics surrounding the occurrence of a specific event, it makes it possible to perform analyses on a predefined list of events.

There are a number of challenges for conducting climate model-based EA. First of all, as we have already seen, extreme weather is (by definition) localized and rare, and therefore appropriately resolving such weather requires large ensembles of high-resolution simulations needed. Of course, this is both difficult and computationally expensive particularly for fully coupled models, which regardless have important biases in means, variability, and the ocean-atmosphere interface. Alternatively, scientists often use uncoupled, atmosphere- and land-only climate simulations with prescribed ocean conditions, for which it is possible to obtain large ensembles (on the order of hundreds). More recently, in order to avoid these challenges, or sometimes simply to buttress the estimates derived by climate model analysis through the use of an independent methodology and data source, EA studies have been conducted instead using observational data only (Risser & Wehner, 2017). In this case, the counterfactual is constructed statistically using a regression-based predictive approach: Usually a regression model is fitted to the historical record of the metric chosen to represent the event (e.g., maximum seasonal daily precipitation amount) that contains a term representative of time, or greenhouse gas concentrations, or global average temperature during the observational period. If the term is statistically significant, the prediction from the regression corresponding to that regressor's preindustrial value is compared to the prediction corresponding to its current value and the RR and FAR are computed accordingly. Some analyses go a step further and project ahead the changes under future conditions (e.g., future years, higher concentrations of atmospheric greenhouse gases, or higher global warming levels), thus connecting EA to future risk characterization (Otto et al., 2018).

For any EA study, both the definition of the event and its framing are extremely important, and the results of the study should be interpreted carefully in light of these choices. For example, critics of EA often point to two analyses of the infamous Russian heat wave in 2003, which came to conflicting conclusions about the role of anthropogenic influence on the occurrence of the heat wave. Otto et al. (2012) resolve this apparent conflict by pointing out that the two studies considered different scientific questions: The first set out to assess the anthropogenic influence on the frequency of a heat wave like the one experienced in Russia, while the second examined the anthropogenic influence on the magnitude or severity of the heat wave. In this particular case, Otto et al. (2012) showed that there was, in fact, no contradiction between these two analyses: A single event like the Russian heat wave can be "mostly internally-generated in terms of magnitude and mostly externally-driven in terms of occurrence-probability"; i.e., human climate change can have no influence on the severity while increasing the probability.

As an example, we utilize simulations of five-day temperature averages over the wet season (November to April) from two 50-member ensembles of simulations of the CAM5.1 global atmosphere/land climate model. The model is run in its conventional ~1° longitude/latitude configuration (Neale et al., 2012) under the experiment protocols of the C20C+ Detection and Attribution Project (Stone & Pall, 2021). The first set of simulations (representing the factual scenario) is driven by observed boundary conditions, while the second set of simulations (the counterfactual) is driven by what observed boundary conditions might have been in the absence of historical anthropogenic emissions; simulations within a scenario differ only in the starting conditions. Here we examine data averaged over a region comprised of California and Nevada, two states in the western United States, during the 2009–2018 period; the data and further details on the simulations are available at http://portal.nersc.gov/c20c. Density plots that show the distribution of the five-day TAS averages for each climate scenario are shown in Fig. 21.3.

Figure 21.3 also nicely demonstrates how the point of an EA study is to quantify changes in the tails of the distribution and specifically not the center of the distribution. As an example, the event of interest for these data might be experiencing a five-day TAS average that exceeds 12.85°C: This threshold is shown in Fig. 21.3 with the black vertical line. Note that we specifically are not interested in how the center of the distribution changes (the dashed vertical lines in Fig. 21.3); instead, we are concerned with how the upper tail of each distribution changes, denoted by the shaded in areas under each density (the tail probabilities are p_F and p_C for the factual and counterfactual, respectively).

Quantifying uncertainty in either the risk ratio or fraction of attributable risk involves an extra degree of complication, in that we must account for uncertainty in two probabilities (p_F and p_C) as well as quantify how these uncertainties are propagated through functional quantities (i.e., p_F/p_C or $1 - p_C/p_F$). Furthermore, calculation of confidence intervals should make use of the fact that both the risk ratio and FAR have bounds: The risk ratio must be greater than zero and the FAR cannot exceed one. Both the bootstrap and the Delta method (see Sect. 21.5) are useful strategies for this task; for more information, see Paciorek et al. (2018).

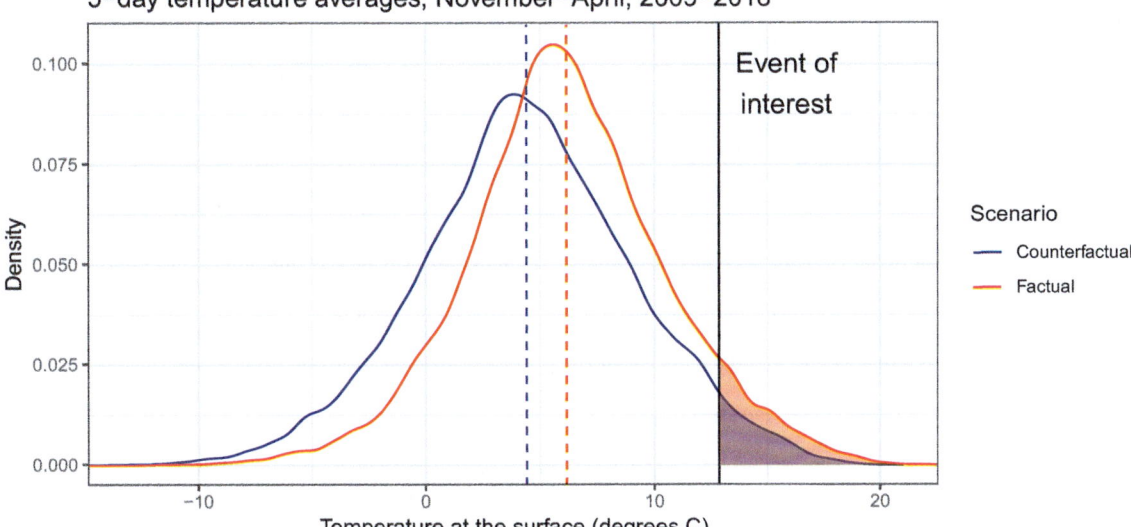

Fig. 21.3 Density plots for five-day temperature at the surface (TAS) averages (°C), averaged over California and Nevada, for November to April, 2009–2018, from a 50-member ensemble of CAM5.1-1degree global climate model simulations. The distributions are shown for both the factual and counterfactual scenarios; the extreme event of interest is defined to be exceeding 12.85°C, and the probabilities of interest (p_F and p_C) are the shaded tails of each distribution

References

Aerenson, T., Tebaldi, C., Sanderson, B., & Lamarque, J.-F. (2018). Changes in a suite of indicators of extreme temperature and precipitation under 1.5 and 2 degrees warming. *Environmental Research Letters, 13*(3), 035009. https://doi.org/10.1088%2F1748-9326%2Faaafd6

Alexander, L. V. (2016). Global observed long-term changes in temperature and precipitation extremes: A review of progress and limitations in IPCC assessments and beyond. *Weather and Climate Extremes, 11*, 4–16. https://doi.org/10.1016/j.wace.2015.10.007

Allen, M. (2003). Liability for climate change. *Nature, 421*, 891–892. https://doi.org/10.1038/421891a

Angélil, O., Stone, D., Wehner, M., Paciorek, C. J., Krishnan, H., & Collins, W. (2017). An independent assessment of anthropogenic attribution statements for recent extreme temperature and rainfall events. *Journal of Climate, 30*(1), 5–16. https://doi.org/http://doi.org/10.1175/JCLI-D-16-0077.1

Arent, D. J., & Coauthors (2014). Key economic sectors and services. In C. B. Field, V. R. Barros, & et al. (Eds.), *Climate Change 2014: Impacts, Adaptation, and Vulnerability. Part A: Global and Sectoral Aspects. Contribution of Working Group II to the Fifth Assessment Report of the Intergovernmental Panel on Climate Change* (pp. 659–708). Cambridge University. https://www.ipcc.ch/site/assets/uploads/2018/02/WGIIAR5-Chap10_FINAL.pdf

Behrens, C. N., Lopes, H. F., & Gamerman, D. (2004). Bayesian analysis of extreme events with threshold estimation. *Statistical Modelling, 4*(3), 227–244. https://doi.org/10.1191/1471082X04st075oa

Coles, S. (2001). *An introduction to statistical modeling of extreme values*. Lecture Notes in Control and Information Sciences. Springer.

Deser, C., & Coauthors (2020). Insights from earth system model initial-condition large ensembles and future prospects. *Nature Climate Change, 10*(4), 277–286. https://doi.org/10.1038/s41558-020-0731-2

Frame, D. J., Wehner, M. F., Noy, I., & Rosier, S. M. (2020). The economic costs of Hurricane Harvey attributable to climate change. *Climatic Change, 160*(2), 271–281. https://doi.org/10.1007/s10584-020-02692-8

Gelman, A., Carlin, J. B., Stern, H. S., Dunson, D. B., Vehtari, A., & Rubin, D. B. (2013). *Bayesian data analysis*. CRC Press.

Hansen, G., Auffhammer, M., & Solow, A. R. (2014). On the attribution of a single event to climate change. *Journal of Climate, 27*, 8297–8301. https://doi.org/10.1175/JCLI-D-14-00399.1

Hewitt, J., Fix, M. J., Hoeting, J. A., & Cooley, D. S. (2019). Improved return level estimation via a weighted likelihood, latent spatial extremes model. *Journal of Agricultural, Biological and Environmental Statistics, 24*(3), 426–443. https://doi.org/10.1007/s13253-019-00354-6

Huang, W. K., Cooley, D. S., Ebert-Uphoff, I., Chen, C., & Chatterjee, S. (2019). New exploratory tools for extremal dependence: χ networks and annual extremal networks. *Journal of Agricultural, Biological and Environmental Statistics, 24*(3), 484–501. https://doi.org/10.1007/s13253-019-00356-4

Huerta, G., & Sansó, B. (2007). Time-varying models for extreme values. *Environmental and Ecological Statistics, 14*(3), 285–299. https://doi.org/10.1007/s10651-007-0014-3

Hüsler, J., & Reiss, R.-D. (1989). Maxima of normal random vectors: Between independence and complete dependence. *Statistics & Probability Letters, 7*(4), 283–286. https://doi.org/10.1016/0167-7152(89)90106-5

Jonkman, S. N., Godfroy, M., Sebastian, A., & Kolen, B. (2018). Brief communication: Loss of life due to Hurricane Harvey. *Natural Hazards and Earth System Sciences, 18*(4), 1073–1078. https://doi.org/10.5194/nhess-18-1073-2018

Kharin, V. V., Zwiers, F. W., Zhang, X., & Hegerl, G. C. (2007). Changes in temperature and precipitation extremes in the IPCC ensemble of global coupled model simulations. *Journal of Climate, 20*(8), 1419–1444. https://doi.org/10.1175/JCLI4066.1

Kunkel, K. E., & Champion, S. M. (2019). An assessment of rainfall from hurricanes Harvey and Florence relative to other extremely wet storms in the United States. *Geophysical Research Letters*. https://doi.org/10.1029/2019GL085034

Menne, M. J., Durre, I., Vose, R. S., Gleason, B. E., & Houston, T. G. (2012). An overview of the Global Historical Climatology Network-daily database. *Journal of Atmospheric and Oceanic Technology, 29*(7), 897–910. https://doi.org/10.1175/JTECH-D-11-00103.1

National Academies of Sciences, Engineering, and Medicine (2016). *Attribution of extreme weather events in the context of climate change*. The National Academies Press. https://doi.org/10.17226/21852

Neale, R. B., & Coauthors (2012). Description of the NCAR community atmosphere model (CAM 5.0). Technical report, NCAR Technical Note NCAR/TN-486+STR. https://www.researchgate.net/publication/224017878_Description_of_the_NCAR_Community_Atmosphere_Model

Otto, F. E., Massey, N., Van Oldenborgh, G., Jones, R., & Allen, M. (2012). Reconciling two approaches to attribution of the 2010 Russian heat wave. *Geophysical Research Letters, 39*(4). https://doi.org/10.1029/2011GL050422

Otto, F. E. L., & Coauthors (2018). Anthropogenic influence on the drivers of the western cape drought 2015–2017. *Environmental Research Letters, 13*(12), 124010. https://doi.org/10.1088/1748-9326/aae9f9

Paciorek, C. J., Stone, D. A., & Wehner, M. F. (2018). Quantifying statistical uncertainty in the attribution of human influence on severe weather. *Weather and Climate Extremes, 20*, 69–80. https://doi.org/10.1016/j.wace.2018.01.002

Pall, P., Aina, T., Stone, D. A., Stott, P. A., Nozawa, T., Hilberts, A. G. J., Lohmann, D., & Allen, M. R. (2011). Anthropogenic greenhouse gas contribution to flood risk in England and Wales in Autumn 2000. *Nature, 470*, 382–385. https://doi.org/10.1038/nature09762

Pall, P., Patricola, C. M., Wehner, M. F., Stone, D. A., Paciorek, C. J., & Collins, W. D. (2017). Diagnosing conditional anthropogenic contributions to heavy Colorado rainfall in September 2013. *Weather and Climate Extremes, 17*, 1–6. https://doi.org/10.1016/j.wace.2017.03.004

Risser, M. D., & Wehner, M. F. (2017). Attributable human-induced changes in the likelihood and magnitude of the observed extreme precipitation during hurricane Harvey. *Geophysical Research Letters, 44*(24), 12–457. https://doi.org/10.1002/2017GL075888

Scarrott, C., & MacDonald, A. (2012). A review of extreme value threshold estimation and uncertainty quantification. *REVSTAT–Statistical Journal, 10*(1), 33–60. https://doi.org/10.57805/revstat.v10i1.110

Sillmann, J., Kharin, V. V., Zwiers, F. W., Zhang, X., & Bronaugh, D. (2013). Climate extremes indices in the CMIP5 multimodel ensemble: Part 2. Future climate projections. *Journal of Geophysical Research: Atmospheres, 118*(6), 2473–2493. https://doi.org/10.1002/jgrd.50188

Smith, K. R., & Coauthors (2014). Human health: Impacts, adaptation, and co-benefits. In C. B. Field, V. R. Barros, & et al. (Eds.), *Climate Change 2014: Impacts, Adaptation, and Vulnerability. Part A: Global and Sectoral Aspects. Contribution of Working Group II to the Fifth Assessment Report of the Intergovernmental Panel on Climate Change* (pp. 709–754). Cambridge University. https://www.ipcc.ch/site/assets/uploads/2018/02/WGIIAR5-Chap11_FINAL.pdf

Stone, D. A., & Allen, M. R. (2005). The end-to-end attribution problem: From emissions to impacts. *Climatic Change, 71*, 303–318. https://doi.org/10.1007/s10584-005-6778-2

Stone, D. A., & Pall, P. (2021). Benchmark estimate of the effect of anthropogenic emissions on the ocean surface. *International Journal of Climatology, 41*(5), 3010–3026. https://doi.org/10.1002/joc.7002

Stott, P. A., Stone, D. A., & Allen, M. R. (2004). Human contribution to the European heatwave of 2003. *Nature, 432*, 610–614. https://doi.org/10.1038/nature03089

Tebaldi, C., Hayhoe, K., Arblaster, J. M., & Meehl, G. A. (2006). Going to the extremes. *Climatic Change, 79*(3), 185–211. https://doi.org/10.1007/s10584-006-9051-4

Tye, M. R., & Cooley, D. (2015). A spatial model to examine rainfall extremes in Colorado's front range. *Journal of Hydrology, 530*, 15–23. https://doi.org/10.1016/j.jhydrol.2015.09.023

Wehner, M., Stone, D., Krishnan, H., AchutaRao, K., & Castillo, F. (2016). The deadly combination of heat and humidity in India and Pakistan in summer 2015. *Bulletin of the American Meteorological Society, 97*(12), S81–S86. https://doi.org/10.1175/BAMS-D-16-0145.1

Open Access This chapter is licensed under the terms of the Creative Commons Attribution 4.0 International License (http://creativecommons.org/licenses/by/4.0/), which permits use, sharing, adaptation, distribution and reproduction in any medium or format, as long as you give appropriate credit to the original author(s) and the source, provide a link to the Creative Commons license and indicate if changes were made.

The images or other third party material in this chapter are included in the chapter's Creative Commons license, unless indicated otherwise in a credit line to the material. If material is not included in the chapter's Creative Commons license and your intended use is not permitted by statutory regulation or exceeds the permitted use, you will need to obtain permission directly from the copyright holder.

22. How Uncertainty Interacts with Ethical Values in Climate Change Research

Casey Helgeson, Wendy Parker, and Nancy Tuana

22.1 Introduction

Much climate change research aims to inform decision-making in one way or another. A common vision of how science and ethics work together in this decision-making has science spelling out the (probable) consequences of different policy options, while ethical judgments determine which option's consequences are most desirable. For example, climate projections and impact studies may suggest the likely consequences of different mitigation pathways, but ethical judgments are required to evaluate how good or bad those consequences are and how preferable one possible future is over another.

While correct as far as it goes, this standard picture can encourage an overly sharp distinction between scientific activities and ethical deliberation. Far from entering only at the policy-making stage, ethical judgments often shape scientific research itself. This is most obvious in the choice of research questions. The choice of what to study ultimately affects what knowledge can be brought to bear in real-world decisions, including consequences for which (and whose) decisions can be made with the benefit of scientific insight. Such considerations are routinely referenced when motivating funding proposals and research articles. Of course, more purely scientific motivations such as fundamental discovery and filling gaps in knowledge are also critical in choosing research questions. In this way, a researcher's choice of what to investigate illustrates a central concept of this chapter: *coupled ethical–epistemic* choices (Tuana 2018).

A little terminology is needed to unpack this jargon. We use the word *values* as a general term for the reasons or perspectives from which one evaluates something as good or bad. Applying this notion of values very broadly, any goal judged worthy of pursuit will be done so on the basis of values. Sometimes these will be *ethical values* such as concern for justice, human welfare, or environmental protection. (The overlapping concept of *social values* includes things valued by communities or individuals—like greenspaces or social services—even if these may not be recognizably ethical in nature. Here we use "ethical values" broadly to also include these social values.) In contrast, scientific findings can be valued for how they advance understanding, and scientific methods or models can be valued for their accuracy, reliability, or generality. These aspects of research are valued because they are thought to promote (or constitute) a central aim of science: gaining knowledge. Such values are often called *epistemic values*.

Many decisions made in the course of scientific research are coupled ethical–epistemic choices in the sense that their consequences can be judged from the perspective of both epistemic values (i.e., what are the contributions to scientific knowledge) and ethical values (i.e., what are the upshots for policy, society, and the environment). Coupled ethical–epistemic choices can be found at any spot along the continuum of research-design choices, from the broad end of choosing and refining research questions to narrower decisions regarding approaches to answering those questions, specific methods, and interpretation of results.

Scientific training tends to focus on epistemic values—especially when it comes to the narrower, finer-grained re-

---

C. Helgeson (✉)
Earth and Environmental Systems Institute, Penn State University, University Park, PA, USA

Department of Philosophy, Penn State University, University Park, PA, USA
e-mail: casey@psu.edu

W. Parker
Department of Philosophy, Virginia Tech, Blacksburg, VA, USA
e-mail: wendyparker@vt.edu

N. Tuana
Department of Philosophy, Penn State University, University Park, PA, USA
e-mail: nat3@psu.edu

© The Author(s) 2025
L. O. Mearns et al. (eds.), *Uncertainty in Climate Change Research*,
https://doi.org/10.1007/978-3-031-85542-9_22

search choices. In this chapter, we draw attention to the ethical values that are often linked to the same choices. Our aim is to encourage more deliberate and more reflective engagement with the ethical components of these choices. The topic of this volume is uncertainty in climate change research, and decisions about how to address sources of uncertainty in research provide a particularly rich arena for interaction between epistemic and ethical values. We present a series of examples of such interaction, followed by a short list of recommendations on how to approach coupled ethical–epistemic choices in research.

22.2 Attribution Methods and Public Communication

Our first example concerns extreme event attribution (National Academies of Sciences, Engineering, and Medicine, 2016). Increasingly, climate scientists are investigating the extent to which particular extreme weather events, such as floods, droughts, and heat waves, can be linked to anthropogenic climate change. Depending on the choice of method, different pictures can emerge regarding what can and cannot be attributed to climate change, with implications for public communication and litigation for damages.

The standard "risk-based" approach has been adapted from epidemiology (Allen, 2003; Haustein et al., 2016; Stott et al., 2016). Researchers attempt to quantify the change in likelihood of a weather event like the one observed, given rising greenhouse gas concentrations. This is done via climate modeling studies that compare the frequency of such event types across simulations driven by different greenhouse gas concentrations. In one set of simulations, historical (i.e., increasing) greenhouse gas concentrations are used; in the other, concentrations are held (counterfactually) at pre-industrial levels.

For a variety of reasons, studies following the risk-based approach can be inconclusive. These reasons include the difficulties and uncertainties in simulating the atmospheric circulation driving some types of extreme events (Shepherd, 2014), use of null-hypothesis significance testing to interpret simulation results, and use of "no change in likelihood" as a null hypothesis (Shepherd, 2014; Lloyd & Oreskes, 2018). Failure to reject such a null hypothesis means that the possibility of no change in likelihood cannot be excluded at the chosen significance level, given available evidence. But careful and cautious statements such as this are sometimes misinterpreted in public discourse as saying something stronger and more conclusive, namely that there is no connection between anthropogenic climate change and the weather event in question.

Uncertainties about circulation, notwithstanding broad thermodynamic changes in the climate system such as rising sea surface temperature and increased moisture content are well understood as anthropogenic. Moreover, it is very plausible that these thermodynamic changes can make weather events, when they do occur, more intense than they would otherwise be. Critics thus worry that the (often inconclusive) risk-based approach to attribution will miss some valuable opportunities to communicate to the public, via salient events such as extreme floods, that climate change is already having negative impacts (Trenberth, 2011; Trenberth et al., 2015). This line of thought has led to a second approach to attribution, sometimes referred to as the "storyline" approach. (Though note that the storyline concept is also used more broadly for communication, uncertainty characterization, and risk management beyond the context of attribution science (Shepherd et al., 2018; Sillmann et al., 2021).

In general terms, the storyline approach to event attribution offers descriptive narratives of specific past events, with emphasis on understanding the driving factors that were involved in those events and that may shape future events as well (Shepherd et al., 2018). Such an approach would typically ask: How did "known" thermodynamic changes in climate make a difference to the intensity of this particular weather event? To address this question, the first step is to simulate the extreme event as it occurred. The second step is to re-simulate the event, removing the human-caused thermodynamic changes, e.g., making the nearby sea surface temperature cooler by a specified amount in the simulations. These studies very often do find a link between anthropogenic climate change and an extreme event of interest—specifically, an increase in intensity. For example, the conclusion might be that rising greenhouse gas concentrations, via their effects on sea surface temperature, increased a flood-causing storm's precipitation by at least 30% (see, e.g., Meredith et al., 2015; Hoerling et al., 2013).

The risk-based and storyline approaches ask different questions (Lloyd & Oreskes, 2018). One asks whether increasing greenhouse gas concentrations have, all things considered, changed the probability of a given event type. The other brackets anthropogenic circulation changes and asks whether the thermodynamic consequences of increasing concentrations affected the intensity of a specific event, holding fixed the actual circulation that led up to the event. When applied to the same case, the two methods can give different answers (e.g., "no" and "yes," respectively) with no logical contradiction.

Given limited time and resources, which approach should attribution researchers prioritize? The considerations that have been aired in discussions contrasting the two approaches include not only aspects subject to epistemic values (different kinds of insights; different degrees of uncertainty in results) but also consequences judged by ethical values. The latter include purported differences in: messaging to the public regarding "links" from climate change to extreme weather; potential for misinterpretation of results; relevance of results for climate risk management; long-term effects on public trust in science; and potential for reputational damage to individual scientists (Otto et al., 2016; Lloyd & Oreskes, 2018).

Each approach to attribution thus comes with a bundle of features and consequences, some of which are important for epistemic reasons and some of which are important for ethical reasons. The ethical and epistemic merits of an approach can be judged separately, yet they are bound together in the same scientific choice. In this way, attribution methods illustrate the concept of coupled ethical–epistemic choices in research.

22.3 Parameter Choices and the Consequences of Error

A second example concerns the way in which method choices can affect the balance of *inductive risk*: the risk of erring in one's scientific conclusions (Douglas, 2000a). The errors at issue could be Type I ("false positives") versus Type II ("false negatives") or could concern overestimating versus underestimating a quantity of interest. A classic example is the choice of significance level used in null-hypothesis significance testing. This significance level (often fixed conventionally at .05) affects the balance between the relative risks of Type I and Type II errors. More broadly, choices between alternative datasets, modeling assumptions, or statistical algorithms can have analogous consequences for the risk of different types of error in the findings of a study (see, e.g., Fujiwara et al., 2017; Flato et al., 2013).

As an example, consider the assignment of numerical values to uncertain parameters in a climate or impacts model (i.e., model calibration). When model output is compared to observations across a suite of performance metrics, some parameter assignments result in better model performance on some important metrics, while other assignments result in better performance on others (Mauritsen et al., 2012). A number of different model versions might fit the observations reasonably well and yet differ substantially in their projections. With different projections come different inductive risk profiles: for a given quantity of interest (e.g., precipitation extremes, heat stress, or crop loss), higher projections come with a greater risk of overestimating that quantity, while lower projections risk underestimation to a greater degree.

One approach to managing inductive risk is to make one's method choices while giving some consideration to the potential consequences of erring in one way versus another. Would overestimating future precipitation extremes or crop losses be *worse* than underestimating them? If so, this could be factored in as the researcher chooses among the scientifically reasonable approaches to addressing the research question. Indeed, it has been argued that doing so helps the researcher fulfill her obligations as a moral agent, which include taking due care to avoid errors with particularly bad consequences (Douglas 2000a, 2009). Of course, the question of which consequences are particularly bad is informed by ethical values, not epistemic ones. In this way, consideration of the risks of error can generate coupled ethical–epistemic choices. (Approaches to transparently incorporating ethical values in the model-calibration example include risk-based calibration (e.g., Pappenberger et al., 2007) and careful definition of loss functions (Jaynes, 2003) when comparing model performance with observations.)

When facing research design choices, instead of choosing a single approach, sometimes several options can be tried, producing a range of results. Ensemble modeling studies, for instance, involve multiple simulations that incorporate different options for modeling equations, parameter assignments, or initial conditions. But ensemble studies can still involve uncertain method choices, such as specifying the boundaries of the "plausible" ranges for the parameters (or model structures) to be sampled. For these choices too, there may be a range of scientifically reasonable options with different associated risks of error. Indeed, it seems likely that almost every modeling study in the climate-change context will involve uncertain method choices with potentially different risks of error.

This does not mean, however, that ethical values ought to influence method choices in every modeling study, even if one is persuaded by the reasoning above. The inductive risk implications of some choices will be unforeseeable in practice (Undorf et al., 2022; Betz, 2017). And there might be overriding reasons for making choices on other grounds. For example, researchers might stick with "default" parameter assignments for the sake of more meaningful model intercomparisons, tractability, or to avoid upsetting an existing "balance of approximations" among model components. The case for ethical values influencing method choices seems most compelling when modeling is done in support of particular decision-making tasks, and where some method options have clear inductive risk implications that align better with the aims and values of stakeholders or clients. Such situations may arise, for instance, in the context of climate services (Adams et al., 2015; Parker & Lusk, 2019). In any case, whenever such precautionary thinking does lead to ethical values shaping method choices, this should be communicated clearly and transparently (Adams et al., 2015; Baldissera Pacchetti et al., 2022).

Ultimately, even if one remains unpersuaded that ethical judgments about potential errors ought to influence method choices, there is a crucial insight here that should not be overlooked: method choices that are not directly influenced by ethical values can nevertheless affect the balance of inductive risk in ways that serve the needs and interests of some stakeholders better than others. That is, even method choices that are not value-*influenced* can, in an important sense, fail to be value-*neutral*.

22.4 Model Complexity and High-Impact Events

High-impact, low-probability events provide another example of interaction between ethical values and the treatment of uncertainties in research. By definition, high-impact events are those that are particularly dangerous or concerning—a judgment based on ethical values. Because they are of such concern, learning about the likelihood of high-impact events can be particularly important for understanding climate change impacts and assessing risk-management strategies. (In terms of specific decision-support frameworks, the probability of extreme, high-impact outcomes can, for example, have an outsized impact on expected damage calculations (Weitzman 2009) and can shape the range of possibilities across which satisfying strategies are sought in robustness-based frameworks (Quinn et al., 2020).)

The highest-impact events also tend to be low-probability occurrences, which can complicate uncertainty assessment (Keller et al., 2021). For example, where uncertainty in projections is characterized through an ensemble of simulations, use of computationally expensive models can limit ensemble size and impede estimation of the small probabilities associated with high-impact outcomes (Lee et al., 2020; Sriver et al., 2012; Wong & Keller, 2017). A state-of-the-art Earth System Model may be the richest and most complete encapsulation of knowledge relevant to, e.g., sea-level rise by century's end. Yet the large number of model runs needed for ensemble-based uncertainty quantification of extreme sea-level rise may be feasible only using faster, more idealized models (Bakker et al., 2016; Helgeson et al., 2021; Wong et al., 2017). In this way, some of the scientific or epistemic merits of models can, in practice, trade off against the *relevance* of the questions that can be addressed using those models, where relevance is a question of ethical values.

22.5 Disaggregation and Distributive Justice

So far, we have discussed examples that specifically concern the treatment of uncertainties. Here we relax this focus somewhat in order to provide an indication of the broader character of coupled ethical–epistemic research choices in climate change research (which need not always link directly to the treatment of uncertainties).

There is a particularly rich and explicit role for ethical values when it comes to designing and assessing climate risk management strategies. To be relevant for decision-makers and stakeholders, such analyses should characterize potential futures in terms that allow those actors to apply their own values to the decision problem (Helgeson et al., 2024). What are these values? Climate change impacts people in many ways, and people care about those impacts from many different perspectives (Tschakert et al., 2017; O'Brien & Wolf, 2010). To give just one example, an interview-based study with community members in the city of New Orleans found that stakeholder views on coastal flood risk encompassed values such as concern for personal safety, property damage, broader economic impacts, sense of place, perception of safety, non-human welfare, distributive justice, intergenerational justice, and having a say in risk management decisions (Bessette et al., 2017). Each of these concerns provides a perspective from which projected outcomes and impacts can be evaluated (except for the last one, which is about *process* rather than outcomes).

Consider one specific concern mentioned above: distributive justice. In the context of local flood risk management, distributive justice addresses the fairness of how flood risk, or related costs and benefits, are distributed across communities and populations. Analysis of adaptation strategies (such as levees, evacuation planning, or funding programs for home elevation) that estimates costs and benefits only in the aggregate—e.g., for a whole city or region—will be blind to differences in the way that alternative strategies distribute risk across smaller units such as neighborhoods or households. For stakeholders who care about distributive justice, a distribution-blind analysis will fail to provide relevant decision support because those stakeholders will be unable to apply their values to the evaluation of the adaptation strategies (Jafino et al., 2021; Vezér et al., 2018). (For related illustrations, see Khosrowi (2019); Parker and Winsberg (2018).)

Estimating the effectiveness of adaptation measures with attention to distributive justice may require a more complex or disaggregated modeling framework that resolves neighborhoods or even households (Jafino et al., 2021). For example, Vezér et al. (2018) contrast two specific models used for coastal flood risk analysis in the state of Louisiana, including the city of New Orleans. Both models take flood hazards and adaptation measures as inputs and project the success of those measures as outputs. But one model (Groves et al., 2014) includes detailed and disaggregated spatial information, while the other (Jonkman et al., 2009) works with a simplified and highly aggregated representation of the study system. The models also differ in their usability, adaptability, and transparency (Vezér et al., 2018). At the same time, model choice is, as always, subject to a range of *epistemic* considerations concerning the accuracy and trustworthiness of a model's representations and projections. Like previous examples, here a single choice in the design of a study can have consequences both for the epistemic or purely scientific side of a study (including but not limited to the treatment of uncertainties) and also for the treatment of ethical values in the analysis.

22.6 Conclusion

We have presented a series of examples illustrating how choices made during the conduct of research can carry implicit value judgments or create side effects and consequences with ethical import. These consequences include what (and whose) questions receive scientific attention, how mitigation and adaptation strategies are evaluated, which impacts are prioritized, how science is communicated, and what kinds of errors are avoided. We have focused on examples in which the research choices in question also shape how uncertainties are addressed: alternative attribution methods can subtly recast the research question and shift the burden of proof; model complexity can enable or constrain the characterization of ethically important uncertainties, and model calibration plays a key role in determining which uncertainties and which types of futures are characterized and how.

Many research choices are like these examples. On the one hand, they have consequences that might be judged from the perspective of ethical values, and on the other hand, they have consequences—regarding, e.g., the depth of insight or reliability of findings—that can be judged by scientific standards that express epistemic values. In other words, many research choices (perhaps even most) are *coupled ethical–epistemic choices* (see (Beck & Krueger, 2016) and (Deitrick et al., 2021) for further illustrations). Scientific training naturally focuses on the epistemic side. Here we have highlighted the ethical side and the coupling of the two sides.

Once this coupling is recognized, many further questions arise, such as: whose or which values should be considered? How should we balance epistemic and ethical considerations when they are in tension? What are the best approaches for representing the tradeoffs between value considerations? How should the connections between epistemic and ethical considerations be discussed in scientific publications? For views on some of these questions, readers can consult (Adams et al., 2015; Elliott, 2017; Hicks, 2014; Baldissera Pacchetti et al., 2022). Here, we close with some brief recommendations on first steps toward engaging with coupled ethical–epistemic choices (see (Pulkkinen et al., 2022) for related, complementary recommendations).

- **Develop an eye for the ethical side of research choices.** Make a habit of thinking through how your findings might be used and by whom. Ask questions like: Whose information needs does my research design serve? What value system does my policy-evaluation framework assume? Whose vulnerabilities does my approach to hazard mapping prioritize? Who might be disadvantaged by my research findings? What kinds of errors have I been most/least careful to avoid?
- **Discuss ethical values explicitly in research outputs.** Answers to questions like those listed under suggestion the previous bullet point can help readers contextualize your findings and assess whether they are useful for a given purpose. Be transparent about your explicit and implicit working assumptions. Briefly explain how your research design balances relevant ethical and epistemic values. Note any tradeoffs between value considerations. Declare any motivating ethical priorities and, especially if the rationale for these priorities is not obvious, defend them.
- **Engage with end users and/or boundary organizations.** While there are many reasons to engage with decision-makers, stakeholders, and boundary organizations, one important reason is to facilitate the alignment of research with stakeholder values and priorities (Adams et al., 2015; Helgeson et al., 2024).

Acknowledgments The authors thank Linda Mearns, Marina Baldissera Pacchetti, Vivek Srikrishnan, Ted Shepherd, and Rob Wilby for comments on a draft of this chapter. This work was supported by the Rock Ethics Institute and Center for Climate Risk Management (CLIMA) at Penn State.

References

Adams, P., E. Eitland, B. Hewitson, C. Vaughan, R. Wilby, & Zebiak, S. (2015). Toward an ethical framework for climate services: A white paper of the climate services partnership working group on climate services ethics. *Climate Services Partnership*, 12 pp. https://hdl.handle.net/10568/68833.

Allen, M. (2003). Liability for climate change. *Nature, 421*, 891–892. https://doi.org/10.1038/421891a

Bakker, A. M. R., Applegate, P. J., & Keller, K. (2016). A simple, physically motivated model of sea-level contributions from the Greenland ice sheet in response to temperature changes. *Environmental Modelling & Software, 83*, 27–35. https://doi.org/10.1016/j.envsoft.2016.05.003

Baldissera Pacchetti, M., Schacher, J., Dessai, S., Soares, M. B., Lawlor, R., & Daron, J. (2022). Toward a UK climate service code of ethics. *Bulletin of the American Meteorological Society, 103*, E25–E32. https://doi.org/10.1175/BAMS-D-21-0137.1

Beck, M., & Krueger, T. (2016). The epistemic, ethical, and political dimensions of uncertainty in integrated assessment modeling. *Wiley Interdisciplinary Reviews Climate Change, 7*, 627–645. https://doi.org/10.1002/wcc.415

Bessette, D. L., Mayer, L. A., Cwik, B., Vezér, M., Keller, K., Lempert, R. J., & Tuana, N. (2017). Building a values-informed mental model for New Orleans climate risk management. *Risk Analysis, 37*, 1993–2004. https://doi.org/10.1111/risa.12743

Betz, G. (2017). Why the argument from inductive risk doesn't justify incorporating non-epistemic values in scientific reasoning. In K. C. Elliott & D. Steel (Eds.), *Current Controversies in Values and Science* (pp. 94–110). Routledge.

Deitrick, A. R., Torhan, S. A., & Grady, C. A. (2021). Investigating the influence of ethical and epistemic values on decisions in the watershed modeling process. *Water Resources Research, 57*, e2021WR030481. https://doi.org/10.1029/2021wr030481

Douglas, H. (2000a). Inductive risk and values in science. *Philosophy of Science, 67*, 559–579. https://doi.org/10.1086/392855

Douglas, H.. (2000b). *Inductive risk and values in science. 2009: Science, policy, and the value-free ideal* (256 pp). University of Pittsburgh Press.

Elliott, K. C. (2017). *A tapestry of values: An introduction to values in science* (208 pp). Oxford University Press.

Flato, G., et al. (2013). Evaluation of climate models. In T. F. Stocker et al. (Eds.), *Climate Change 2013: The physical science basis. Contribution of Working Group I to the Fifth Assessment Report of the Intergovernmental Panel on Climate Change*. Cambridge University Press.

Fujiwara, M., et al. (2017). Introduction to the SPARC Reanalysis Intercomparison Project (S-RIP) and overview of the reanalysis systems. *Atmospheric Chemistry and Physics, 17*, 1417–1452. https://doi.org/10.5194/acp-17-1417-2017

Groves, D. G., Fischbach, J. R., Knopman, D., Johnson, D. R., & Giglio, K. (2014). *Strengthening coastal planning: How coastal regions could benefit from Louisiana's planning and analysis framework*. Rand Corporation.

Haustein, K., et al. (2016). Real-time extreme weather event attribution with forecast seasonal SSTs. *Environmental Research Letters, 11*, 064006. https://doi.org/10.1088/1748-9326/11/6/064006

Helgeson, C., Srikrishnan, V., Keller, K., & Tuana, N. (2021). Why simpler computer simulation models can be epistemically better for informing decisions. *Philosophy of Science, 88*, 213–233. https://doi.org/10.1086/711501

Helgeson, C., Keller, K., Nicholas, R. E., Srikrishnan, V., Cooper, C., Smithwick, E. A. H., & Tuana, N. (2024). Integrating values to improve the relevance of climate-risk research. *Earth's Future, 12*, e2022EF003025. https://doi.org/10.1029/2022EF003025

Hicks, D. J. (2014). A new direction for science and values. *Synthese, 191*, 3271–3295. https://doi.org/10.1007/s11229-014-0447-9

Hoerling, M., et al. (2013). Anatomy of an extreme event. *Journal of Climate, 26*, 2811–2832. https://doi.org/10.1175/jcli-d-12-00270.1

Jafino, B. A., Kwakkel, J. H., & Taebi, B. (2021). Enabling assessment of distributive justice through models for climate change planning: A review of recent advances and a research agenda. *Wiley Interdisciplinary Reviews Climate Change, 12*. https://doi.org/10.1002/wcc.721

Jaynes, E. T. (2003). *Probability theory: The logic of science*. Cambridge University Press.

Jonkman, S. N., Kok, M., van Ledden, M., & Vrijling, J. K. (2009). Risk-based design of flood defence systems: A preliminary analysis of the optimal protection level for the New Orleans metropolitan area. *Journal of Flood Risk Management, 2*, 170–181. https://doi.org/10.1111/j.1753-318x.2009.01036.x

Keller, K., Helgeson, C., & Srikrishnan, V. (2021). Climate risk management. *Annual Review of Earth and Planetary Sciences, 49*. https://doi.org/10.1146/annurev-earth-080320-055847

Khosrowi, D. (2019). Trade-offs between epistemic and moral values in evidence-based policy. *Economics & Philosophy, 35*, 49–78. https://doi.org/10.1017/S0266267118000159

Lee, B. S., Haran, M., Fuller, R. W., Pollard, D., & Keller, K. (2020). A fast particle-based approach for calibrating a 3-D model of the Antarctic ice sheet. *The Annals of Applied Statistics, 14*, 605–634. https://doi.org/10.1214/19-aoas1305

Lloyd, E. A., & Oreskes, N. (2018). Climate change attribution: When is it appropriate to accept new methods? *Earth's Future, 6*, 311–325. https://doi.org/10.1002/2017EF000665

Mauritsen, T., et al. (2012). Tuning the climate of a global model. *Journal of Advances in Modeling Earth Systems, 4*. https://doi.org/10.1029/2012ms000154

Meredith, E. P., Semenov, V. A., Maraun, D., Park, W., & Chernokulsky, A. V. (2015). Crucial role of Black Sea warming in amplifying the 2012 Krymsk precipitation extreme. *Nature Geoscience, 8*, 615–619. https://doi.org/10.1038/ngeo2483

National Academies of Sciences, Engineering, and Medicine. (2016). *Attribution of extreme weather events in the context of climate change* (186 pp). The National Academies Press.

O'Brien, K. L., & Wolf, J. (2010). A values-based approach to vulnerability and adaptation to climate change: A values-based approach. *Wiley Interdisciplinary Reviews Climate Change, 1*, 232–242. https://doi.org/10.1002/wcc.30

Otto, F. E. L., van Oldenborgh, G. J., Eden, J., Stott, P. A., Karoly, D. J., & Allen, M. R. (2016). The attribution question. *Nature Climate Change, 6*, 813–816. https://doi.org/10.1038/nclimate3089

Pappenberger, F., Beven, K., Frodsham, K., Romanowicz, R., & Matgen, P. (2007). Grasping the unavoidable subjectivity in calibration of flood inundation models: A vulnerability weighted approach. *Journal of Hydrology, 333*, 275–287. https://doi.org/10.1016/j.jhydrol.2006.08.017

Parker, W. S., & Lusk, G. (2019). Incorporating user values into climate services. *Bulletin of the American Meteorological Society, 100*, 1643–1650. https://doi.org/10.1175/BAMS-D-17-0325.1

Parker, W. S., & Winsberg, E. (2018). Values and evidence: How models make a difference. *European Journal for Philosophy of Science, 8*, 125–142. https://doi.org/10.1007/s13194-017-0180-6

Pulkkinen, K., et al. (2022). The value of values in climate science. *Nature Climate Change, 12*, 4–6. https://doi.org/10.1038/s41558-021-01238-9

Quinn, J. D., Hadjimichael, A., Reed, P. M., & Steinschneider, S. (2020). Can exploratory modeling of water scarcity vulnerabilities and robustness be scenario neutral? *Earth's Future, 8*(104), 699. https://doi.org/10.1029/2020EF001650

Shepherd, T. G. (2014). Atmospheric circulation as a source of uncertainty in climate change projections. *Nature Geoscience, 7*, 703–708. https://doi.org/10.1038/ngeo2253

Shepherd, T. G., et al. (2018). Storylines: An alternative approach to representing uncertainty in physical aspects of climate change. *Climatic Change, 151*, 555–571. https://doi.org/10.1007/s10584-018-2317-9

Sillmann, J., Shepherd, T. G., van den Hurk, B., Hazeleger, W., Martius, O., Slingo, J., & Zscheischler, J. (2021). Event-based storylines to address climate risk. *Earth's Future, 9*. https://doi.org/10.1029/2020ef001783

Sriver, R. L., Urban, N. M., Olson, R., & Keller, K. (2012). Toward a physically plausible upper bound of sea-level rise projections. *Climatic Change, 115*, 893–902. https://doi.org/10.1007/s10584-012-0610-6

Stott, P. A., et al. (2016). Attribution of extreme weather and climate-related events. *Wiley Interdisciplinary Reviews Climate Change, 7*, 23–41. https://doi.org/10.1002/wcc.380

Trenberth, K. E. (2011). Attribution of climate variations and trends to human influences and natural variability. *Wiley Interdisciplinary Reviews Climate Change, 2*, 925–930. https://doi.org/10.1002/wcc.142

Trenberth, K. E., Fasullo, J. T., & Shepherd, T. G. (2015). Attribution of climate extreme events. *Nature Climate Change, 5*, 725–730. https://doi.org/10.1038/nclimate2657

Tschakert, P., et al. (2017). Climate change and loss, as if people mattered: Values, places, and experiences. *Wiley Interdisciplinary Reviews Climate Change, 8*, e476. https://doi.org/10.1002/wcc.476

Tuana, N. (2018). Understanding coupled ethical-epistemic issues relevant to climate modeling and decision support science. In L. C. Gundersen (Ed.), *Scientific integrity and ethics in the geosciences* (pp. 155–173). American Geophysical Union and John Wiley & Sons.

Undorf, S., Pulkkinen, K., Wikman-Svahn, P., & Bender, F. A.-M. (2022). How do value-judgments enter model-based assessments of climate sensitivity? *Climatic Change, 174*, 19. https://doi.org/10.1007/s10584-022-03435-7

Vezér, M., Bakker, A., Keller, K., & Tuana, N. (2018). Epistemic and ethical trade-offs in decision analytical modelling. *Climatic Change, 147*, 1–10. https://doi.org/10.1007/s10584-017-2123-9

Weitzman, M. L. (2009). On modeling and interpreting the economics of catastrophic climate change. *The Review of Economics and Statistics, 91*, 1–19. https://doi.org/10.1162/rest.91.1.1

Wong, T. E., & Keller, K. (2017). Deep uncertainty surrounding coastal flood risk projections: A case study for New Orleans. *Earth's Future, 5*, 1015–1026. https://doi.org/10.1002/2017EF000607

Wong, T. E., Bakker, A. M. R., Ruckert, K., Applegate, P., Slangen, A. B. A., & Keller, K. (2017). BRICK v0.2, a simple, accessible, and transparent model framework for climate and regional sea-level projections. *Geoscientific Model Development, 10*, 2741–2760. https://doi.org/10.5194/gmd-10-2741-2017

Open Access This chapter is licensed under the terms of the Creative Commons Attribution 4.0 International License (http://creativecommons.org/licenses/by/4.0/), which permits use, sharing, adaptation, distribution and reproduction in any medium or format, as long as you give appropriate credit to the original author(s) and the source, provide a link to the Creative Commons license and indicate if changes were made.

The images or other third party material in this chapter are included in the chapter's Creative Commons license, unless indicated otherwise in a credit line to the material. If material is not included in the chapter's Creative Commons license and your intended use is not permitted by statutory regulation or exceeds the permitted use, you will need to obtain permission directly from the copyright holder.

Expert Judgment and Communication of Uncertainty

Stephen B. Broomell, Emily Ho, Daniel M. Benjamin, and David V. Budescu

23.1 Introduction

Uncertainty is ever present in research, and climate research is no exception. With vast time scales, atmospheric complexities, and multiple policies that can be implemented, expert judgment is often required to make informed decisions based on climate predictions and their uncertainties. As outlined elsewhere, uncertainty in climate science applies not only to projections of climate (Chap. 15, this volume) and sea level rise (Chap. 19, this volume), but also to impacts on human health (Chap. 13, this volume) and infrastructure (Chap. 11, this volume). Uncertainty is a complex topic, and there exist differences between experts in their views and approaches to uncertainty. Moreover, experts, the public, and policymakers vary in their understanding and interpretation of why climate predictions are uncertain and how they should affect policy and decision-making. Turning science into action requires communication of the aspects important to those unfamiliar with the science (Fischhoff & Scheufele, 2013).

Psychological research focusing on human judgment and decision-making under uncertainty provides a theoretical framework that can help climate researchers understand how their message is understood and interpreted (and, sometimes, misunderstood and misinterpreted) and develop effective ways to communicate the uncertainty in their research to the public. Applications of this approach to the study of uncertainty and scientific communication have revealed some prominent features that drive public perceptions and choices and have identified cases where these features differ from current guidelines (and normative frameworks) employed by researchers to define and communicate uncertainty.

We first provide a brief review and introduction to the psychological study of decision-making under uncertainty. Second, we discuss expert conceptions of uncertainty along with research that tests whether the public differs in its interpretation of uncertainty. We review differences between approaches for handling uncertainty, finding both effective and ineffective communication methods. Third, we review recent psychological advances that can benefit the development and testing of effective communications moving forward. We conclude with a discussion of how the generation of communication guidance and empirical testing of their efficacy can proceed as an iterative process.

23.2 Judgment and Decision-Making Under Uncertainty

The study of judgment and decision-making can be applied to analyze decisions (e.g., prescribing choices), summarize decisions (e.g., model natural choice processes), and aid decisions (e.g., design interventions to help decision makers). For a more detailed review, see Fischhoff and Broomell (2020). Each of these branches of study can help in building effective communications of uncertain expert judgment.

We analyze decisions by applying *normative models* to the study of choice. These models typically start from a set of mathematical and logical axioms and establish rationality as a benchmark for optimal decision-making. Rational choices are perfectly consistent with the axioms and can, typically, be represented by models that translate the DM's subjective and personal preferences into actions and choices. Specific examples include (subjective) expected utility theory and Bayes' theorem.

We summarize such decisions by applying *descriptive models* to the study of choice. Typically, these models start from a set of psychological principles and empirical regularities and seek to develop theories for how a person's choices are affected by different features that are often not part of normative models. These features may include the presentation and description of the choice options, their framing, uncertainty, and more.

While there are many definitions of uncertainty unique to scientific disciplines (henceforth, scientific uncertainty; see Chaps. 11, 13, 15, and 19, this volume), we discuss uncertainty as it pertains to behavioral science focusing on judgment and decision-making (henceforth, decision uncertainty). Knight (1921) categorized decisions as being made in contexts labeled as certainty, risk, and uncertainty. Knight defined decisions under *certainty* to have known outcomes that will be realized; decisions under *risk* to have known outcomes that are realized with known probabilities; and decisions under *uncertainty* to have known outcomes that are realized with completely unknown probabilities. One can think of these three cases as salient, and easy to analyze, points on a continuum. Other points on this continuum involve decisions under *vagueness* (or imprecision) where the outcomes and probabilities are only partially or imprecisely known (e.g., Budescu & Wallsten, 1995).

Decision-making under Knight's (1921) categories of risk and uncertainty has a long history of research from the normative perspective, resulting in different models of choice depending on how we interpret probability. The classical "Frequentist" definition of probability assumes that the relevant random processes are repeatable, and risky choices are defined by a shared understanding of the "objective" probabilities governing these processes (e.g., an expected utility model). Subjective (or Bayesian) probabilities reject this view and allow more flexibility. Decision makers can have their own personal beliefs about the probabilities of target outcomes, and these can be mapped into probabilities provided that the DMs are consistent and coherent (e.g., a subjective expected utility model).

Recent research in psychology has focused on decision-making under vagueness, expanding Knight's categories to allow for a richer psychological study of decision-making under uncertainty, where decision uncertainty varies in its degree and source. Choice options can range from complete uncertainty (where any outcome or probability is possible) to more certain (where the range of possible outcomes and probabilities can be narrowed down). A DM's understanding of decision uncertainty cannot be fully captured by standard normative frameworks because it may also depend on (a) context, (b) its source, (c) if it arises from inherent randomness/chance (aleatory) or from a lack of complete or reliable knowledge (epistemic) (Ülkümen et al., 2016). Therefore, perceptions of decision uncertainty require deeper investigations of judgments of uncertainty (summarized in the last section of this chapter).

We apply this approach to judgment and decision-making under vagueness and uncertainty to investigate the efficacy of methods for communicating uncertain expert judgment. In the next section, we discuss some existing guidelines for communicating the various forms of scientific uncertainty related to climate as decision uncertainty. We also discuss several empirical tests of the efficacy of these approaches.

23.3 Scientific Conceptions of Uncertainty

Scientific conceptions of uncertainty are rooted in the methodology and statistical procedures that underlie scientific research. This includes concepts such as probability distributions, confidence intervals, and fuzzy sets. Expert guidance for communicating scientific uncertainty is often generated from the perspective of research methodology. However, these concepts are abstract, require training to understand and use, and may not be fully, or well, understood by the public. If the public needs to use this information to form judgments and make decisions, it is critical to analyze the effectiveness of scientific uncertainty communications based on how it looks through the lens of decision uncertainty introduced above.

The Intergovernmental Panel on Climate Change (IPCC) provides an excellent case study for understanding expert conceptions of uncertainty. Climate science is an inherently multidisciplinary field, ranging from the natural and physical sciences studying the physical scientific bases of climate systems to the social sciences studying medical, environmental, and social impacts of systemic changes. This convergence of disciplines results in different discipline-specific treatments of (and even different languages for describing) uncertainty such as estimated probabilities in the natural sciences, which are qualitatively different from that in the social sciences (Grübler & Nakicenovic, 2001). For example, uncertainty in climate models tends to arise from limited knowledge about model assumptions and parameter values (e.g., standard errors of complex systems; Curry, 2011), whereas uncertainty in social sciences tends to arise from small samples, inaccurate and less than fully reliable measurements, and limited generalizability across contexts.

Recent work seeks to model and account for contributions of scientific uncertainty in climate metamodels (van Vuuren et al., 2020), which combine the contributions of multiple models such as Integrated Assessment Models and other relevant information from both natural and social sciences to determine the likelihood of reaching certain temperature targets by the year 2100. Among the findings in this exercise is identifying and, where possible, quantifying the relative magnitude of uncertainty across the different components of the model. Van Vuuren et al. (2020) find that compared with natural science inputs, socioeconomic sources of uncertainty are more prominent, suggesting the importance of policy decision-making in setting appropriate climate targets.

The IPCC therefore faces a difficult task in creating guidance for the communication of scientific uncertainty that can be equally useful and informative for all the different scientific disciplines represented in this organization. The goal of the IPCC is to, clearly and accurately, communicate the most current science on climate change so that policymakers and other stakeholders can act decisively on such information. The IPCC consists of three Working Groups (WGs), which assess, respectively: the quality of scientific information on climate change (WG I); the environmental and socioeconomic impacts of climate change (WG II); and the response strategies to mitigation and adaptation of climate change impacts (WG III).

The three WGs produce and release periodical Assessment Reports (ARs). The first two ARs did not include any definition of uncertainty which may have led to miscommunications, particularly in the media portrayal of the available scientific evidence (Pidgeon, 2012; Segnit & Ereaut, 2007). This led to a summit to develop clearer recommendations for communicating uncertainty (Moss & Schneider, 1999). However, these recommendations were not fully followed, and the guidelines were applied unsystematically across the three WGs (Patt & Schrag, 2003; Swart et al., 2009; Bjurström & Polk, 2011). Over the years, IPCC guidelines have explored several ways of communicating uncertainty, including (a) uncertainty of exact outcomes defined by probabilities (e.g., 36%), (b) uncertainty in probabilities defined by ranges of probabilities (e.g., 24–41%) or verbal probability phrases (e.g., very unlikely), (c) separate disclosures of uncertainty defined by high/medium/low confidence, quality of evidence, and expert disagreement, and (d) qualitative uncertainty terms such as "speculative" (Moss & Schneider, 1999). For a historical account of IPCC treatment of uncertainty across ARs, refer to Ho et al. (2017).

Most recently, AR5 follows the (Mastrandrea et al., 2010) guidance for uncertainty communication, and AR6 follows the same approach with updates described in (Mach et al., 2017). This guidance has evolved over time and settled on communicating two key metrics for the degree of certainty: confidence and likelihood. Confidence communications are determined by the quality of the evidence and degree of expert agreement. Likelihood communications are generated when the author team has sufficient confidence and evidence to do so using specific likelihood statements defined by pre-specified probability ranges (see Box 1.1 in AR6 Ch. 1, Chen et al. (2021)).

As with any type of policy guideline, they should be continually tested for effectiveness and updated accordingly. Research from judgment and decision-making has examined several communication approaches for their effect on judgments and choices and documented several empirical regularities. First, there are several important limitations to communicating risks to the public using probabilities. Basic statements of statistical information such as percentages or probabilities are difficult to understand and can have less influence on decisions than anecdotal information and evidence presented as a narrative (Fagerlin et al., 2005). Additionally, DMs have trouble combining probabilities, leading to systematic biases in subjective judgments of compound events (Bar-Hillel, 1973). Probabilities can be interpreted only based on comparisons to understood entities because "all probabilities are conditional" (Genest & Zidek, 1986, p. 117), and cannot be properly interpreted without understanding the reference class used to compute the probability (e.g., the probability of pregnancy for women, married women, or teenage women) (Gigerenzer et al., 1991).

Much work has focused on the use of verbal probability phrases to communicate scientific uncertainty currently used in the IPCC AR6. Researchers found that verbal probabilities are perceived differently (a) depending on perceived outcome severity (Harris & Corner, 2011), (b) for positive and negative phrases (Smithson et al., 2012), and (c) by different people creating a large heterogeneity of interpretations of risk (Budescu et al., 2009, 2012, 2014). Verbal probabilities can therefore create an "illusion of communication" (Budescu et al., 2009; Wallsten & Budescu, 1995) where consumers of communications perceive a different message than that intended by the communicator, but neither party is fully aware of this communication failure. Dhami and Mandel (2021) propose replacing verbal probabilities with numerical probabilities, particularly for domains where failure to communicate can have heavily negative downstream consequences, such as in climate science. However, this recommendation seems to ignore the reluctance of experts to commit to precise numerical probabilities based on less than perfect information and observations. Budescu and Wallsten (1995) suggested uncertainty should be communicated using statements that reflect faithfully the degree of their underlying uncertainty. For example, people may expect precise estimates of weather forecasts for short time horizons (e.g., tomorrow's temperature) but will expect more uncertainty for longer time horizons (e.g., the temperature a month from now).

There is some evidence that climate science communications using verbal probabilities can be improved by (a) using a verbal-numerical format that combines ranges of probabilities with verbal probability phrases (Budescu et al.,

2009; Wintle et al., 2019), (b) including explicit statements of the lower or upper bound implied by the expression (Harris et al., 2017), and (c) empirically generating guidelines for translating words to probabilities that match the target audience's general interpretations (Ho et al., 2015). In general, the empirical evidence is overwhelming that the verbal probability phrases used in the AR5 and AR6 should be avoided if clear actionable communication is the goal.

There can be stark differences in how different groups of people perceive uncertainty (experts vs. laypeople, experts from different disciplines, and communicators vs. audiences). Communicators should not be satisfied by just setting clear guidelines for describing uncertainty but should revisit, test, and revise those guidelines as evidence evolves. Communicators can never be cognizant enough about how their audience will interpret the uncertainties in technical communications, no matter how clear originators think they are being. Both over- and under-precision can damage impressions of the content and alter decision-making. Communicators should be transparent, avoiding vagueness when possible, and make sure to understand how communication design choices might systematically affect judgment and decision-making.

23.4 Public Perceptions of Scientific Uncertainty

As described above, communication guidelines are often developed by (and from) the experts' view of uncertainty and, often, without investigating how the public understands the uncertainty conveyed. For example, the boundaries defining "likely" and "very likely" in the IPCC guidance are based on prominent values in the Normal (Gaussian) distribution instead of empirical research that elicits prominent values among the public. However, effective communication requires, by definition, a common understanding by all parties involved in the communication. Therefore, communication of expert judgment to the general public requires an understanding of how people perceive uncertain evidence, research, and scientific findings. Such an approach would require the IPCC to generate boundaries based on public perceptions instead of standard distributions. Ho et al. (2015) illustrate the impact of this approach. Additionally, uncertain evidence ultimately leads to disagreement and conflict, which also affects public perceptions of expert judgment. Here we describe more general psychological research on public perceptions of scientific evidence and disagreement that can inform the development of effective communication of scientific uncertainty in climate science.

23.4.1 Public Perception of Scientific Evidence

Broomell and Kane (2017) present three studies looking at how scientific definitions of uncertainty relate to public perceptions of scientific uncertainty. They find that the public's perceptions of scientific uncertainty are based mostly on their perceptions of precision in any given scientific field. When judging the uncertainty of individual research findings, participant responses did not significantly differ across conditions that framed the result as coming from scientific fields with high versus low perceived uncertainty. This suggests that participants may not judge individual results in light of the scientific field that created them, but instead judge the scientific field in light of the individual results that they associate with that field. Therefore, this evidence suggests that laypersons' perceptions of the uncertainty of a scientific field are based on experiences with the science's results and predictions, and not on the uncertainty and precision inherent in the methods and topic of study.

For example, seismic data cannot be used to predict earthquakes, but atmospheric data can be used to predict weather and climate. Yet Broomell and Kane (2017) find that the public perceives seismology as more precise than climate science. While seismology makes few predictions that can be refuted empirically (although see Cartlidge, 2012), climate science may be perceived as being highly uncertain (and imprecise) by the public because of the perceptions of the accuracy of climate change predictions. This can be amplified by systematic efforts to use uncertainty strategically to discredit work on the effects of climate change (Oreskes & Conway, 2011) and by message framing designed to manufacture uncertainty (McCright et al., 2016).

Broomell and Kane (2017) created a perceptual map of scientific uncertainty by collecting ratings of 16 research fields on 14 dimensions of uncertainty. As shown in Fig. 23.1, Broomell and Kane (2017) used principal component analysis to plot the 16 fields of research based on two prominent dimensions of perceived uncertainty: precision (explaining 73% of judgment variance) and mathematical abstraction (explaining 10% of judgment variance). Climate science is perceived to have low precision (along with psychology). The public expects to see this low precision in communications of climate projections. For example, Howe et al. (2019) and Joslyn and LeClerc (2016) found that providing uncertainty in the form of bounds around an estimate tended to increase public trust in the estimate. Indeed, Du et al. (2011) found that in finance, managers actually prefer imprecise advice (up to a point) because uncertainty is expected and is perceived as inherent to financial predictions. However, Howe et al. (2019) also found that fully acknowledging the irreducible

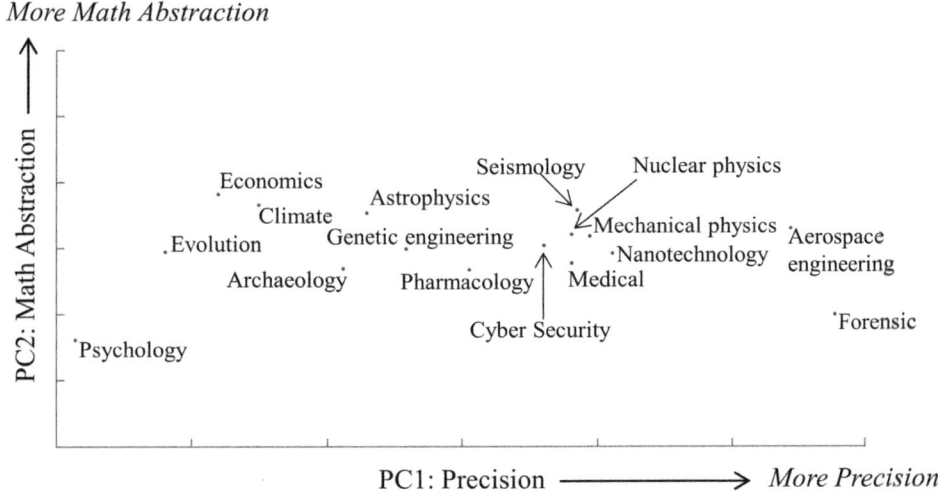

Fig. 23.1 Reproduction of Fig. 1 from Broomell and Kane (2017) displaying a perceptual map of scientific uncertainty as a function of perceptions of precision and mathematical abstraction related to each field

uncertainty around a scientific estimate undermined public trust and acceptance of the information. This may be due to a general expectation that science resolves uncertainties and provides clear answers, limiting the public's receptiveness to information describing events that have never happened within recent history where the extent of uncertainty is inherently unknowable (e.g., the number of feet the sea level may rise in the year 2100; Ho & Budescu, 2019).

How does the general public evaluate the accuracy of scientific predictions? People may use their perceptions of their local environment (Fiedler, 2000; Galesic et al., 2016), expecting features of their local environment to generalize to global predictions (Broomell, 2020; Broomell & Kane, 2021). However, such judgment processes are theorized to lead to systematic misperceptions of precision in scientific predictions of global problems due to global-local incompatibility (Broomell, 2020). This problem arises when local environments appear stable and reliable but fail to reflect the true trend of a more global phenomenon being considered.

Climate change is a perfect example because individual experiences of localized weather rarely reveal information about climate trends. Global trends require data on larger geographical and temporal scales to be fully revealed. In fact, due to natural variance between locations, experiences with localized weather patterns can lead to increased error in judgments of populations of people all affected by the same global risk (Broomell, 2020; Broomell & Kane, 2021; Kane & Broomell, 2020). This happens when the risk is the average of all possible experiences, but few experiences reflect the average. Despite the low information content of local experiences in the climate setting, psychological research has documented that people accurately attend to local weather (Broomell et al., 2017) and update their beliefs about climate based on their localized experiences (Li et al.,

2011; Sugerman et al., 2021). Indeed, Konisky et a;/ (2016) find that extreme weather experiences were associated with increased public concern about climate change, but the effect decayed over time such that only the most recent experiences had any link to public concern. The overall effect is that many severe and extreme manifestations of climate change may fail to generate sustained collective action because they affect only a fragment of the population and at different times. Additionally, as variance in weather increases, many impacts on the climate seem counterintuitive to a warming narrative (e.g., colder than normal temperatures; Rawlins, 2022), making sustained collective action to combat climate change from geographically dispersed populations difficult.

Because of the natural variance in weather experiences and the uncertainty surrounding climate science, even individuals with the same experiences can draw different conclusions about the degree to which these experiences support (or oppose) climate predictions (Broomell et al., 2017). For example, Budescu et al. (2012) found that in a US nationally representative sample, Democrats reported higher levels of personal experience with climate change than Republicans. Different interpretations of the same weather events can be explained using Signal Detection Theory, a model that incorporates a decision process into the detection of noisy stimuli (Swets et al., 1961). Broomell et al. (2017) provide evidence of a positive association between individuals interpreting weather extremes as evidence of climate change and prior beliefs about whether they have already observed the effects of climate change. Those who believed climate change could not be experienced had stricter decision thresholds, requiring larger extremes before linking temperatures to global warming.

Overall, these studies suggest several general results regarding public perceptions of uncertainty in scientific evidence. First, perceptions of scientific uncertainty are highly influenced by personal experiences and may not be directly linked to weaknesses in scientific methodology, data, or measurements. Second, science is expected to be precise, and admissions of uncertainty (especially large levels) can be perceived as scientific failures, rather than natural limitations of knowledge of a topic. Third, local experiences with scientific claims about climate are heterogeneous, unreliable, and ambiguous. Different individuals have different experiences, and depending on their prior beliefs, may also interpret their experiences differently.

These results have several implications for communicating expert judgment. As outlined in the previous section, verbal communications of probability are intended to express decision uncertainty but require context to interpret. If everyone's context is different due to natural variability in experiences and beliefs and there is a general expectation for precision in scientific predictions, then verbal communications may not increase uncertainty, but instead increase heterogeneity in interpretations of expert judgments. Researchers may also consider analyzing the natural variability of their predictions and focus communications on important predictions that are more likely to create shared experiences among the public, so that predictions are experienced similarly by most of the target audience.

23.4.2 Public Perception of Conflict and Disagreement

Scientific uncertainty also leads to conflict and disagreement among experts. Evidence can conflict and experts can disagree, and these disagreements need to be resolved to form consensus judgments. Psychological research suggests that conflict and disagreement affect how uncertainty is perceived by the public. How the public understands and reckons with scientific disagreement is also a topic of prescriptive concern, as it impacts the call to collective action and policy decisions.

Conflict is a specific source of uncertainty that arises from, "disagreement over states of reality that cannot hold true simultaneously" (Smithson, 1999, p. 180) and its presence can alter decision-making. For example, a survey of insurance adjusters found they would charge higher prices when experts' risk estimates disagreed, especially for catastrophic risk (Cabantous et al., 2011). Additionally, conflict can lead to over-sensitivity to imprecision and ambiguity (Baillon et al., 2012). This can impede objectivity. For instance, participants provided estimates that were more similar to the experts after viewing sets of agreeing imprecise expert projections compared to conflicting precise projections (Benjamin & Budescu, 2018).

Public reactions to conflict are a function of cognitive ability and topic knowledge. Individuals with lower ability and knowledge attribute disagreement to incompetence, while higher ability and knowledge attribute disagreement to bias. Higher ability individuals find complexity and randomness to play a vital role (Dieckmann et al., 2017). A more complete analysis of reasons behind expert disagreement found that explanations load onto three factors: interests and values, process and competence, and topic complexity (Johnson & Dieckmann, 2018). This study employed a latent class analysis—a method to classify respondents into groups based on observed heterogeneity—of a survey of US adults. The study identified four classes–one high on all three factors (45%), one moderate on all three (29%), one high on topic complexity (20%), and one low on topic complexity (5%). Positive and credible views of science were two of the better predictors of class membership.

Conflict is psychologically distinct from imprecision and uniquely contributes to perceptions of uncertainty (Smithson, 1999). Judges view conflict as less desirable than a similar degree of imprecision. Stakeholders must determine whether partially agreeing estimates, such as overlapping interval forecasts, are considered agreeing or conflicting. The intersecting portion can simultaneously hold true while the disjoint portions cannot. Participants found disagreeing, precise estimates to be less credible than imprecise intervals (Smithson, 1999). When both sources of disagreement are present, the impacts of conflict and imprecision are task-dependent; they are additive for estimation tasks and averaged for preferential judgments (Benjamin & Budescu, 2018).

Reactions to disagreement are also affected by source credibility, which is determined by a decision maker's perceptions of two factors: expertise and bias (Birnbaum & Stegner, 1979). Perceptions of expertise are theorized to be derived from the perceived accuracy of the source and increases credibility. Perceptions of bias are theorized to be derived from knowledge of the source's point of view, and non-neutral points of view serve to undermine credibility. Birnbaum and Stegner (1979) asked research participants to integrate information from multiple sources. They found that participants placed more weight on sources with higher expertise, but less weight on expert sources that were perceived to be biased (i.e., not having a neutral point of view). Indeed, situations with disagreeing sources that are each perceived to be highly credible are the most difficult to resolve (Einhorn & Hogarth, 1985).

Decision makers resolve disagreement by combining multiple conflicting judgments through aggregation. When decision makers combine forecasts, the resulting judgment is sensitive to the structure of the information and the nature of their cognitive processes (Wallsten et al., 1997). The process of aggregating information can generally be expressed as a function of confidence in the individual projections plus randomness (Erev et al., 1994; Wallsten et al., 1997). Confidence can help determine how to weigh different inputs. Confidence when combining inputs is a function of the number and (in)accuracy of the judges, the number of total and common cues and correlation between them, and the (un)predictability of the event (Budescu, 2006; Budescu & Rantilla, 2000; Budescu et al., 2003).

Typically, some form of averaging most accurately describes the aggregation process and outperforms a randomly selected individual forecaster (e.g., Budescu & Rantilla, 2000; Budescu et al., 2003; Clemen & Winkler, 1987; Clemen, 1989; Lorge et al., 1958; Wallsten et al., 1997; Winkler, 1971). Therefore, the presence of conflicting expert judgments can have the paradoxical effect of improving aggregate judgment accuracy while also reducing the public's confidence in the aggregated judgment.

The results of these studies have several implications for communicating expert judgment. As outlined in the previous section, IPCC guidance has considered the possibility of communicating scientific uncertainty in terms of high/medium/low quality of evidence and expert disagreement. Although expert disagreement is present in all scientific enterprises, the expert disagreement regarding climate science is much lower than is often portrayed in the media (Boykoff & Boykoff, 2004). Experts agree on the general causes and consequences of climate change but may disagree on their quantitative estimates of the impacts and their timing. Perceptions of uncertainty derived from disagreement tend to be less desirable and lead to more violations of rational choice. While the accuracy of aggregating conflicting predictions may benefit from independent and disagreeing points of view, expressions of disagreement and conflict may do more to undermine public confidence than to increase transparency of the science.

23.5 Conclusion

Expert judgment plays an important role in communicating the current state of scientific knowledge to the public. We provided a review from the perspective of human judgment and decision-making of several proposed methods for communicating uncertain expert judgment along with empirical evaluations of the efficacy of these approaches. We focus our conclusion on future guideline development by building from current psychological theory.

Finding a shared definition of uncertainty is not a trivial task. Our understanding of uncertainty differs with both discipline (i.e., different understanding of the data, methods, and their limitations) and occupation (e.g., scientists differ from policymakers). Even if communicators are aware of the vast number of ways uncertainty can be conceived, consumers of scientific information may not be aware. If definitions of uncertainty are too complicated to be conveyed clearly, communicators could find alternative cognitively compatible ways of adding uncertainty to projections, and empirically test them for effectiveness within each application and context. We propose that empirical tests can focus on decision uncertainty as a shared representation of how scientific uncertainty constrains the imprecision of the outcomes and their likelihood for each choice option. For example, Fagerlin et al. (2005) found that pictograms of probabilities reduced the biasing effect of anecdotes in medical decisions.

Expressions of scientific uncertainty can sometimes be at odds with the perceived usefulness of communications. A full disclosure of uncertainty surrounding a simple projection can make it look far more complicated. The future states of the world that can be ruled out are at times minimal, with error bars being so wide that they appear to be wholly uninformative. Yet, the appeal of verbal expressions of probability is that they may appear less complex to consumers. We reviewed work on verbal descriptions of probability, finding that their usefulness is limited outside of conveying uncertainty through their inherent ambiguity. However, ambiguity is not the same as uncertainty, and readers are left guessing which probability applies based on their understanding of the context instead of understanding that a range of probabilities might apply. Therefore, we propose that future psychological work is needed to expand the toolbox for communicating uncertainty in cognitively compatible ways beyond the use of verbal probabilities. Novel effective methods will become apparent with the development of theories that can explain the role and representation of uncertainty in cognition.

Expressions of scientific uncertainty are sometimes given to increase trust, as withholding this information and providing only the best estimates can be misleading. If not disclosed, detractors can use this omission to further reduce trust. However, we find that the sources of uncertainty themselves can sometimes have the unintended consequence of undermining trust, such as expressions of conflict and disagreement. Furthermore, it is natural for people to seek individual experts to find answers and to discredit certain experts when disagreements arise. In deeply uncertain settings, like climate science, we argue that it is imperative to convey the pool of expert opinion as credible and state of the art.

Finally, personal experiences play an important role in how the public interprets vague and ambiguous information. Single instances (or anecdotes) can quickly undermine the perceived accuracy of probabilistic projections. For example, weather forecasters provide probabilities of precipitation that may be the most calibrated and accurate probabilities conveyed to the public. Yet the public may still perceive these probabilities as inaccurate by observing recent "failures" (e.g., no rain on a day with a 60% chance of rain). Because of the heterogeneity in the probabilities of precipitation across a single viewing area, there is heterogeneity in the perceptions of forecast accuracy. One solution adopted by broadcast meteorologists is to provide projections that will appear the most accurate to the largest number of people in their viewing area (e.g., by weighting probabilities by population density). Thus, one potential strategy for climate communications is to analyze how the public will interpret different climate predictions within their own context in both scientifically accurate and inaccurate ways. For example, a prediction of increased atmospheric warmth might lead people to expect less snow. This may be true in some regions, but other regions might experience increased snowfall due to increases in evaporation. We recommend analyzing differences in interpretations and the various experiences the public might associate with different predictions and guiding communication strategies toward communications that minimize such differences. When possible, globally distributed communications could emphasize predictions that are expected to appear more uniform to everyone. Targeted communications could inform specific locations with essential information to prepare them for their experiences, such as extreme precipitation or drought.

Given the vast heterogeneity of the audience and the choices they are facing, generating effective communications of expert judgment will require its own dedicated empirical study. Developing communication strategies based on scientific representations of uncertainty is not likely to resonate with large swaths of the public. Finding approaches that will resonate requires collaborative research efforts between climate and behavioral scientists, broadening our theoretical understanding of risk communication and improving our ability to apply these theories to help the public make informed decisions in deeply uncertain contexts.

References

Baillon, A., Cabantous, L., & Wakker, P. P. (2012). Aggregating imprecise or conflicting beliefs: An experimental investigation using modern ambiguity theories. *Journal of Risk and Uncertainty, 44*(2), 115–147. https://doi.org/10.1007/s11166-012-9140-x

Bar-Hillel, M. (1973). On the subjective probability of compound events. *Organizational Behavior and Human Performance, 9*(3), 396–406. https://doi.org/10.1016/0030-5073(73)90061-5

Benjamin, D. M., & Budescu, D. V. (2018). The role of type and source of uncertainty on the processing of climate models projections. *Frontiers in Psychology, 9*. https://doi.org/10.3389/fpsyg.2018.00403

Birnbaum, M. H., & Stegner, S. E. (1979). Source credibility in social judgment: Bias, expertise, and the judge's point of view. *Journal of Personality and Social Psychology, 37*(1), 48–74. https://doi.org/10.1037/0022-3514.37.1.48

Bjurström, A., & Polk, M. (2011). Physical and economic bias in climate change research: A scientometric study of IPCC Third Assessment Report. *Climatic Change, 108*(1–2), 1–22. https://doi.org/10.1007/s10584-011-0018-8

Boykoff, M. T., & Boykoff, J. M. (2004). Balance as bias: Global warming and the US prestige press. *Global Environmental Change, 14*(2), 125–136. https://doi.org/10.1016/j.gloenvcha.2003.10.001

Broomell, S. B. (2020). Global–local incompatibility: The misperception of reliability in judgment regarding global variables. *Cognitive Science, 44*(4), e12831. https://doi.org/10.1111/cogs.12831

Broomell, S. B., & Kane, P. B. (2017). Public perception and communication of scientific uncertainty. *Journal of Experimental Psychology: General, 146*(2), 286–304. https://dx.doi.org.libproxy1.usc.edu/10.1037/xge0000260

Broomell, S. B., & Kane, P. B. (2021). Perceiving a pandemic: Global–local incompatibility and COVID-19 superspreading events. *Decision, 8*(4), 227–236. https://doi.org/10.1037/dec0000155

Broomell, S. B., Winkles, J.-F., & Kane, P. B. (2017). The perception of daily temperatures as evidence of global warming. *Weather, Climate, and Society, 9*(3), 563–574. https://dx.doi.org.libproxy1.usc.edu/10.1175/WCAS-D-17-0003.1

Budescu, D. V. (2006). Confidence in aggregation of opinions from multiple sources. In K. Fiedler & P. Juslin (Eds.), *Information sampling and adaptive cognition* (pp. 327–352). Cambridge University Press.

Budescu, D. V., & Rantilla, A. K. (2000). Confidence in aggregation of expert opinions. *Acta Psychologica, 104*(3), 371–398. https://doi.org/10.1016/S0001-6918(00)00037-8

Budescu, D. V., & Wallsten, T. S. (1995). Processing linguistic probabilities: General principles and empirical evidence. In J. Busemeyer, R. Hastie, & D. L. Medin (Eds.), *Psychology of learning and motivation* (Vol. 32, pp. 275–318). Academic Press. https://doi.org/10.1016/S0079-7421(08)60313-8

Budescu, D. V., Rantilla, A. K., Yu, H.-T., & Karelitz, T. M. (2003). The effects of asymmetry among advisors on the aggregation of their opinions. *Organizational Behavior and Human Decision Processes, 90*(1), 178–194. https://doi.org/10.1016/S0749-5978(02)00516-2

Budescu, D. V., Broomell, S., & Por, H.-H. (2009). Improving communication of uncertainty in the reports of the Intergovernmental Panel on Climate Change. *Psychological Science, 20*(3), 299–308. https://doi.org/10.1111/j.1467-9280.2009.02284.x

Budescu, D. V., Por, H.-H., & Broomell, S. B. (2012). Effective communication of uncertainty in the IPCC reports. *Climatic Change, 113*(2), 181–200. https://doi.org/10.1007/s10584-011-0330-3

Budescu, D. V., Por, H.-H., Broomell, S. B., & Smithson, M. (2014). The interpretation of IPCC probabilistic statements around the world. *Nature Climate Change, 4*(6), 508–512. https://doi.org/10.1038/nclimate2194

Cabantous, L., Hilton, D., Kunreuther, H., & Michel-Kerjan, E. (2011). Is imprecise knowledge better than conflicting expertise? Evidence from insurers' decisions in the United States. *Journal of Risk and Uncertainty, 42*(3), 211–232. https://doi.org/10.1007/s11166-011-9117-1

Cartlidge, E. (2012). Aftershocks in the courtroom. *Science, 338*(6104), 184–188. https://doi.org/10.1126/science.338.6104.184

Chen, D., and et al., 2021: Framing, context, and methods. Climate Change 2021: The physical science basis. Contribution of Working Group I to the Sixth Assessment Report of the Intergovernmental Panel on Climate Change, V. P. Masson-Delmotte, A. Zhai, S. Pirani, C. Connors, S. Péan, N. Berger, Y. Caud, L. Chen, M. Goldfarb, M. Gomis, K. Huang, E. Leitzell, J. Lonnoy, T. Matthews, T. Maycock, O. Waterfield, R. Yelekçi, Yu, and B. Zhou, Eds., Cambridge University Press., https://www.ipcc.ch/report/ar6/wg1/downloads/report/IPCC_AR6_WGI_Chapter_01.pdf.

Clemen, R. T. (1989). Combining forecasts: A review and annotated bibliography. *International Journal of Forecasting, 5*(4), 559–583. https://doi.org/10.1016/0169-2070(89)90012-5

Clemen, R. T., & Winkler, R. L. (1987). Calibrating and combining precipitation probability forecasts. In R. Viertl (Ed.), *Probability and Bayesian statistics* (pp. 97–110). Springer. https://doi.org/10.1007/978-1-4613-1885-9_10

Curry, J. (2011). Reasoning about climate uncertainty. *Climatic Change, 108*(4), 723–732. https://doi.org/10.1007/s10584-011-0180-z

Dhami, M. K., & Mandel, D. R. (2021). Words or numbers? Communicating probability in intelligence analysis. *American Psychologist, 76*(3), 549–560. https://doi.org/10.1037/amp0000637

Dieckmann, N. F., Johnson, B. B., Gregory, R., Mayorga, M., Han, P. K. J., & Slovic, P. (2017). Public perceptions of expert disagreement: Bias and incompetence or a complex and random world? *Public Understanding of Science, 26*(3), 325–338. https://doi.org/10.1177/0963662515603271

Du, N., Budescu, D. V., Shelly, M. K., & Omer, T. C. (2011). The appeal of vague financial forecasts. *Organizational Behavior and Human Decision Processes, 114*(2), 179–189. https://doi.org/10.1016/j.obhdp.2010.10.005

Einhorn, H. J., & Hogarth, R. M. (1985). Ambiguity and uncertainty in probabilistic inference. *Psychological Review, 92*(4), 433–461. http://dx.doi.org.libproxy1.usc.edu/10.1037/0033-295X.92.4.433

Erev, I., Wallsten, T. S., & Budescu, D. V. (1994). Simultaneous over- and under confidence: The role of error in judgment processes. *Psychological Review, 101*(3), 519–527. https://doi.org/10.1037/0033-295X.101.3.519

Fagerlin, A., Wang, C., & Ubel, P. A. (2005). Reducing the influence of anecdotal reasoning on people's health care decisions: Is a picture worth a thousand statistics? *Medical Decision Making, 25*(4), 398–405. https://doi.org/10.1177/0272989X05278931

Fiedler, K. (2000). Beware of samples! A cognitive-ecological sampling approach to judgment biases. *Psychological Review, 107*(4), 659–676. https://doi.org/10.1037/0033-295X.107.4.659

Fischhoff, B., & Broomell, S. B. (2020). Judgment and decision making. *Annual Review of Psychology, 71*(1), 331–355. https://doi.org/10.1146/annurev-psych-010419-050747

Fischhoff, B., & Scheufele, D. A. (2013). The science of science communication. *Proceedings of the National Academy of Sciences, 110*(3), 14031–14032. https://doi.org/10.1073/pnas.1312080110

Galesic, M., Kause, A., & Gaissmaier, W. (2016). A sampling framework for uncertainty in individual environmental decisions. *Topics in Cognitive Science, 8*(1), 242–258. https://doi.org/10.1111/tops.12172

Genest, C., & Zidek, J. V. (1986). Combining probability distributions: A critique and an annotated bibliography. *Statistical Science, 1*(1), 114–135. https://doi.org/10.1214/ss/1177013825

Gigerenzer, G., Hoffrage, U., & Kleinbolting, H. (1991). Probabilistic mental models: A Brunswikian theory of confidence. *Psychological Review, 98*(4), 506–528. https://doi.org/10.1037/0033-295X.98.4.506

Grübler, A., & Nakicenovic, N. (2001). Identifying dangers in an uncertain climate. *Nature, 412*(6834), 15. http://www.nature.com/nature/journal/v411/n6833/full/411017a0.html

Harris, A. J. L., & Corner, A. (2011). Communicating environmental risks: Clarifying the severity effect in interpretations of verbal probability expressions. *Journal of Experimental Psychology: Learning, Memory, and Cognition, 37*(6), 1571–1578. https://doi.org/10.1037/a0024195

Harris, A. J. L., Por, H.-H., & Broomell, S. B. (2017). Anchoring climate change communications. *Climatic Change, 140*(3–4), 387–398. https://doi.org/10.1007/s10584-016-1859-y

Ho, E. H., & Budescu, D. V. (2019). Climate uncertainty communication. *Nature Climate Change, 9*(11), 802–803. https://doi.org/10.1038/s41558-019-0606-6

Ho, E. H., Budescu, D. V., Dhami, M. K., & Mandel, D. R. (2015). Improving the communication of uncertainty in climate science and intelligence analysis. *Behavioral Science & Policy, 1*(2), 43–55. https://doi.org/10.1353/bsp.2015.0015

Ho, E. H., Budescu, D. V., & Por, H.-H. (2017). Psychological challenges in communicating about climate change and its uncertainties. *Oxford Research Encyclopedia of Climate Science*. https://doi.org/10.1093/acrefore/9780190228620.013

Howe, L., MacInnis, B., Krosnick, J. A., Markowitz, E. M., & Socolow, R. (2019). Acknowledging uncertainty impacts public acceptance of climate scientists' predictions. *Nature Climate Change, 9*, 863–867. https://doi.org/10.1038/s41558-019-0587-5

Johnson, B. B., & Dieckmann, N. F. (2018). Lay Americans' views of why scientists disagree with each other. *Public Understanding of Science, 27*(7), 824–835. https://doi.org/10.1177/0963662517738408

Joslyn, S. L., & LeClerc, J. E. (2016). Climate projections and uncertainty communication. *Topics in Cognitive Science, 8*(1), 222–241. https://doi.org/10.1111/tops.12177

Kane, P. B., & Broomell, S. B. (2020). Applications of the bias–variance decomposition to human forecasting. *Journal of Mathematical Psychology, 98*, 102417. https://doi.org/10.1016/j.jmp.2020.102417

Knight, F. H. (1921). *Risk, uncertainty and profit*. Houghton Mifflin.

Konisky, D. M., Hughes, L., & Kaylor, C. H. (2016). Extreme weather events and climate change concern. *Climatic Change, 134*, 533–547. https://doi.org/10.1007/s10584-015-1555-3

Li, Y., Johnson, E. J., & Zaval, L. (2011). Local warming: Daily temperature change influences belief in global warming. *Psychological Science, 22*(4), 454–459. https://doi.org/10.1177/0956797611400913

Lorge, I., Fox, D., Davitz, J., & Brenner, M. (1958). A survey of studies contrasting the quality of group performance and individual performance, 1920-1957. *Psychological Bulletin, 55*(6), 337–372. https://doi.org/10.1037/h0042344

Mach, K. J., Mastrandrea, M. D., Freeman, P. T., & Field, C. B. (2017). Unleashing expert judgment in assessment. *Global Environmental Change, 44*, 1–14. https://doi.org/10.1016/j.oenvcha.2017.02.005

Mastrandrea, M. D., et al., 2010: Guidance note for lead authors of the IPCC Fifth Assessment Report on consistent treatment of uncertainties. Intergovernmental Panel on Climate Change, https://www.ipcc.ch/site/assets/uploads/2017/08/AR5_Uncertainty_Guidance_Note.pdf.

McCright, A. M., Charters, M., Dentzman, K., & Dietz, T. (2016). Examining the effectiveness of climate change frames in the face of a climate change denial counter-frame. *Topics in Cognitive Science, 8*(1), 76–97. https://doi.org/10.1111/tops.12171

Moss, R., and S. H. Schneider, 1999: Uncertainties in the IPCC TAR: Recommendations to lead authors for more consistent assessment and reporting. Unpublished document.

Oreskes, N., & Conway, E. M. (2011). *Merchants of doubt: How a handful of scientists obscured the truth on issues from tobacco smoke to global warming*. Bloomsbury Publishing.

Patt, A. G., & Schrag, D. P. (2003). Using specific language to describe risk and probability. *Climatic Change, 61*(1–2), 17–30. http://link.springer.com/article/10.1023/A:1026314523443

Pidgeon, N. (2012). Public understanding of, and attitudes to, climate change: UK and international perspectives and policy. *Climate Policy, 12*(1), S85–S106. http://search.proquest.com/pqrl/docview/1321403763/abstract/AB2E5918701C4FDAPQ/5

Rawlins, M. A., 2022: Why a warming climate can bring bigger snowstorms. http://theconversation.com/why-a-warming-climate-can-bring-bigger-snowstorms-176201.

Segnit, N., & Ereaut, G. (2007). *Warm words II: How the climate story is evolving and the lesson we can learn for encouraging public action*. IPPR & Energy Saving Trust. http://www.ippr.org/files/images/media/files/publication/2011/05/warmwordsfull_1596.pdf?noredirect=1

Smithson, M. (1999). Conflict aversion: Preference for ambiguity vs conflict in sources and evidence. *Organizational Behavior and Human Decision Processes, 79*(3), 179–198. https://doi.org/10.1006/obhd.1999.2844

Smithson, M., Budescu, D. V., Broomell, S. B., & Por, H.-H. (2012). Never say "not": Impact of negative wording in probability phrases on imprecise probability judgments. *International Journal of Approximate Reasoning, 53*(8), 1262–1270. https://doi.org/10.1016/j.ijar.2012.06.019

Sugerman, E. R., Li, Y., & Johnson, E. J. (2021). Local warming is real: A meta-analysis of the effect of recent temperature on climate change beliefs. *Current Opinion in Behavioral Sciences, 42*, 121–126. https://doi.org/10.1016/j.cobeha.2021.04.015

Swart, R., Bernstein, L., Ha-Duong, M., & Petersen, A. (2009). Agreeing to disagree: Uncertainty management in assessing climate change, impacts and responses by the IPCC. *Climatic Change, 92*(1–2), 1–29. https://doi.org/10.1007/s10584-008-9444-7

Swets, J. A., Tanner, W. P., Jr., & Birdsall, T. G. (1961). Decision processes in perception. *Psychological Review, 68*(5), 301–340. https://doi.org/10.1037/h0040547

Ülkümen, G., Fox, C. R., & Malle, B. F. (2016). Two dimensions of subjective uncertainty: Clues from natural language. *Journal of Experimental Psychology: General, 145*(10), 1280–1297. https://doi.org/10.1037/xge0000202

van Vuuren, D. P., van der Wijst, K.-I., Marsman, S., van den Berg, M., Hof, A. F., & Jones, C. D. (2020). The costs of achieving climate targets and the sources of uncertainty. *Nature Climate Change, 10*(4), 329–334. https://doi.org/10.1038/s41558-020-0732-1

Wallsten, T. S., & Budescu, D. V. (1995). A review of human linguistic probability processing: General principles and empirical evidence. *The Knowledge Engineering Review, 10*(1), 43–62. https://doi.org/10.1017/S0269888900007256

Wallsten, T. S., Budescu, D. V., Erev, I., & Diederich, A. (1997). Evaluating and combining subjective probability estimates. *Journal of Behavioral Decision Making, 10*(3), 243–268. https://doi.org/10.1002/(SICI)1099-0771(199709)10:3<243::AID-BDM268>3.0.CO;2-M

Winkler, R. L. (1971). Probabilistic prediction: Some experimental results. *Journal of the American Statistical Association, 66*(336), 675–685. https://doi.org/10.1080/01621459.1971.10482329

Wintle, B. C., Fraser, H., Wills, B. C., Nicholson, A. E., & Fidler, F. (2019). Verbal probabilities: Very likely to be somewhat more confusing than numbers. *PLoS One, 14*(4), e0213522. https://doi.org/10.1371/journal.pone.0213522

Open Access This chapter is licensed under the terms of the Creative Commons Attribution 4.0 International License (http://creativecommons.org/licenses/by/4.0/), which permits use, sharing, adaptation, distribution and reproduction in any medium or format, as long as you give appropriate credit to the original author(s) and the source, provide a link to the Creative Commons license and indicate if changes were made.

The images or other third party material in this chapter are included in the chapter's Creative Commons license, unless indicated otherwise in a credit line to the material. If material is not included in the chapter's Creative Commons license and your intended use is not permitted by statutory regulation or exceeds the permitted use, you will need to obtain permission directly from the copyright holder.

Uncertainty in the Economic Appraisal of Adaptation

Paul Watkiss

24.1 Introduction

Economic appraisal—as used by Governments, Multi-lateral Donor Banks (MDBs) and public sector organisations—is based on the principles of welfare economics and aims to assess the ability of a policy, programme or project to improve social welfare or well-being (HMT, 2020). It is therefore carried out from the perspective of society and includes the economic valuation of non-market effects, such as environmental benefits and costs. This enables assessment of the socio-economic costs, benefits and other dimensions of alternative choices.

With the scaling-up of climate finance, there is an increasing need to apply economic appraisal to potential adaptation investments. Information from the Climate Policy Initiative (CPI, 2019) reports that global finance flows for adaptation in developing countries were US$30 billion/year in 2017–2018. Almost all of this was from the public sector, and a large proportion ($7.4 billion) was funding from the MDBs and International Financial Institutions (IFIs) (MDBs, 2017).

There is an abundance of existing guidance for economic appraisal, produced by national finance ministries to harmonise appraisal within countries (e.g., HMT, 2020) or to harmonise economic appraisal within organisations such as within MDBs (e.g., EIB, 2013). However, there are challenges in applying conventional economic appraisal to adaptation. These include the greater difficulty in the quantification and valuation of benefits, as well as the high uncertainty associated with future climate change and thus adaptation benefits (OECD, 2015). This has led to a greater emphasis on decision-making under uncertainty, including for economic appraisal of adaptation (Watkiss et al., 2014: Dittrich et al., 2016).

This chapter investigates these issues and discusses, in particular, how to address uncertainty in the economic appraisal of adaptation. It starts with a number of framing issues, to set the context, and includes a discussion of the challenges involved with the economic appraisal of adaptation. It then provides a discussion of methodological approaches for different types of adaptation problems, including examples of applications in economic analysis. To finish, it provides a summary and identifies a number of key future priorities.

24.2 Framing and Methods

An economic appraisal is usually undertaken through some form of cost-benefit analysis. This assesses a policy, programme or project by estimating the economic benefits it produces over time and compares these to the costs (from capital, operating and maintenance costs over time) from a societal perspective.

In economic appraisal, costs and benefits are assessed in terms of present values through the use of discount rates (HMT, 2020) to allow analysis on a consistent basis.[1] As well as considering environmental and social costs and benefits, economic appraisal often takes a long-term perspective as it uses a social discount rate. The results of an economic appraisal are expressed as the Net Present Value (NPV),[2]

P. Watkiss (✉)
Paul Watkiss Associates, Oxford, UK

---

[1] Costs and benefits in appraisal are estimated in 'real' base year prices, which means the effects of inflation are removed. Subsequently, costs and benefits that arise in different future periods are adjusted to provide equivalent values using some form of discount scheme. Many governments and organisations use a 'social time preference rate' (STPR), reflecting the fact generally people (and society) prefer to receive goods and services now rather than later, though some schemes use alternatives, such as the social opportunity cost of capital.

[2] The Net Present Value (NPV) is the sum of future values (in real prices) that have been discounted to bring them to today's value (HMT, 2020) and is estimated as the total present value (discounted) benefits divided by the total present value of costs.

the benefit-to-cost ratio,³ or the Economic Internal Rate of Return (EIRR).⁴

It is highlighted that an economic appraisal differs from a financial appraisal (or analysis). A financial appraisal considers the incremental revenues and costs generated by an investment or project, and the ability of the project to generate cash flows, recover the financial costs and generate profits. It is therefore carried out from the perspective of an investor, not the perspective of society.⁵

While most attention has been on the economic appraisal of interventions or options, economic analysis has a much wider role in public decision-making. In the government public policy context (e.g., HMT, 2020), economic principles are used early in the policy process to articulate the rationale for intervention, i.e. why should the public sector act? For example, economic analysis helps to identify the presence of market failures that may require public policy action or investment, as well as to explore the type of intervention that might be appropriate, notably whether to directly intervene or create the enabling environment for others to act (such as the private sector, households). A more formalised economic appraisal—and the analysis of costs and benefits—is then used later in the process, to assess alternative choices, relative to a counterfactual.

MDBs use economic analysis and appraisal in a similar way. To illustrate, the Asian Development Bank (ADB) (2017) uses economic analysis to inform investment decisions. First, this is to ensure that there is a strong rationale for the public sector to intervene and thus the MDB to finance the project. Second, that the selected project represents the most efficient option among the feasible alternatives for achieving the intended project benefits and outcomes. Third, that it generates a positive economic net present value (assessed through economic appraisal).

This chapter focuses on the application of economic appraisal within a climate adaptation (or climate resilience⁶) context. In order to develop this, a number of framing and methodological issues are first discussed.

Types of Adaptation It is important to note that there are different types of adaptation. An important differentiation is around the objective of adaptation. For investments, this includes (ADB, 2020):

- Adaptation of projects (climate proofing⁷). These aim to improve the climate resilience of existing/new projects or investments, such as including climate adaptation in the design of a new major road, in order to help address future climate risks. This focuses on the additional adaptation response–and the marginal costs and benefits–to tackle climate risks or take advantage of opportunities. This is sometimes known as 'adaptation [or resilience] *in* projects'.
- Projects for adaptation. These are primarily intended to address climate change risks that will affect people, investments, and economic activity. They involve targeted adaptation investments, such as a new coastal defence project to reduce the impacts of sea-level rise. This is sometimes termed 'adaptation [or resilience] *through* projects'.

For the first of these, adaptation is a secondary objective, since the primary goal will be associated with the investment itself, such as the building of a new road to enhance access, reduce congestion etc. This is important because it means that there will usually be an existing economic appraisal cycle, and the task is to undertake an economic appraisal of adaptation alongside or within this. The extra steps are, therefore, to examine the additional options for adaptation and assess the costs and benefits of such action. This may include decisions over the timing of the adaptation investment, whether now or later. There are some examples of economic guidance for these types of projects (e.g., ADB, 2015). Moreover, the same concepts are relevant for economic appraisal of policies (e.g., regulatory impact assessment), when the objective is to mainstream climate adaptation into new policies or plans.

For the second of these, the primary objective is adaptation per se. This is likely to necessitate a greater focus on quantifying the economic benefits of adaptation and considering uncertainty, since these could have a greater effect on benefit estimates. Again, the same concepts apply to new policies or plans (e.g., for new adaptation policies).

Methodological Challenges Economic appraisal of adaptation may be difficult for several reasons (Chambwera et al., 2014). First, the future impacts of climate change, and thus

---

³Total present value of benefits divided by total present value of costs.

⁴The rate at which the NPV is zero, which can be compared with the discount rate to assess if a project generates a sufficient return on investment to be viable.

⁵A financial analysis only uses market prices—it excludes environmental or social benefits. The financial attractiveness of a project is usually expressed in terms of an internal rate of return (IRR), the annual return that makes the net present value equal to zero, or a payback period. This generally takes a short-term perspective and uses (higher) discount rates that reflect a required rate of return or the opportunity cost of capital, noting that commercial or private investors typically expect much higher returns than public investments.

⁶In recent years, there has been an increase use of the term climate resilience as an alternative to adaptation. However, there are many definitions of resilience, framed from different disciplines, and so we use the term adaptation in this paper, centred on the IPCC fifth Assessment Report (AR5) Working Group 2 core conceptions and definitions (IPCC, 2014).

⁷While typically referred to as climate 'proofing', this term is not recommended, as it is neither practical nor economically efficient to seek to eliminate all climate risk (to climate-proof).

the benefits of adaptation, are uncertain. Indeed, it is often characterised as a case of deep uncertainty because key forces that shape the future are not known, or probabilities cannot be assigned (Hallegatte et al., 2012). This is due to uncertainty around future emission pathways (such as whether the world is on a 2 °C or 4 °C trajectory) as well as uncertainty associated with different climate model outputs for a given emissions trajectory. Second, there are challenges around the timing of adaptation costs and benefits (OECD, 2015). A standard approach to dealing with an immediate risk in an investment project would be to change the design, noting this would likely incur a cost, though this would also provide an immediate benefit. In contrast, the risks of climate change are likely to increase over time, especially in future decades. The full benefits of adapting to these future impacts are, therefore, expected to arise in the longer term as well. In economic appraisal, the timing of costs and benefits matters, and this is captured through the use of social discount rates (see above).[8] These implicitly assign lower weight to benefits in the future (in present value terms), reflecting the principle that people (and society) generally prefer to receive goods and services now rather than later. Implementing an approach that over-designs for adaptation may not, therefore, pass a cost-benefit test. This is because implementing a costly investment now for benefits in say 20 years may generate low present values.

This has led to a focus on frameworks for helping to prioritise adaptation decisions, and alongside this, the development of methods (decision-support tools) to address the particular challenge of uncertainty—decision-making under uncertainty (DMUU). These frameworks are worth considering separately because they have relevance at different points in the policy or investment cycle (Watkiss et al., 2019):

- There are various frameworks for early adaptation prioritisation and sequencing over time (Watkiss et al., 2019). These are broad typologies that help with developing adaptation policy and programmes, as well as initial scoping of options at the project level, such as part of an initial economic review to short-list options.
- Decision-support methods or tools for adaptation (Watkiss & Hunt, 2013). These are more formalised methods that can be used in appraisal. For some adaptation options, conventional decision-support tools can be used; but for others, there is a need for DMUU. While these can be applied as part of policy or programme applications, they are most relevant and applicable for project appraisal. These frameworks can also be used as part of economic appraisal.

These are discussed in turn below.

Frameworks for Early Adaptation One of the challenges for adaptation, and adaptation economics, is that much of the analysis to date has been undertaken using stylised models. These typically use a science-led, impact assessment framework, to quantify future impacts, and then assess the quantified benefits and costs of adaptation (ECONADAPT, 2017). These are extremely valuable in raising awareness around the economic potential of adaptation. However, their future orientation towards specific periods, such as the 2040s, and the application of if-then (predict-then-act) frameworks do not inform immediate adaptation decisions or appraisal.

As a consequence, there has been the development of framing concepts for the economics of adaptation that focus on short-term and immediate decisions. This builds on the fact that there are types of early adaptation decisions that make sense in the next decade or so from an economic perspective. Over time, the literature has built up a reasonable idea of what these are (e.g., Fankhauser et al., 1999; Ranger et al., 2010) and they can be grouped into three areas (Warren et al., 2018)[9]:

- Address <u>current</u> climate impacts—and the current adaptation gap[10]—by implementing 'no-regret' or 'low-regret' actions[11] to reduce risks associated with current climate variability as well as building future climate resilience. These generate immediate economic benefits.
- Ensure that adaptation is considered in near-term decisions (now) that have long lifetimes and involve potential <u>future</u> climate risks, such as major infrastructure devel-

---

[8] The use of discount rates when calculating the social cost of carbon, or the costs and benefits of mitigation policy, has been very contentious. This is because of issues around inter-generational wealth transfers. However, most adaptation is associated with shorter-term interventions and thus uses existing discounting approaches. It is noted that the use of such rates might need to be re-considered when looking at larger scale risks and transformational adaptation.

[9] These look similar, but they have different decision characteristics related to the timing of the investment decision and the timing of the climate risk. They are, in turn, now-now (e.g., upgrading an existing road to address current risks), now-future (building a new hydropower plant today and making it climate resilient to future climate change) and future-future (developing a plan to potentially build a river barrage in the 2050s).

[10] The adaptation gap is defined by UNEP (2020) as the difference between implemented adaptation and a societally set goal.

[11] No-regret adaptation is defined as options that generate net social and/or economic benefits irrespective of whether or not climate change occurs. This can include win-win options, which have positive co-benefits. These are differentiated from low-regret options, which may have low costs or high benefits, or low levels of regret, or may be no-regret options that have some opportunity or transaction costs in practice.

opments or land-use change, to avoid 'lock-in'.[12] This may involve use of DMUU concepts including economic analysis.
- Fast-track early adaptive management activities when there are long lead times or possible major future changes involving tipping points. This involves monitoring and learning to generate evidence for forthcoming future decisions (but note the actual interventions will be in future years, not in the short term).

It is stressed that at the national level, all three of these adaptation interventions/strategies are needed, and this is likely to require portfolios. Indeed, the three approaches above can be part of an overall adaptive management process or roadmap. What is also important is that the exact economic appraisal method needs to be different for each of them, i.e. there is not a single approach that can be applied to all adaptation types.

The sections below expand on how economic appraisal is being applied to these areas to take account of uncertainty.

24.3 The Economic Appraisal of Early No- and Low-Regret Measures

Much of the focus of early practical adaptation, especially in developing countries, has been directed towards no- and low-regret measures (ECONADAPT, 2017). These are interventions that tend to have net economic benefits, even without future climate change, because they reduce the impacts of current climate variability or extreme events and, at the same time, could help reduce the increasing risks of future climate change. These may be implemented on their own, and/or as part of a package of adaptation options, or an early stage of an adaptation pathway.

Economic appraisal of these measures can be undertaken using conventional approaches, because the decision lifetime is short, and the focus is on the current climate and emerging trends, so uncertainty does not dominate. The most commonly used methods in economic appraisal are cost-benefit analysis (CBA), cost-effectiveness analysis (CEA), and multicriteria analysis (MCA).

Social CBA was described earlier: it values all relevant costs and benefits to society (including non-market effects), in present value terms, and then estimates a net present value and/or a benefit-to-cost ratio. There is existing guidance on economic appraisal and cost-benefit analysis (e.g., ADB, 2017; HMT, 2020), and this can be used to prioritise options and provide an overall economic case for the preferred option. There is a growing literature on the economic appraisal of no- and low-regret options for adaptation (e.g., Shreve & Kelman, 2014; Mechler 2016; ECONADAPT, 2017; GCA, 2019). This shows that such adaptation has high benefit-to-cost ratios. However, there are some additional aspects that can be considered to address uncertainty. For instance, it is good practice to check that options remain low- or no-regret under a changing climate, by sensitivity testing with future climate scenarios to check if these alter the benefit-to-cost ratio. An example of such an approach was undertaken for an adaptation project focused on climate smart agriculture, where sensitivity runs were used to assess whether future climate change would alter the economic return of proposed measures (GCF, 2019).

However, CBA requires the quantification of all costs and benefits, and the latter is often difficult for adaptation, notably when there are non-market benefits. It is also difficult to undertake CBA for non-technical adaptation options such as capacity building, as benefits are hard to identify in quantitative terms. As a result, there are other methods that can be used: two of these are routinely included for decision support in government and MDB economic appraisal guidance already.

Cost-effectiveness analysis (CEA) compares options by assessing the cost per unit of benefit in order to identify the options that are the most cost-effective (highest benefit for lowest cost). It thus avoids monetary valuation of benefits and quantifies benefits in physical terms. The cost-effectiveness score is then used to rank alternative options. While CEA has become the default appraisal method for mitigation, it is sparingly used for adaptation, although there is guidance (Watkiss & Hunt, 2012), and there are some applications, especially in the water sector (e.g., ASC, 2011). Part of the reason for the low uptake of CEA is that it optimises on one metric only and thus does not prioritise more integrated, multi-sectoral, solutions. It also does not lend itself to uncertainty analysis. Nonetheless, as with CBA, a simple solution is to apply sensitivity analysis which repeats the cost-effectiveness assessment for a set of different climate scenarios.

When valuation is particularly challenging, it is also possible to use multi-criteria analysis (MCA), which considers quantitative and qualitative data together for ranking options. This method assesses and scores options against a range of decision criteria, some of which are physical or monetary, whereas others are qualitative. Various criteria can then be weighted and scores aggregated to provide an overall ranking of different options. The method has been applied quite widely to adaptation, and there is guidance (Van Ierland et al., 2013) and examples (e.g., De Bruin et al., 2009). A particular advantage is that MCA provides a structured framework for combining expert judgement and stakeholder preferences.

---

[12] Early actions or decisions that involve long lifetimes or path dependency, which will potentially increase future risk or vulnerability, and that are difficult or costly to reverse later (quasi-irreversibility). This can be from an action or decision that is 'business-as-usual', from a lack of an action or decision, or from a mal-adaptative action or decision.

It can also introduce criteria to score options in terms of their ability to address uncertainty, though this is usually subjective and qualitative.

24.4 The Economics of Decision-Making Under Uncertainty

Moving beyond the low- and no-regret options above, a more detailed set of appraisal methods are applicable, which extend the principles of CBA (Watkiss et al., 2014: Dittrich et al., 2016). These approaches have been applied mostly at the project investment level. They are particularly applied for near-term decisions that have a long lifetime. This might be new infrastructure projects where there are risks from future climate change, and where it could be costly or difficult to make change later (quasi-irreversibility). In such cases, there is a one-off opportunity to incorporate adaptation in the design from the outset. However, some approaches can also be used to look at longer-term decisions, as part of iterative adaptation pathways, enabling learning and changes over time.

The description of the most common approaches for use in economic appraisal is presented in Fig. 24.1. Interestingly, while they all seek to address uncertainty, they apply different principles to do this (see right-hand column). Some centre on DMUU (covered elsewhere in this book) and can be extended to economic appraisal. For example, robust decision-making (Groves & Lempert, 2007) involves testing options or strategies across a large number of plausible 'futures' to identify which perform well over the range, rather than making choices that are optimal to one central (or average) scenario. It is possible to include an analysis of potential costs and benefits in such analysis (e.g., Nassopoulos et al., 2012; Mereu et al., 2018). The same also applies to decision-scaling (Ray and Brown, 2015), not least because this can select economic or financial criteria as a key performance indicator to stress test (e.g., Bonzanigo et al., 2015). Similarly, dynamic adaptation pathways (Haasnoot et al., 2013) provide an approach for iterative risk management, and these have been extended to dynamic cost-benefit analysis (Eijgenraam et al., 2014).

Other methods such as real options analysis (ROA) are more grounded in the economic literature and are extensions of cost-benefit methods. ROA is sometimes included in standard economic appraisal guidance anyway (e.g., EIB, 2013). It is particularly useful when considering the value of flexibility in investments. This includes the ability to adjust

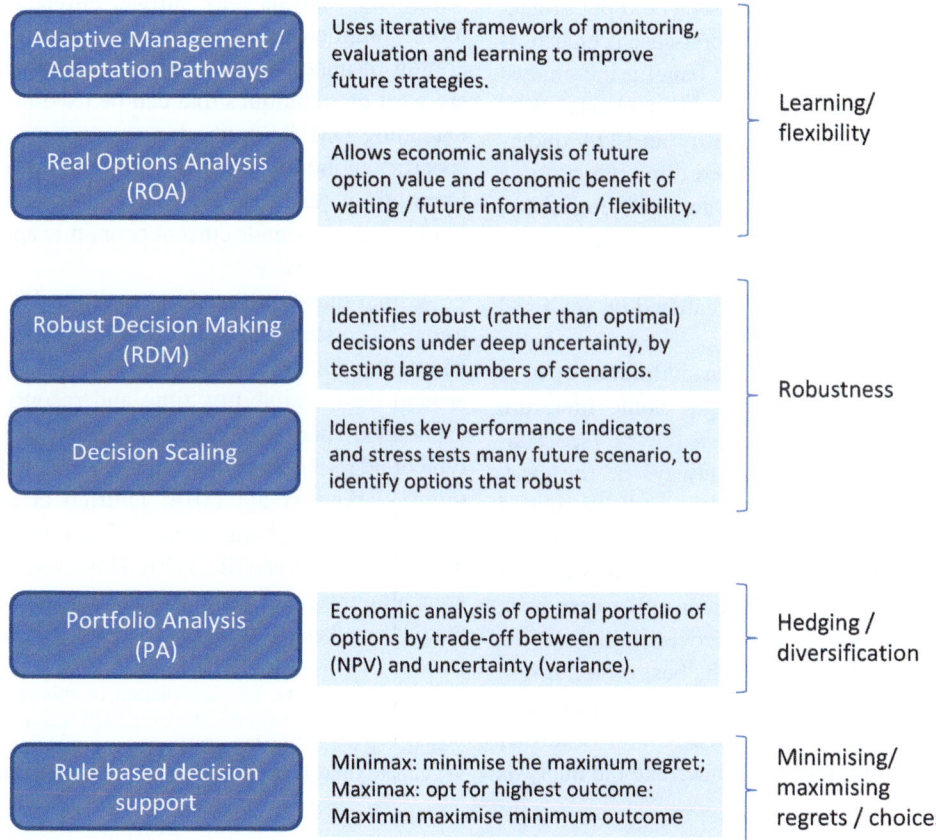

Fig. 24.1 Decision-making under uncertainty in appraisal. Adapted from Watkiss et al. (2014)

the timing of the capital investment or to include flexibility for later changes in the investment, i.e. allowing a project to adapt, expand or scale-back in response to unfolding events. ROA has been quite extensively applied to adaptation, but mainly for sea-level rise and coastal impacts (e.g., Linquiti & Vonortas, 2012; Woodward et al., 2014; Dawson et al., 2018). There is also portfolio analysis (PA) (Hunt and Watkiss 2013). This appears to be a highly promising method for economic appraisal of adaptation, given the focus on designing portfolios, which (together) are effective over a range of possible future climates, rather than a single option best suited to one possible future. However, while there have been applications (Aerts et al., 2008), uptake is much lower than for the other methods discussed above. This may be because it is resource-intensive, requires a high degree of expert knowledge, and relies on the availability of quantitative data.

ROA and PA applications are more likely to provide detailed analyses of the economic costs and benefits of addressing uncertainty, whilst aligning well with economic appraisal guidance. However, they are quite complex to implement and tend to require probabilistic inputs (with accompanying assumptions when dealing with deep uncertainty). One way of utilising these analytical frameworks whilst recognising that objective probabilities of future climate change scenarios do not exist, is to allow the decision analyst to impose probabilities that reflect their attitudes to risk. For example, a risk-averse decision-maker (and analyst) would put greater weight (i.e., higher probability) on a worst-case scenario. If the CBA then produces a net economic benefit, the analyst can be assured that the adaptation investment is robust to their attitude to risk. This was demonstrated by an ROA undertaken by Dawson et al. (2018). They used three risk attitude archetypes—the Pessimist, the Optimist and the Neutralist—to test alternative probability distributions that characterise these archetypes in the context of building defences to protect a railway line in Devon, UK, that is threatened by sea-level rise.

A key point is that there is often an additional cost of addressing uncertainty in project design. Some literature reports that this is often low, i.e. only a few % (see Hallegatte et al., 2019), but this will not always be the case. Adaptation interventions should, therefore, still be subject to an economic appraisal, even for options that are robust or flexible. Indeed, sometimes the most economically efficient adaptation option will be to fix things later rather than now. This might be the case if there is little irreversibility and the costs of retrofit are low. Alternatively, the best option might be to use other approaches to manage risks, such as insurance, rather than increased resilience of infrastructure. A further principle of economic appraisal is to undertake the analysis compared with a do-nothing (counterfactual), noting that in cases where adaptation does not generate net present values, the preferred option may be to just live with the risk (unless there are particular downside risks of failure).

Finally, there is now supplementary economic appraisal guidance for adaptation emerging, which complements the standard approaches and includes many of these aspects outlined above. This includes advice for national economic appraisal (Defra, 2024) as well as within MDBs (ADB, 2015). This points towards the greater uptake of these approaches and their integration within core economic appraisal methods.

24.5 Discussion and Conclusions

This chapter focuses on the economic appraisal of adaptation and the consideration of uncertainty. It identifies a number of key issues relating to the types of adaptation, the challenges for analysis and the potential for frameworks for prioritisation of early adaptation using economic principles.

It then went on to discuss the economic appraisal of adaptation. This highlights that for some adaptation decisions—including no- and low-regret options—it may be possible to use traditional economic appraisal methods. This precludes the need to consider uncertainty in depth, and when simple sensitivity analysis may suffice. Alongside this, there is a more complicated set of decisions that do require consideration of uncertainty as part of economic appraisal. There are a set of techniques that can be used in such cases, either extending DMUU methods to economic appraisal or using specific economic DMMU methods. There is a growing literature of case studies of such applications and emerging guidance that extends current economic appraisal to consider these approaches.

A final question that follows from this is which method to use? The short answer is that this depends on the type of adaptation decision, as well as other practical considerations around data availability, time and resources, and technical capacities. There has been some analysis to try and match the type of adaptation problem to the relevant decision approach (e.g., Watkiss et al., 2014; Dittrich et al., 2016), and to identify project characteristics that might necessitate more detailed analysis (ADB, 2020). However, further work in this area, along with the need to build and share portfolios of case studies, are key priorities.

Acknowledgements The completion of this paper received funding from the European Union's Horizon 2020 research and innovation programme, as part of the COACCH project (CO-designing the Assessment of Climate Change costs) (grant agreement No 776479).

References

Aerts, J. C. J. H., Botzen, W., van der Veen, A., Krywkow, J., & Werners, S. (2008). Dealing with uncertainty in flood management through diversification. *Ecology and Society, 13*, 1–17. http://www.ecologyandsociety.org/vol13/iss1/art41/

ASC. (2011). *Research to identify potential low-regrets adaptation options to climate change in the residential buildings sector*. Commissioned by the Adaptation Sub Committee, July 2011. London, UK.

Asian Development Bank. (2015). *Economic analysis of climate proofing investment projects*. Accessed 17 September 2024. https://www.adb.org/sites/default/files/publication/173454/economic-analysis-climate-proofing-projects.pdf.

Asian Development Bank. (2017). *Guidelines for the economic analysis of projects*. Accessed 17 September 2024. https://www.adb.org/documents/guidelines-economic-analysis-projects.

Asian Development Bank. (2020). *Principles of climate risk management for climate-proofing projects*. Accessed 17 September 2024. https://www.adb.org/publications/climate-risk-management-climate-proofing-projects.

Bonzanigo, L., Brown, C., Harou, J., Hurford, A., Karki, P., Newmann, J., & Ray, P., 2015: South Asia investment decision making in hydropower: Decision tree case study of the Upper Arun Hydropower Project and Koshi Basin Hydropower Development in Nepal. GEEDR South Asia, the World Bank. Report no.: AUS 11077.

Chambwera, M., et al. (2014). Economics of adaptation. In C. B. Field et al. (Eds.), *Climate change 2014: Impacts, adaptation, and vulnerability. Part A: Global and sectoral aspects. Contribution of Working Group II to the Fifth Assessment Report of the Intergovernmental Panel on Climate Change*. Cambridge University Press.

CPI. (2019). *Global landscape of climate finance 2019*. Climate Policy Initiative.

Dawson, D., Hunt, A., Shaw, J., & Gehrels, R. (2018). The economic value of climate information in adaptation decisions: Learning in the sea-level rise and coastal infrastructure context. *Ecological Economics, 150*, 1–10. https://doi.org/10.1016/j.ecolecon.2018.03.027

De Bruin, K., et al. (2009). Adapting to climate change in The Netherlands: An inventory of climate adaptation options and ran-king of alternatives. *Climatic Change, 95*, 23–45. https://doi.org/10.1007/s10584-009-9576-4

Defra, 2024: Accounting for the effects of climate change. Supplementary guidance to HMT Green book.. https://www.gov.uk/government/publications/green-book-supplementary-guidance-environment.

Dittrich, R., Wreford, A., & Moran, D. (2016). A survey of decision-making approaches for climate change adaptation: Are robust methods the way forward? *Ecological Economics, 122*, 79–89. https://doi.org/10.1016/j.ecolecon.2015.12.006

ECONADAPT. (2017). The costs and benefits of adaptation: Results from the ECONADAPT Project. Watkiss, P., (ed.). Accessed 17 September 2024. https://econadapt.eu/sites/default/files/docs/Econadapt-policy-report-on-costs-and-benefits-of-adaptaiton-july-draft-2015.pdf.

Eijgenraam, C., et al. (2014). Economically efficient standards to protect The Netherlands against flooding. *Interfaces, 44*, 7–21. https://doi.org/10.1287/inte.2013.0721

European Investment Bank. (2013). Economic Appraisal of Investment Projects at the EIB. Accessed 17 September 2024. http://www.eib.org/attachments/thematic/economic_appraisal_of_investment_projects_en.pdf.

Fankhauser, S., Smith, J. B., & Tol, R. (1999). Weathering climate change: Some simple rules to guide adaptation decisions. *Ecological Economics, 30*, 67–78. https://doi.org/10.1016/S0921-8009(98)00117-7

GCA. (2019). Adapt Now. Global Commission on Adaptation. Accessed 17 September 2024, https://gca.org/reports/adapt-now-a-global-call-for-leadership-on-climate-resilience/.

GCF. (2019). Economic appraisal of the Green Climate Fund Project Rwanda, Rural Green Economy and Climate Resilient Development Project. Preparatory studies completed on behalf of the Ministry of Natural Resources Rwanda and funded through the Project Preparation Facility of the Green Climate Fund.

Groves, D. G., & Lempert, R. J. (2007). A new analytic method for finding policy-relevant scenarios. *Global Environmental Change, 17*, 73–85. https://doi.org/10.1016/j.gloenvcha.2006.11.006

Haasnoot, M., Kwakkel, J. H., Walker, W. E., & ter Maat, J. (2013). Dynamic adaptive policy pathways: A method for crafting robust decisions for a deeply uncertain world. *Global Environmental Change, 23*, 485–498. https://doi.org/10.1016/j.gloenvcha.2012.12.006

Hallegatte, S., Shah, A., Lempert, R., Brown, C., & Gill, S. (2012). Investment decision making under deep uncertainty—Application to climate change. World Bank Policy Research Working Paper 6193. Accessed 17 September 2024. https://doi.org/10.1596/1813-9450-6193

Hallegatte, S., Rentschler, J., & Rozenberg, J. (2019). *Lifelines: The resilient infrastructure opportunity. Sustainable Infrastructure Series*. World Bank. Accessed 17 September 2024. http://hdl.handle.net/10986/31805

HMT. (2020). The Green Book: Central government guidance on appraisal and evaluation. Accessed 17 September 2024. https://www.gov.uk/government/publications/the-green-book-appraisal-and-evaluation-in-central-governent.

Hunt, A., & Watkiss, P.. (2013). Portfolio analysis: Decision support methods for adaptation. MEDIATION Project, Briefing Note 5. Accessed 17 September 2024. https://www.sei.org/mediamanager/documents/Publications/sei-mediation-briefing5-portfolio-analysis.pdf.

IPCC. (2014). Glossary. Annex II. Impacts, adaptation, and vulnerability, Working Group 2 of the IPCC 5th assessment report. Accessed 17 September 2024. https://www.ipcc.ch/site/assets/uploads/2018/02/WGIIAR5-AnnexII_FINAL.pdf.

Linquiti, P., & Vonortas, N. (2012). The value of flexibility in adapting to climate change: A real options analysis of investments in coastal defence. *Climate Change Economics, 3*, 1250008. https://doi.org/10.1142/S201000781250008X

MDBs, 2017: Joint report on multilateral development banks climate finance. Multilateral development banks. Accessed 17 September 2024. https://www.eib.org/attachments/press/2017-joint-report-on-mdbs-climate-finance-48p.pdf

Mechler, R. (2016). Reviewing estimates of the economic efficiency of disaster risk management: Opportunities and limitations of using risk-based cost–benefit analysis. *Natural Hazards, 81*, 2121–2147. https://doi.org/10.1007/s11069-016-2170-y

Mereu, V., Santini, M., Cervigni, R., Augeard, B., Bosello, F., Scoccimarro, E., Spano, D., & Valentini, R. (2018). Robust decision making for a climate-resilient development of the agricultural sector in Nigeria. In L. Lipper, N. McCarthy, D. Zilberman, S. Asfaw, & G. Branca (Eds.), *Climate smart agriculture, natural resource management and policy* (Vol. 52, pp. 277–306).

Nassopoulos, H., Dumas, P., & Hallegatte, S. (2012). Adaptation to an uncertain climate change: Cost benefit analysis and robust decision making for dam dimensioning. *Climatic Change, 114*, 497–508. https://doi.org/10.1007/s10584-012-0423-7

OECD. (2015). *Climate change risks and adaptation: Linking policy and economics*. OECD Publishing, . Accessed 17 September 2024. https://doi.org/10.1787/9789264234611-en

Ranger, N., Millner, A., Dietz, S., Fankhauser, S., Lopez, A., Ruta, G., & G. (2010). Adaptation in the UK: A decision-making process. In *Policy briefing note for the adaptation sub-Committee of the Climate Change Committee*.

Ray, P. A., & Brown, C. M. (2015). *Confronting climate uncertainty in water resources planning and project design: The decision tree framework*. World Bank. Accessed 17 September 2024. https://doi.org/10.1596/978-1-4648-0477-9

Shreve, C. M., & Kelman, I. (2014). Does mitigation save? Reviewing cost-benefit analyses of disaster risk reduction. *International Journal of Disaster Risk Reduction, 10*, 213–235. https://doi.org/10.1016/j.ijdrr.2014.08.004

UNEP. (2020). *The adaptation gap report*. Published by the United Nations Environment Programme. Accessed 17 September 2024. https://www.unep.org/resources/adaptation-gap-report-2020

Van Ierland, E.C. , de Bruin, K., & Watkiss, P.. (2013). Multi-criteria analysis: Decision support methods for adaptation. MEDIATION Project, Briefing Note 6. Accessed 17 September 2024. https://www.sei.org/publications/decision-support-methods-for-climate-change-adaptation-multi-criteria-analysis/.

Warren, R. F., Wilby, R. L., Brown, K., Watkiss, P., Betts, R. A., Murphy, J. M., & Lowe, J. A. (2018). Advancing national climate change risk assessment to deliver national adaptation plans. *Philosophical Transactions of the Royal Society A, 376*, 20170295. https://doi.org/10.1098/rsta.2017.0295

Watkiss, P. and A. Hunt, 2012: Cost-effectiveness analysis: Decision support methods for adaptation. MEDIATION Project, Briefing Note 2. Accessed 17 September 2024. https://mediamanager.sei.org/documents/Publications/sei-mediation-briefing2-cost-effective-analysis.pdf.

Watkiss, P. and A. Hunt, 2013: Method overview: Decision support methods for adaptation, briefing note 1. Summary of methods and case study examples from the MEDIATION Project. Accessed 17 September 2024. https://www.sei.org/mediamanager/documents/Publications/sei-mediation-briefing1-method-overview.pdf

Watkiss, P., Hunt, A., Blyth, W., & Dyszynski, J. (2014). The use of new economic decision support tools for adaptation assessment: A review of methods and applications, towards guidance on applicability. *Climatic Change, 132*, 401–416. https://doi.org/10.1007/s10584-014-1250-9

Watkiss, P., Ventura, A., & Poulain, F. (2019). *Decision-making and economics of adaptation to climate change in the fisheries and aquaculture sector* (FAO Fisheries and Aquaculture Technical Paper No. 650). FAO. Accessed 17 September 2024. https://openknowledge.fao.org/server/api/core/bitstreams/32239839-8f34-4bd8-8a8b-ebba5c1dceeb/content

Woodward, M., Kapelan, Z., & Gouldby, B. (2014). Adaptive flood risk management under climate change uncertainty using real options and optimization. *Risk Analysis, 34*, 75–92. https://doi.org/10.1111/risa.12088

Open Access This chapter is licensed under the terms of the Creative Commons Attribution 4.0 International License (http://creativecommons.org/licenses/by/4.0/), which permits use, sharing, adaptation, distribution and reproduction in any medium or format, as long as you give appropriate credit to the original author(s) and the source, provide a link to the Creative Commons license and indicate if changes were made.

The images or other third party material in this chapter are included in the chapter's Creative Commons license, unless indicated otherwise in a credit line to the material. If material is not included in the chapter's Creative Commons license and your intended use is not permitted by statutory regulation or exceeds the permitted use, you will need to obtain permission directly from the copyright holder.

Uncertainty Management in a Decision Context: What We Learned from the Decision Center for a Desert City

Patricia Gober and Howard Wheater

25.1 Water Scarcity Is an Existential Threat

Water is central to the desert city of Phoenix's past, present, and future development. The city enjoys a favorable geographic location at the base of two large watersheds of the Salt and Verde Rivers in Central Arizona (Fig. 25.1). The modern site of Phoenix once hosted a large and complex prehistoric population based on irrigated agriculture, from 0 to 1400 AD. At its peak, the Hohokam civilization numbered 40,000 residents and irrigated some 110,000 acres. The Hohokam story is, however, a cautionary tale about how *not* to respond to climatic uncertainty, as the civilization collapsed after 1250. Archeological evidence indicates that the Hohokam experienced changes in climate regime and responded with unsustainable strategies such as overplanting their good fields and expanding into more marginal lands. As drought conditions persisted, the Hohokam all but disappeared by 1400 AD (Redman, 1999; Abbott, 2000). The present city of Phoenix is named after the mythic Phoenix bird that rose from the ashes of its predecessor.

Modern Phoenix developed a diverse and extensive water portfolio including surface water from the Salt and Verde Rivers, groundwater from deep alluvial aquifers, and allocations from the Colorado River via a 336-mile aqueduct from the Colorado River to Central Arizona cities (Fig. 25.1). Until recently, water resource management was a matter of juggling these three sources to meet the needs of the environment, municipal users, industry, and farmers. The dominant management paradigm was to "predict and plan," in other words to anticipate future demand and then plan for it by securing water rights and building infrastructure (Quay et al., 2010). This management paradigm uses the past as a guide to the future, to provide reliable water services to the urban population. This model assumes stationarity—that past empirical observations are a valid basis upon which to estimate future supply and demand (Milly et al., 2008).

The onset of climate-change impacts, poor governance, and uncertainties associated with urban water demand threaten Phoenix's traditional predict-and-plan model of urban water management. Climate scientists have long warned that the Colorado River Basin is likely to be warmer and drier in the future than in the past with negative consequences for water resources, hydropower production, and environmental flows (Christensen et al., 2004; Sheikh & Stern, 2019). In August 2021, observed data from the US Bureau of Reclamation showed that Lake Mead, the largest reservoir in the United States, was at its lowest level since the Hoover Dam was constructed in the 1930s (Fig. 25.2). The lake level was at 1067 feet above sea level, and the lake volume was only 35% full (US Bureau of Reclamation, 2024). At this level, pre-agreed upon shortages to Arizona, Nevada, and Mexico came into effect reducing their allocations. California was not subject to immediate water cutbacks but had to consider the possibility that hydroelectric power generation at Hoover Dam could cease if Lake Mead falls below 1025 feet (Sheikh & Stern, 2019). Uncertainties about Colorado River flows and Lake Mead water levels present significant challenges for Phoenix as some urban communities are completely dependent upon the Colorado River for their water supplies whereas Colorado River water is used to replenish groundwater recharge for others. As can be seen from Fig. 25.2, these low levels persist

P. Gober
School of Geographical Sciences and Urban Planning, Arizona State University, Tempe, AZ, USA
e-mail: gober@asu.edu

H. Wheater (✉)
Department of Civil and Environmental Engineering, Imperial College London, London, UK

School of Environment and Sustainability and Global Institute for Water Security, University of Saskatchewan, Saskatoon, SK, Canada
e-mail: howard.wheater@usask.ca

Fig. 25.1 Phoenix water supplies. (Source: Decision Center for a Desert City)

Arizona passed the Groundwater Management Act of 1980 to sustainably manage its groundwater resource. The law accounts for the gradual retirement of agricultural water use and shift in water allocation to meet urban needs. This transition requires developers to demonstrate a 100-year water supply for new home construction in areas that rely on groundwater. Unfortunately, political forces have gotten in the way of this common-sense legislation, and the Arizona Legislature now allows developers to use Central Arizona Project (Colorado River) water to meet the 100-year requirement, even though the Bureau of Reclamation anticipates near-term shortages on the river (Hirt et al., 2008; Sheikh & Stern, 2019). There are also profound uncertainties associated with the groundwater resource where it is possible for one community to undermine the water sustainability of neighboring communities because they share seven interconnected aquifers that underlie the metropolitan region.

25.2 National Science Foundation Decision-Making Under Uncertainty Program

In 2004, the National Science Foundation (NSF) put out a request for proposals to address climate change and adaptation from a social science perspective. The call asserted that society had allocated hundreds of millions of dollars to address the uncertainties associated with future climate impacts, but little had been spent in the US on the social sciences to address decision-making for climate adaptation as a research problem. NSF called for decision-making under uncertainty (DMUU) projects structured around the social science of climate adaptation. Researchers were expected to develop new strategies to facilitate DMUU for climate change. The Arizona State University (ASU) proposal was to deploy DMUU strategies for water management in Phoenix and then disseminate what was learned to other cities facing similar water challenges.[1]

The water story in Phoenix dovetailed nicely with NSF's vision of a complex system problem amenable to DMUU. The prevailing paradigm of predict-and-plan would need to evolve in the face of deep uncertainties about the climate. Research questions would have to shift from asking what the optimal solution is, to identifying robust decisions and/or strategies that work well across a range of climate futures. A new scientific and social infrastructure would be required to support DMUU for water management in Phoenix. Decision scientists had established protocols for DMUU research, but

to the present (2024) and continue to challenge allocations. Temporary reductions were agreed by Arizona, California, and Nevada in 2023, but only until 2026—state and federal discussions continue.

Possible lower flows in the Colorado River are just one of the many uncertainties that challenge water planning in Phoenix. Water is allocated and managed at the municipal level leaving each of 120 local water companies to make individual decisions about water supplies, rates, and conservation programs (Gober, 2018). Older cities in the urban core have more secure water supplies than rapidly growing suburbs. Thus, it is possible to envision a future in which residents of older communities continue high use levels with lower price structures while urban fringe communities face water shortages and higher prices. The idea of sharing risk to manage uncertainty generally is not part of Phoenix's cultural heritage.

---

[1] DMUU is used here to represent problems of deep uncertainty. Deep uncertainty characterizes situations in which analysts do not know or cannot agree upon the key drivers that will shape the future, probability functions that represent uncertainty, and how to value gains and losses from key outcomes (Lempert 2003).

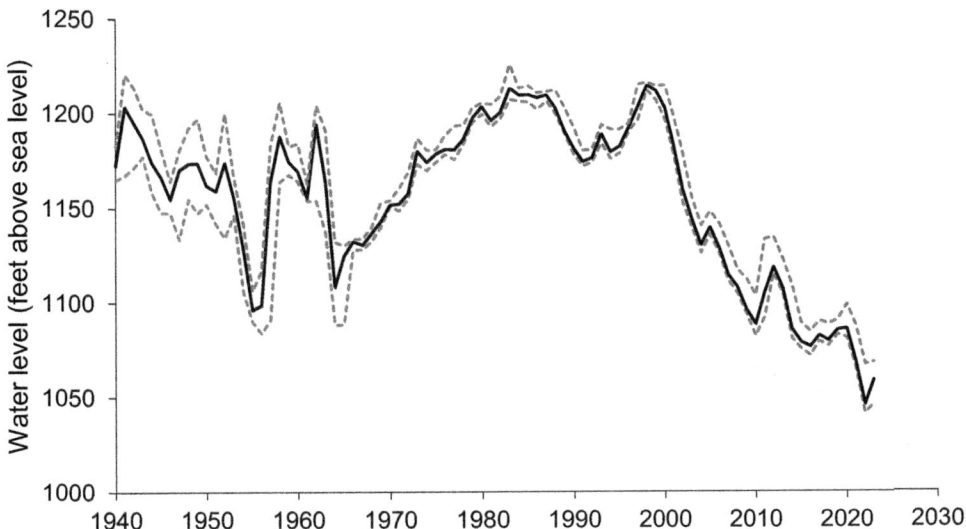

Fig. 25.2 Lake Mead water levels, 1940–2023. The solid line is the annual median monthly level; dashed lines denote the range between annual monthly maximum and minimum levels. (Data source: Bureau of Reclamation)

not how to build a boundary organization to link scientists with local decision makers for climate adaptation research and action. The Decision Center for a Desert City (DCDC) proposal (1) reframed water management in Phoenix into a DMUU problem, (2) assembled a team of interdisciplinary environmental scientists led by social scientists, and (3) secured critical support and promises of collaboration from community partners. The goal was to nudge water planning toward scenarios, stakeholder engagement, and robust decision-making.

There was always an aim to share results and experiences with other cities. DCDC initiated the Urban Water Demand Roundtable to foster collaboration among North American cities. The aim was to share lessons learned in Phoenix with other cities, exchange data, and tease out national trends in urban water demand. Topics included urban green spaces, pools, outdoor water use, forecasting and management, and falling revenues as fixtures and appliances became more efficient with the Energy Policy Act of 2005. The DCDC hosted three Urban Water Demand Roundtables between 2009 and 2012. And worked to build a community of North American cities experiencing similar water management challenges and invite them to share their experiences and data across the wider network.

The DCDC aspired to be a boundary organization sitting between water science and local and regional decision makers. Monthly colloquia brought together local water managers with university scientists and students. An internship program placed students in local water agencies and engaged water managers in the scientific process and in student mentoring. While the circumstances of Phoenix's water issues were not necessarily transferable to other cities, the process of building a boundary organization for climate adaptation was of significant interest to NSF, The Water Demand Roundtable, and the many visitors who observed DCDC's work over the years.

Also significant from an institutional perspective was that the university hosted the Central Arizona Phoenix Long-Term Ecological Research (CAP LTER) project, launched in 1997 as one of two urban LTERs in NSF's network of place-based ecological monitoring and scientific research network. When the DMUU initiative appeared in 2004, there already existed working teams of interdisciplinary scientists and social scientists with ongoing collaborative research projects on urban heat, landscape design, water use, and urban fringe development. The LTER collected data, established relationships with community partners, and built a solid research portfolio about environmental change in Phoenix. In addition, the university had already developed the intellectual and administrative infrastructure needed to muster a large and complex environmental science and social science research program.

It took a fair amount of effort early on to deploy DCDC's assets directly for DMUU research and outreach. Initially many internal investigators sought research funding for their ongoing disciplinary projects. It helped enormously that the University President was a policy scientist and advocate for DMUU. He attended site visits and shared his vision of the role of higher education in building more sustainable cities and the importance of interdisciplinary research. He contemporaneously sought funding from a local donor to build what he called a Decision Theater in which scientific products, including DCDC's WaterSim model would be visualized for policy discussion and decision-making (Fig. 25.3). The co-location of the Decision Theater and DCDC on neutral ground in a suburban downtown location adjacent to the

Fig. 25.3 The Decision Theater at Arizona State University

ASU campus conveyed the institution's deep commitment to linking scientists and practitioners in boundary science at the intersection of science and decision-making.

DMUU acknowledges that many aspects of the future are unknowable and that predictions and forecasts only partially capture the many possible futures. Herman Kahn (1985) first articulated forecasting problems of deep uncertainty several decades ago. There are some problems like nuclear war, climate change adaptation, and sustainable development that are so complex, intractable, and controversial that it is unproductive to try to predict future outcomes. Standard methods of risk analysis work only when system behavior is predictable, and relevant probability distributions are widely used and accepted. Some decision scientists refer to DMUU problems as "wicked problems" because solutions are not true or false but good or bad in the eyes of stakeholders. Problem definitions are ambiguous, and solutions are open-ended. Wicked problems are managed, not solved (Head, 2010).

The idea of DMUU for water planning initially ran counter to the training and worldviews of Phoenix area water stakeholders who were trained in hydrology and water engineering, where the emphasis was on narrow analytical problems and solutions. They were taught to reduce risk and emphasize the reliability of water resources (Lach et al., 2004; Gober, 2018). At the same time, they came to the DMUU table with a sophisticated understanding of deep uncertainty based on their professional experiences. In an early study of stakeholders' views of uncertainty (White et al., 2008), local managers explained that they dealt with political, scientific, and climatic uncertainty almost every day. They described uncertainty as a defining characteristic of water management. Sources of uncertainty included the lack of a definitive historical record of river flows, precipitation levels and drought, and problems with long-term records.

When asked about uncertainties in the water resources management context, decision makers in Phoenix described their concerns about science-based predictions (White et al. 2015). Scientists are primarily concerned with the accuracy of input data, the models used, and the assumptions made. They often speak in probabilistic terms. Policy makers are more likely to follow a political model to characterize uncertainty based on the specific decision context. They evaluate uncertainty not only in statistical terms but also around the perceived political costs of being wrong. Hence, it is easier to take risks when the political environment is settled and the cost of making mistakes is low. Political uncertainties make it more difficult for water managers to follow the science and ignore public opinion.

Political uncertainty jeopardizes Phoenix's groundwater resource. Unsustainable groundwater practices in the 1950s, 1960s, and 1970s led to the Groundwater Management Act of 1980. The Act mandates that new development is required to have a 100-year assured supply for new development to proceed. The state agency responsible for groundwater management has failed to adequately ensure a sustainable supply, and some outlying communities run the risk of water shortages in the mid-term future. In Buckeye, for example, on the far west side of the metropolitan Phoenix, the state agency has allowed the city to use groundwater and to replenish it with unsustainable Central Arizona Project Colorado River water. This sleight of hand robs Peter to pay Paul to sustain rapid growth on the urban fringe 25 miles from central Phoenix. These "borrowed" CAP water supplies are highly vulnerable given what we already know about declining flows on the Colorado River. They are not a reliable source to support long-term urban development. Many new home buyers in Buckeye are not aware that their water supplies are highly vulnerable to CAP supplies although rapidly increasing water tariffs communicate to them the fragility of the city's water supplies. Water rates in Buckeye are among the highest in the metropolitan region.

25.3 Boundary Organization

DCDC was framed from the onset as boundary science, straddling the fence between science and policy. Boundary science includes scientists and decision makers in research projects and other activities. Cash et al. (2003) described the need for new organizational structures to face the challenges of sustainability and manage the boundary of knowledge and action. They stressed the importance of salience (the relevance of assessments to the needs of decision makers), credibility (the scientific adequacy of the technical evidence and arguments), and legitimacy (the perception that knowledge production has been respectful of stakeholders' divergent values and unbiased in orientation). DCDC took these principles to heart and queried stakeholders about DCDC's performance relative to them. White et al. (2010) evaluated the knowledge embedded in the production of one of DCDC's early simulation modeling products, WaterSim. In structured interviews with 62 water resource decision makers from diverse backgrounds, the study found that stakeholders were critical of the model's validity, relevance, and bias. They were skeptical about the reliability and validity of the data that were used to calibrate the model, and they believed they were not given sufficient detail about the equations underlying the model's operations. They wanted to know more about what was "under the hood" in terms of detail. They were skeptical that the model would meet their needs as a water resource decision tool and recommended that the model be downscaled to the city level to deal with the geographic scale at which water decision-making in Phoenix occurs. Stakeholder views were more mixed about the model's legitimacy, with policy makers seeing it as potentially useful but criticizing the overemphasis on supply at the expense of demand and the failure to integrate environmental flows into the model design.

We learned a lot from this initial boundary work about the need to engage early and often with stakeholders, not to fine-tune results but to determine what questions they are interested in from outset. We learned also that boundary work is iterative and thus not conducive to the 1- or 2-year timelines of traditional research projects. We came to acknowledge Dilling and Lemos' (2011) observation that scientists and stakeholders have differing perceptions of what is and is not useful, given their cultural settings and reward structures.

One of our most successful initiatives was a monthly series of Water-Climate Briefings—a 15-year endeavor in building common knowledge and trust across scientists and regional water managers. The briefings typically involved knowledge sharing from a four-member panel of scientists and water managers followed by a question-and-answer period and audience discussion. These briefings involved a catered lunch, free parking, and an opportunity to stay abreast of a large range of local and global water issues. These spanned Native American water rights settlements; status of Colorado River agreements; the energy-water nexus; innovative stormwater management; extreme fires and post-fire flooding; through to Latin American water policy; and Mexico's perspective on the Colorado River management. There was something for everyone on the agenda, and the opportunity to get away from the work desk for an hour and a half, see old friends in the regional water community and meet new ones, plus learn about competing perspectives on a range of water issues. The Water-Climate Briefings socialized students into DCDC's boundary work and facilitated side talk about collaboration.

25.4 WaterSim

DCDC's premier "boundary object" was WaterSim—a simulation model for water planning in Greater Phoenix (Gober et al., 2011). It was not mentioned in the original DCDC proposal to the NSF but came from the need to have a signature product that reflected our aims and work. WaterSim first appeared in 2008 to enable investigations into how alternative climate conditions and policies might interact to affect future water supply and demand conditions in Phoenix. It was a quintessential boundary-science product in the sense that it drew on the scientific expertise of ASU faculty members and students and on insights and data from the local policy community partners. WaterSim represented supply from surface and groundwater sources and demand for residential, commercial, and agricultural users. It included the deep uncertainties about future climate and its impacts on the watersheds that support growth and commerce in Phoenix-area cities. Initially, WaterSim was developed as a region-wide representation of water resources but was quickly downscaled to the municipal level, where most water decisions are made (Sampson et al., 2016).

WaterSim facilitated scenario planning and robust decision-making. It simulated the supply and demand for water under different hydro-climatic conditions, including the historical record, the recent past, or mega drought conditions of the prehistoric Hohokam. It included policy options designed for users to ask: what are the consequences of making particular choices in a complex system under given climate conditions? Output metrics focused on groundwater levels because of their critical role as a bank from which Phoenix draws water when surface water is in short supply. Social-science-based discourse analysis identified five potential options that users can manipulate to explore the consequences of policy decisions on their water security. They included (1) population growth management, (2) municipal and industrial water conservation, (3) groundwater banking and recharge, (4) direct potable use of reverse osmosis reclaimed water, and (5) supply augmentation (Gober et al., 2016). Users explored the consequences of specific policy decisions and combinations of those decisions on long-term water security. Results showed that it will be challenging but possible for Phoenix to continue to grow, even under mega-drought conditions, with sustainable groundwater use. A suite of policies implemented now would go a long way in helping the region manage a severe, prolonged drought. Modest reductions in growth, coupled with continued conservation and reuse, would maintain years of available supply at sustainable levels; banking and augmentation would reduce the negative impacts of high inter-annual variability on years of available supply.

The ability to demonstrate WaterSim in the Decision Theater was initially a powerful draw for community partners and visitors to the university. WaterSim brought positive attention to boundary science and was a useful vehicle to talk about climate change and adaptation. Emphasis was on strategies for adaptation and policy analysis and topics that the water managers expressed interest in and over which they felt they had some control. Ultimately, however, the novelty of the Decision Theater wore out, and decision makers were less interested in traveling to a bustling suburban downtown with scarce car parking. WaterSim gradually transitioned from the Decision Theater to a mobile kiosk, which favors bringing the science to decision makers over bringing the decision makers to the science, and a web browser version that provides an educational tool, available to teachers for classroom use.

25.5 Lessons Learned

Decades of work at DCDC have confirmed that climate adaptation research is an interdisciplinary exercise in uncertainty research as both a social science and science problem. Stakeholders told us early and often that they wanted to talk about policy because that was something over which they had control. Early efforts to explain the climate models and their associated uncertainties fell on deaf ears. As the original creators of the DMUU program imagined, there was a dearth of material about how to make decisions in the face of inevitable uncertainty about the climate. Our journey into boundary science left us skeptical of easy answers and one-size-fits-all solutions to uncertainty problems but enthusiastic about the importance of boundary science.

Resolving the challenge of replicability remained elusive. Initially, the goal was to figure out climate adaptation in the water sector in Phoenix and then transfer what we learned to other problems and other cities. We discovered what adaptation scientists had learned before us—that adaptation is a place-specific process, and each place is different. To be sure, the challenges of building a boundary organization and a boundary product like WaterSim have broader application, but less so for some of the individual science projects that focused on Phoenix. DCDC enjoyed national success as convener of events like an annual water demand roundtable for cities in North America to share insights about trends in urban water demand and policy experiments.

We learned firsthand the value of scenario development and robust decision-making and the need for social learning across science and decision-making. The 15-year+ time horizon gave us an opportunity to develop long-term relationships and to move past some early mistakes. We hired one of our stakeholders to be an associate director of DCDC, and he was not shy in giving advice about how to produce usable science and interact successfully with the local water community.

In moving the conversation from climate change to climate adaptation, we linked the social sciences with climate science and nudged the water management paradigm from predict-and-plan to DMUU. There is still more work to be done in Phoenix and elsewhere, but there is reservoir of good will, awareness that water issues transcend climate science to include social science, and a growing acknowledgment that adaptation is an ongoing process that needs to be managed, not a one-off problem to be solved.

References

Abbott, D. (2000). *Ceramics and community organization among the Hohokam* (280 pp). University of Arizona Press.

Cash, D. W., et al. (2003). Knowledge systems for sustainable development. *PNAS, 100*(14), 8086–8091. https://doi.org/10.1073/pnas.1231332100

Christensen, N. S., et al. (2004). The effects of climate change on the hydrology and water resources of the Colorado River Basin. *Climatic Change, 62*, 337–363. https://doi.org/10.1023/B:CLIM.0000013684.13621.1f

Dilling, L., & Lemos, M. C. (2011). Creating usable science: Opportunities and constraints for climate knowledge use and their implications for science polity. *Global Environmental Change, 21*, 680–689. https://doi.org/10.1016/j.gloenvcha.2010.11.006

Gober, P. (2018). *Building resilience for uncertain water futures* (213 pp). Palgrave Macmillan.

Gober, P., et al. (2011). WaterSim: A simulation model for urban water planning in Phoenix, Arizona USA. *Environment and Planning B, 38*, 197–215. https://doi.org/10.1068/b36075

Gober, P., et al. (2016). Urban adaptation to mega-drought: Anticipatory water modeling, policy, and planning for the urban southwest. *Sustainable Cities and Society, 27*, 497–504. https://doi.org/10.1007/s11269-015-1205-6

Head, B. (2010). *Wicked problems in water governance: Paradigm changes to promote water sustainability and address planning uncertainty* (Urban Security Technical Alliance, Technical Report No. 38). http://www.urbanwateralliance.org.au/publications/UWSRA-tr38.pdf. Accessed 6 November 2019.

Hirt, P., et al. (2008). The mirage in the valley of the sun. *Environmental History, 13*, 482–514.

Kahn, H. (1985). *Thinking the unthinkable in the 1980s* (250 pp). Simon & Schuster.

Lach, D., et al. (2004). Maintaining the status quo: How institutional norms and practices create conservative water organizations. *Texas Law Review, 83*, 2027–2053.

Lempert, R. J. (2003) *Shaping the next one hundred years: New methods of quantitative, Long-term Policy Analysis* (187 pp). Rand.

Milly, P. C. D., et al. (2008). Stationarity is dead: Whither water management. *Science, 319*(5863), 573–574.

Quay, R., et al. (2010). Anticipatory governance: A tool for climate change adaptation. *Journal of the American Planning Association, 76*(4), 496–511. https://doi.org/10.1080/01944363.2010.508428

Redman, C. L. (1999). *Human Impact on ancient environments* (256 pp). University of Arizona Press.

Sampson, D. A., et al. (2016). Anticipatory modeling for water supply sustainability in Phoenix, Arizona. *Environmental Science & Policy, 55*, 36–46.

Sheikh, P. A., & Stern, C. V. (2019). *Management of the Colorado River: Water allocation, drought, and the Federal Role*. Congressional Research Service, https://csreports.congress.gov Report R5546.

US Bureau of Reclamation. (2024). https://www.usbr.gov/lc/region/g4000/hourly/mead-elv.html. Accessed September 2024.

White, D. D., Corley, E. A., & White, M. S. (2008). Water managers' perceptions of the science-policy interface in Phoenix, Arizona: Implications for an emerging boundary organization. *Society & Natural Resources, 21*(3), 1–14. https://doi.org/10.1080/08941920701329678

White, D. D., et al. (2010). Credibility, salience, and legitimacy of boundary objects: Water managers' assessment of a simulation model in an immersive decision theater. *Science and Public Policy, 37*(3), 219–232.

White, D. D., et al. (2015). Water management decision makers' evaluations of uncertainty in a decision support system: The case of WaterSim in the Decision Theater. *Journal of Environmental Planning and Management, 58*(4), 616–630. https://doi.org/10.1080/09640568.2013.875892

Open Access This chapter is licensed under the terms of the Creative Commons Attribution 4.0 International License (http://creativecommons.org/licenses/by/4.0/), which permits use, sharing, adaptation, distribution and reproduction in any medium or format, as long as you give appropriate credit to the original author(s) and the source, provide a link to the Creative Commons license and indicate if changes were made.

The images or other third party material in this chapter are included in the chapter's Creative Commons license, unless indicated otherwise in a credit line to the material. If material is not included in the chapter's Creative Commons license and your intended use is not permitted by statutory regulation or exceeds the permitted use, you will need to obtain permission directly from the copyright holder.

Acting with Uncertainty: Reflecting on a Decade of Rapid Progress in Climate Policy, Research and Practice

Linda O. Mearns and Robert L. Wilby

26.1 Introduction

A lot has changed since the 2014 Advanced Study Program Colloquium on *Uncertainty in Climate Change Research: An Integrated Approach*. Between 2014 and 2023, annual global CO_2 emissions rose by 6% from 35.47 to 37.55 billion tonnes;[1] the atmospheric concentration of CO_2 increased from 399 to 421 ppm;[2] and in 2023 (the hottest year on record at that time), the global mean temperature increase reached 1.48 °C above the 1850–1900 pre-industrial level.[3] Real-time attribution analyses show that anthropogenic greenhouse gases are increasing the likelihood of extreme events such as floods, droughts, heatwaves and storms globally.[4] When writing, Hurricane Milton was about to make landfall in Florida. This cyclone was noteworthy for a pressure drop to 897 mb in just 10 h, coinciding with abnormally warm waters in the Gulf of Mexico.[5] Similarly, higher sea temperatures in the North Atlantic are favouring the formation of severe atmospheric rivers which then impact Europe. For example, 'Super' Storm Desmond in December 2015 brought powerful winds and record 24 and 48 h rainfall across northwest England (Matthews et al., 2018). The associated total runoff from Great Britain smashed the previous maximum by more than 30% (Barker et al., 2016).

Notable temperature-related events include 16 August 2020, when the weather station at Furnace Creek, Death Valley, recorded a peak of 54.4 °C (Matthews 2020); in summer 2021, the Pacific Northwest experienced a 2-week-long heatwave unprecedented in the last millennium (Heeter et al., 2023); and in February–April 2022, millions of people in South Asia were severely affected by the most intense (mega) heat wave in the observed period (Aadhar & Mishra, 2023). Meanwhile, longer term reductions in snow and ice stores of the 'Third Pole' in Asia pose existential threats to the water, food and energy security of nearly two billion people (Kumar et al., 2021; Huggel et al., 2022). There is also growing evidence of the profound vulnerability of small island states and coastal megacities to sea level rise, combined with subsidence, storm surges and waves (De Dominicis et al., 2020; Strauss et al., 2021; Yin et al., 2020).

The following sections reflect on cross-cutting responses to the above threats made by climate policy, research and practice over the last decade. Each section begins with a list of key advances, followed by brief examples of where uncertainty still persists, then others where uncertainty has been reduced. We have scoped the most significant developments using Artificial Intelligence (AI)—an assistive technology that was not widely available in 2014 but now plays important roles in education and research (Wilby & Esson, 2024). We briefly reflect on some ethical and moral challenges ahead for controversial and deeply uncertain technologies such as geo-engineering. Finally, we flag topics that fell outside the span of this text yet, nonetheless, merit attention.

Linda O. Mearns has died before the publication of this book.

[1] https://www.statista.com/statistics/276629/global-co2-emissions/#:~:text=Global%20carbon%20dioxide%20emissions%20from,by%20more%20than%2060%20percent

[2] https://gml.noaa.gov/ccgg/trends/data.html

[3] https://climate.copernicus.eu/copernicus-2023-hottest-year-record

[4] https://www.worldweatherattribution.org/

[5] https://wmo.int/media/news/hurricane-milton-threatens-destruction-florida

---

L. O. Mearns (deceased)

R. L. Wilby (✉)
Department of Geography & Environment, Loughborough University, Loughborough, UK
e-mail: r.l.wilby@lboro.ac.uk

26.2 Advances in Climate Policy

According to ChatGPT, the six most significant advances in climate policy during the last 10 years were chronologically the: (1) Paris Agreement (2015); (2) European Green Deal (2019); (3) China's Carbon Neutrality Pledge (2020); (4) re-entry of the United States into the Paris Agreement and the Inflation Reduction Act (2021–2022); (5) general rise of net zero pledges globally; and (6) the COP26 promise by 40 countries to phase out coal. Further contenders overlooked by the AI might include the Glasgow Leaders Declaration on Forests and Land Use (2021) by which 141 countries pledged to end deforestation by 2030; the establishment of a new 'Loss and Damage' fund for vulnerable countries at COP27 (2022); the rise in climate change litigation suits against governments, corporations and individuals;[6] and a doubling of global climate finance,[7] which reached US$ 1.3 trillion in 2021/2022. Because of the Paris Agreement, 195 countries have now tendered national emission targets (UNFCCC, 2024); and 48 countries have submitted at least one National Adaptation Plan (UNCC, 2024). In the last decade, there were also two synthesis reports for policymakers published by the Intergovernmental Panel on Climate Change (IPCC, 2014, 2023).

The 2015 Paris Agreement set the overarching goal to hold *"the increase in the global average temperature to well below 2 °C above pre-industrial levels"* and pursue efforts *"to limit the temperature increase to 1.5 °C above pre-industrial levels."* The likelihood of achieving the 1.5 °C target is now vanishingly slim—some assert that existing net zero pledges may stabilize global warming at just below 2 °C (Meinshausen et al., 2022). Others claim that 1.5 °C is still feasible provided that 2030 net-zero pledges are better aligned, broadened and strengthened (Dafnomilis et al., 2024). However, the required mix of technologies and emissions pathways required to achieve this, and other targets still hinge critically on the prescribed climate sensitivity and climate modules within Integrated Assessment Models (IAMs) (Chap. 5, this text). Although the best (high confidence) estimate of climate sensitivity narrowed from 1.5 to 4.5 °C in AR5 to 2.5 to 4 °C in AR6, the global temperature responses to a given climate sensitivity vary between IAMs, especially for high sensitivities (Wang et al., 2022). There is also ambiguity about the amount of future emissions that convert to atmospheric greenhouse gas concentrations because of uncertainty around the functioning of global land and ocean carbon sinks (Chap. 14, this text). This is aside from attendant uncertainties in socio-economic development pathways that might deliver the technologies and policies needed for specified global warming targets (Morris et al., 2022). Ultimately, policy making under uncertainty is about generating and selecting from such alternative options (Chap. 2, this text). It is also important to establish who owns the responsibility for managing the uncertainty (Chap. 10, this text).

China's Five-Year Plan for Low Carbon Development (2021–2025) is an example where climate policy uncertainty has reduced. This is because the plan lays out clearer goals for transitioning to a low-carbon economy, including setting peak carbon emissions by 2030 and then achieving carbon neutrality by 2060. This will require a more sustainable approach to growth (Stern & Xie, 2023). Such commitments, along with sector-specific roadmaps, have reduced uncertainty for industry and investors in China, creating a more predictable policy landscape for climate-related projects. Similarly, the UK Climate Change Act (2008)—which legally binds the government to reduce greenhouse gas emissions by 80% (later updated to 100%) by 2050—provides a clear legal and policy framework for long-term decarbonization, reducing policy uncertainty and providing a clear timeline for businesses and energy providers. For instance, the UK government has set a minimum electric car sales target for manufacturers of 80% by 2030 and increased public funds available to incentivize heat pump uptake (Climate Change Committee, 2024). Unfortunately, there were backward steps such as scrapping Energy Performance Certificates for the private-rental sector in 2023 (but now back under review), and failure to incentivize additional offshore wind energy capacity.

26.3 Advances in Research and Technology

When asked about the six most significant advances in climate research in the last 10 years, ChatGPT suggests: (1) higher resolution Earth system and regional climate models; (2) attribution of extreme events to climate change; (3) improved understanding of the carbon cycle and sinks; (4) tipping points and climate feedback mechanisms; (5) new satellite and remote sensing technologies, supported by cloud computing; and (6) new renewable energy and decarbonization technologies, such as direct air capture, carbon capture and storage, and green hydrogen. Again, this list may be extended to cover other innovations in multi-century reconstructions and analysis of mega droughts (Williams et al., 2022); evaluation and prediction of unprecedented climate extremes (Thompson et al., 2017); improved measurement and understanding of methane sources and sinks (Lan et al., 2021a, b); advances in high performance- and cloud-computing resources enabling generation of quicker, longer, higher-resolution, probabilistic projections (Chen et al., 2017); multi-model convection-permitting ensembles (Coppola et al., 2020); plus use

---

[6] https://climatecasechart.com/non-us-climate-change-litigation/
[7] https://www.climatepolicyinitiative.org/publication/global-landscape-of-climate-finance-2023/

of AI to emulate climate models (Eyring et al., 2024), parameterization of small-scale processes (Schneider et al., 2024), or for hybrid dynamical-statistical forecasting (Slater et al. 2023). There have also been advances in understanding of land-surface and land-use change feedbacks on climate extremes (Seneviratne et al., 2018).

Areas where scientific uncertainty has increased or remained largely unchanged include aerosol climate forcing (Kahn et al., 2023), net greenhouse gas budgets and climate feedbacks from permafrost degradation (Ramage et al., 2024). There is also uncertainty around the likelihood, timing and impacts of tipping points, such as a collapse of the Atlantic Meridional Overturning Circulation (Gong et al., 2022). This is partly due to the coarse resolution of climate models and incomplete representation of ocean-atmosphere coupling. Another area of persistent scientific uncertainty surrounds ice sheet dynamics and stability (Bassis et al., 2024). Recent observations of rapid changes in the Antarctic and Greenland ice sheets—including previously poorly understood processes like marine ice sheet instability—have added uncertainty to projections of sea-level rise (see Chap. 19, this text). Moreover, statistical modelling and satellite observations have revealed that the El Niño Southern Oscillation and the Southern Annular Mode contribute substantially to variations in the height of the Antarctic Ice Sheet, especially in coastal areas (King & Christofferson, 2024). Although satellite altimetry and ice sheet models are improving, these new discoveries indicate that the melting of ice sheets could be more nonlinear and unpredictable than previously understood, raising questions about the pace and scale of future contributions to sea-level rise. If marine ice sheets become unstable, the range of possible sea level rise scenarios widens (Robel et al., 2019), with concomitant implications for coastal management and infrastructure (Chap. 12, this text).

The past decade has witnessed further degradation of ground-based hydrometeorology networks, offset by a marked proliferation of gridded data sets blended from point meteorological observations, satellite images, and reanalysis products. The strengths, limitations and key uncertainties associated with the underpinning data are now better understood—especially for daily precipitation (Sun et al., 2018), monthly temperature (Osborn et al., 2021), and river flows (Wilby et al., 2017). However, efforts to compile global observation-based sub-daily indices for precipitation have only just begun (Pritchard et al., 2023). It is further recognized that no single source of gridded climate data is universally the best—unsurprisingly, regions with greater station density tend to have higher quality gridded estimates. For example, one global assessment of 60 gridded data sets showed that reanalysis-based information can be superior to ground-based data in places with low station density (<1 per 1000 km^2), such as mountainous and humid regions (Mankin et al., 2024). Preserving covariance between gridded climate variables continues to be challenging at local scales and in areas with complex topography. Nonetheless, growing availability of high-resolution, high-quality gridded climate data is supporting advances in other research areas such as process-based evaluations of convection-permitting climate models (Kendon et al., 2021), or detection and attribution of non-stationary climate extremes (Slater et al., 2021). These, in turn, contribute to more certain projections of extreme events (e.g., Fosser et al., 2024). Even so, more high-resolution information about soil moisture and the vertical properties of the atmosphere would still be welcome.

26.4 Advances in Science-Informed Practice

ChatGPT lists the following as the six most significant advances in climate practice over the last 10 years: (1) rapid growth and widespread deployment of renewable energy technologies—especially solar and wind power; (2) electrification of public and private transport; (3) corporate commitments to improve sustainability based on Environmental, Social and Governance (ESG) criteria; (4) sector mainstreaming of climate adaptation and resilience planning; (5) use of nature-based methods to enhance carbon sinks and for climate adaptation; and (6) rise of the circular economy concept. The last decade has also witnessed a rapid growth in climate services seeking to translate climate science and information into climate action and to support uptake by decision-makers (Hewitt et al., 2017). Allied with this, there have been marked improvements in accessibility to hydrometeorological data archives and climate model outputs via platforms such as the KNMI Climate Explorer,[8] the WMO Climate Information[9] platform, and the World Bank Climate Change Knowledge Portal.[10] Sectors such as water and health have made steady progress in applying such resources to improve the climate resilience of critical public services (see Chaps. 6 and 9, this text). Although uptake of nature-based methods has been strongly advocated by many (Cohen-Shacham et al., 2019), others call for greater candor when communicating their limited effectiveness in reducing the impacts of extreme events (Dadson et al., 2017).

As mentioned previously, surface hydromet networks and monitoring systems are degrading, even in places that are already data sparse and highly vulnerable—such as the Hindu Kush-Karakoram-Himalayan (HKKH) system (Lehner & Formayer, 2023). This 'Third Pole' holds the largest global store of frozen water outside the poles and is

---

[8] https://climexp.knmi.nl/start.cgi
[9] https://climateinformation.org/
[10] https://climateknowledgeportal.worldbank.org/

essential to the water supply of nearly two billion people (Yao et al., 2022). However, the future of this 'water tower' is highly uncertain due to wide-ranging climate model projections of precipitation, different hydrological model structures and parameterizations, land-use changes and thereby estimates of glacier melt and runoff across the HKKH (Su et al., 2022). Resulting uncertainty in the timing of peak water—which could occur in a few decades—has major ramifications for hydropower generation and downstream water users. Thereafter, river regimes are expected to become more rain-dominated as the influence of glacier melt diminishes, and the risk of glacial-lake outburst floods increases (Nie et al., 2021). Arguably, the uncertain fate of the HKKH system remains one of the most significant unresolved socio-hydrological challenges ahead.

The above example underlines the need for management strategies that can deliver sustainable human development *despite* uncertainty around climate variability and change. These practical actions are low regret, reversible, use safety margins, are time-limited, and/or apply integrated measures (Hallegatte 2009). 'Low regret' adaptations are often favoured because they are designed to bring intended benefits now and, in the future, regardless of climate uncertainty. For example, a low-regret adaptation for natural systems would be the removal of human-related pressures (see Reid et al., 2019). Others include forecasting and early warning systems, which contribute to better preparedness for emergencies with lead times of a few hours to several seasons ahead. For instance, AI and Machine Learning (ML) algorithms are improving heatwave forecasts across all timescales by extracting useful patterns hidden in historical weather data, satellite imagery, and climate model output (Materia et al., 2024). Emulators can also be trained on weather prediction models with 50 members to then generate ensembles with thousands of members. This allows probabilistic forecasting of the likelihood of previously unseen extreme events (Li et al., 2024). By training ML-based methods on historic reanalysis data, systems like Graph Castcan now outperform dynamical forecast systems up to 10 days ahead (Lam et al., 2023). The benefits of more accurate, timely and actionable forecasts are truly realized when embedded in (heat) action plans and combined with investments in more climate-resilient healthcare, water and energy services.

26.5 Challenges and Opportunities Ahead

Previous sections have cited examples of where there is now more or less policy, research and practitioner uncertainty than a decade ago. However, it is important to recognize that the way researchers approach epistemic uncertainty can have important ethical implications (Chap. 22, this text). Here, maybe inductive risk is most pertinent where there is a danger of either overestimating or underestimating a quantity of interest, depending on the choice of method(s) or parameter set(s). For example, a climate policy advisor who runs IAMs and downplays the possibility of uncertain catastrophic ice sheet instability or ecosystem collapse—evident in some simulations—is likely failing to adhere to expected ethical standards (Frank, 2019). Surely, the ethical advisor would adopt a precautionary approach and explain the consequences of being wrong in various ways?

Such considerations are possibly most extreme in the context of geo-engineering—a raft of technologies that is unpalatable to many climate scientists and only mentioned in passing within this text. The profound uncertainties and safety concerns around deployment are simply too great to contemplate, even alongside conventional mitigation options (e.g., Sovacool et al., 2023). Some are now even calling for an International Non-Use Agreement on Solar Geoengineering (Biermann et al., 2022). This is aside from moral hazard arguments which are themselves ambiguous (Tsipiras & Grant, 2022). Will citizens be less inclined to accept economically painful cuts in emissions if a partial technological fix to some aspects of climate change is available? Are policy bandwidth and research funds being misdirected to geo-engineering away from mitigation? With a growing climate crisis, the case will likely be built for more radical mitigation policies, despite deep scientific and governance uncertainties.

This is a bleak topic to close a collection devoted to uncertainty in climate change research, but it is hard to imagine a better case where economic, social, ethical, scientific, technical and policy uncertainties converge. Perhaps similar discussions about uncertain outcomes from proliferating AI technologies have something to offer us (Ulnicane et al., 2021). Other topics warranting more research include the climate mitigation, impacts and adaptation of sectors such as education, energy, housing, sport and leisure. Although outside the present contributions, there are ample opportunities to investigate uncertainty, risk, and decision-making in each case.

If there is one overarching message from this sketch of advances—in climate policy, research and practice—it is that uncertainty should never be an excuse for inaction. Rather, by recognizing and acknowledging socio-scientific uncertainties, we actually deepen understanding of system vulnerabilities and hence can focus resources on what really matters. This is acting with uncertainty.

Acknowledgements The authors thank Rachel Warren and Hayley Fowler for providing constructive feedback and advice on this chapter. Thanks also to Robert Lempert for the chapter title suggestion!

References

Aadhar, S., & Mishra, V. (2023). The 2022 mega heatwave in South Asia in the observed and projected future climate. *Environmental Research Letters*, 18, 104011. https://doi.org/10.1088/1748-9326/acf778

Barker, L., et al. (2016). The winter 2015/2016 floods in the UK: A hydrological appraisal. *Weather*, 71, 324–333. https://doi.org/10.1002/wea.2822

Bassis, J. N., et al. (2024). Stability of ice shelves and ice cliffs in a changing climate. *Annual Review of Earth and Planetary Sciences*, 52, 221–247. https://doi.org/10.1146/annurev-earth-040522-122817

Biermann, F., et al. (2022). Solar geoengineering: The case for an international non-use agreement. *Wiley Interdisciplinary Reviews: Climate Change*, 13, e754. https://doi.org/10.1002/wcc.754

Chen, X., Huang, X., Jiao, C., Flanner, M. G., Raeker, T., & Palen, B. (2017). Running climate model on a commercial cloud computing environment: A case study using Community Earth System Model (CESM) on Amazon AWS. *Computers and Geosciences*, 98, 21–25. https://doi.org/10.1016/j.cageo.2016.09.014

Climate Change Committee. (2024). *2024 Progress report to parliament*. Accessed 28 September 2024. https://www.theccc.org.uk/publication/progress-in-reducing-emissions-2024-report-to-parliament/.

Cohen-Shacham, E., et al. (2019). Core principles for successfully implementing and upscaling nature-based solutions. *Environmental Science and Policy*, 98, 20–29. https://doi.org/10.1016/j.envsci.2019.04.014

Coppola, E., et al. (2020). A first-of-its-kind multi-model convection permitting ensemble for investigating convective phenomena over Europe and the Mediterranean. *Climate Dynamics*, 55, 3–34. https://doi.org/10.1007/s00382-018-4521-8

Dadson, S. J., et al. (2017). A restatement of the natural science evidence concerning catchment-based 'natural' flood management in the UK. *Proceedings of the Royal Society A*, 473, 20160706. https://doi.org/10.1098/rspa.2016.0706

Dafnomilis, I., den Elzen, M., & van Vuuren, D. (2024). Paris targets within reach by aligning, broadening and strengthening net-zero pledges. *Communications Earth and Environment*, 5, 48. https://doi.org/10.1038/s43247-023-01184-8

De Dominicis, M., Wolf, J., Jevrejeva, S., Zheng, P., & Hu, Z. (2020). Future interactions between sea level rise, tides, and storm surges in the world's largest urban area. *Geophysical Research Letters*, 47, e2020GL087002. https://doi.org/10.1029/2020GL087002

Eyring, V., Gentine, P., Camps-Valls, G., Lawrence, D. M., & Reichstein, M. (2024). AI-empowered next-generation multiscale climate modelling for mitigation and adaptation. *Nature Geoscience*, 17, 963. https://doi.org/10.1038/s41561-024-01527-w

Fosser, G., et al. (2024). Convection-permitting climate models offer more certain extreme rainfall projections. *npj Climate and Atmospheric Science*, 7, 51. https://doi.org/10.1038/s41612-024-00600-w

Frank, D. M. (2019). Ethics of the scientist qua policy advisor: Inductive risk, uncertainty, and catastrophe in climate economics. *Synthese*, 196, 3123–3138. https://doi.org/10.1007/s11229-017-1617-3

Gong, X., Liu, H., Wang, F., & Heuzé, C. (2022). Of Atlantic meridional overturning circulation in the CMIP6 project. *Deep Sea Research Part II: Topical Studies in Oceanography*, 206, 105193. https://doi.org/10.1016/j.dsr2.2022.105193

Hallegatte, S. (2009). Strategies to adapt to an uncertain climate change. *Global Environmental Change*, 19, 240–247. https://doi.org/10.1016/j.gloenvcha.2008.12.003

Heeter, K. J., Harley, G. L., Abatzoglou, J. T., Anchukaitis, K. J., Cook, E. R., Coulthard, B. L., Dye, L. A., & Homfeld, I. K. (2023). Unprecedented 21st century heat across the Pacific northwest of North America. *npj Climate and Atmospheric Science*, 6, 5. https://doi.org/10.1038/s41612-023-00340-3

Hewitt, C. D., Stone, R. C., & Tait, A. B. (2017). Improving the use of climate information in decision-making. *Nature Climate Change*, 7, 614–616. https://doi.org/10.1038/nclimate3378

Huggel, C., Bouwer, L. M., Juhola, S., Mechler, R., Muccione, V., Orlove, B., & Wallimann-Helmer, I. (2022). The existential risk space of climate change. *Climatic Change*, 174, 8. https://doi.org/10.1007/s10584-022-03430-y

IPCC. (2014). Summary for policymakers. In *Climate change 2014: Impacts, adaptation, and vulnerability. Contribution of Working Groups II to the Fifth Assessment Report of the Intergovernmental Panel on Climate Change*. IPCC. https://www.ipcc.ch/site/assets/uploads/2018/03/ar5_wgII_spm_en-1.pdf

IPCC. (2023). Summary for policymakers. In *Climate change 2023: Synthesis report. Contribution of working groups I, II and III to the sixth assessment report of the intergovernmental panel on climate change*. IPCC. https://doi.org/10.59327/IPCC/AR6-9789291691647.001

Kahn, R. A., et al. (2023). Reducing aerosol forcing uncertainty by combining models with satellite and within-the-atmosphere observations: A three-way street. *Reviews of Geophysics*, 61, e2022RG000796. https://doi.org/10.1029/2022RG000796

Kendon, E. J., Prein, A. F., Senior, C. A., & Stirling, A. (2021). Challenges and outlook for convection-permitting climate modelling. *Philosophical Transactions of the Royal Society A*, 379, 20190547. https://doi.org/10.1098/rsta.2019.0547

King, M. A., & Christoffersen, P. (2024). Major modes of climate variability dominate nonlinear Antarctic ice-sheet elevation changes 2002-2020. *Geophysical Research Letters*, 51, e2024GL108844. https://doi.org/10.1029/2024GL108844

Kumar, A., Nagar, S., & Anand, S. (2021). In S. Singh, P. Singh, R. Selvasembian, & K. K. Srivastava (Eds.), *Climate change and existential threats. Global climate change*. Elsevier. https://doi.org/10.1016/B978-0-12-822928-6.00005-8

Lam, R., et al. (2023). Learning skillful medium-range global weather forecasting. *Science*, 382, 1416–1421. https://doi.org/10.1126/science.adi2336

Lan, X., et al. (2021a). Improved constraints on global methane emissions and sinks using $\delta 13C$-CH_4. *Global Biogeochemical Cycles*, 35, e2021GB007000. https://doi.org/10.1029/2021GB007000

Lan, X., Nisbet, E. G., Dlugokencky, E. J., & Michel, S. E. (2021b). What do we know about the global methane budget? Results from four decades of atmospheric CH_4 observations and the way forward. *Philosophical Transactions of the Royal Society A*, 379, 20200440. https://doi.org/10.1098/rsta.2020.0440

Lehner, F., & Formayer, H. (2023). Insights on the climate of Bhutan from a new daily 1 km gridded data set for temperature and precipitation. *International Journal of Climatology*, 43, 4927–4943. https://doi.org/10.1002/joc.8125

Li, L., Carver, R., Lopez-Gomez, I., Sha, F., & Anderson, J. (2024). Generative emulation of weather forecast ensembles with diffusion models. *Science Advances*, 10, eadk4489. https://doi.org/10.1126/sciadv.adk4489

Mankin, K. R., Mehan, S., Green, T. R., & Barnard, D. M. (2024). Review of gridded climate products and their use in hydrological analyses reveals overlaps, gaps, and need for more objective approach to model forcings. *Hydrology and Earth System Sciences Discuss*. https://doi.org/10.5194/hess-2024-58. in review.

Materia, S., et al. (2024). Artificial intelligence for climate prediction of extremes: State of the art, challenges, and future perspectives. *Wiley Interdisciplinary Reviews: Climate Change*, 15, e914. https://doi.org/10.1002/wcc.914

Matthews, T. (2020). Death valley: World-beating temperatures, no sweat. *Weather*, 75, 347–348. https://doi.org/10.1002/wea.3858

Matthews, T., Murphy, C., McCarthy, G., Broderick, C., & Wilby, R. L. (2018). Super storm Desmond: A process-based assessment. *Environmental Research Letters, 13, 014024. https://doi.org/10.1088/1748-9326/aa98c8

Meinshausen, M., Lewis, J., McGlade, C., Gütschow, J., Nicholls, Z., Burdon, R., Cozzi, L., & Hackmann, B. (2022). Realization of Paris Agreement pledges may limit warming just below 2°C. Nature, 604, 304–309. https://doi.org/10.1038/s41586-022-04553-z

Morris, J., Reilly, J., Paltsev, S., Sokolov, A., & Cox, K. (2022). Representing socio-economic uncertainty in human system models. Earth's Futures, 10, e2021EF002239. https://doi.org/10.1029/2021EF002239

Nie, Y., et al. (2021). Glacial change and hydrological implications in the Himalaya and Karakoram. Nature Reviews Earth & Environment, 2, 91–106. https://doi.org/10.1038/s43017-020-00124-w

Osborn, T. J., Jones, P. D., Lister, D. H., Morice, C. P., Simpson, I. R., Winn, J. P., Hogan, E., & Harris, I. C. (2021). Land surface air temperature variations across the globe updated to 2019: The CRUTEM5 data set. Journal of Geophysical Research: Atmospheres, 126, e2019JD032352. https://doi.org/10.1029/2019JD032352

Pritchard, D., Lewis, E., Blenkinsop, S., Patino Velasquez, L., Whitford, A., & Fowler, H. J. (2023). An observation-based dataset of global sub-daily precipitation indices (GSDR-I). Scientific Data, 10, 393. https://doi.org/10.1038/s41597-023-02238-4

Ramage, J., et al. (2024). The net GHG balance and budget of the permafrost region (2000-2020) from ecosystem flux upscaling. Global Biogeochemical Cycles, 38, e2023GB007953. https://doi.org/10.1029/2023GB007953

Reid, A. J., et al. (2019). Emerging threats and persistent conservation challenges for freshwater biodiversity. Biological Reviews, 94, 849–873. https://doi.org/10.1111/brv.12480

Robel, A. A., Seroussi, H., & Roe, G. H. (2019). Marine ice sheet instability amplifies and skews uncertainty in projections of future sea-level rise. Proceedings of the National Academy of Sciences, 116, 14887–14892. https://doi.org/10.1073/pnas.1904822116

Schneider, T., Leung, L. R., & Wills, R. C. (2024). Opinion: Optimizing climate models with process knowledge, resolution, and artificial intelligence. Atmospheric Chemistry and Physics, 24, 7041–7062. https://doi.org/10.5194/acp-24-7041-2024

Seneviratne, S. I., et al. (2018). Climate extremes, land–climate feedbacks and land-use forcing at 1.5 C. Philosophical Transactions of the Royal Society A, 376, 20160450. https://doi.org/10.1098/rsta.2016.0450

Slater, L. J., et al. (2021). Nonstationary weather and water extremes: A review of methods for their detection, attribution, and management. Hydrology and Earth System Sciences, 25, 3897–3935. https://doi.org/10.5194/hess-25-3897-2021

Slater, L. J., et al. (2023). Hybrid forecasting: Blending climate predictions with AI models. Hydrology and Earth System Sciences, 27, 1865–1889. https://doi.org/10.5194/hess-27-1865-2023

Sovacool, B. K., Baum, C., & Low, S. (2023). The next climate war? Statecraft, security, and weaponization in the geopolitics of a low-carbon future. Energy Strategy Reviews, 45, 101031. https://doi.org/10.1016/j.esr.2022.101031

Stern, N., & Xie, C. (2023). China's new growth story: Linking the 14th five-year plan with the 2060 carbon neutrality pledge. Journal of Chinese Economic and Business Studies, 21, 5–25. https://doi.org/10.1080/14765284.2022.2073172

Strauss, B. H., Kulp, S. A., Rasmussen, D. J., & Levermann, A. (2021). Unprecedented threats to cities from multi-century sea level rise. Environmental Research Letters, 16, 114015. https://doi.org/10.1088/1748-9326/ac2e6b

Su, F., et al. (2022). Contrasting fate of western Third Pole's water resources under 21st century climate change. Earth's Futures, 10, e2022EF002776. https://doi.org/10.1029/2022EF002776

Sun, Q., Miao, C., Duan, Q., Ashouri, H., Sorooshian, S., & Hsu, K. L. (2018). A review of global precipitation data sets: Data sources, estimation, and intercomparisons. Reviews of Geophysics, 56, 79–107. https://doi.org/10.1002/2017RG000574

Thompson, V., Dunstone, N. J., Scaife, A. A., Smith, D. M., Slingo, J. M., Brown, S., & Belcher, S. E. (2017). High risk of unprecedented UK rainfall in the current climate. Nature Communications, 8, 107. https://doi.org/10.1038/s41467-017-00275-3

Tsipiras, K., & Grant, W. J. (2022). What do we mean when we talk about the moral hazard of geoengineering? Environmental Law Review, 24, 27–44. https://doi.org/10.1177/14614529211069839

Ulnicane, I., Knight, W., Leach, T., Stahl, B. C., & Wanjiku, W. G. (2021). Framing governance for a contested emerging technology: Insights from AI policy. Policy and Society, 40, 158–177. https://doi.org/10.1080/14494035.2020.1855800

United Nations Climate Change (UNCC). 2024. NAP Central. Accessed 28 September 2024. https://napcentral.org/.

United Nations Framework Convention on Climate Change (UNFCCC). 2024. NDC Registry. Accessed 28 September 2024. https://unfccc.int/NDCREG

Wang, T., Teng, F., Deng, X., & Xie, J. (2022). Climate module disparities explain inconsistent estimates of the social cost of carbon in integrated assessment models. One Earth, 5, 767–778. https://doi.org/10.1016/j.oneear.2022.06.005

Wilby, R. L., & Esson, J. (2024). AI literacy in geographic education and research: Capabilities, caveats, and criticality. The Geographical Journal, 190, e12548. https://doi.org/10.1111/geoj.12548

Wilby, R. L., et al. (2017). The 'dirty dozen' of freshwater science: Detecting then reconciling hydrological data biases and errors. Wiley Interdisciplinary Reviews: Water, 4, e1209. https://doi.org/10.1002/wat2.1209

Williams, A. P., Cook, B. I., & Smerdon, J. E. (2022). Rapid intensification of the emerging southwestern north American megadrought in 2020–2021. Nature Climate Change, 12, 232–234. https://doi.org/10.1038/s41558-022-01290-z

Yao, T., et al. (2022). The imbalance of the Asian water tower. Nature Reviews Earth & Environment, 3, 618–632. https://doi.org/10.1038/s43017-022-00299-4

Yin, J., et al. (2020). Flood risks in sinking delta cities: Time for a reevaluation? Earth's Futures, 8, e2020EF001614. https://doi.org/10.1029/2020EF001614

Open Access This chapter is licensed under the terms of the Creative Commons Attribution 4.0 International License (http://creativecommons.org/licenses/by/4.0/), which permits use, sharing, adaptation, distribution and reproduction in any medium or format, as long as you give appropriate credit to the original author(s) and the source, provide a link to the Creative Commons license and indicate if changes were made.

The images or other third party material in this chapter are included in the chapter's Creative Commons license, unless indicated otherwise in a credit line to the material. If material is not included in the chapter's Creative Commons license and your intended use is not permitted by statutory regulation or exceeds the permitted use, you will need to obtain permission directly from the copyright holder.

Index

A
Adaptation, 2, 12, 19, 32, 47, 61, 76, 94, 112, 118, 126, 141, 145, 164, 185, 199, 225, 232, 239, 247, 256, 265
Adaptation options, 2, 5, 56, 118, 131–133, 142, 143, 192, 194, 249, 250, 252
Adaptive capacity, 53, 56, 61, 62, 123, 145, 147, 148
Adaptive management, 101, 121, 123, 250, 251
Added value, 185, 192–194
Aerosols, 49, 75, 155, 157, 160, 161, 167, 169, 171, 177, 186, 265
Agricultural Model Intercomparison and Improvement Project (AgMIP), 73
Agro climatic hazards, 71
Agroclimatic risks, 72
Ambiguity, 4, 19, 23–24, 94, 132, 146, 149–151, 242, 243, 264
Appraisal, 5, 99, 101, 247–252
Artificial intelligence (AI), 2, 3, 193, 263–266
Atmospheric circulation, 155, 181, 182, 202
Attribution, 3, 11, 218, 225–227, 230–231, 233, 263–265

B
Bayesian statistics, 207, 211
Benchmarking, 4, 185, 193, 194
Benefits, 5, 12, 14, 15, 20–22, 35, 49–53, 56, 57, 66, 67, 76, 90, 96, 97, 111, 123, 126, 130–133, 145, 164, 174, 191, 192, 200, 203, 204, 229, 232, 237, 243, 247–252, 266
Biodiversity, 47–49, 52, 53, 55–57, 137
Block Maxima, 220–222, 224
Boulder CO, 150–151, 213, 217
Boundary forcing, 191, 192
Boundary science, 258–260

C
Calibration, 73–75, 78, 88, 191, 192, 203, 204, 208–210, 231, 233
Carbon budget, 14, 171
Certainty, 3, 40, 61, 62, 110, 127–130, 133, 146, 238, 239
Climate adaptation, 21–22, 62, 145, 146, 149, 150, 248, 256, 257, 260, 265
Climate assessments, 160
Climate attribution, 230, 231, 265
Climate change, 1, 9, 19, 32, 47, 61, 71, 81, 94, 105, 115, 131, 137, 145, 155, 163, 182, 185, 210, 225, 229, 239, 247, 255, 264
Climate change adaptation, 20, 53, 61–68, 119, 146, 258
Climate change allowance, 96, 97, 101
Climate change mitigation, 47, 49, 53, 122, 149, 169
Climate change risk assessment (CCRA), 50, 56, 164
Climate model, 2, 20, 74, 76, 77, 94–96, 99, 101, 116, 118, 141, 155–161, 168–170, 172, 178, 182, 189, 190, 194, 199, 200, 202, 204, 207, 208, 210, 212, 213, 220, 221, 225, 226, 238, 249, 260, 265, 266
Climate model projection, 95, 157, 220, 266
Climate model weighting, 212–213
Climate projections, 4, 19, 20, 22–23, 25, 26, 50, 99, 142, 157, 163–166, 168, 178–180, 182, 185–194, 240
Climate risk management, 231, 232
Climate sensitivity, 4, 52, 75, 156–158, 167, 188, 264
Climate services, 161, 192, 231
Cloud computing, 264
Coastal hazard, 128, 204
Coastal management, 128, 265
Coastal risk, 96
Colorado River, 255, 256, 258, 259
Communication, 4–5, 31, 40–41, 57, 89, 96, 151, 159, 178, 230–231, 237–244
Computer experiments, 207–208
Concentrations, 3, 23, 49, 71–73, 75, 93, 120, 140, 156–158, 163–174, 177, 178, 226, 230, 263, 264
Conflicts, 52–53, 55, 57, 137, 138, 149, 170, 226, 240, 242–243
Conservation, 47, 52, 53, 55, 63, 65, 81, 89, 90, 149, 202, 260
Coping capacity, 61, 62
Costs, 5, 12–15, 21, 27, 35, 38–41, 49–52, 55, 63, 65–67, 81, 94, 95, 97–99, 110, 116, 119–123, 126, 129, 130, 132, 133, 145, 150, 151, 156, 158–160, 167, 169, 171, 172, 174, 200, 232, 247–252, 258
Coupled ethical-epistemic choices, 229–233
COVID-19, 93, 100, 116, 139, 141
Crop management, 76, 77
Crop simulation models, 73

D
Decision analysis, 19–27, 32–33, 38, 117, 118
Decision Center for a Desert City (DCDC), 255–260
Decision-making, 1, 9, 20, 31, 52, 61, 71, 81, 94, 109, 116, 125, 137, 146, 164, 182, 192, 200, 229, 237, 247, 256, 266
Decision making under deep uncertainty (DMDU), 31, 32, 36, 42, 120, 131
Decision making under uncertainty (DMUU), 2, 94, 110, 118, 237–238, 247, 249–252, 256–260
Decision-support, 32, 36, 37, 41, 42, 110, 142, 193, 232, 249, 251
Decision Theater, 257, 258, 260
Decision uncertainty, 238, 242

Deep uncertainty, 1, 4, 33, 36, 38, 41, 119–121, 131, 132, 174, 200–203, 249, 251, 252, 256, 258, 259
Disaster, 66, 72, 105, 106, 117, 128, 139, 145–147, 149, 225
Distributive justice, 232
Domain, 9, 10, 15, 38, 77, 94, 160, 185–188, 191, 192, 194, 200, 207, 223, 239
Drought, 3, 53, 62–64, 71, 93, 94, 97–99, 101, 106, 116, 137, 217, 230, 244, 255, 258–260, 263, 264
Dynamical downscaling, 185–188, 191, 192
Dynamic and thermodynamic drivers, 180, 181
Dynamic crop simulation models, 75

E
Early warning, 72, 142, 266
Earth system models, 155, 157, 159, 160, 172, 173, 203, 204, 207, 232
Ecological Niche, 81
Ecology, 5, 138, 147
Economics, 4, 11, 35, 48, 61, 71, 98, 106, 126, 137, 146, 166, 199, 232, 247, 266
Emissions, 2, 11, 20, 38, 42, 62, 71, 89, 93, 111, 115, 125, 138, 157, 163, 178, 185, 199, 210, 218, 249, 263
Emissions drivers, 167, 168, 172
Emissions scenario, 2, 74, 94, 96, 118, 126, 165, 166, 168, 172, 178, 179, 182, 185, 187, 191, 192, 205, 210
Emulators, 203, 204, 207–210, 266
Endangered Species, 49
Energy, 4, 11–13, 35, 38–40, 48, 52, 53, 55, 56, 71, 101, 122, 138, 146, 149–151, 155, 156, 159, 166, 167, 169–173, 177, 257, 259, 263–266
Engineering, 5, 33, 76, 94, 96, 97, 111, 115–122, 170, 221
Ensemble, 2, 20, 34, 50, 73, 86, 94, 122, 159, 165, 178, 186, 202, 207, 221, 231, 264
Environmental flow, 101, 255, 259
Environmental loads, 117–118
Environmental Niche model, 83
Epistemic uncertainty, 15, 111, 146, 147, 149–151, 266
Epistemic values, 229, 230, 233
Equity, 13, 137, 146
ETCCDI indices, 220, 224
Ethical values, 229–233
Expert judgment, 26, 37, 130, 200, 237–244
Exploratory modeling, 34, 37
Exposure, 105, 106, 108–112, 117, 137, 138, 140, 142, 145–147
Extreme event attribution, 11, 218, 225–227, 230
Extreme events, 3, 68, 71, 73, 98, 108, 116–118, 121, 151, 182, 186, 187, 190, 217, 219, 221–223, 225, 230, 263, 264, 266
Extreme value analysis, 221–223, 225

F
Fire, 52, 62, 63, 65, 67, 96, 116, 117, 151, 259
Flood, 3, 27, 63, 93, 105, 115, 128, 137, 145, 199, 230, 263
Flooding, 27, 53, 66, 67, 71, 105–112, 115, 118, 119, 126, 129, 137, 139, 145, 150, 190, 202, 204, 217, 221, 259
Food security, 137, 138, 170
Forecasting by analogy, 72
Fraction of attributable risk, 225
Free-air carbon dioxide enrichment (FACE) experiments, 72

G
Gaussian processes, 208–210, 223
Generalized extreme value (GEV) distribution, 219–222
Generalized Pareto distribution, 222–223
Geo-engineering, 48, 170, 263, 266

Greenhouse gases (GHGs), 3, 4, 14, 15, 20, 26, 38, 39, 41, 48, 50, 52, 55, 71, 75, 111, 125, 126, 132, 138, 140, 141, 163, 164, 166–169, 171, 177, 178, 180, 181, 186, 199, 202–205, 208, 210, 225, 226, 230, 263–265
Gridded data, 265

H
Harm, 11, 67, 106, 112, 131, 150
Hazard, 63, 64, 71, 72, 93, 94, 101, 105, 106, 108–112, 116–118, 128–130, 137, 145–147, 149, 151, 200, 202–204, 232, 266
Health data, 138–140, 142
Health outcomes, 137, 138, 140–142
Health risks, 137–142
Hierarchical models, 204, 211
High-impact events, 232
Human health, 2, 4, 137–143, 220, 221
Human system, 48, 49, 138, 163, 168, 204
Hydrological model, 93–96, 98, 266
Hydrologic engineering, 94–96

I
Ice sheets, 1, 49, 52, 111, 125, 155, 161, 199–202, 204, 265, 266
Impacts, 2, 9, 19, 32, 48, 62, 71, 81, 93, 106, 115, 126, 137, 145, 156, 163, 180, 185, 199, 207, 217, 229, 237, 248, 255, 263
Infrastructure, 19, 32, 53, 67, 72, 93, 94, 96–99, 106, 112, 115–123, 129, 131, 137, 138, 141–143, 145–147, 149–151, 192, 204, 217, 237, 249, 251, 252, 256, 257, 265
Initial conditions, 24, 39, 76, 116, 158, 177, 178, 182, 185–188, 191, 202, 203, 210, 213, 221, 231
Input uncertainty, 74–76, 163
Institutions, 41, 108–111, 123, 132, 150, 170, 247
Integrated assessment, 3, 51, 53, 163, 168, 238, 264
Integrated assessment modelling, 5, 51, 163, 238, 264
Intercomparison, 73, 160, 164, 179, 192, 193, 202
Internal variability, 22, 159, 163, 164, 177–182, 212, 213

L
Land management, 81
Land use, 3, 53, 55, 77, 83, 116, 121, 137, 138, 151, 160, 166, 167, 171, 200, 250, 265, 266
Lock-in, 5, 56, 133, 250
Loss, 47–49, 51, 52, 57, 71, 81, 89, 106, 109, 125–131, 137, 138, 147, 150, 173, 200, 202, 205, 217, 231, 264

M
Metropolitan Phoenix, 258
Mitigation, 2–5, 47–57, 71, 112, 122, 141, 146, 149, 164, 166, 169–171, 173, 174, 202, 205, 229, 233, 239, 249, 250, 266
Mobility, 116
Model, 2, 10, 19, 31, 49, 62, 73, 81, 93, 110, 115, 126, 137, 146, 155, 163, 177, 185, 199, 207, 218, 229, 237, 249, 255, 264
Model structure uncertainty, 73–74, 77
Multidisciplinary research, 140

N
National Mall, Washington, DC, 105–109, 111
Natural History, 108, 111
Net-zero, 5, 50, 53, 55, 56, 157, 171, 264
Net-zero emissions, 55, 164, 167, 168, 171, 172
No-regret, 249–251

O
Ontological uncertainty, 15, 146

P
Parameter uncertainty, 74, 76, 77, 95
Paris Agreement, 13, 47, 50, 52, 55, 111, 264
Pathways, 3–5, 47, 50, 52, 53, 55, 57, 74, 75, 93, 111, 121, 122, 130, 132, 137, 138, 140, 141, 167–173, 202, 229, 249–251, 264
Pattern, 4, 41, 49, 50, 61, 63, 71, 75, 83, 86, 87, 93, 110, 116, 121, 137, 138, 141, 142, 147, 164, 166, 170, 177, 178, 181, 185, 189–191, 199, 200, 202, 210, 241, 266
Perceptions of climate change, 68, 240
Performance, 13–15, 21–25, 35, 37, 39–41, 82, 84, 86–88, 96, 116–118, 121, 123, 190, 192, 210, 212, 231, 251, 264
Plan, 11, 32, 33, 35, 37, 40, 41, 50, 56, 61, 63, 65–67, 93, 94, 96–99, 101, 111, 118, 119, 121, 122, 126, 128, 129, 138, 142, 151, 248, 249, 255, 264, 266
Policy, 1, 9, 31, 47, 105, 121, 128, 138, 145, 161, 164, 204, 225, 229, 237, 247, 257, 263
Policy analysis, 2, 9, 10, 13–15, 110, 260
Policy design, 12
Policy process, 9–15, 109, 110, 112, 145, 172, 248
Policy sphere, 9–15, 110
Politics, 5, 9, 14, 110, 149, 172
Prediction uncertainty, 77, 78, 89, 159
Prioritization, 56
Process-based crop models, 73

R
Rational decision-making, 127–131, 133
Realized niche, 83
Reducing uncertainty, 2, 74, 77, 78, 87, 94, 123, 147, 151, 160, 192
Regional climate model (RCM), 185–192, 194, 208, 213, 264
Regional projections, 97
Representative agricultural pathways (RAPs), 76
Representative concentration pathways (RCPs), 76, 120, 168–170, 202
Research design, 229, 231, 233
Researcher decisions, 89
Resilience, 5, 53, 62, 66, 67, 94, 96, 98, 99, 101, 128, 129, 137, 138, 142, 145–147, 150, 151, 248, 252, 265
Return level, 222
Return period, 94, 96–98, 222–224
Risk, 2, 12, 22, 31, 47, 62, 72, 93, 105, 115, 126, 137, 145, 157, 164, 182, 192, 199, 217, 230, 238, 248, 256, 266
Risk-based frameworks, 118
Risk communication, 130, 244
Risk perception, 232
Risk ratio, 225, 226
Robust decision-making (RDM), 2, 27, 31–42, 121, 131, 132, 173, 182, 200, 204, 251, 257, 259, 260

S
Scaling, 19–27, 86, 142, 156, 161, 185, 189–192, 210, 247, 251
Scenario discovery, 27, 33, 35, 37, 39, 41, 173, 174
Scenarios, 2–4, 14, 15, 19–27, 31–38, 41, 55, 56, 64, 67, 68, 74–76, 84, 88, 89, 94–100, 110, 112, 116, 118–120, 122, 126, 129, 130, 141, 146, 148, 149, 157, 160, 163–174, 179, 185–187, 189, 191, 192, 200–205, 210, 225, 227, 250–252, 257, 265
Science-policy interface, 133, 134
Scientific uncertainty, 127–131, 133, 140, 141, 164, 238–243, 265
Sea level change, 96, 199–202, 204, 205
Sea level projections, 199–205
Sea level rise, 1, 49, 52, 66, 67, 97, 100, 110–112, 118, 125, 126, 131, 133, 137, 145, 150, 167, 201–204, 232, 248, 252, 265
Shared socioeconomic pathways, 168
Social cost of carbon, 39, 40, 51, 249
Socioeconomics, 52, 56, 62, 67, 71, 76, 93, 106, 127, 137, 141, 145–152, 163–166, 168–170, 172, 173, 178, 202, 239
Socioeconomic vulnerability, 145–152
Spatial statistics, 211
Species distribution models (SDMs), 81–84, 86–90
Stakeholder engagement, 24, 257
Stakeholder involvement, 62, 67
Statistical design, 208, 213
Statistical downscaling, 94, 142, 185, 188–192, 194
Storylines, 2, 3, 15, 94–96, 98, 100, 168, 172–174, 230
Stress test, 4, 19, 23–27, 32–37, 39, 96, 101, 190, 192, 251
Sub-grid processes, 185
Subject knowledge, 9, 14, 15
Successful decision making, 41
Sustainability, 42, 47, 53, 56, 121, 170, 172, 256, 259, 265
Synergies, 49, 50, 52–55, 57, 122, 132

T
Temperature, 3, 20, 23, 26, 27, 47, 49, 50, 52, 61, 64, 66, 67, 71, 73–75, 78, 98, 100, 108, 110, 111, 117, 122, 139, 141, 142, 149, 156–160, 164–167, 169, 171, 177–182, 185–187, 189–191, 202, 210–213, 217, 219–221, 226, 227, 230, 233, 239, 241, 263–265
Tipping points, 1, 93, 123, 202, 250, 264, 265
Top-down and bottom-up, 2
Transparency, 13, 96, 97, 132, 141, 145, 147, 192, 194, 231–233, 240, 243
Transportation, 4, 12, 63, 65, 66, 115–123, 150, 151, 166, 167

U
Uncertainty, 1, 9, 20, 31, 47, 61, 71, 81, 93, 105, 115, 125, 137, 145, 155, 163, 177, 185, 199, 207, 217, 229, 237, 247, 255, 263
Uncertainty characterization, 203
Uncertainty partitioning, 187
Uncertainty quantification, 172–174, 207–213, 223, 224, 232
Up-scaling uncertainty, 77
Urban water, 255, 257, 260

V
Values, 1, 12, 19, 37, 50, 64, 74, 83, 95, 105, 116, 129, 146, 157, 165, 177, 185, 209, 217, 229, 238, 247, 256
Vulnerability, 2, 5, 16, 19, 26, 27, 32, 33, 35–37, 41, 52, 56, 64, 95, 99, 100, 105, 106, 108–111, 116–118, 120, 121, 137, 142, 145–152, 170, 200, 204, 233, 250, 263, 266
Vulnerability indices, 147–149, 151

W
Water resources, 4, 24, 33, 50, 93–101, 220, 255, 256, 258, 259
Water resources management, 33, 93, 94, 98, 100
Water resources planning, 24
WaterSim, 257, 259–260
Water supply system, 24, 101

The manufacturer's authorised representative in the EU is Springer Nature Customer Service Centre GmbH, Europaplatz 3, 69115 Heidelberg, Germany. If you have any concerns regarding our products, please contact ProductSafety@springernature.com

Printed and bound by CPI Group (UK) Ltd, Croydon, CR0 4YY

26/03/2026

02079001-0001